Two Manuals Combined into One Book

Training Ranges

And

Military Marksmanship, all M16 and M4 Series Weapons

Department of Defense

Pentagon
Publishing

Training Circular
No. 25-8

HEADQUARTERS
DEPARTMENT OF THE ARMY
Washington, DC, 5 April 2004

TRAINING RANGES

Contents

DISTRIBUTION RESTRICTION: Approved for public release; distribution is unlimited.

*This publication supersedes TC 25-8, 15 March 2004.

FIGURES

Page

ARMY RANGES

TABLES

Page

PREFACE

This circular provides information about Army ranges. With AR 210-21, *Army Ranges and Training Land Program,* this circular provides guidance for developing and operating Army ranges. It is a working guide for trainers, range and mobilization planners, engineers, coordinators, and mandated range project review boards at all levels of the Active Army, Army National Guard, and Army Reserve. It is the primary guide for installation and major Army command range development plans and for developing the Army Master Range Plan.

Army Regulation 385-63, *Policies and Procedures for Firing Ammunition for Training Target Practice and Combat,* provides criteria and instructions for developing surface danger zones for weapons fired on Army and Marine Corps ranges. The US Army Training Support Center is the executive agent for the Deputy Chief of Staff for Operations and Plans, Headquarters Department of the Army, for the Army Range and Training Land Program. As such, the Army Training Support Center is the proponent for this circular.

Submit comments and recommendations on Department of the Army Form 2028, (Recommended Changes to Publications and Blank Forms) directly to Commander, US Army Training Support Center, ATTN: ATIC-ATMD, Fort Eustis, Virginia 23604-5166.

"The Division fought to train under realistic conditions in preparation for combat. The priorities were clear and subordinate leaders executed taining that produced a disciplined, lethal and flexible force capable of accomplishing any tactical task.

- The requirement for tough realistic training has not changed in the past 227 years. The Division lived under the "train as you fight" motto for the 12 months that preceded the war. The training proficiency, lethality, and maturity of the Division serves as an example for the Army to follow.

- Lessons Learned:

- **Key leader and soldier stabilization produced a mature fighting force.**
- **Combined arms force on force and live-fire exercises prepared units at all levels for combat.**
- **Command and control training under offensive conditions at task force through division level was essential to 3ID(M)'s success."** ***Operation Iraqi Freedom, Third Infantry Division (Mechanized) "Rock of the Marne",*** **After Action Report Final Draft, 12 May 2003**

Chapter 1
Introduction

1-1. Overview

a. Effective live training, carried out to a high doctrinal standard, is the cornerstone of operational success. The training of the critical tasks that individual, crew, platoon and companies have to accomplish to be combat ready is directly related to the availability and capability of live-fire ranges and maneuver areas. The continued improvement of live-fire ranges and facilities is key to the development of the go-to-war skills. Live-fire ranges and facilities will be even more important for units in the future as they must be able to be deploy and be operational within ninety-six hours in a combat zone. The successful implementation of the business end of the Future Combat System (FCS) Unit of Action is when the sharing of information in the sensor array allows for precision targeting and grouping of precision fires for target kills. This capability can only be validated on a live fire range or complex.

b. The Army modernization of the combat force from the Current Force to the Future Force will increase the lethality and precision of weapon systems over greater distances. As these weapons are introduced and combined with a net centric digital command and control the achievement of overwhelming tactical dominance and control of the battlefield will be obtained. The precision of the weapon system will allow for the determinate of appropriate force, preventing unnecessary loss of life, as well as laying the foundation for the future operational environment.

1-2. Army Ranges

a. The Army's family of training ranges provides training opportunities to develop and improve soldier and team proficiency and competence in the use of sophisticated weaponry. Individual soldier proficiency and collective training ranges realistically portray combat conditions to mold the team into an effective fighting unit.

b. The computer-controlled ranges of today allow trainers to develop scenarios and control targets and battlefield simulation devices. This permits soldiers and units to practice mission essential tasks in a stressful environment. Computerized systems also provide performance feedback. After-action reviews (AAR), using data recorded during training, permits the commander to assess the unit's performance. The accurate feedback allows leaders to assess the mission status of their units and design training programs to overcome the identified shortcomings. The performance feedback highlights positive actions to reinforce correct procedures and to foster soldiers' confidence, enabling soldiers and leaders to recognize and correct their shortcomings.

1-3. Challenge

a. As weapon systems become more lethal and capable of delivering greater firepower over increased distances Army ranges must change. Current training ranges are required to support soldiers using their weapons through live-fire, subcaliber devices, and laser and simulation technology. The ranges of the future must serve as the focal point of training as we integrate the Live-Virtual-Constructive linkups and add digital command and control elements. At company level and below soldiers train and hone their combat skills in live-fire and maneuver. In an era of intense resource competition, each dollar spent to develop, mitigate, or restore training ranges must deliver the maximum return in effective training and combat readiness.

b. With greater firepower and maneuver capabilities, the requirement for range land will continue to grow. Acquiring additional range lands will be difficult, so sustaining our current range lands becomes a critical task. Challenges to range land acquisition include: encroachment of commercial and private development, protection of threatened and endangered species, loss of wetlands, prevention of soil, surface and ground water contamination, deterioration of air quality and noise effects.

1-4. Approaches

a. One approach to meet the training challenge is to develop ranges capable of supporting training for multiple purposes, weapons, and combined arms. For example, training on an automated field-fire (AFF) range and an automated record-fire (ARF) range can be provided on a single modified record-fire (MRF) range, a land savings when compared to construction of separate ranges. The qualification training range (QTR) provides a similar capability to combine many other types of ranges, see Figure D-5.

b. Risk management integration is integral to successful range design, construction, use and sustainment. Early application of risk management in the planning process provides the planners and decision makes with alternatives that reduce safety hazards, limit adverse environmental impact and minimize operational constraints. Information and data developed during the risk management process will facilitate and expedite the development of environmental impact documentation and accelerate the approval process.

1-5. Planning

a. Range development projects require careful, deliberate planning by a team who coordinates the Range and Training Land Program (RTLP) process in accordance with Army Regulation (AR) 210-21, *Army Ranges and Training Land Program,* and this circular. The members of this interdisciplinary team are collectively called "range planners." Range planners should seek innovative alternatives to fulfill training requirements. At the same time, they consider the need for ranges to support surge requirements on short notice in the event of mobilization. Command interest and support are essential to each step of the development process. Accurate, complete justification ensures that resource expenditures provide an optimum return.

b. The range planning team consists of:

(1) trainers

(2) installation real property master planners

(3) environmental and natural/cultural resource managers

(4) financial and manpower resource managers

(5) range officers

(6) safety and occupational health managers

(7) force developers

(8) facility engineers

Chapter 2
Range Modernization

2-1. The Range Modernization Program

a. The increased accuracy and lethality of weapons was one driving force to change Army ranges. The other was the need to train soldiers in more realistic environments. The recognition that urban combat is unavoidable has resulted in new types of ranges such as Live Fire Exercise Shoothouses, Urban Assault Course, Battle Area Complex (BAX), and Combined Arms Collective Training Facility (CACTF). As the proponent schools identify training requirements, the Army will continue to modify existing ranges and build new types of ranges as required. The focus for ranges will be interoperability, standardization, improved targetry, digital capability and multi-purpose utility.

b. The Department of the Army (DA), Office of the Deputy Chief of Staff, G3 (ODCS, G-3), manages the RTLP. The ODCS, G-3 coordinates training range activities with the U.S. Army Training and Doctrine Command (TRADOC), Corps of Engineers (COE), Army Materiel Command (AMC) IAW 350-38, *Training Device Policies and Management*, and other major Army commands (MACOM) to achieve operational readiness.

c. Army doctrine requires combined arms teamwork and synchronization. Units must train for wartime combined arms operations. Combined arms proficiency results from regular practice of combat missions and tasks in the live domain. It starts with the development of individual skills. Individual skills, when combined and practiced build unit proficiency from crew through brigade task force.

d. The modernization of Army ranges under the RTLP supports this doctrine. Range design and construction must support the development of soldier skills in individual weapons and crew-served weapon systems. The ranges also support unit training to standards established in Army Training and Evaluation Program (ARTEP) and Mission Training Plans (MTP). Multiple-use ranges meet these requirements and reduce construction and operating costs by permitting training with a variety of weapons on the same range. Many of these ranges support collective training for small units.

e. Computer technology can equip ranges with target remote control. These devices can depict realistic battlefield conditions under a variety of offensive and defensive scenarios. Computer-recorded hits and misses enable trainers to analyze performance, provide corrective instruction, and provide accurate AARs. Computer technology combines with other training devices to create stressful, challenging scenarios for soldiers to train as they will fight.

f. Prudent implementation of the standardization program:

(1) Establishes a family of ranges for compatible weapons to provide training of one or more well-defined requirements.

(2) Base range size and configuration on the units that the range will support. The quantity of firing positions and target emplacements can be tailored to user needs, making the most effective use of the range.

(3) Provides training to meet standard weapons qualification and sustainment training requirements.

(4) Permits commanders to assess combat readiness and prepare individuals and units for advanced targeted training.

(5) Fosters standard operating procedures leading to a common understanding of force employment.

(6) Provides accurate throughput capabilities of ranges for mobilization planning. This lets mobilization planners determine the number of ranges needed to meet training requirements.

g. Standard design manuals are available for ranges included in the RTLP. These manuals provide layouts of firing positions, lanes, and target arrangements as guidelines for range planners, architects, and construction contractors. The manuals provide flexibility for range development projects to adapt to local throughput, terrain, surface danger zone (SDZ) placement, and environment. These conditions may warrant variances in the number of lanes and targets and in the configuration of ranges from one location to another.

h. The RTLP established automated ranges both in the continental United States (CONUS) and Outside CONUS (OCONUS) and includes active Army, Army Reserve, and Army National Guard installations. The complexities of ranges are from the Automated Field-Fire (AFF) Range which provides entry-level familiarization with the M16-series rifle to the highly sophisticated Digital Multipurpose Range Complexes (DMPRC) that supports collective training at the platoon level. In some cases, DMPRCs support companies in combined arms live-fire exercises (CALFEX) when augmented with adjacent ranges, maneuver areas, and additional targetry.

2-2. Future Outlook

a. Weapon systems currently planned or under development will require land resources to include expanded SDZs. Costs of ordnance used for training and qualifying will likely increase. Increased operating tempos may incur greater costs. Increased reliance on weapon system training simulators, embedded training devices, and the tactical engagement simulation (TES) training system can avoid or offset many of these costs. Planning for range development and range upgrades must anticipate the impact of new weapon systems and training aids, devices, simulators, and simulations (TADSS).

b. Ranges will evolve to support both live-fire and TES using non-live-firing systems similar to the Multiple Integrated Laser Engagement System (MILES). Although the emphasis on multiple-use ranges will continue, constraints on land and fiscal resources will limit construction of large range complexes. Long-range, direct-fire gunnery training will concentrate in areas capable of containing projectiles in existing SDZs or with land available to expand SDZs. Where expansion is not possible, other answers must be found, such as shared use of ranges or use of non-firing training systems.

c. Target control and scoring will continue to improve through more powerful computer systems. These systems will provide more flexibility to program training scenarios and permit event-driven target acquisitions. Voice-recognition systems linked with range-control computers will permit trainers and instructors to play a direct role in activating targets to keep pace with the evolving battle situation. Range instrumentation systems, incorporating position-location technology, will identify units and track their relative location to targets. Improved hit sensors, thermal imagery, and round discriminators will provide more accurate scoring and feedback on shooter performance. Through-sight video systems will confirm gun-laying accuracy and provide additional performance feedback. Battlefield effects will improve as targets are equipped with "shootback" simulators. Casualty-assessment systems will determine the effects of indirect fire. Observer-controllers (OC) equipped with electronic clipboards containing training and evaluation outlines (T&EO) will provide more effective training performance feedback, and input to lessons learned.

Chapter 3
Training Areas and Centers

3-1. Introduction

a. Army training doctrine provides for the progressive achievement and sustainment of proficiency in individual and collective tasks. The Army classifies ranges and training areas by level of training from individual soldier qualification skills, through integrated live-fire and maneuver unit training, to large-scale force-on-force exercises using TADSS including progressive excellence in markmanship both in training areas and on ranges.

b. Three types of training areas support progressively higher levels of proficiency—local training areas (LTAs), major training areas (MTAs), and combat training centers (CTCs).

c. LTAs and MTAs support individual, crew and unit collective training using a combination of standard ranges that provide the flexibility to support a variety of weapons and weapon systems. Stationary and moving targets, remotely controlled and supplemented with battlefield weapons-effects simulators, provide opportunities to train under challenging conditions. The target arrays portray a threat environment. Units conduct training on LTAs facilities under their installation's direct purview. These are usually limited in size. Training areas and ranges at home station and Army schools use a combination of ranges and TADSS to develop soldier and crew proficiency and to exercise small-unit mission essential tasks.

d. CTCs offer stressful exercises for maneuver battalions and command and control headquarters. These exercises focus on missions and tasks that enable the unit to capitalize on its strengths and improve its performance.

e. Occasionally units in overseas locations arrange for maneuver rights areas (MRAs) with local jurisdictions and the MACOM. This provides opportunities for periodic maneuver training on land not directly controlled by the Army.

3-2. Local Training Areas (LTA)

a. For CONUS-based active units, homestation training for individual and crew weapons proficiency training typically occurs in the LTA. Local training area facilities allow familiarization, qualification, and sustainment training with minimum impact on resources for travel to and from other types of training areas. Training focuses on individual through platoon weapons proficiency and battalion maneuver requirements. Tactical engagement simulation systems and other weapon system training simulators support maneuver and live-fire exercises.

b. Training land and range availability at overseas LTAs and installations are usually more constrained than those available for CONUS-based active units. Reserve Component (RC) centers, armories, and weekend training sites, face limited LTAs as well. Training proceeds to the extent available resources will support including individual and limited collective weapons proficiency and small-unit maneuver training.

3-3. Major Training Areas (MTA)

a. An MTA usually has enough ranges and training land resources to support:

(1) Collective live-fire proficiency.

(2) Combined arms live-fire exercises.

b. The MTAs are usually geographically separate from LTAs. For example:

(1) CONUS – Yakima Firing Center, subinstallation and approximately 130 miles southeast of Fort Lewis, Washington.

(2) OCONUS – Grafenwoehr Training Area (GTA) and Hohenfels Training Area (HTA), Germany; Pohakaloa Training Area (PTA), Hawaii.

c. Units training at MTAs concentrate on large unit collective live-fire training (platoon through battalion) and maneuver training (battalion or brigade) according to doctrine and standards. This builds on the training proficiency achieved at LTAs. An MTA provides training to sustain lessons learned from CTC experiences that LTAs could not tactically or doctrinally accommodate. The size of some installations supports simultaneous individual weapons proficiency and both small and large unit collective training (e.g., Fort Hood, Texas).

3-4. Combat Training Centers (CTC)

a. The CTCs are the Army's premier training centers. Instrumentation systems collect data to support training feedback to units. The Army derives lessons learned from the collected data and uses the lessons learned to improve doctrine, training, leader development, organization, materiel, personnel, and facilities (DTLOMPF). The four CTCs are:

(1) National Training Center (NTC), Fort Irwin, California.

(2) Combat Maneuver Training Center (CMTC), GTA (live-fire) and HTA (maneuver), Germany.

(3) Joint Readiness Training Center (JRTC), Fort Polk, Louisiana.

(4) Battle Command Training Program (BCTP), Fort Leavenworth, Kansas.

b. The three maneuver CTCs (NTC, CMTC, and JRTC) focus on brigade task force training using a combination of live-fire ranges and maneuver training land. The Army's goal is for battalions to experience the rigorous and realistic training environment at a maneuver CTC during a commander's tour. Organic to these CTCs are:

(1) A dedicated opposing force unit.

(2) Observer-controllers.

(3) Instrumentation systems not available at LTA and MTA sites.

c. The maneuver CTCs require sufficient land to accommodate doctrinal fire and maneuver training using multiple scenarios over varied terrain. This allows rehabilitation of the land and precludes over-familiarization with the terrain, which detracts from training realism.

3-5. Training Area Functional Uses

The Army's force modernization program has increased the demand for collective training.

a. As a result, greater reliance must be placed on the use of LTAs for individual training and the initial stages of crew training. Local training areas include cantonment areas at installations, RC centers, and armories. Tactical engagements can be simulated using appropriate TADSS to overcome the space limitations required for live-fire exercises. Training at LTAs achieves and sustains soldier and crew proficiency and prepares for advanced collective training at MTAs and CTCs.

(1) The collocation of an MPRC, a multipurpose training range (MPTR), and an infantry squad battle course (ISBC) is desirable at an MTA. This accommodates CALFEX and other forms of combined arms training using a wide variety of weapon systems and units. Collocating these facilities affords the land area and target arrays required to train platoon through company live-fire engagements. Although primarily designed for M1 Abrams tank and M2/M3 Bradley/Cavalry Fighting Vehicle (BFV/CFV) training, all division weapon systems can be trained on the MPRC. The Stryker vehicle can train on a MPRC/digital MPRC (DMPRC) though the BAX is preferred. Air defense artillery, field artillery, and attack helicopters can support combined arms scenarios. The MPRC supports up to company-size light infantry, airborne, and air assault forces. The DMPRC enhances the capabilities of the standard MPRC. The DMPRC increases target count, maneuver area size, and supports the collection of digital data to support the assessment of training.

(2) The MTA maneuver areas provide limited battalion and brigade maneuver force-on-force capabilities.

b. Combat training centers provide an advanced collective training experience for combat, CS, and CSS units. In contrast to MTAs, CTCs have a dedicated opposing force (OPFOR) trained in threat tactics, observer controllers, and instrumentation systems. During force-on-force engagements, an instrumentation system collects, reports, and stores engagement data from TES-equipped participants. OC's use this information to prepare AARs and post exercise take home packages (THP). Arrays of automated, radio-controlled targets are distributed over a 5-kilometer by 10-kilometer area for the offensive and defensive live-fire ranges. On the offensive live-fire range, the battalion task force employs maneuver and live fire to attack a defending force represented by an appropriate threat target array. On the defensive range, the task force engages a target array that represents multiple threats.

(1) The NTC provides CONUS-based maneuver battalion task forces the experience of a demanding high-intensity force-on-force engagement against an OPFOR organized and trained to replicate a threat motorized rifle regiment. During NTC rotations, battalion task forces also train on offensive and defensive live-fire ranges using organic and supporting direct- and indirect-fire weapon systems, artillery, helicopter gunship, and U.S. Air Force (USAF) close-air support.

(2) The JRTC provides training for contingency forces (light infantry, ranger, airborne, air assault, and special operations) in joint operations. Scenarios include forced and non-forced entry and special operations missions for low- and mid-intensity conflicts worldwide. These scenarios employ elements of the USAF Air Combat Command and Air Mobility Command. The JRTC supports maneuver in day and night exercises. The well-trained OPFOR represents a threat based on the scenario. Planned exercises include urban operations, small boat operations, river crossing operations, and offensive air support. The instrumentation system supports a two-battalion brigade task force with its supporting CS and CSS units.

(3) Combat forces in U.S. Army Europe (USAREUR) conduct combined arms, force-on-force training at the CMTC. The CMTC training takes place at two locations. Live-fire training takes place at GTA. GTA provides MPRC, MPTR and ISBC training ranges. HTA supports maneuver training. Units training at HTA use the TES training system with a mature instrumentation system. Infantry/armor battalions and cavalry squadrons train at the CMTC annually. A well-trained OPFOR portrays major elements of a scenario driven enemy force and are trained in the appropriate threat tactics and doctrine. The OPFOR provides opportunities for realistic exercises from low intensity conflicts to full combat. The CMTC instrumentation systems support exercise control and collects real time position/location and engagement data. This data is used to conduct AARs and prepare THPs.

Chapter 4
Range Requirements

4-1. Introduction

a. Training ranges are major contributors to keeping the Army ready to accomplish its missions. They also represent a considerable investment in time, land, and fiscal resources. Once a range training need is established, range planners must examine other alternatives to a new range such as:

(1) The required range may be available at another installation.

(2) An existing range might be converted to support multiple training requirements.

(3) A simulation, subcaliber training device, or reduced-scale range might satisfy the need.

(4) Existing ranges might increase operational hours.

b. Clearly defined requirements for individual, collective, or combined arms training must support decisions to initiate range development projects.

c. This chapter provides guidance on assessing needs for range development. AR 210-21 establishes the requirement for this assessment and lists many of the factors that require consideration.

d. Throughput is an important factor in deciding whether to undertake a major expenditure for range modification or construction. Required throughput is the number of individuals, crews, or units requiring a particular type of training. Throughput capacity is the number of soldiers, crews, or units a given range can support during a specific period. Throughput calculations help evaluate alternatives and determine how many firing points, lanes, or ranges are needed to satisfy a training requirement. They must include both peacetime and mobilization requirements.

e. This chapter presents formulae for calculating throughput and examples of applying the formulae to different weapon systems. It also provides a matrix of standard ranges. Appendix A provides tables to help unit trainers select appropriate ranges for training and weapons qualification. It further identifies ranges for the current and future forces including the SBCT.

4-2. Determining Training Requirements

a. Field Manuals (FM) 7.0, *Training the Force* and FM 7.1, *Battle Focused Training* are primary references for determining training requirements. As described in the FMs, battle focus enables unit leaders to develop a Mission Essential Task List (METL). The METL comes from the unit's war plans and external directives. It links the unit's collective mission essential tasks with the individual tasks that support them. It is also a checklist to identify requirements for collective training of these tasks.

b. Applicable ARTEP MTPs identify conditions and standards for collective training tasks. The MTPs show the relationships among wartime missions, collective tasks, battle drills, and individual tasks. They give examples and guidance for developing situational training exercises (STX) based on mission requirements. The MTPs provide training criteria for critical wartime mission tasks. For additional sources of individual and collective training objectives, see:

(1) DA Pamphlet (Pam) 350-38, *Standards in Weapons Training.*

(2) Soldier's manuals.

(3) Applicable gunnery manuals.

(4) Mobilization plans.

(5) Force modernization plans.

(6) General defense plans.

(7) Army Regulations.

4-3. Assessing Range Needs

a. DA Pamphlet 350-38 provides weapon system qualification standards and suggested training strategies to attain and sustain the standards. It establishes prescribed standards and training resources. For each weapon system the pamphlet outlines the number of rounds allocated per training event and the frequency of events for each weapon.

b. To assess range needs, match the METL, individual and collective task training objectives, and associated training range requirements to the suitable weapon system. DA Pamphlet 350-38 provides frequency requirements for each event and can be accessed from the Army Training Support Center (ATSC) website at http://www.atsc.army.mil/atmd/STRAC/stracweb.asp

c. In selecting ranges, consider:

(1) The number of troops and weapon systems at the installation.

(2) Range throughput capacity.

(3) External training support requirements; for example, those of the RC.

(4) Availability of Army Target System (ATS) ranges to support the gunnery tables and tasks outlined in weapon system field manuals.

(5) Types of ATS targets available.

(6) Overlapping Surface Danger Zones (SDZs).

(7) Environmental restrictions.

(8) Availability of training devices, simulations, and subcaliber munitions.

(9) Installation priorities.

d. Factors to consider when assessing range needs include:

(1) Unit METL.

(2) Weapon system quantities.

(3) Required gunnery qualification frequencies.

(4) STRAC standards.

(5) Availability and throughput capacity of ranges.

e. This assessment may identify a requirement for a new range or an upgrade to an existing range.

4-4. Range Availability

Subtract scheduled and unscheduled maintenance days, holidays, and inclement weather days from 365 calendar days to determine range availability. Master training schedules for tenant and regional units will impact range availability and must be considered by range managers. Table 4-1 indicates typical range availability.

Table 4-1. Typical range availability (days)

Factors	AFF/ARF	MRF	MPTR	MPRC	DMPRC
Scheduled Maintenance	22	25	55	59	59
Unscheduled Maintenance	16	11	15	17	17
Holidays	10	10	10	10	10
Inclement Weather	5	5	5	5	5
Total Non-Available Days	53	51	85	91	91
Available Days	312	314	280	274	274

4-5. Scheduling And Utilization

a. The Army's goal is to schedule ranges to be used 80 percent of the days available for training. For example, the MPTR shown in Table 4-1 should be scheduled for use at least 224 days (80% of 280 days) annually. Use of existing facilities is critical to the requirements identification and needs assessment phase discussed in AR 210-21. The Army's utilization goal for ranges and training land is 90 percent of scheduled days. Therefore, if the MPTR in Table 4-1 is scheduled for 224 days, it should be used at least 202 days of the year.

b. When calculating range requirements for a unit, consider:

(1) The number of training elements (e.g., soldiers, crews, and squads) assigned to the unit.

(2) The number of lanes available on the range.

(3) The average time required for one training element to complete a course of firing.

c. For example, experienced soldiers in an infantry company may have a high level of marksmanship proficiency completing the course of firing with greater efficiency than soldiers in other types of units. The analysis described below estimates both day and night firing to require 30 minutes per soldier. Although night firing requires more time because of operations in the dark, other factors compensate. Fewer rounds are fired at night (30 versus 40 for day firing), and the distances fired are shorter (25 and 50 meters versus 50 to 300 meters for day firing).

d. Calculate the time required to cycle one 200-soldier company through an MRF range as follows:

NUMBER OF LANES: 16

NUMBER OF SOLDIERS IN COMPANY: 200

TIME REQUIRED FOR ONE SOLDIER TO COMPLETE FIRING: 0.5 HOUR

NUMBER OF SOLDIERS PER HOUR = 16 ÷ 0.5 = 32

NUMBER OF HOURS = 200 ÷ 32 = 6.25 HOURS PER COMPANY

e. Calculate the total time required for zeroing, sustainment firing, and qualification firing by one company during the course of a year as follows:

Zeroing:

NUMBER OF LANES: 110

NUMBER OF SOLDIERS IN COMPANY: 200

TIME REQUIRED FOR ONE SOLDIER TO COMPLETE FIRING: 0.5 HOUR

FREQUENCY OF ZEROING PER YEAR (DAY ONLY): 2

NUMBER OF SOLDIERS PER HOUR = 100 ÷ 0.5 = 200

NUMBER OF HOURS = 2(200 ÷ 200) = 2 HOURS PER COMPANY PER YEAR

Sustainment:

NUMBER OF LANES: 16

NUMBER OF SOLDIERS IN COMPANY: 200

TIME REQUIRED FOR ONE SOLDIER TO COMPLETE FIRING: 0.5 HOUR

FREQUENCY OF SUSTAINMENT FIRING PER YEAR: 2 DAY AND 2 NIGHT

NUMBER OF SOLDIERS PER HOUR = 16 ÷ 0.5 = 32

NUMBER OF HOURS = 2(200 ÷ 32) = 12.5 DAY HOURS AND 12.5 NIGHT HOURS PER COMPANY PER YEAR

Qualification:

SAME FACTORS AS FOR SUSTAINMENT

NUMBER OF HOURS = 12.5 DAY AND 12.5 NIGHT HOURS PER YEAR

Total time (hours) required for a single company to complete M16-series/M4 firing per year:

ZEROING	2.0 DAY HOURS	
SUSTAINMENT	12.5 DAY HOURS	12.5 NIGHT HOURS
QUALIFICATION	12.5 DAY HOURS	12.5 NIGHT HOURS
Total	27.0 DAY HOURS	25.0 NIGHT HOURS

f. Given the foregoing, the annual range requirement (days) for M16-series/M4 rifle training for one company is 3.4 (27.0 hours ÷ 8 hours per day) days and 3.1 (25.0 hours ÷ 8 hours per night) nights.

g. The process shown above can be used to calculate usage requirements for other weapon systems, training requirements, and training elements.

4-6. Quantifying Range Requirements

a. Department of the Army Pamphlet 350-38 identifies weapon system training requirements, frequency of range events, and ammunition resources allocated to training. Using the formula in the throughput calculations paragraph below, installation range personnel can quantify the number of days required to conduct training. They can use this information to determine the number of ranges needed to support training requirements. For example, a mechanized infantry company must qualify its soldiers semiannually with the M16-series and M4 rifle. To meet this requirement, the unit conducts zeroing, practice, and day and night record firing. Two types of ranges support this firing:

(1) 25-meter zero range.

(2) 16-lane MRF range.

b. A mechanized infantry division includes five mechanized infantry battalions with four companies each. Based on single-company annual usage requirements of 3.4 days and 3.1 nights, compute one battalion's annual usage requirements as follows:

DAY: 3.4 DAYS PER COMPANY X 4 COMPANIES PER BATTALION = 13.6 OR 14 DAYS PER BATTALION

NIGHT: 3.1 NIGHTS PER COMPANY X 4 COMPANIES PER BATTALION = 12.4 OR 13 NIGHTS PER BATTALION

To determine battalion, and above, range throughput requirements, round up fractional days or hours to the next whole number.

c. Compute the total divisional annual M16-series and M4 rifle range requirement for mechanized battalions as follows:

DAY: 14 DAYS PER BATTALION X 5 BATTALIONS PER DIVISION = 70 DAYS PER DIVISION

NIGHT: 13 NIGHTS PER BATTALION X 5 BATTALIONS PER DIVISION = 65 NIGHTS PER DIVISION

d. An installation's M16-series and M4 rifle range requirement is determined by totaling the firing requirements of:

(1) Combat units (infantry, armor, field artillery, engineer, air defense, aviation, and cavalry).

(2) CS and CSS units.

(3) Installation support personnel.

(4) Reserve Component units.

(5) Reserve Officers' Training Corps units.

(6) Others requiring range support.

e. A comparison of this total with the range availability determines if the requirement can be met. If not, an availability shortfall exists that indicates the need for additional range capability.

f. Another example is a tank company must qualify its 14 crews semiannually on Gunnery Table VIII-A and B. Before firing for qualification, the unit practices on Table VII-A and B. Either the DMPRC or the DMPTR will support this training requirement.

g. Table 4-2 shows the approximate times for tank units to complete firing all offensive (Off) and defensive (Def) engagements on the tank tables shown. Time data include the time allowed for each engagement, five minutes between engagements, and ten minutes for the element to clear the range before the next element starts downrange. The time required to cycle a tank crew through a DMPRC or DMPTR is 74 minutes for Table VII-A and B and 76 minutes for Table VIII-A and B. Calculate the hours required for one company of 14 crews to train on the tables as follows:

Table 4-2. Tank gunnery tables - time requirements (minutes)

Table	Crew	Section	Platoon
Table V-A and B	95+		
Table VI-A and B	104+		
Table VII-A and B	74+		
Table VIII-A and B	76+		
Table XII-A and B (Off and Def)			120+
+ = Approximate times in minutes			

TABLE VII-A and B: 74 MINUTES PER CREW X 14 CREWS PER COMPANY DIVIDED BY 60 MINUTES TO GET HOURS OR (74 x 14) ÷ 60 = 17.3 HOURS PER COMPANY.

TABLE VIII-A and B: 76 MINUTES PER CREW X 14 CREWS PER COMPANY DIVIDED BY 60 MINUTES OR (76 X 14) ÷ 60 = 17.7 HOURS PER COMPANY

h. Obtain the time requirement for a single battalion consisting of three tank companies plus the two headquarter's tanks to train semiannually is as follows:

TABLE VII-A and B: 17.3 HOURS PER COMPANY X 3 COMPANIES + 2 HQ TANKS PER BATTALION = 54.4 HOURS PER BATTALION

TABLE VIII-A and B: 17.7 HOURS PER COMPANY X 3 COMPANIES + 2 HQ TANKS PER BATTALION = 55.6 HOURS PER BATTALION

i. The annual requirement for the battalion is twice the semiannual requirement, or 104 hours for Table VII-A and B and 107 hours for Table VIII-A and B.

j. Obtain the total divisional usage requirements for five battalions:

TABLE VII-A and B: 104 HOURS PER BATTALION X 5 BATTALIONS PER DIVISION = 520 HOURS PER DIVISION

TABLE VIII-A and B: 107 HOURS PER BATTALION X 5 BATTALIONS PER DIVISION = 535 HOURS PER DIVISION

k. Based on an average of 20 hours per day available for training on DMPTRs and DMPRCs, the total division requirement is approximately 26 days for Table VII-A and B and 27 days for Table VIII-A and B:

TABLE VII-A and B: 520 HOURS PER DIVISION ÷ 20 HOURS PER DAY = 26 DAYS PER DIVISION

TABLE VIII-A and B: 535 HOURS PER DIVISION ÷ 20 HOURS PER DAY = 26.75 or 27 DAYS PER DIVISION

l. Range control personnel determine the remaining tank gunnery usage requirements (Tables IX, X, XI, and XII) and BFV gunnery requirements for units that will fire on the DMPRC or DMPTR. Once again, the projected number of available training days must be established and compared to the number of required range days. If the required days exceed the number of available days, a shortfall exists.

4-7. Throughput Capacity

a. Throughput refers to the total number of soldiers or units (e.g., crews, sections, squads, platoons, and companies) to be trained in a given period of time. The throughput capacity of a range is the number of soldiers or units that it can accommodate in a given period of time. For example, a rifle range might have a capacity of one company averaging 200 soldiers over a period of eight hours. This capacity would support at least one day-firing cycle and one night-firing cycle per day. Assuming the range is available 314 days per year, the range would have an annual throughput capacity of 314 companies, or 62,800 soldiers (314 companies x 200 soldiers per company), for day and night firing.

b. Calculation of throughput capacity is based on the type of training, the time required for a single individual or unit to complete a training event or series of events, and the period of time (day, week, month, year) which applies. Throughput calculations for some types of ranges may also include the number of soldiers or units that can train simultaneously.

c. For example, develop the daily throughput capacity of an MRF range with 16 lanes as follows:

NUMBER OF LANES: 16

TIME REQUIRED FOR ONE SOLDIER TO COMPLETE FIRING: 30 MINUTES OR 2 SOLDIERS PER HOUR

TIME AVAILABLE FOR TRAINING: 8 HOURS PER DAY

NUMBER OF SOLDIERS PER HOUR: 16 LANES X 2 SOLDIERS PER LANE PER HOUR = 32 SOLDIERS PER HOUR

THROUGHPUT CAPACITY = 32 SOLDIERS PER HOUR X 8 HOURS PER DAY = 256 SOLDIERS PER DAY

d. Obtain the annual capacity by multiplying the daily capacity by the number of days available for training:

DAILY THROUGHPUT CAPACITY = 256 SOLDIERS PER DAY

DAYS AVAILABLE FOR TRAINING: 314 (ESTIMATED)

ANNUAL THROUGHPUT CAPACITY = 256 SOLDIERS PER DAY X 314 DAYS PER YEAR = 80,384 SOLDIERS PER YEAR

4-8. Throughput Requirements

a. The number of required training iterations is included in the calculation to compute annual throughput requirements. For example, annual M16-series and M4 rifle training for infantrymen consists of:

(1) Two iterations of daytime zeroing with the weapon.

(2) Two day and two night iterations of qualification firing.

(3) Two day and two night iterations of sustainment firing.

b. Compute the total rifle training throughput requirement for a mechanized infantry division as follows:

NUMBER OF SOLDIERS IN COMPANY: 200

NUMBER OF COMPANIES: 30

NUMBER OF ITERATIONS REQUIRED: 6 DAY AND 4 NIGHT ITERATIONS OR 10 ITERATIONS PER YEAR

NUMBER OF SOLDIERS PER COMPANY PER YEAR = 200 SOLDIERS PER COMPANY PER ITERATION X 10 ITERATIONS PER YEAR = 2000 SOLDIERS PER COMPANY PER YEAR

THROUGHPUT REQUIREMENT = 2000 SOLDIERS PER COMPANY PER YEAR X 30 COMPANIES = 60,000 SOLDIERS PER YEAR

c. Comparing the throughput capacity to the throughput requirement (80,384 versus 60,000) reveals an excess capacity of 20,384 soldiers per year.

4-9. Armored Vehicle Gunnery Requirements

a. Tables 4-2 can be used to develop throughput requirements for firing gunnery Tables V through XII for tank battalions. Tables 4-3 and 4-4 can be used to develop throughput requirements for firing gunnery Tables V through XII mechanized battalions. Calculate the time required for a unit to fire a gunnery table by using the formula:

$H = N (M \div 60)$, where:

H is the number of hours required to complete the gunnery table.

N is the number of vehicle crews required to fire the table.

M is the number of minutes required for one crew to complete day and night firing events (divide by 60 to convert minutes to hours).

b. For example, the time required for a tank company consisting of 14 tank crews to complete Tank Table VIII-A and B (day and night) is:

N = 14 CREWS

M = 76 MINUTES (FROM TABLE 4-2)

H = 14 (76 ÷ 60) = 17.7 HOURS (17 HOURS AND 42 MINUTES)

c. Determine the number of days required for training by dividing the required hours by the number of hours that the range is available for daily training. In the example above if the Table VIII range were available 16 hours per day, the number of days required for a company to complete firing would be:

H ÷ 16 = 17.7 ÷ 16 = 1.1 DAYS (1 DAY, 1 HOUR, AND 36 MINUTES)

This can be expressed in a formula as:

D = N(M ÷ 60) ÷ A, where:

D is the number of days required to complete the table.

A is the number of hours that the range is available on a daily basis.

d. Throughput capacity for an MPRC or MPTR, where only one training element (tank) can complete a gunnery table at a time, is developed as follows:

NUMBER OF LANES: 1

TIME REQUIRED FOR ONE TRAINING ELEMENT TO COMPLETE TABLE: FROM TABLE 4-2

TIME AVAILABLE FOR TRAINING: 20 HOURS PER DAY (AVERAGE)

DAYS AVAILABLE FOR TRAINING: 277 (AVERAGE)

NUMBER OF TRAINING ELEMENTS PER DAY: NUMBER OF CREWS, SECTIONS, OR PLATOONS

THROUGHPUT CAPACITY = NUMBER OF LANES x ((TIME REQUIRED FOR TRAINING ELEMENT ÷ 60) x TIME AVAILABLE FOR TRAINING) x DAYS AVAILABLE FOR TRAINING

e. For example, a gunnery table requires an average of 87 minutes to complete. Therefore:

THROUGHPUT CAPACITY = 1((87 ÷ 60) X 20) X 277 = 8033 CREWS, SECTIONS, OR PLATOONS PER YEAR

f. Develop the tank gunnery training throughput requirement for an armored division assigned 15 tank companies with 14 crews per company as follows:

For crews:

NUMBER OF CREWS: 210 (15 TANK COMPANIES X 14 CREWS PER COMPANY)

NUMBER OF CREW TABLES: 4 TABLES PER CREW

NUMBER OF ITERATIONS: 2 PER TABLE PER YEAR

THROUGHPUT REQUIREMENT (TR) = 210 CREWS X 4 TABLES PER CREW X 2 PER TABLE PER YEAR = 1680 CREWS PER YEAR

For sections:

NUMBER OF SECTIONS: 105 SECTIONS (15 TANK COMPANIES X 7 SECTIONS PER COMPANY)

NUMBER OF SECTION TABLES: 2 TABLES PER SECTION

NUMBER OF ITERATIONS: 2 PER TABLE PER YEAR

TR = 105 SECTIONS X 2 TABLES PER SECTION X 2 PER TABLE PER YEAR = 420 SECTIONS PER YEAR

For platoons:

NUMBER OF PLATOONS: 60 (15 TANK COMPANIES X 4 PLATOONS PER COMPANY)

NUMBER OF PLATOON TABLES: 2 TABLES PER PLATOON

NUMBER OF ITERATIONS: 2 PER PLATOON PER YEAR

TR = 60 PLATOONS X 2 TABLES PER PLATOON X2 PER TABLE PER YEAR = 240 PLATOONS PER YEAR

Total tank gunnery training:

TOTAL TR = 1680 CREWS + 420 SECTIONS + 240 PLATOONS = 2,340 TRAINING ELEMENTS PER YEAR.

g. Comparing the throughput requirement to the capacity (2,340 versus 8,033) indicates an excess capacity of 5,693 training elements per year. This excess capacity can help meet BFV gunnery requirements of the division's mechanized battalions. Trainers develop the throughput requirement for these battalions in the same manner.

h. Table 4-3 displays the average time required to complete BFV Tables V-A and B through VIII-A and B and squad firing exercises. These times are not consistent worldwide. Trainers should adjust for local conditions including the level of range automation and the distance traveled down range between engagements.

Table 4-3. BFV team/squad gunnery tables and exercises – time requirements (AP only) (minutes)

Table	Team/Squad
Table V-A and B	68
Table VI-A and B	69
Table VII-A and B	81
Table VIII-A and B	81
Infantry Squad Combat Exercise (Day and Night)	120

i. Bradley Tables IX, X, XI and XII involve gunnery tasks associated with missions. Day and night engagements occur in one or more missions, as determined by the unit commander. Timelines for day and night engagements of these missions follow:

MOVEMENT TO CONTACT: 3.8 HOURS

ATTACK: 3.7 HOURS

RAID: 3.4 HOURS

AMBUSH: 3.3 HOURS

RECON/SECURITY: 3.6 HOURS

DEFENSE: 3.4 HOURS

RETROGRADE: 3.4 HOURS

AVERAGE: 3.5 HOURS

j. Table 4-4 displays the average time required for day and night firing on BFV Tables IX, X, XI, and XII.

Table 4-4. BFV section/platoon gunnery tables – time requirements (hours)

Table	Section	Platoon
Table IX (Day and Night)	3.5	
Table X (Day and Night)	3.5	
Table XI (Day and Night)		3.5
Table XII (Day and Night)		3.5

k. Scout sections perform Tables IX-A and B and X-A and B. Timing for these tables depends on terrain, weather, and distance between engagements. The estimated time for completion of these tables is eight hours for Table IX-A and B and eight hours for Table X-A and B.

4-10. Range Selection

Refer to Appendix A for the types of ranges needed to accomplish training events for selected weapon systems.

Chapter 5
Range Development

5-1. Introduction

This chapter provides procedures to implement the range modernization and standardization policies in AR 210-21. It discusses the range design and construction process and describes the quality assurance features in the range modernization program.

5-2. Planning

a. As identified in paragraph 1-5, the "range planners" coordinate the RTLP planning process in accordance with AR210-21.

b. Planning an Army training range at installation level must be comprehensive, coordinated, and systematic. The first step is to determine facility needs based on the training and unit throughput requirements discussed in Chapter 4. The next step – range planning and location (the pre-design phase) – consists of project initiation, site analysis and selection, and documentation.

c. Planners must consider the resources available to support the development of training ranges. Project planning must follow the Army's planning, programming, budgeting, and execution system (PPBES) to ensure sufficient funds are allocated for construction and maintenance. Installation resource managers are familiar with the PPBES and can provide assistance to project planners.

d. The installation master plan must consider current and future use of installation and adjoining land when identifying suitable land for ranges and maneuver training areas. Maneuver areas should accommodate and be configured for units that will use them. Force structure, weapon systems, applicable doctrine, and METLs developed per 7-0 *Training the Force* and appropriate type unit Field Manuals. Training Circular (TC) 25-1, *Training Land*, is the guide to determining maneuver land requirements.

e. Planners evaluate requirements based on annual schedules, throughput capacity based on range or training land availability, and actual use. Requirements for additional range and maneuver area capabilities are valid where throughput requirements exceed the capacity of existing facilities.

f. Adverse environmental impact, safety and occupational health, and industrial hygiene aspects of range construction and operations must be taken into account. Contaminated impact areas will not be created unless an exception to policy has been approved per AR 200-1, *Environmental Protection and Enhancement,* and AR 210-20, *Master Planning for Army Installations.* Undertaking construction in an existing impact area will require expensive ordnance clearance to ensure the safety of construction personnel. Planned weapons impact areas must meet requirements for current weapon systems and those programmed for fielding. Planners should consider the effects of impact areas and SDZs on current ranges and maneuver areas; this avoids reducing the availability or capability of existing training areas. Planning should also address airspace requirements for combined arms training.

g. Planners must locate facilities and equipment storage to meet the needs of the proposed training and support planned land use. For example, grouping small-arms ranges with close-in training facilities reduces the need for separate support facilities (e.g., latrines, instruction buildings, and parking areas). Planners must also consider fuel consumption. Fuel is a critical resource so maneuver areas and ranges should be located as close as practical to cantonment and bivouac areas to minimize travel and provide expansion potential.

h. Sub-caliber ranges and TADSS should be used to conserve full caliber ammunition for zero, sustainment, and qualification firing.

i. Operations and maintenance plans for new facilities, including ranges and training areas must achieve manpower and other resource savings.

j. Documentation for new facilities should include information on the size, condition, and capacity of land, as well as the ranges and training facilities available locally or at other accessible training areas. Safety requirements, environmental considerations, and terrain profiles for proposed range locations should also be included. Much of this information is in the catalog of installation firing ranges required by the local range regulation. Land condition and capability data can be obtained from the installation directorate of public works (DPW) natural resources office. These data are based on land-condition trends determined by using the Integrated Training Area Management (ITAM) program. The ITAM program includes the geographic information system (GIS). Information on the ITAM program is available from Headquarters, Department of the Army (DAMO-TRS) .

k. Doctrine and training strategies, information on types of units, personnel, frequency, specific weapon systems, and assigned missions provide the basis for the range/training area planning phase. Use Standards in Training Commission (STRAC) guidance as a baseline to determine alternatives to construction or land acquisition.

l. Information on the arrival of a new weapon system and the associated facility requirements is contained in Army Modernization Reference Data (AMRD), distributed on compact disc – read only memory (CD-ROM) by HQDA (DAMO-FDR), and a support facility annex (SFA) published by the U.S. Army Corps of Engineers (USACE). An SFA-assessment of the facility impact of weapons development and fielding is accomplished through logistics management for major and directed acquisition programs per AR 700-127, *Integrated Logistical Support (ILS) Manager's Guide.* Support facility annex data are updated in coordination with the materiel system program executive officer (PEO) or program manager (PM) and TRADOC system manager (TSM) as weapon system development matures.

m. Engineering expertise is provided by:

(1) The divisional engineer staff officer.

(2) The local DPW or US Army National Guard facility management officer.

(3) The Corps of Engineers division or district engineer.

(4) The USACE RTLP Mandatory Center of Expertise (RTLP-MCX).

(5) The USACE Ordnance and Explosives Center of Expertise (OE-CX)

(6) COE laboratories.

n. The installation DPW includes training facilities requirements in the installation real property master plan.

o. The installation safety manager provides range safety expertise. This expertise is also available at each MACOM safety office. The technical advisor for range safety (TRADOC Safety Office) provides additional assistance. Special-use airspace matters are processed and coordinated under the provisions of AR 95-2, *Air Traffic Control, Airspace, Airfields, Flight Activities and Navigation Aids.* In addition to the installation air traffic and airspace officer and the DA regional representative (DARR) to the Federal Aviation Administration, policy and strategy guidance can be obtained from the U.S. Army Aeronautical Services Office, Fort Belvoir, VA.

p. Health and safety risk assessment and design review assistance may be obtained from the installation medical activity and safety offices, respectively. Additional assistance is available from:

(1) U.S. Army Center for Health Promotion and Preventive Medicine, ATTN: HPPM-MIL, Aberdeen Proving Ground, MD 21010-5422 (health assessment).

(2) U.S. Army Environmental Awareness Resource Center, Huntsville, AL (environmental support services and products).

(3) U.S. Army Environmental Center, ATTN: CETHA-TS-S, Edgewood Area, Building E4460, Aberdeen Proving Ground, MD 21010-5401 (design review assessment).

q. Military construction projects (both live-fire range construction and land acquisition) are long lead time requirements (five-to-seven years). Planners must consider practical alternatives. When evaluating alternatives or determining if existing training facilities can support new weapons systems or missions, planners should consider the:

(1) Number and types of training facilities available.

(2) Use rates.

(3) User density.

(4) Ammunition authorized.

(5) Serviceability or remaining life spans of facilities.

(6) Current and projected throughput requirements.

(7) Current use and potential for new or multiple uses.

(8) Adequacy to meet training standards.

(9) Life-cycle, operating, and maintenance costs.

(10) Adverse environmental impacts, including contamination of surface and ground water, bio-diversity, soil, cultural resources, deterioration of air quality and environmental noise impact.

(11) Contamination of impact areas and the potential for creating new contaminated impact areas.

(12) Land-use impact and compatibility, including current and future use of installation and adjoining land.

r. For additional assistance, contact:

(1) The MACOM.

(2) U.S. Army Training Support Center, ATTN: ATIC-ATMD, Fort Eustis, VA.

(3) U.S. Army Corps of Engineers, ATTN: CEMP-EA, Washington, DC.

(4) Range – Mandatory Center of Expertise (RNG-MCX), U.S. Army Engineering and Support Center, Huntsville, ATTN: CEHNC-IS-SP, Huntsville, AL.

(5) PEO-STRI, (Program Executive Office - Simulations, Training and Instrumentation Command (STRICOM)) Orlando, FL in accordance with AR 350-38.

5-3. Project Initiation

a. Project initiation includes:

(1) Identifying the training requirement.

(2) Identifying training facility needs based on adequacy and availability of existing facilities.

(3) Considering alternatives to facility modification or new construction.

(4) Developing a project milestone schedule.

(5) Collecting resource data.

b. Project initiation is a coordinated effort of planners from:

(1) The directorate of plans, training, and mobilization (DPTM)/G3 (trainers, range officers, force developers).

(2) The DPW (facility engineers; installation master planners; environmental, natural, and cultural resource managers).

(3) Major using units.

(4) Installation resource management office.

(5) Security and safety agencies.

c. Representatives from the Directorate of Reserve Components, Staff Judge Advocate, and Public Affairs Office should participate as required.

d. Planners:

(1) Review training requirements.

(2) Consider alternatives to facility modification or new construction.

(3) Evaluate resources available to support development of a training range.

(4) Determine the specific type of range required, including weapon systems and ammunition.

(5) Consider construction impact on both the installation master plan and the RTLP.

5-4. Range Land

Land must be identified that accommodates the range layout, the SDZ of applicable weapon systems, and impact area requirements. An environmental analysis determines the impact of range construction and operations.

5-5. Facilities and Equipment

Resources must meet the needs of the proposed range and support nearby ranges. For example, collocating small-arms ranges with an obstacle or infiltration course would conserve land and resources by reducing common support facilities (e.g., latrines, instruction buildings, and parking areas).

5-6. Training Schedules

Range construction projects may impact training schedules for nearby ranges. Coordination with scheduling activities of the installation DPTM/G-3 must take place early in the planning process.

5-7. Training Aids, Devices, Simulators, Simulations, and Sub-Caliber Ranges

Use of these devices should be maximized to save full-caliber ammunition for zero, qualification, and sustainment firing. The acquisition of TADSS should be accomplished through AMC in accordance with AR 350-38.

5-8. Resource Management

a. Efficient use of manpower and resources for operations and maintenance should be planned.

b. Planners must consider the impact on operating costs when selecting a range location. Travel time and fuel are critical resources.

5-9. Host Nations

Range projects in overseas areas require coordination with host nation officials throughout the development process. Requirements will vary from nation to nation. The unit G-3 or local area commander should have information concerning any host nation agreement or legalities.

5-10. Alternatives and Justification

a. The Master Planning Board (MPB) approves decisions of installation planners to build new range facilities.

(1) Sharing use of an existing facility at another installation.

(2) Converting an older or specialized range to a multipurpose range.

(3) Using TADSS or additional portable targets on existing facilities.

(4) Increasing the hours of operation of existing ranges.

b. When deciding whether to share use of an existing range at another installation, commanders and planners consider:

(1) Availability of the other range.

(2) Distance and cost of travel to the alternate range.

(3) Frequency of firing required to meet weapons training standards.

(4) Possible limitations on the use of ammunition required for training.

(5) Obsolescence of equipment, facilities, or range design.

(6) Potential restrictions and waivers on range use.

c. Factors bearing on justification of a facility include:

(1) The overall range facility needs to meet valid training requirements and throughput.

(2) Fielding needs to meet unit training requirements and throughput levels.

(3) A change in the installation mission or force structure requiring an increased capability to conduct training.

(4) Capability of standard ranges to meet homestation peacetime and mobilization training requirements. For example, modernization of one range could satisfy training needs and allow the installation to close other ranges to free land for other training needs.

d. The construction and use of multipurpose training ranges ensure better management of training activities, facilities, and land.

5-11. Site Planning Process

a. Once the decision to develop a range is made and approved according to AR 210-21, the site planning process begins. The steps in the procedure are:

(1) Analyze training task requirements.

(2) Determine the type of range needed.

(3) Identify candidate sites.

(4) Analyze sites.

(5) Select the best site.

b. Site planning is crucial to successful training facility planning and development. Site planning must comply with requirements outlined in AR 200-1, AR 200-2, *Environmental Effects of Army Actions*, AR 210-20, TC 25-1, and this training circular.

c. The trainer works with the installation DPW (including environmental, natural, and cultural resource managers) and safety officials to coordinate range-site decisions that enable effective training. A new or modified range may have environmental impact, especially noise, on the installation and nearby communities. Planners should minimize the impact, but they must maintain a balance between environmental concerns and training requirements. For example, keeping maximum natural vegetation promotes realism and reduces environmental damage. Information on planting programs that enhance cover and concealment for training realism is available from the U.S. Army Construction Engineering Research Laboratory's pamphlet, *Tactical Concealment Area*, August 1999.

d. Planners consider the location of the proposed range in relation to the rest of the installation. The site should not isolate useful maneuver terrain, cut off impact areas, or make clearance operations difficult. When the proposed range requires frequent maintenance, the site should provide easy access when adjacent ranges are active. Unusable terrain should be incorporated into impact areas. Moreover, the range should have potential for growth and improvement to accommodate new weapon systems. The key to efficient operations is a site that is easily accessible to using units. Planners need to consider overall training requirements and the flow to and from ranges when determining site locations. Training areas, such as small-arms ranges and bayonet courses, should be close to garrisons. Tracked-vehicle and artillery ranges may be located further away from unit housing. When ranges are remote, planners should consider overnight parking and bivouac facilities.

e. Due to the shortage of lands for live-fire training, ranges are sited over existing ranges or extended to encroach on existing impact areas. The trainer must consider the increased financial impact and construction time required to clear the site of unexploded ordnance to ensure safe construction operations.

5-12. Terrain

a. Terrain configuration will influence the extent of construction and the cost required to transform land into a usable range. Contour profile, soil content, and trafficability requirements affect the extent of moving and other clearing work needed. Sites requiring minimal earthwork to create level trails, visible targets, and good drainage are preferred.

b. Terrain has a major effect on direct-fire weapon systems. It influences target locations, target visibility, and a soldier's ability to fire on moving targets. Training in challenging terrain with well-planned scenarios can improve a soldier's ability to hit a moving target. Terrain features that influence target location are:

(1) Topography.

(2) Soil characteristics.

(3) Surface composition.

(4) Hydrology.

(5) Vegetation.

(6) Man-made obstacles.

c. These features also directly affect range maintenance and the environmental impact of construction and operations.

d. A terrain profile of the proposed range should be developed to support feasibility estimates, review safety requirements, and determine target locations.

e. Range sites must be selected to ensure that terrain allows maximum use of weapon systems. Planners attempt to develop ranges in areas with different types of terrain to promote weapon-firing skills under varying conditions. The site should slope downward from the firing points to the target area. This assists drainage, provides visibility, and helps contain projectiles within the SDZs. Fire-and-movement ranges should have both natural and manmade cover and concealment islands. Vegetation, however, should be removed only to provide a clear view of targets. Planners must balance the need for realism with unit training objectives. A land maintenance plan provides for regrading, revegetation, and erosion control to maintain training realism and reduce long-term maintenance costs.

f. Water areas contribute to training and increase combat realism. Reflections and temperature variations over or near water create conditions that can only be learned through experience. However, unexploded (dud) ordnance embedded in shallow water could hamper future land clearing and decontamination operations. Precautions must also be exercised if lasers are used on a range; the reflection of laser beams on water is a potential eye hazard, and open water directly in front of or behind targets could render the range unusable with laser devices. Water areas should be sited as part of the overall storm water runoff and erosion control without detrimentally affecting the use of the range.

5-13. Range Impact Area Requirements

Impact area requirements vary with training requirements and with SDZs. New contaminated impact areas will not be created unless an exception to policy is approved per ARs 200-1, 210-20, and 210-21. Conveniently located hill masses or other natural terrain features may significantly reduce real estate requirements for SDZs. Ranges should be sited so they do not restrict maneuver areas. Refer to AR 385-63, *Range Safety*, and DA PAM 385-63, *Range Safety*. Impact area requirements are based on:

a. The number and type of ranges required.

b. The SDZ for each type of weapon.

c. The use of common impact areas, if possible.

5-14. Gun-Target Lines

a. The gun-target line (GTL), also known as the "line of fire," is an imaginary line drawn from a weapon system (firing position) to a target. For efficient small-arms range operations, it is best to orient the GTL to the northeast in the Northern Hemisphere. This places light on the face of the target for the longest part of the day. Orient small-arms marksmanship ranges so that soldiers are not required to fire into the rising or setting sun. However, direction of fire is not as important as accessibility, security, and safety.

b. Consider terrain configuration when orienting the GTL in order to reduce the area covered by the SDZ. If a site is on rough terrain, the GTL should be perpendicular to high ground. Since small-arms ranges are usually on flat terrain, the GTL should be horizontal or slightly below horizontal. Range planners must avoid sites requiring the GTL for direct-fire, line-of-site weapon systems to pass over public roads or navigable waters. If firing over installation roads is unavoidable, it is mandatory that these firing areas be marked with appropriate signs and separate warnings and closed during firing periods. Indirect artillery firing over public access areas must comply with procedures in AR 385-63 and DA Pam 385-63.

c. Planning GTLs for vehicle-mounted weapon systems presents a greater challenge. Vehicular weapon systems such as tanks, Bradleys, and Strykers require that GTLs be planned for multiple targets from stationary firing points and for single and multiple targets from a moving tank or Bradley Fighting Vehicle.

d. Firing on the move requires planners to analyze a series of GTLs from various firing points along the path of the moving vehicle or within designated "firing boxes." Multiple firing scenarios for combat vehicle ranges that include both stationary and moving gun platforms and stationary and moving targets require a series of GTLs with their respective SDZs superimposed over common terrain.

e. Set safety requirements for moving vehicle firing exercises by designing a course with sequential firing points perpendicular to the target area. This arrangement may require several kilometers to accommodate training with high-speed vehicles.

f. The result is a range that needs more depth than one with the vehicle firing from a fixed point. A course on which a vehicle is required to change direction frequently more closely represents tactical conditions. It also increases land requirements since it requires larger SDZs. Refer to AR/DA Pam 385-63 for procedural guidance on this type of GTL and SDZ planning.

5-15. SDZ Requirements

a. In the most economical layout for a range complex the impact areas of SDZs overlap, but the maneuver or firing areas themselves do not. This reduces the overall acreage needed for the range system, which conserves land and minimizes the size of impact areas that are contaminated.

b. Some ranges should be sited close together, such as 25-meter zero ranges and record-fire ranges. Some ranges are incompatible for contiguous siting. For example, field artillery firing points should not be located next to rifle ranges since the field artillery noise detracts from rifle marksmanship training.

5-16. Target Areas

a. Once the range layout is approved, planners must design and develop the target area. Target areas on ranges used for individual and collective training should provide realistic training and meet gunnery qualification requirements.

b. Target arrays must:

(1) Incorporate valid training requirements.

(2) Meet size, quantity, and distance requirements of gunnery manuals.

(3) Make logical use of terrain for target arrangement.

(4) Provide a combination of fire and maneuver when necessary to meet training requirements.

5-17. Target Visibility

Firing points and targets must be located to maximize flexibility for scenario development and to meet training criteria. Planners must position targets within safe limits and be sure they are visible from firing points, positions, or lanes. To meet training requirements, 90 percent of the target must be visible from the firing position. Simulation of combat situations is important. When threat forces are portrayed on a range, groups of targets that may be visible to personnel in several firing positions will appear at the same time. If visual obstacles are used, their placement should allow reasonable target exposure time. The key is to provide an environment in which correct gunnery techniques can be exercised using realistic battlefield obstacles and target speeds.

5-18. Range Analysis and Design System

a. The Range Analysis and Design System (RADS) provides a graphical and analytical profile from each firing position to each target on a range. This stored data can be sorted to produce a target catalog. This catalog displays the engagement pairs and distance between them, meeting the criteria for the training requirement found in armor, Bradley and Stryker gunnery FMs. A line of sight (LOS) analysis by RADS should be performed concurrently with the concept design review of the range design and again at the final review. This process ensures that the targets are visible to meet training requirements. RADS assists the master gunners or equivalent in the initial layout of a range and the development of training scenarios. The LOS analysis is required for all ranges, if possible. The design engineer normally has the capability to accomplish this function.

b. The LOS analysis is performed by the U.S. Army Engineering and Support Center, Huntsville (USAESCH). Approximately 30 days is required to complete the analysis and a master gunner is required in Huntsville for approximately one week to assist with the analysis. The following data are required to perform an LOS analysis:

(1) Digital 3D topographical data of the range site.

(2) Proposed layout of the range (including digital design data, XY Z coordinates).

5-19. Target Accessibility and Protection

a. Complex target-elevating mechanisms are maintained most economically in central maintenance facilities equipped for the task. Range maintenance trails are needed for access to target-elevating mechanisms so that they can be removed from the range for maintenance.

b. Targets should be placed at distances from firing points that match training guidance, taking advantage of visibility, cover, and concealment. Landmarks at the target location, such as tree lines, manmade objects, or streams and roads, help soldiers estimate distance to a target. Gunnery manuals, this training circular, and (USAESCH) CEHNC1110-23 series standard design manuals provide guidance and distance requirements for target placement by the type of range.

c. Stationary target-elevating mechanisms and moving target carriers need protection from damage. Target emplacements prepared during construction protect target mechanisms. Natural terrain can provide additional protection and reduces the extent of the earth-moving operations needed to construct a large protective berm.

5-20. Gun-Target Engagement Sequence

Develop and plot gun-target engagement sequence capability to ensure that weapons can be contained within the designated impact area without endangering personnel or creating a hazard to property or facilities. Although primarily used on multi-purpose range complexes, a gun-target engagement sequence matrix is a valuable tool that helps determine target array options on any range. The matrix lists all targets that can be engaged from each firing position and the distance to each target. It identifies the primary targets for each scenario.

5-21. Preliminary Documentation

a. After site selection, the range-planning team prepares a preliminary range site layout. The layout shows a draft diagram (rough sketch format) of the site outline, downrange area, firing line and maneuver boxes, and support facilities. The team prepares perimeter descriptions and layout maps following guidance in AR 210-20, AR 415-15, *Army Military Construction Program Development and Execution,* and AR 415-10. Perimeter descriptions and layout maps show the proposed range area with existing and planned facilities. A site description may also be needed for environmental impact assessments and statements per AR 200-1 and AR 200-2. The planning team coordinates requirements incorporated into the development project, such as construction, equipment, facilities, and safety, with the installation:

(1) Director of Public Works.

(2) Safety Officer.

(3) Public Affairs Officer.

(4) Other appropriate staff offices.

b. Perimeter descriptions and site maps portray the proposed training area with existing and planned facilities. Planners assist in preparing the project summary. A preliminary construction layout is included in the project summary to show the proposed training area in relation to troop housing areas, road networks, existing ranges, and maneuver areas. Layouts will use current mapping data and may be supplemented with aerial photographs.

5-22. Project Summary

a. Once the target area is designed and developed, it is added to the preliminary layout and project summary. Trainers assist engineers to prepare the project summary. It includes a preliminary construction layout to show the proposed range in relation to troop housing areas, road networks, existing ranges, and maneuver areas.

b. The project summary includes general information concerning design and construction. Materials specified for use must conform to local construction policies. Grading, normally held to a minimum, is used to provide adequate drainage, increase target visibility, prevent ricochets, and meet other engineering requirements. A program for seeding, sod and other soil erosion control measures must be established to reduce maintenance costs and enhance operations. Compaction requirements such as embankment, drainage, and general cleanup must meet Corps of Engineers standards. The next step is approval of the project summary by the installation Master Planning Board (MPB).

Chapter 6
Range Operation

6-1. Introduction

a. Installation or community commanders will establish range-control and safety programs according to AR 210-21 and AR 385-63. They will appoint range-control personnel to supervise weapons firing on the installation and enforce safety and operational requirements. The safety manager provides oversight responsibility and monitors the effectiveness of the installation range safety program.

b. This chapter describes the range-control organization and its responsibilities to schedule and maintain positive control over range activities. It describes procedures to follow during live-fire and lasing operations. Refer to Appendix C for a checklist to assist these operations.

6-2. The Range Organization

a. The installation or community range management function is normally found in the Directorate for Plans, Training, and Mobilization (DPTM) or G-3. If the installation range and training functions were consolidated under a single division, it would be titled the "Range, Training, and Training Support Division"; however, the functions are typically assigned to separate training and range divisions. If a range division is not warranted, a training support division performs range functions and is the point of contact between the installation and the U.S. Army Training Support Center (USATSC).

b. The range division/training support division plans, controls, and supervises weapons firing on the installation. Duties of a range division include action to:

(1) Direct and enforce risk management controls pertaining to firing on ranges.

(2) Plan, schedule, and operate the installation's ranges and perform organizational maintenance.

(3) Plan, schedule, supervise use of facilities, training areas (including restricted airspace), and supporting weapons.

(4) Develop, publish, maintain, and enforce range and training area regulations, standing operating procedures (SOP), airspace procedures, and firing bulletins.

(5) Requisition, store, and issue targets, target materials, and equipment.

(6) Ensure that ranges are maintained and capable of meeting training needs in accordance with the Army's ITAM Program. The ITAM program establishes procedures to achieve optimum, sustainable use of training ranges and lands by implementing a uniform land management program.

c. The DPTM or G-3 training division plans, coordinates, and supervises the installation military training program. Many duties relate directly to the range program to include:

(1) Implement and supervise training programs that increase training effectiveness through substitution, simulation, and miniaturization.

(2) Plan and coordinate implementation of competitive marksmanship programs.

(3) Implement, supervise, and evaluate soldier and collective training for installation units.

(4) Supervise and manage battlefield simulation exercises.

(5) Prepare the installation range development plan.

d. Another DPTM or G-3 division with duties relating to the range program is the mobilization division. This division plans for range and training area requirements during mobilization. This division also assists the range division to develop the installation's land use requirements study (LURS) for post-mobilization training requirements (see TC 25-1).

6-3. Range Control and Scheduling

a. Coordination and integration of installation training and firing activities requires effective range control and scheduling procedures. Communications augment these procedures to ensure safe, efficient live-firing activities. Range control begins with development and publication of the installation range regulation. This regulation establishes policies and procedures for range activities consistent with safety and environmental regulations. Installations supplement the safety and environmental regulations with SOPs applicable to individual facilities within the installation-training complex. AR 210-21 describes the range regulation subjects and the minimum content requirements for an SOP.

b. An important feature of a range-control program is an effective range-scheduling process. Scheduling decisions consider range-throughput capacities, usage rates, applicable limitations or restrictions, special training needs identified by unit commanders, and training priorities. A critical scheduling consideration is the elimination of safety conflicts between ranges and maneuver areas.

c. Range Facility Management Support System (RFMSS) assists range-management operations and automates the scheduling process. The database maintains useful data on ranges, training areas, ammunition, range conflicts, and using units. Most Army installations, including RC sites, use RFMSS.

(1) RFMSS is an automated system that will simplify and enhance the current process of range and training facility management and scheduling functions at an installation. It will provide the capability for a unit to electronically submit requests for reservations. Authorized users will be able to view available and scheduled facilities for approval, disapproval, and rescheduling of results; create two and three dimensional SDZs; and detect scheduling, environmental, and safety conflicts. RFMSS will track, collect, monitor, and display range specific information.

(2) Remote access to RFMSS permits unit training personnel to identify available resources and submit schedule requests.

(3) Range-control personnel review and approve these requests and ensure that required information has been provided and ensure scheduling, safety, or environmental conflicts are resolved.

(4) The RFMSS produces a variety of useful reports such as daily and weekly scheduling bulletins, range-facility utilization reports, range-cancellation and no-show reports, ammunition-expenditure and ADHOC reports. RFMSS also uses a Geographic Information System to provide a graphic fire desk and SDZs as overlays to range maps.

d. An effective communications system is essential to range operations. Range Control is the Network Control Station and any orders it issues, such as a general cease fire order must be obeyed immediately. Range officials use the communications system to control firing, coordinate requests for medical assistance, and announce unsafe conditions. Two-way communications must be maintained between the range control organization and using units during firing and weapons training periods. If communications are disrupted, the unit must cease firing and suspend operations until they are reestablished. A range-control communications network normally consists of:

(1) A frequency-modulated (FM) radio range-control firing net.

(2) An FM radio range-control administrative net.

(3) A radio or telephone backup range-control firing net.

e. The network provides communications between the range-control network and each firing range, weapons training facility, and training area within the installation-training complex. The primary communications system may be either FM radio or telephone. The using unit must report firing position openings, closings, and changes in firing status immediately to the range-control communications center. If the RFMSS is used, a communications program is available to record event data. The range-control communications operator can enter training data into the RFMSS database, such as range openings, closings, and firings and, if necessary, print out to provide a detailed reconstruction of all calls and events.

f. Plots of SDZs (described in DA Pam 385-63) play an important part in range scheduling. They are the primary determinant of potential safety conflicts between ranges. Vertical danger zones should also be computed and considered to accommodate the ballistic trajectory and ricochet characteristics of weapons and ammunition that will be used. The scheduling process should also incorporate checks to ensure that two or more units are not scheduled on the same firing range at the same time without a co-use agreement between the commanders concerned. Periodic scheduling conferences may assign major training activities to available training facilities and reconcile non-training requests for range use with the range schedule.

6-4. Live-Fire Operations

a. Safety is paramount during live-fire and laser operations. Using-unit commanders (battalion or squadron) will ensure that the officer-in-charge (OIC), the range safety officer (RSO), and the laser range safety officer (LRSO) receive an installation or community safety instruction before obtaining battalion or squadron commander safety certification. The safety certification programs will train and qualify personnel in OIC, RSO, and LRSO duties for firing exercises or maneuver operations. AR 385-63 and DA PAM 385-63 are the primary source for range safety responsibilities.

b. The OIC is responsible for overall safety of training and proper use of the training facility. Range operations and personnel requirements will be in accordance with AR 210-21, AR 385-63, DA Pam 385-63 and local safety SOP. OIC qualifications and duties are set forth in DA PAM 385-63.

c. The RSO/LRSO will be a commissioned officer, warrant officer, NCO, or a civilian in the rank/grade established in DA PAM 385-63. The RSO/LSRO makes the final determination to fire. The RSO/LSRO must be weapon-system-qualified and have completed a program of instruction in RSO/LSRO duties. The RSO/LSRO must be present on the training complex and have no other duties that period of training, except for aviation weapon systems where instructor pilots may assume RSO/LSRO duties. Assistant range safety officers (ARSO) may be appointed as required. RSO/LSRO qualifications and duties are set forth in DA PAM 385-63.

Appendix A
Range Selection

This appendix identifies primary and alternate ranges used for training and qualification with specific weapon systems and weapons, based on applicable field and gunnery manuals. Table A-1 lists primary and alternate ranges to satisfy weapon-system training events. The Army's Stryker Brigade Combat Team (SBCT) will require ranges to train soldiers and crews. Table A-2 reflects the estimated SBCT training support infrastructure requirements and ranges to support training.

Table A- 1. Range selection

WEAPON SYSTEM	TRAINING EVENT	RANGE PRIMARY	 ALTERNATE
M16-Series and M4	Zero	25m Range	AFF/ARF/MRF with 25m boots
	Sustainment/Record	AFF/ARF/MRF	25m Range
	Qualification	ARF/MRF	25m Range
	Night	AFF/ARF/MRF Night	25m Range
	NBC Conditions	AFF/ARF/MRF	25m Range
	Squad/Platoon ARTEP LFX	ISBC/IPBC/ MPTR/MPRC	MPTR/MPRC
9mm	Instructional Fire	CPQC	25m Alternate Pistol Qualification Course
	Qualification	CPQC	25m Alternate Pistol Qualification Course
	NBC Conditions	CPQC	25m Alternate Pistol Qualification Course
	Military Police Qualification	MPFQC/CPQC	25m Alternate Pistol Qualification Course
	Military Police Night Fire Arms Sustainment	MPFQC/CPQC	25m Alternate Pistol Qualification Course
M21 and M24	Zero	Sniper Field Fire	None
	Qualification	Sniper Field Fire	None
	Squad/Platoon ARTEP LFX	Sniper Field Fire	MPTR/MPRC
	CALFEX/STX	MPRC	None
	Night Fire	Sniper Field Fire	None

Table A-1. Range selection (cont)

WEAPON SYSTEM	TRAINING EVENT	RANGE PRIMARY	RANGE ALTERNATE
M249 SAW/M240B	10m Practice	Machine Gun 10m	None
	10m Record	MPMG	None
	Transition Practice	MPMG	None
	Transition Record	MPMG	None
	NBC Conditions	MPMG	None
	Squad/Platoon ARTEP LFX/STX	ISBC/IBPC/ MPTR/MPRC	None
	CALFEX	MPRC	None
M60/M2	10m Record	Machine Gun 10m	None
	Transition Record	MPMG	None
	Limited Visibility/ Predetermined Fire	MPMG	None
	Squad/Platoon ARTEP LFX	ISBC/IPBC/ MPTR/MPRC	None
	CALFEX	MPRC	None
M203/M79	Instructional	Grenade Launcher	None
	Qualification	Grenade Launcher	None
	NBC	Grenade Launcher	None
	Squad/Platoon ARTEP LFX	ISBC/IPBC MPTR/MPRC	None
	CALFEX	MPRC	None
Hand Grenade	Instructional	HG Qualification Course	None
	Qualification	HG Qualification Course	None
	Live-Grenade Throw	HG Live Fire Range	None
MG/MK-19/Sniper Wpns	Practice	MPMG/Sniper FF	None
	Record Fire	MPMG/Sniper FF	None
	NBC Condition	MPMG/Sniper FF	None
	Night Firing	MPMG/Sniper FF	None
	HE	MPMG/Sniper FF	None
	MG Sustainment	MPMG/Sniper FF	None
	LFX	MPMG/Sniper FF	None
	CALFEX MPRC	MPMG/Sniper FF	None
60mm Mortar	Crew/Section/Platoon Training	Mortar	None
	Evaluated ARTEP	Mortar	None
	CALFEX	Mortar	None
81mm Mortar	Crew/Section/Platoon Training	Mortar	SMR*
	Crew/Section/Platoon Training	Mortar	None
	Evaluated ARTEP	Mortar	None
	CALFEX	Mortar	None
? Scaled mortar range			

Table A-1. Range selection (cont)

WEAPON SYSTEM	TRAINING EVENT	RANGE PRIMARY	RANGE ALTERNATE
120mm Mortar	Crew/Section/Platoon Training	Mortar	SMR
	External Evaluation	Mortar	None
	CALFEX	MPRC	None
M1-Series	Table I	UCOFT*	None
	Table II	UCOFT*	None
	Table III	SGR***	Maneuver Area Day-fire
	Table IV**	MPTR/DMPTR/	MPRC/DMPRC
	Table V	MPTR/DMPTR	MPRC/DMPRC
	Calibration	Screening/Zero Range	None
	Table VI	MPTR/DMPTR	MPTR/MPRC
	Table VII	MPTR/DMPTR	MPRC/DMPRC
	Table VIII	MPRC/DMPRC	MPTR/DMPTR
	Table IX	MPRC/DMPRC	MPTR/DMPTR
	Table X	MPRC/DMPRC	MPTR/DMPTR
	Table XI	MPRC/DMPRC	None
	Table XII	MPRC/DMPRC	None

* Unit Conduct of Fire Trainer
** Tank crew proficiency course (TCPC)
*** Scaled gunnery range
**** Tactical engagement simulation

WEAPON SYSTEM	TRAINING EVENT	RANGE PRIMARY	RANGE ALTERNATE
Bradley Fighting Vehicle (Linebacker and Bradley Engineer Vehicle	Tables I-IV	MPTR/DMPTR	MPRC/DMPRC
	Calibration	Screening/Zero Range	None
	Table V	MPTR/DMPTR	MPRC/DMPRC
	Table VI-A/B	MPTR/DMPTR	MPRC/DMPRC
	Table VII-A/B	MPTR/DMPTR	MPRC/DMPRC
	Table VIII-A/B	MPRC/DMPRC	MPTR/DMPTR
	Infantry Squad Combat Exercise	MPRC/DMPRC	MPTR/DMPTR
	Table IX*	MPRC/DMPRC	MPTR/DMPTR
	Table IX-A/B**	MPRC/DMPRC	MPTR/DMPTR
	Table X-A/B**	MPRC/DMPRC	MPTR/DMPTR
	Table XI	MPRC/DMPRC	MPTR/DMPTR
	Table XII	MPRC/DMPRC	None
	CALFEX	MPRC/DMPRC	None

* Vehicle section gunnery

** Scout section gunnery

There must be installation-level training infrastructure to support the Army's transformation initiative. The Stryker Brigade Combat Team installation training requirements and ranges are shown in Table A-2. These requirements were developed by Training and Doctrine Command and accepted by Headquarters Department of the Army through the Requirements Review Prioritization Board process. The requirements were also adopted for the Assistant Chief of Staff for Installation Management template to identify infrastructure requirements for Transformation.

Table A-2. Estimated Stryker Brigade Combat Team training requirements and ranges

TRAINING ASSET REQUIREMENTS
Simulator building (motion-based).
Simulator building (non-motion based).
Moving target simulator building.
Simulations center.
Maneuver/training area (complex terrain) - battalion 600 KM^2 and brigade 2500 KM^2.
Robust inter- and intra-installation C4I network (which includes a fixed tactical internet capability).
Integrated automation tools for development, storage, distribution, and management of training products.
Sufficient Army Distance Learning Program equipped classrooms for institutional and self-development training support.
MILES 2000 and future One TES.
Home station instrumentation system.
TWGSS/PGS-like system for gunnery training.
Engagement Skills Trainer.
Conduct of fire or artillery fire integration support team gunnery trainer.
CCTT-like system for the Interim Armored Vehicle.

NUMBER	**RANGES**	
	PRIMARY	**ALTERNATE**
1	25m Zero Range	None
1	Modified Record-Fire Range	None
1	Combat Pistol Qualification Course	25m Alternate Pistol Course
1	Multipurpose Machine Gun & Sniper Field Fire	None
1	Grenade Launcher Range	None
1	Sniper Field Fire Range	SPMG/MPTR/MPRC
1	MK-19 Range (for Anti-Tank (AT)-4, Javelin training also)	None
1	Hand Grenade Qualification Course	None
1	Anti-Armor Tracking Range	MPTR/DMPTR
1	Mortar	None
1	Multipurpose Training Range	None
1	Aerial Harmonization Range	None
1	Aerial Gunnery Range	MPRC
1	Digital Multipurpose Training Range	BAX
1	Battle Area Complex (BAX)	Digital Multipurpose Range Complex (DMPRC)
1	Infantry Squad Battle Course	None
1	Infantry Platoon Battle Course	None
1	Urban Assault Course (contains Live Fire Breach Facility and Live Fire Shoot House)	None
1	Combined Arms Collective Training Facility	None

Appendix B
Range Equipment

Figures B-1 through B-69 illustrate targets and targetry equipment used to support ranges. For additional information, contact USATSC ATTN: ATIC-ATMD, Fort Eustis, Virginia 23604-5166, DSN 827-2320.

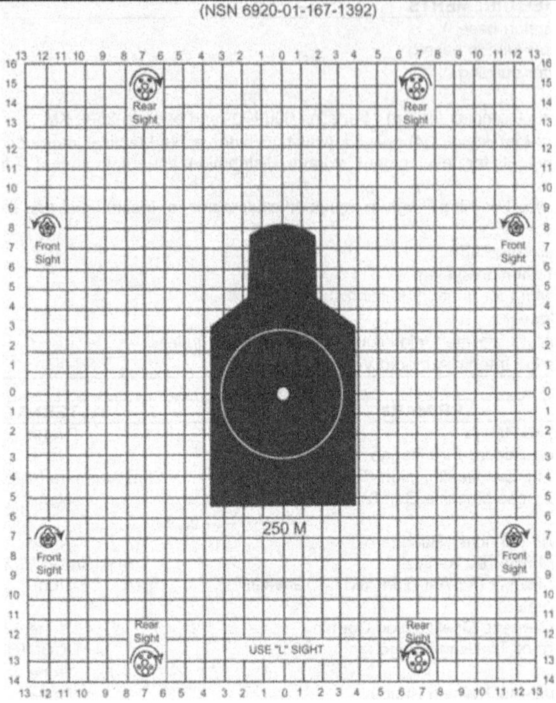

1. **Aim at target center. Adjust sights to move shot group center as close as possible to white dot.**

2. **At completion of zero, rotate rear sight to unmarked aperture and weapon will be battlesight zero for 250 meters.**

Figure B-1. 25-meter zeroing target for M16A1 rifle (standard sights)

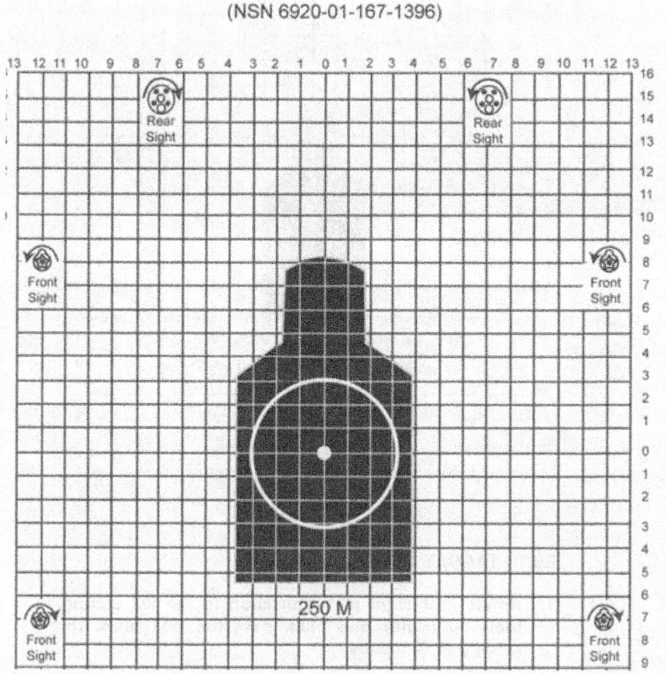

NOTES:

1. **Aim at target center. Adjust sights to move shot group center as close as possible to white dot.**
2. **Use unmarked aperture with standard sights.**
3. **Use "L" sight with low light level sight system.**

Figure B-2. 15-meter qualification zero target

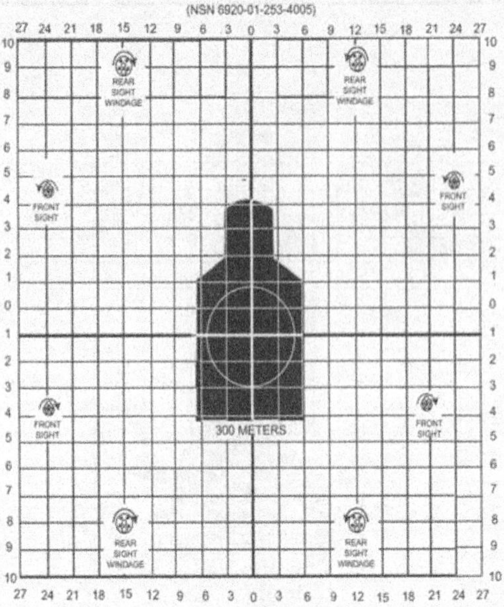

ZERO TARGET DATA:

1. **Rotate rear sight elevation knob to the 8/3 setting then up (right) one click past the 300 mark for zeroing at 25 meters.**

2. **Aim at target center. Adjust sights b move shot group center as close as possible to the center of the target.**

3. **After completion of the 25-meter zero rotate the rear sight elevation knob back one click to the 300/800 meter mark. The weapon will be zeroed for 300 meters.**

Figure B-3. 25-meter zeroing target M16A2

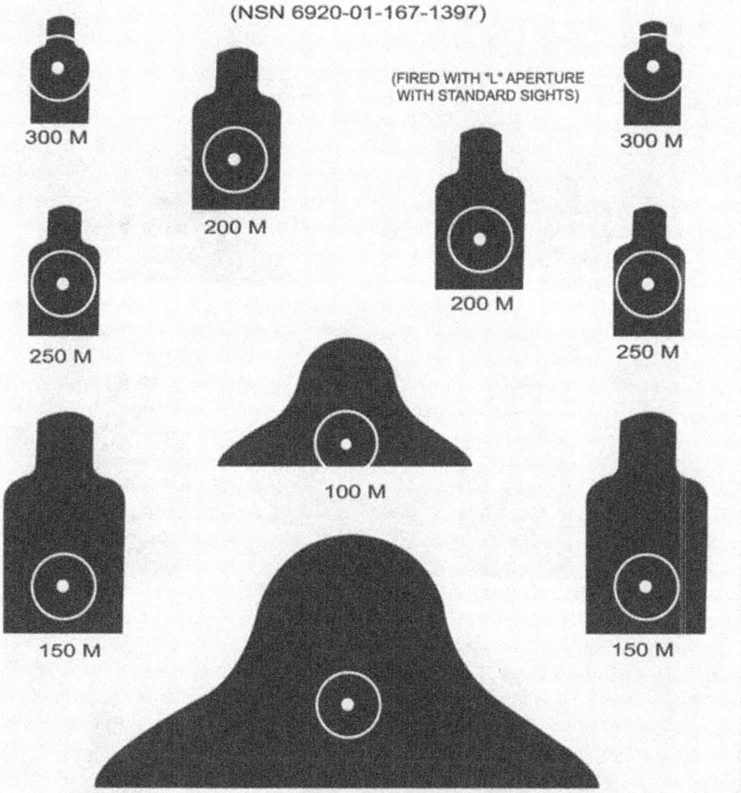

NOTE:

Use long-range sight. The white dot on each target shows the best aiming point for targets at actual distances. If adjusted aiming point is used at 25 meters, bullets should hit within the circles, but are scored as hits if they hit anywhere in the silhouette.

Figure B-4. 25-meter scaled silhouette timed-fire target

(NSN 6920-01-167-1391)

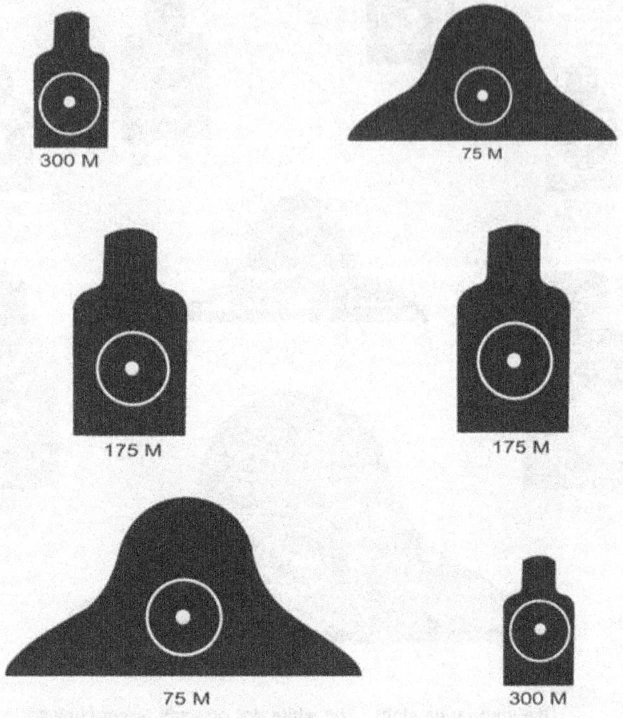

NOTE:
Use long-range sight. The white dot on each target shows the center-of-mass aiming point. Bullets should hit within the circles, but are scored as hits if they hit any part of the silhouette.

Figure B-5. 25-meter scaled silhouette slow-fire target

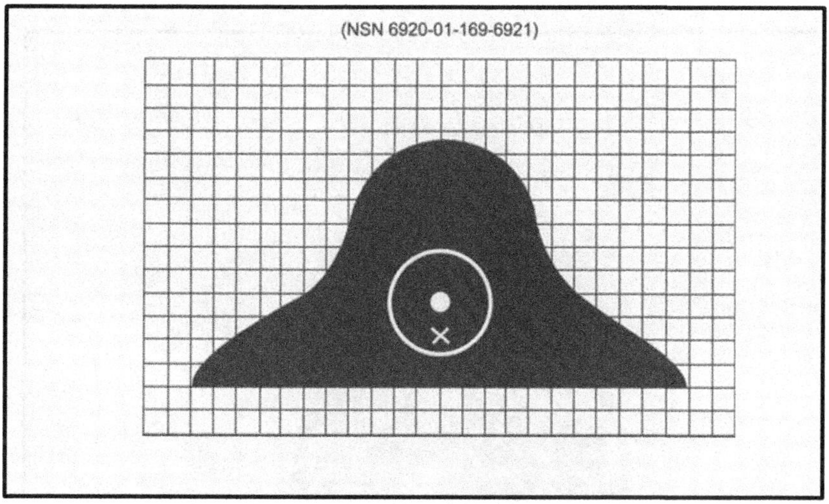

Figure B-6. 75-meter feedback target

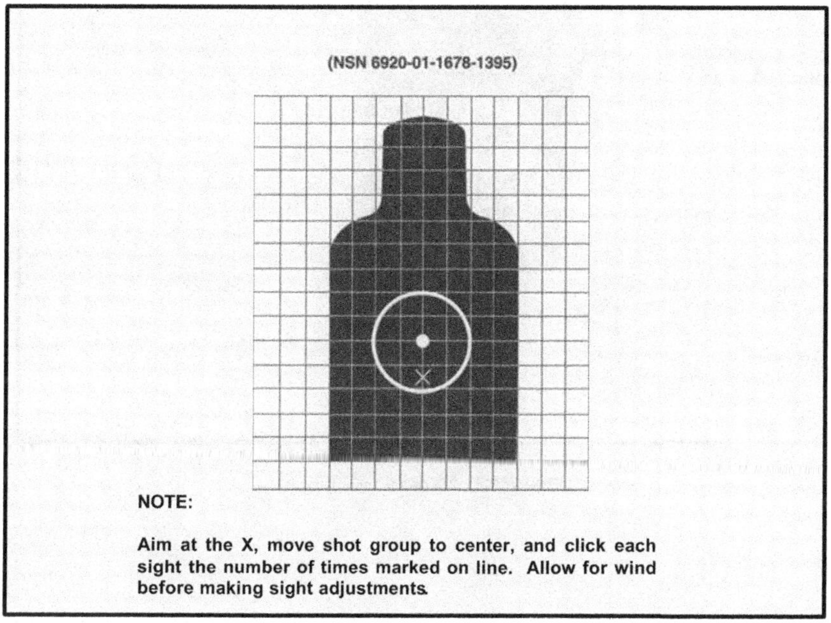

Figure B-7. 175-meter feedback target

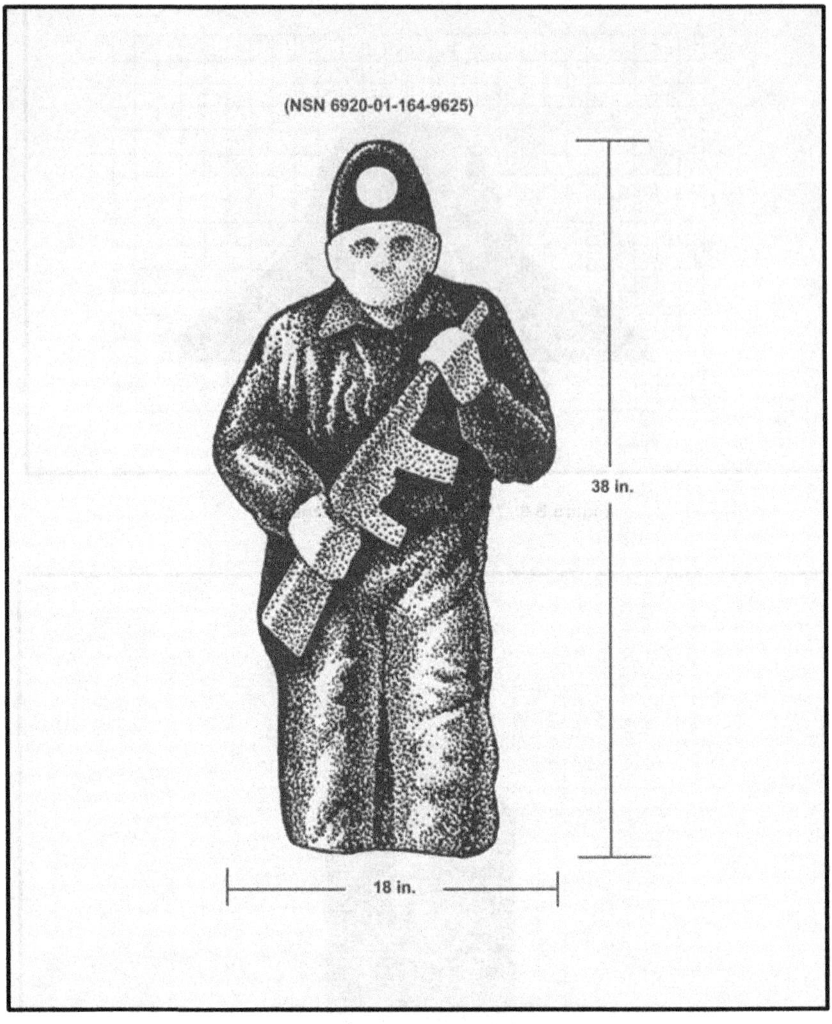

Figure B-8. 3D personnel target[1]

[1] Training requires 90% of the target be visible from the firing position.

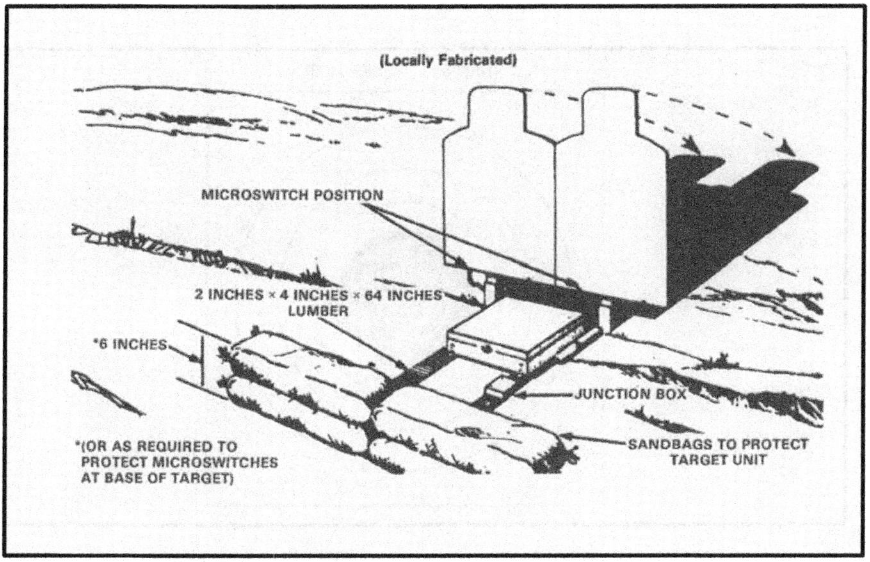

Figure B-9. Small area target

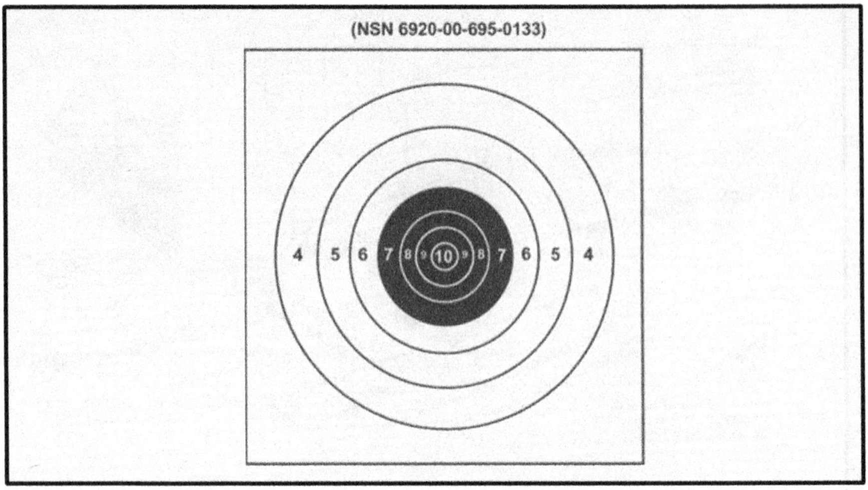

Figure B-10. 50-foot slow-fire pistol bull's eye paper target

Figure B-11. 50-foot timed and rapid-fire pistol bull's eye paper target

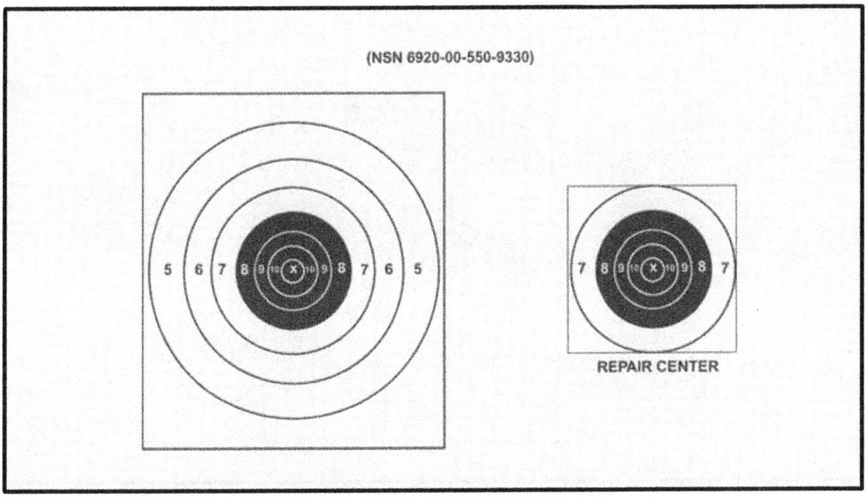

Figure B-12. 50-foot pistol target

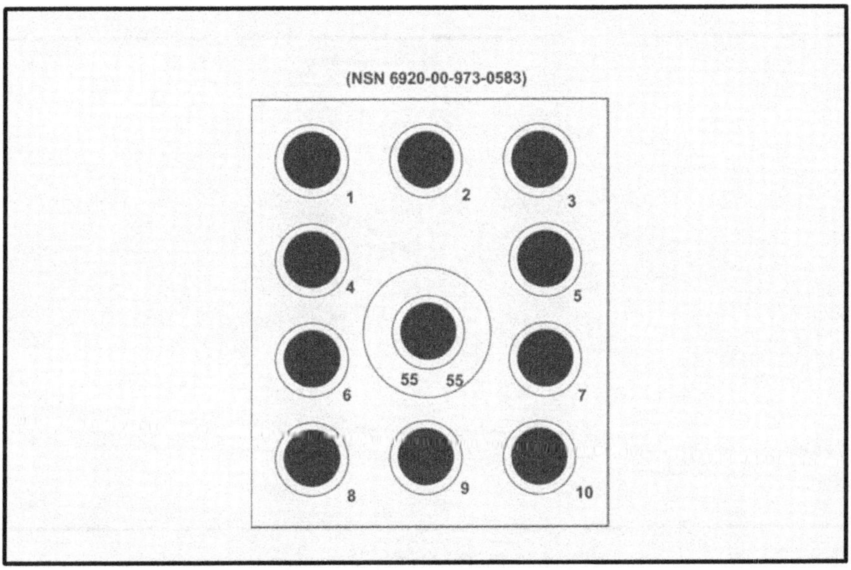

Figure B-13. 50-foot indoor rifle target

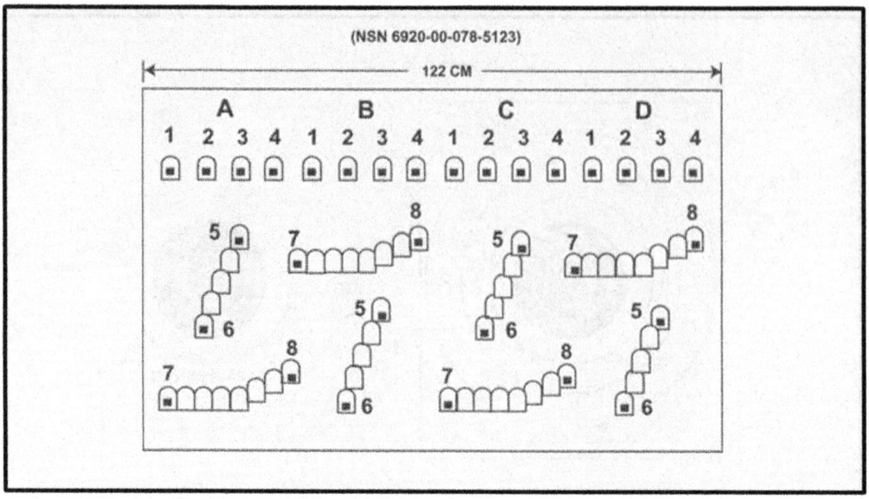

Figure B-14. Basic M60 machine gun target

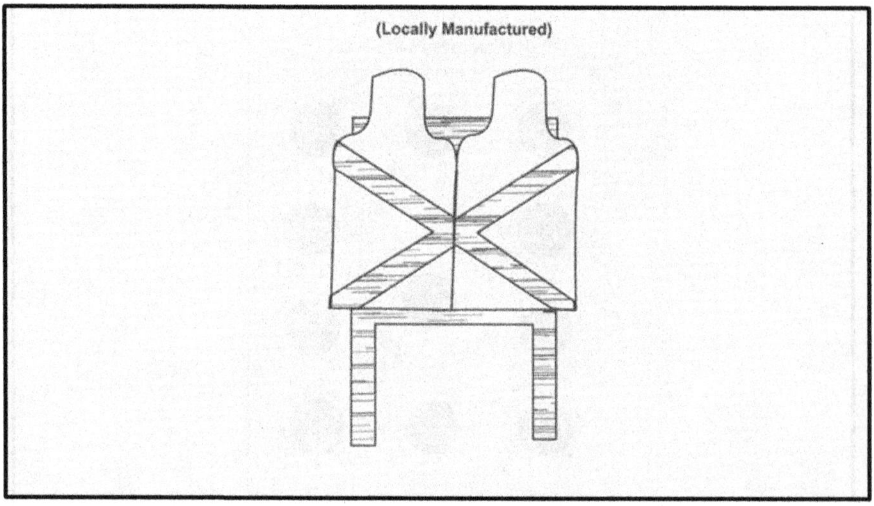

Figure B-15. Double E-type silhouette

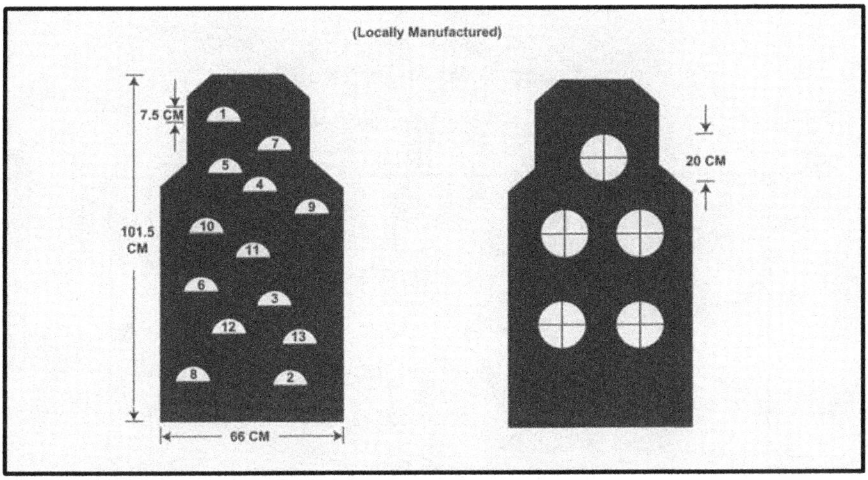

Figure B-16. Target for zeroing exercises (Table I)

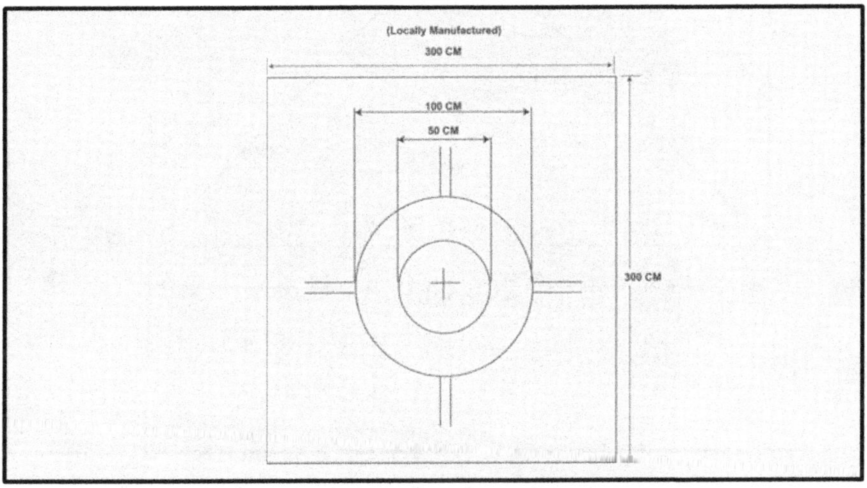

Figure B-17. Zero and bore sight panel

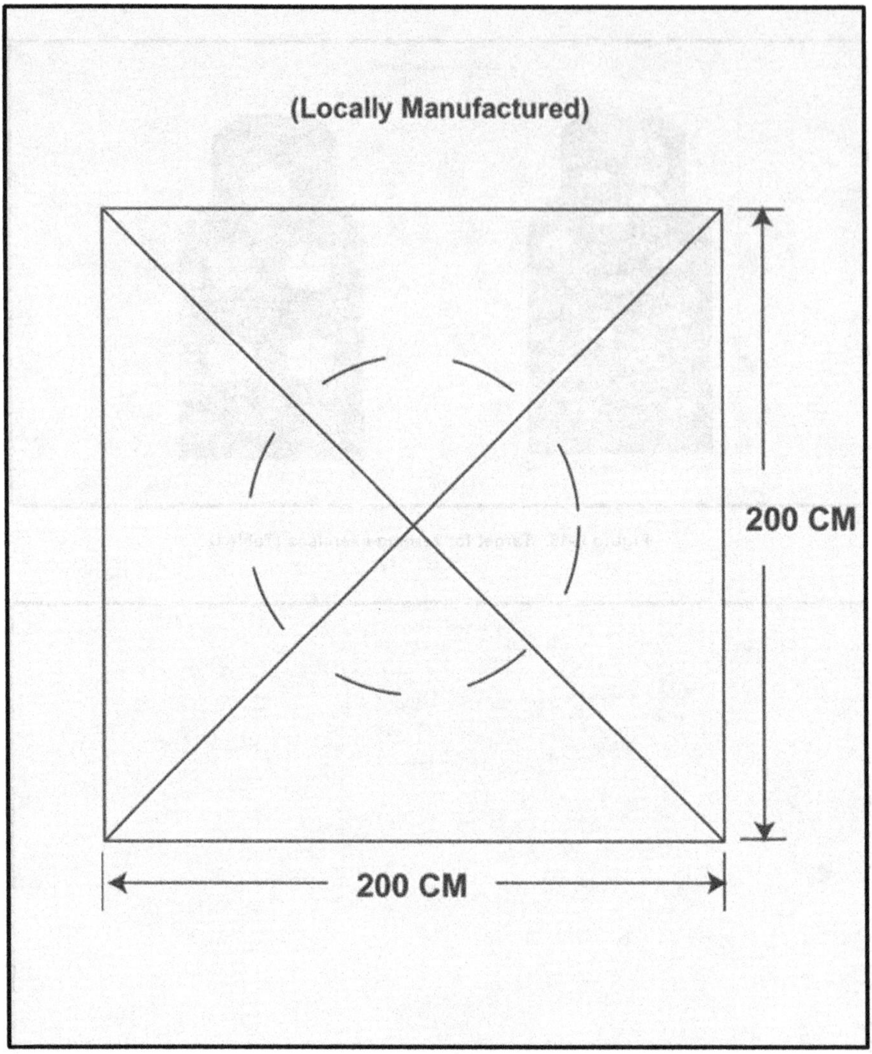

Figure B-18. Boresight and zero target for the cupola-mounted machine gun

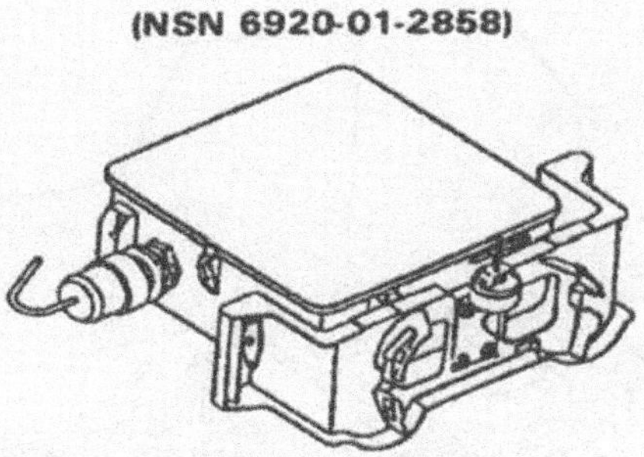

NOTE:
This mechanism mounts either two- or three-dimensional E and F silhouette targets. It raises a target in less than a second. It is automatically or manually controlled from the Range Control Station. It has laser interface and mounts on RETS moving target carrier or stationary emplacement.

Figure B-19. Stationary Infantry Target (SIT)

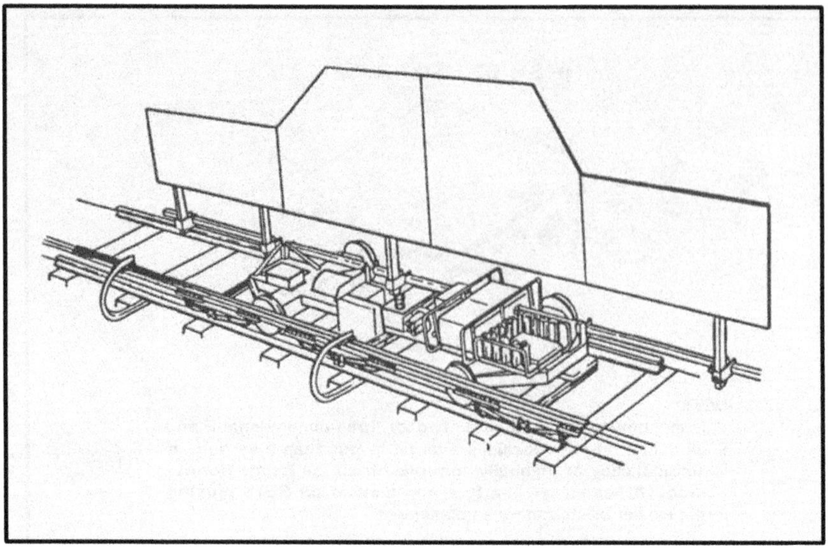

Figure B-20. Moving Armor Target (MAT)

Figure B-21. Stationary Armor Target (SAT)

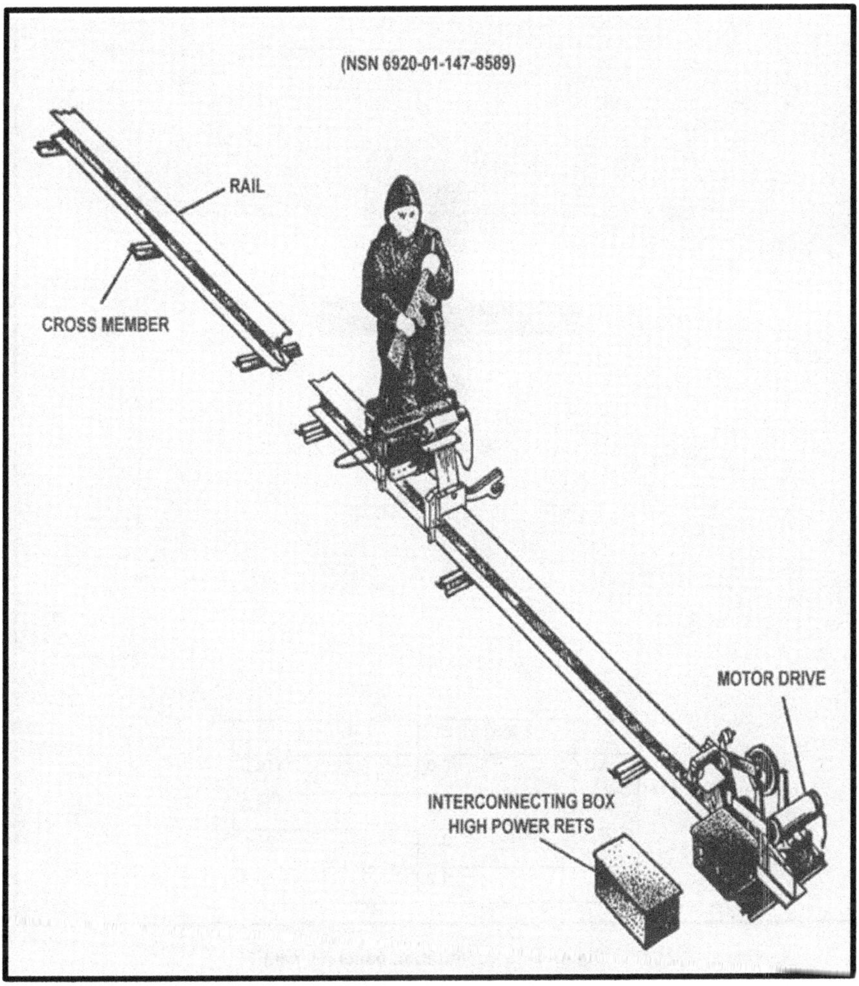

Figure B-22. Moving Infantry Target (MIT)

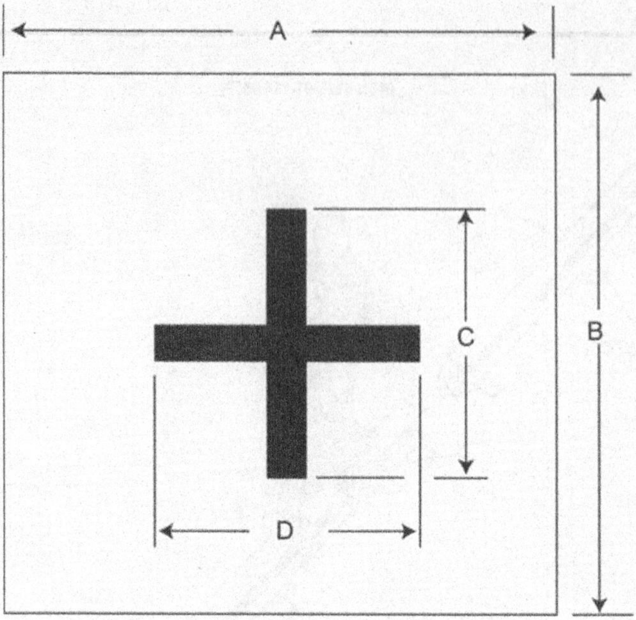

	1-2 METERS	INCHES
A	3.0	118.0
B	3.0	118.0
C	0.6	23.6
D	0.6	23.6

Figure B-23. Calibration panel (50 cal.)

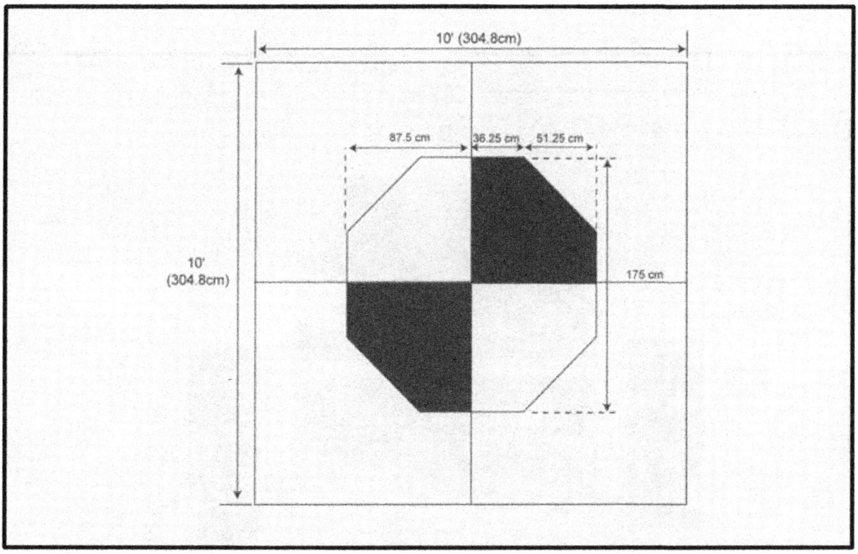

Figure B-24. Screening test target no. 4 (120mm gun)

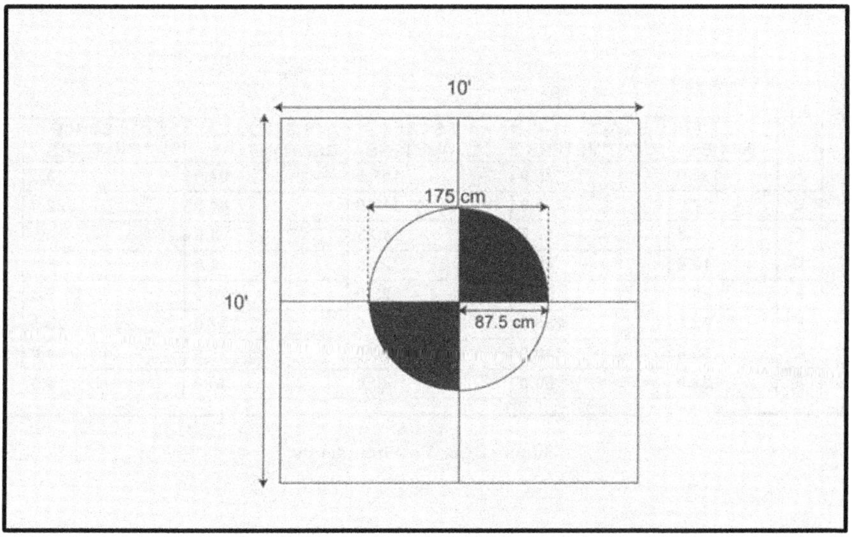

Figure B-25. Screening test target no. 5 (120mm gun)

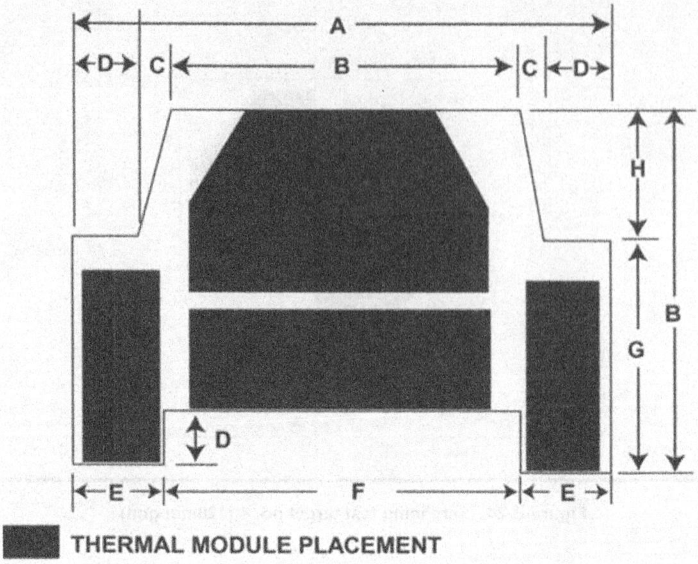

TARGET DIMENSIONS

	FULL-SCALE INCHES	FULL-SCALE CENTIMETERS	1:2 SCALE CENTIMETERS	1:5 SCALE CENTIMETERS	1:10 SCALE CENTIMETERS
A	132.0	335.0	167.5	67.0	33.5
B	87.0	220.0	110.0	44.0	22.0
C	6.5	17.5	8.75	3.5	1.75
D	16.0	40.0	20.0	8.0	4.0
E	20.0	50.0	25.0	10.0	5.0
F	92.0	235.0	117.5	47.0	23.5
G	55.0	140.0	70.0	28.0	14.0
H	32.0	80.0	40.0	16.0	8.0

Figure B-26. T-72 front target

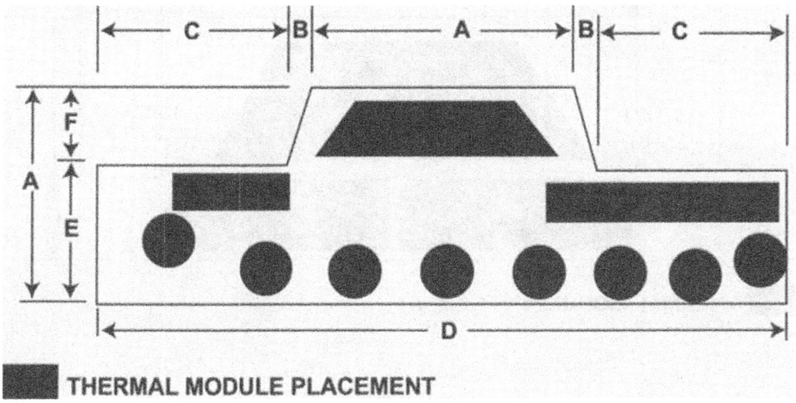

TARGET DIMENSIONS

	FULL-SCALE INCHES	FULL-SCALE CENTIMETERS	1:2 SCALE CENTIMETERS	1:5 SCALE CENTIMETERS	1:10 SCALE CENTIMETERS
A	90.5	230.0	115.0	46.0	23.0
B	16.0	40.0	20.0	8.0	4.0
C	79.0	200.0	100.0	40.0	20.0
D	276.0	700.0	350.0	140.0	70.0
E	55.0	140.0	70.0	28.0	14.0
F	35.0	90.0	45.0	18.0	9.0

Figure B-27. T-72 flank target

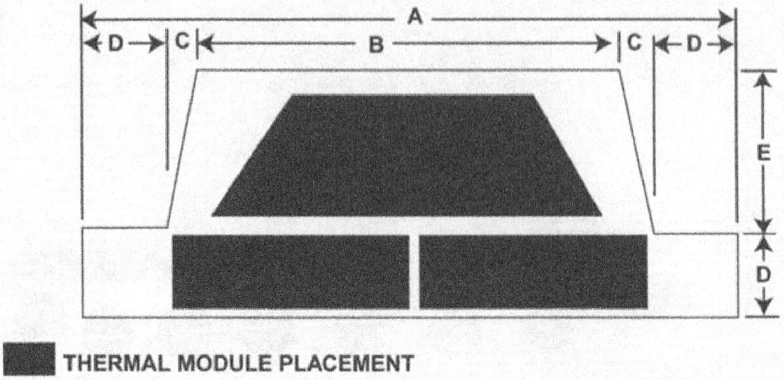

TARGET DIMENSIONS

	FULL-SCALE		1:2 SCALE	1:5 SCALE	1:10 SCALE
	INCHES	CENTIMETERS	CENTIMETERS	CENTIMETERS	CENTIMETERS
A	132.0	335.0	167.5	67.0	33.5
B	87.0	220.0	110.0	44.0	22.0
C	6.5	17.5	8.75	3.5	1.7
D	16.0	40.0	20.0	8.0	4.0
E	35.0	90.0	45.0	18.0	9.0

Figure B-28. T-72 partial defilade target

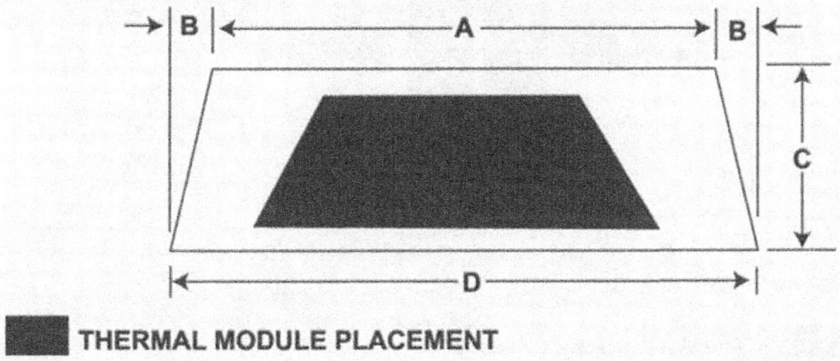

TARGET DIMENSIONS

	FULL-SCALE		1:2 SCALE	1:5 SCALE	1:10 SCALE
	INCHES	CENTIMETERS	CENTIMETERS	CENTIMETERS	CENTIMETERS
A	87.0	220.0	110.0	44.0	22.0
B	5.5	15.0	7.5	3.0	1.5
C	35.0	90.0	45.0	18.0	9.0
D	98.0	250.0	125.0	50.0	25.0

Figure B-29. T-72 Turret Target

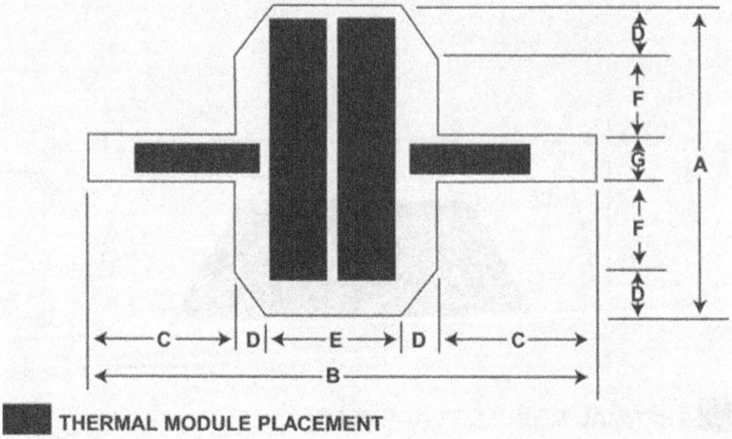

TARGET DIMENSIONS

	FULL-SCALE		1:2 SCALE	1:5 SCALE	1:10 SCALE
	INCHES	CENTIMETERS	CENTIMETERS	CENTIMETERS	CENTIMETERS
A	120.0	305.0	152.5	61.0	30.5
B	192.0	488.0	244.0	97.6	48.8
C	60.0	152.0	76.0	30.4	15.2
D	12.0	31.0	15.5	6.2	3.1
E	48.0	122.0	61.0	24.4	12.2
F	36.0	91.0	45.5	18.2	9.1
G	24.0	61.0	30.5	12.2	6.1

Figure B-30. HIND front target

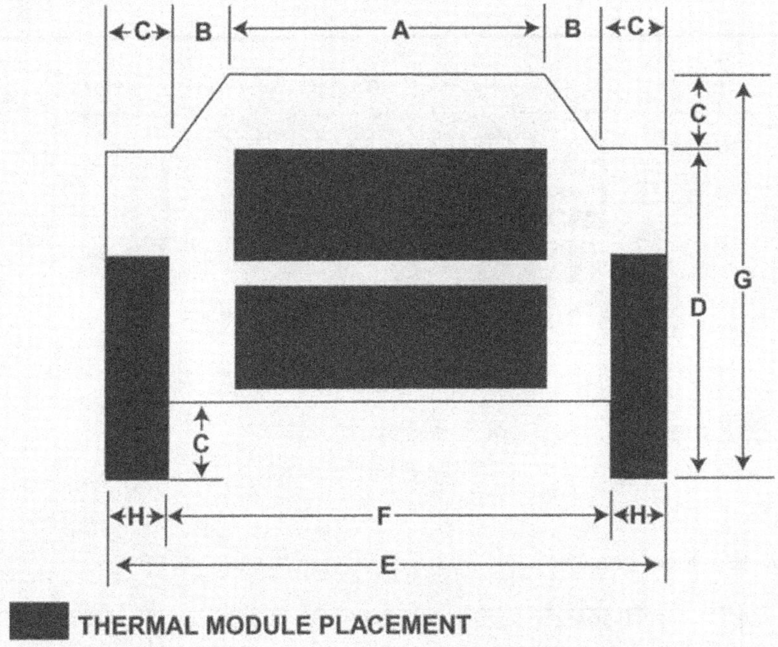

TARGET DIMENSIONS

	FULL-SCALE		1:2 SCALE	1:5 SCALE	1:10 SCALE
	INCHES	CENTIMETERS	CENTIMETERS	CENTIMETERS	CENTIMETERS
A	70.0	179.0	89.5	35.8	17.9
B	10.0	25.5	12.75	5.1	2.6
C	16.0	40.0	20.0	8.0	4.0
D	71.0	180.0	90.0	36.0	18.0
E	122.0	310.0	155.0	62.0	31.0
F	98.0	250.0	125.0	50.0	25.0
G	87.0	220.0	110.0	44.0	22.0
H	12.0	30.0	15.0	6.0	3.0

Figure B-31. BMP 1981 front target

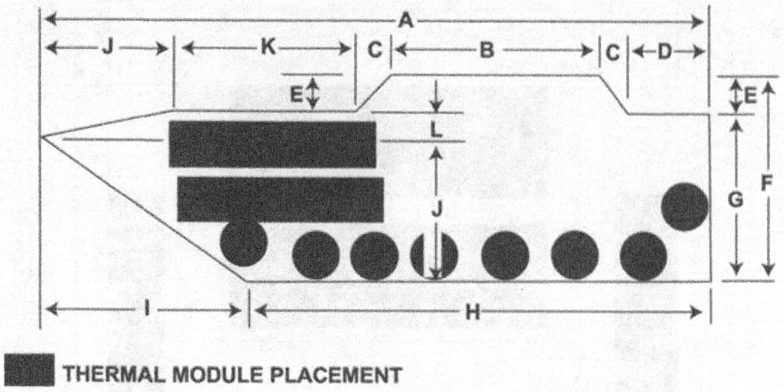

TARGET DIMENSIONS

	FULL-SCALE		1:2 SCALE	1:5 SCALE	1:10 SCALE
	INCHES	CENTIMETERS	CENTIMETERS	CENTIMETERS	CENTIMETERS
A	268.0	680.0	340.0	136.0	68.0
B	67.0	170.0	85.0	34.0	17.0
C	12.0	30.0	15.0	6.0	3.0
D	39.0	100.0	50.0	20.0	10.0
E	18.0	45.0	22.5	9.0	4.5
F	87.0	220.0	110.0	44.0	22.0
G	69.0	175.0	87.5	35.0	17.5
H	173.0	440.0	220.0	88.0	44.0
I	95.0	240.0	120.0	48.0	24.0
J	61.0	155.0	77.5	31.0	15.5
K	77.0	195.0	97.5	39.0	19.5
L	8.0	20.0	10.0	4.0	2.0

Figure B-32. BMP 1981 flank target

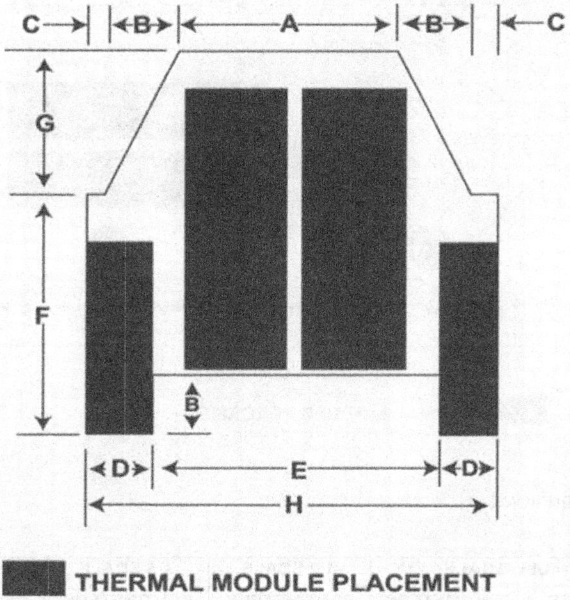

TARGET DIMENSIONS

	FULL-SCALE		1:2 SCALE	1:5 SCALE	1:10 SCALE
	INCHES	CENTIMETERS	CENTIMETERS	CENTIMETERS	CENTIMETERS
A	47.0	120.0	60.0	24.0	12.0
B	16.0	40.0	20.0	8.0	4.0
C	4.0	10.0	5.0	2.0	1.0
D	12.0	30.0	15.0	6.0	3.0
E	63.0	160.0	80.0	32.0	16.0
F	59.0	150.0	75.0	30.0	15.0
G	35.0	90.0	45.0	18.0	9.0
H	87.0	220.0	110.0	44.0	22.0

Figure B-33. BRDM front target

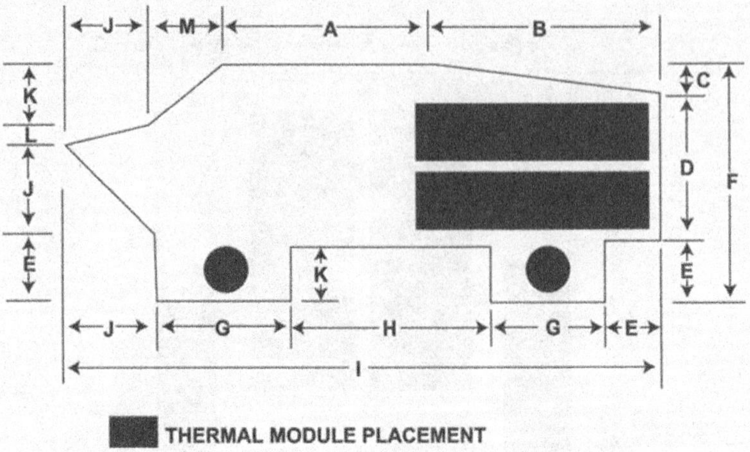

TARGET DIMENSIONS

	FULL-SCALE		1:2 SCALE	1:5 SCALE	1:10 SCALE
	INCHES	CENTIMETERS	CENTIMETERS	CENTIMETERS	CENTIMETERS
A	87.0	221.0	110.5	44.2	22.1
B	81.0	206.0	103.0	41.2	20.6
C	11.0	28.0	14.0	5.6	2.8
D	48.0	122.0	61.0	24.4	12.2
E	24.0	61.0	30.5	12.2	6.1
F	83.0	211.0	105.5	42.2	21.1
G	47.0	120.0	60.0	24.0	12.0
H	75.0	190.0	95.0	38.0	19.0
I	224.0	570.0	285.0	114.0	57.0
J	31.0	79.0	39.5	15.8	7.9
K	20.0	51.0	25.5	10.2	5.1
L	8.0	20.0	10.0	4.0	2.0
M	25.0	64.0	32.0	12.8	6.4

Figure B-34. BRDM flank target

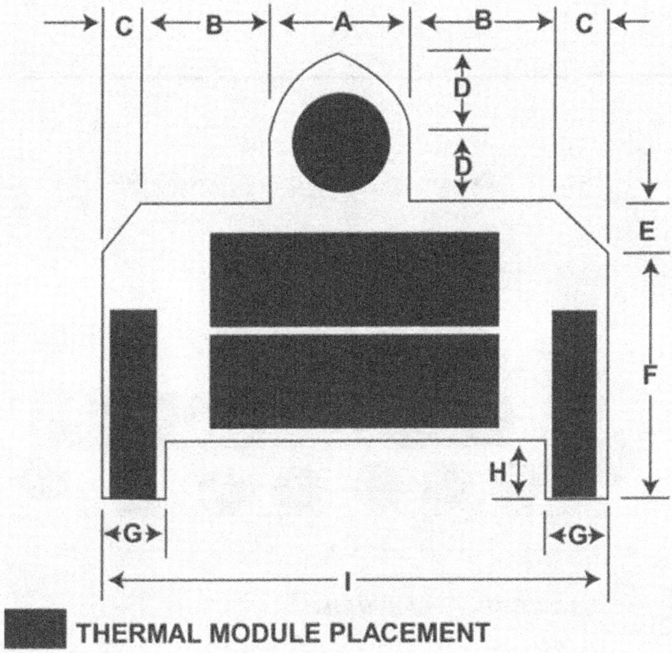

TARGET DIMENSIONS

	FULL-SCALE		1:2 SCALE	1:5 SCALE	1:10 SCALE
	INCHES	CENTIMETERS	CENTIMETERS	CENTIMETERS	CENTIMETERS
A	35.0	90.0	45.0	18.0	9.0
B	33.5	85.0	42.5	17.0	8.5
C	8.0	20.0	10.0	4.0	2.0
D	20.0	50.0	25.0	10.0	5.0
E	12.0	30.0	15.0	6.0	3.0
F	59.0	150.0	75.0	30.0	15.0
G	14.0	35.0	17.5	7.0	3.5
H	16.0	40.0	20.0	8.0	4.0
I	118.0	300.0	150.0	60.0	30.0

Figure B-35. ZSU-23/4 front target

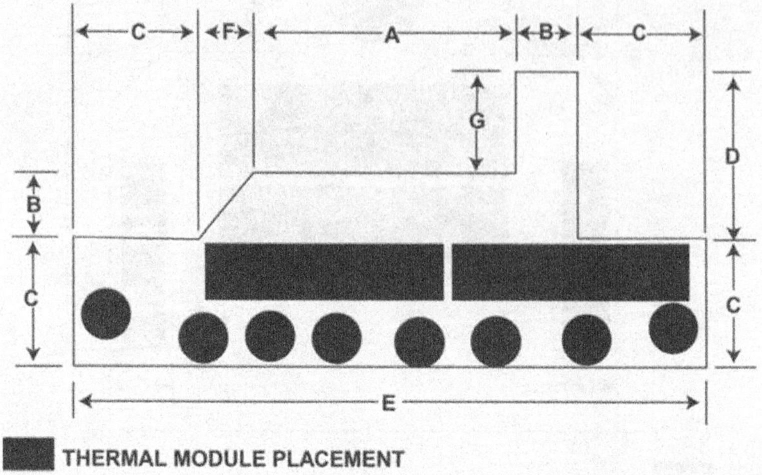

TARGET DIMENSIONS

	FULL-SCALE		1:2 SCALE	1:5 SCALE	1:10 SCALE
	INCHES	CENTIMETERS	CENTIMETERS	CENTIMETERS	CENTIMETERS
A	98.0	250.0	125.0	50.0	25.0
B	24.0	60.0	30.0	12.0	60.0
C	47.0	120.0	60.0	24.0	120.0
D	63.0	160.0	80.0	32.0	16.0
E	236.0	600.0	300.0	120.0	60.0
F	20.0	50.0	25.0	10.0	5.0
G	39.0	100.0	50.0	20.0	10.0

Figure B-36. ZSU-23/4 flank target

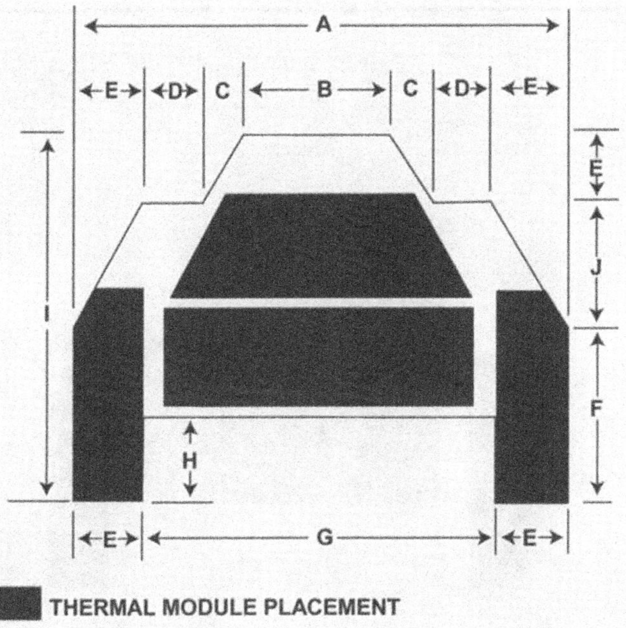

TARGET DIMENSIONS

	FULL-SCALE		1:2 SCALE	1:5 SCALE	1:10 SCALE
	INCHES	CENTIMETERS	CENTIMETERS	CENTIMETERS	CENTIMETERS
A	110.0	280.0	140.0	56.0	28.0
B	30.0	800.0	40.0	16.0	8.0
C	10.0	25.0	12.5	5.0	2.5
D	14.0	35.0	17.5	7.0	3.5
E	16.0	40.0	20.0	8.0	4.0
F	39.0	100.0	50.0	20.0	10.0
G	78.0	200.0	100.0	40.0	20.0
H	20.0	50.0	25.0	10.0	5.0
I	86.0	220.0	110.0	44.0	22.0
J	31.0	80.0	40.0	16.0	8.0

Figure B-37. BTR-series front target

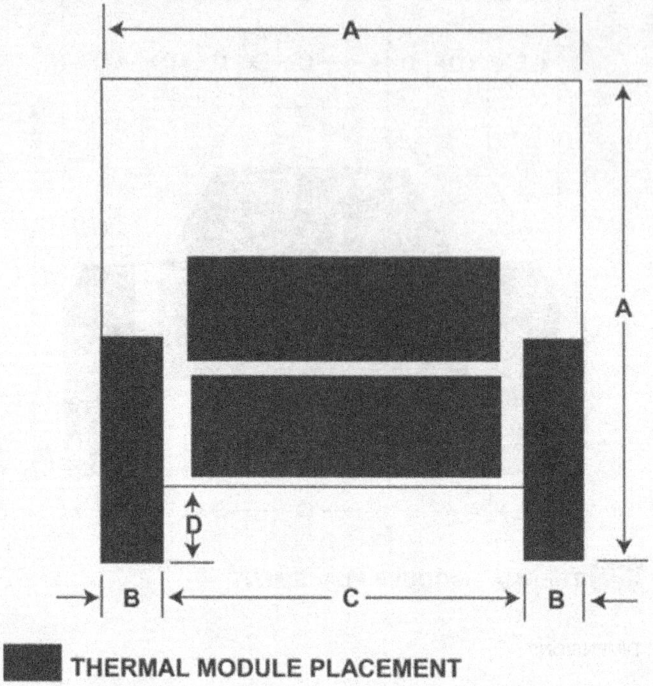

TARGET DIMENSIONS

	FULL-SCALE		1:2 SCALE	1:5 SCALE	1:10 SCALE
	INCHES	CENTIMETERS	CENTIMETERS	CENTIMETERS	CENTIMETERS
A	102.0	260.0	130.0	52.0	26.0
B	12.0	30.0	15.0	6.0	3.0
C	78.0	200.0	100.0	40.0	20.0
D	16.0	40.0	20.0	8.0	4.0

Figure B-38. Truck U-375 front target

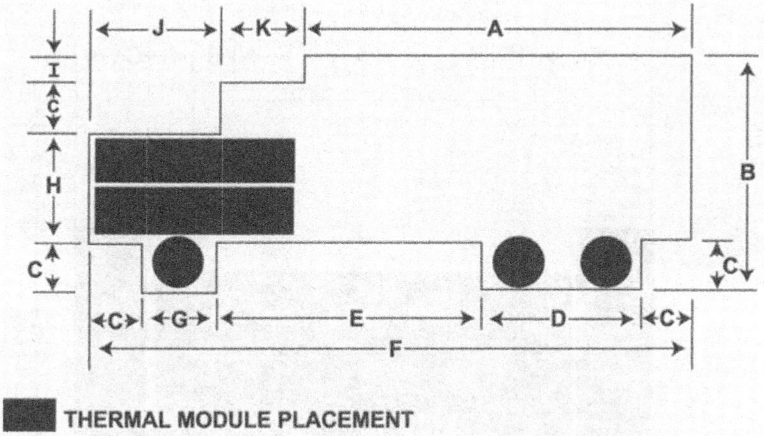

TARGET DIMENSIONS

	FULL-SCALE		1:2 SCALE	1:5 SCALE	1:10 SCALE
	INCHES	CENTIMETERS	CENTIMETERS	CENTIMETERS	CENTIMETERS
A	177.0	450.0	225.0	90.0	45.0
B	102.0	260.0	130.0	52.0	26.0
C	24.0	60.0	30.0	12.0	6.0
D	71.0	180.0	90.0	36.0	18.0
E	126.0	320.0	160.0	64.0	32.0
F	276.0	700.0	350.0	140.0	70.0
G	31.0	80.0	40.0	16.0	8.0
H	47.0	120.0	60.0	24.0	12.0
I	7.0	20.0	100.0	4.0	2.0
J	59.0	150.0	750.0	30.0	15.0
K	40.0	100.0	50.0	20.0	10.0

Figure B-39. Truck U-375 flank target

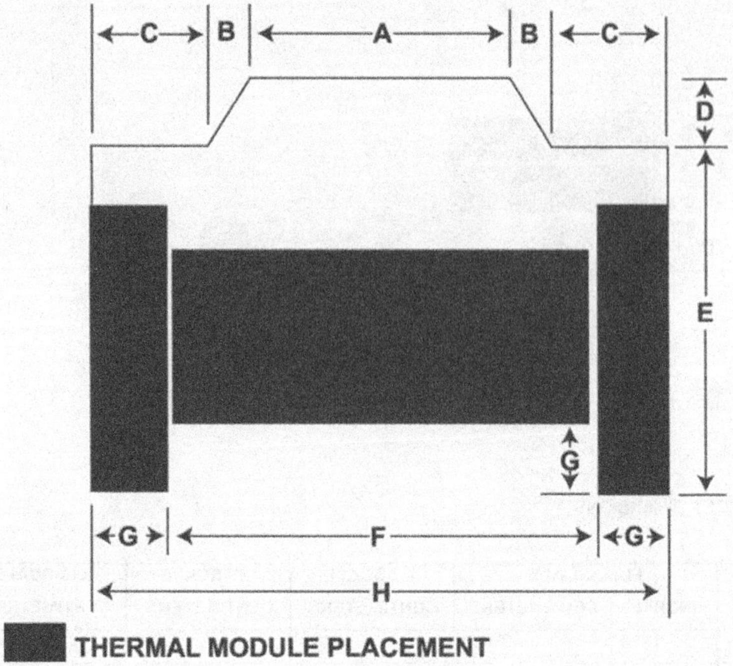

TARGET DIMENSIONS

	FULL-SCALE		1:2 SCALE	1:5 SCALE	1:10 SCALE
	INCHES	CENTIMETERS	CENTIMETERS	CENTIMETERS	CENTIMETERS
A	43.0	110.0	55.0	22.0	11.0
B	80.0	20.0	10.0	4.0	2.0
C	20.0	50.0	25.0	10.0	5.0
D	12.0	30.0	15.0	6.0	3.0
E	55.0	140.0	70.0	28.0	14.0
F	67.0	168.0	84.0	33.6	16.8
G	16.0	40.0	20.0	8.0	4.0
H	99.0	250.0	125.0	50.0	25.0

Figure B-40. BMD front target

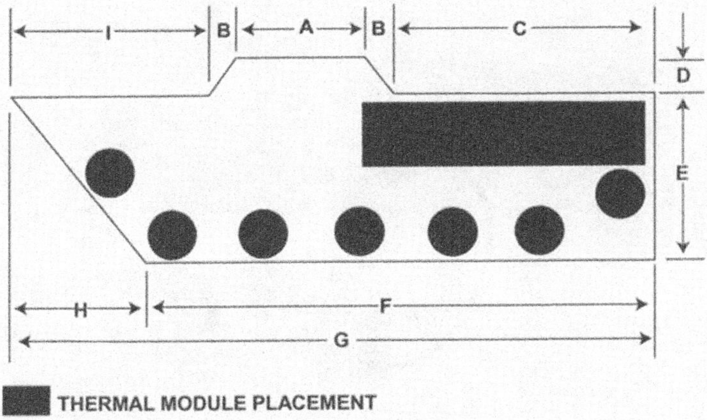

TARGET DIMENSIONS

	FULL-SCALE		1:2 SCALE	1:5 SCALE	1:10 SCALE
	INCHES	CENTIMETERS	CENTIMETERS	CENTIMETERS	CENTIMETERS
A	45.0	114.0	57.0	22.8	11.4
B	7.0	18.0	9.0	3.6	1.8
C	87.0	220.0	11.0	44.0	22.0
D	12.0	30.0	15.0	6.0	3.0
E	55.0	140.0	70.0	28.0	14.0
F	173.0	438.0	219.0	87.6	43.8
G	213.0	540.0	270.0	108.0	54.0
H	400.0	102.0	51.0	20.4	10.2
I	67.0	1700.0	85.0	34.0	17.0

Figure B-41. BMD flank target

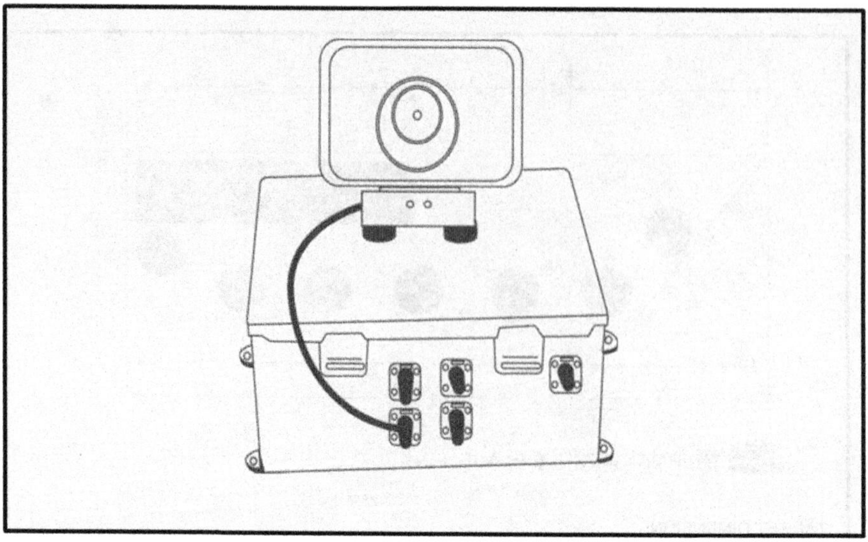

Figure B-42. Sound effects simulator (SES)

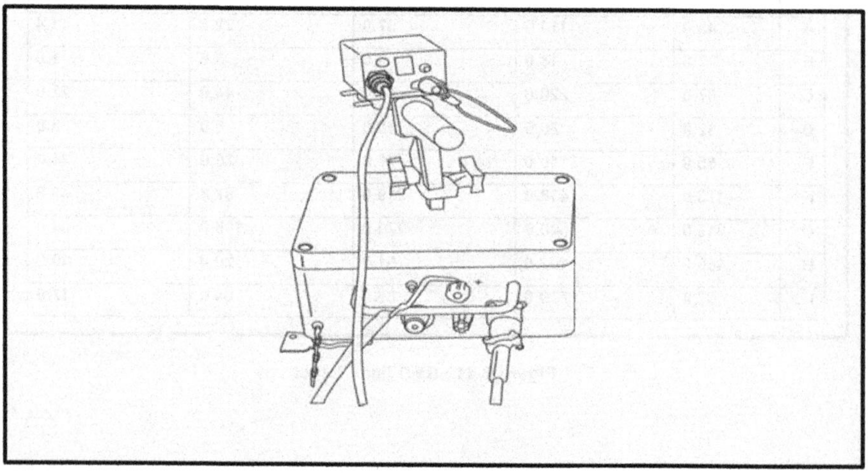

Figure B-43. MILES shootback device (MSD)

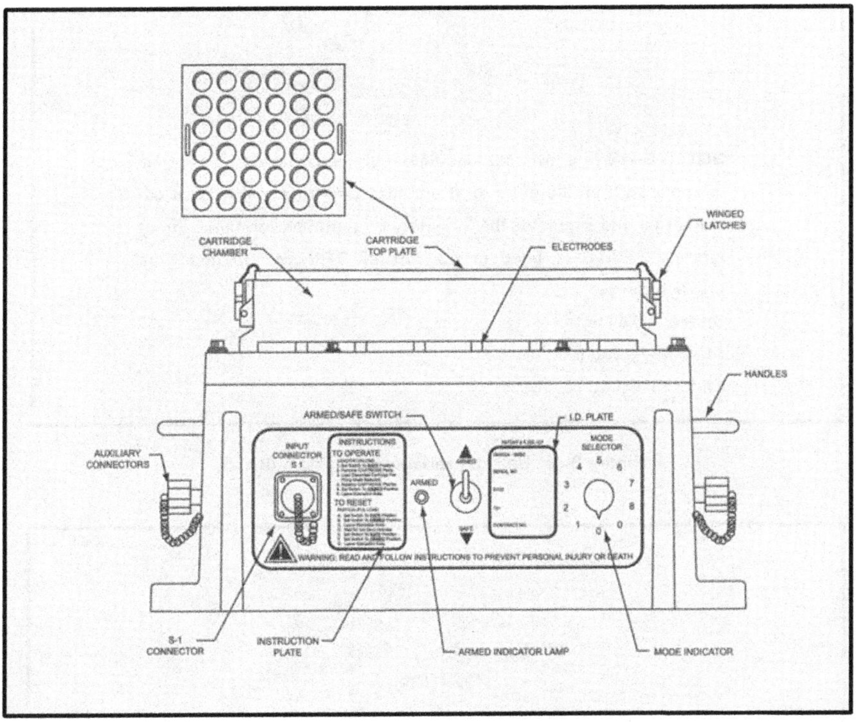

Figure B-44. Battle effects simulator (BES)

NOTE: BATS is a low-cost, one-time flight (expendable) target that is operated from the ground. It provides the firing unit with an on-call target and simulates the trajectory of a high-performance threat aircraft. BATS is used in support of STINGER training. Its characteristics:

Speed – 550 knots

Altitude – 8,000 feet

Endurance – 47 seconds

Figure B-45. Ballistic aerial target system (BATS)

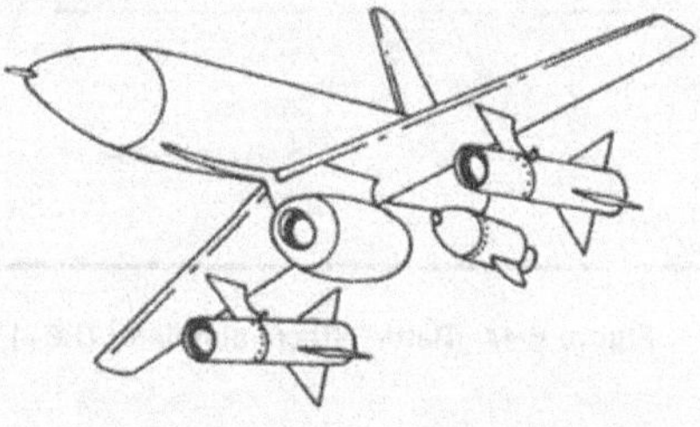

NOTE: The Streaker is a drone target that provides the altitude, speed, and maneuver capability required for supporting the training of Hawk units. It is a low-cost, air transportable system which can be used in both domestic and overseas training areas. The target has the following characteristics:

Speed – 250 to 500 knots

Altitude – 100 to 12,200 meters

Figure B-46. MQM-107 variable-speed training target Streaker

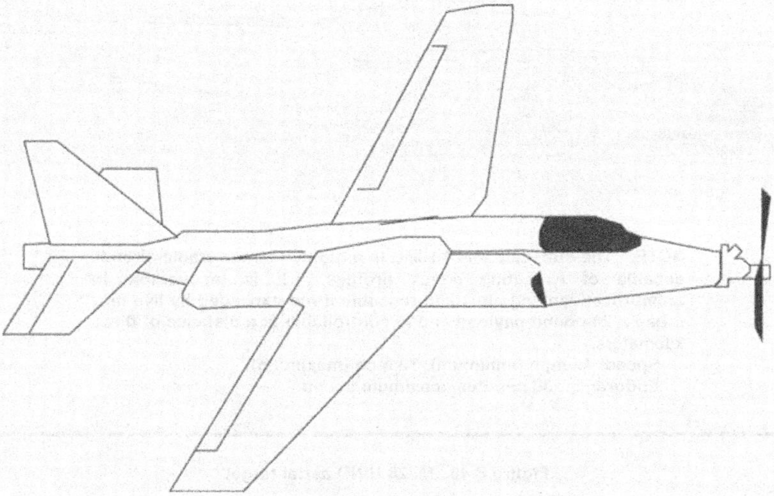

NOTE: The subscale Su-25 Frogfoot is a highly maneuverable aircraft capable of executing attack profiles. It is recoverable in unimproved landing sites and reusable if not expended by live fire. It has a 20-pound payload and is controllable at a distance of three kilometers.

Speed: 45 mph (minimum), 100 mph (maximum)

Endurance: 30 minutes, maximum speed

Figure B-47. Su-25 Frogfoot aerial target

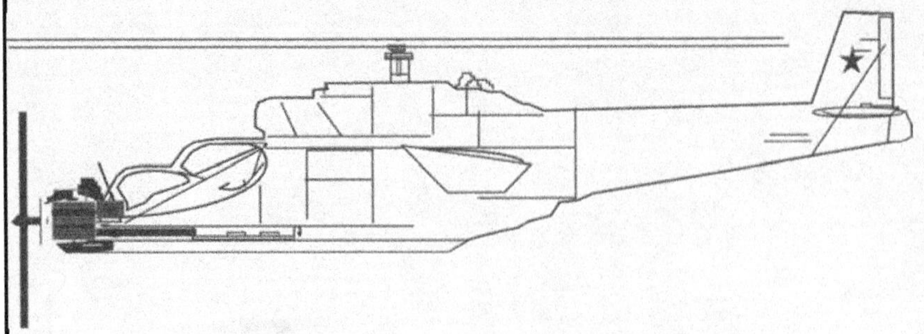

NOTE: The subscale Mi-24 HIND is a highly maneuverable aircraft capable of executing attack profiles. It is recoverable in unimproved landing sites and reusable if not expended by live fire. It has a 20-pound payload and is controllable at a distance of three kilometers.

Speed: 45 mph (minimum), 70 mph (maximum)
Endurance: 30 minutes, maximum speed

Figure B-48. Mi-24 HIND aerial target

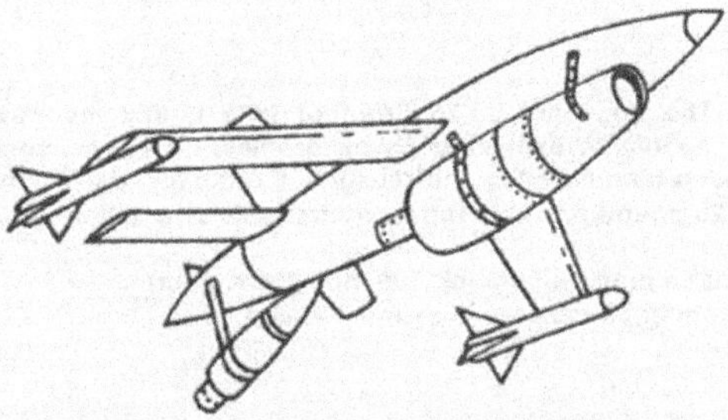

NOTE: The Firebee is a mid-wing, high-speed, subsonic, remotely controlled guided missile. It is propelled by a single turbojet engine. During flight, the missile is controlled from the ground. It is recovered by a parachute system. The mission of the Firebee is to support the evaluation of air defense artillery missile and gun systems. It has the following characteristics.

Speed – 500 knots
Altitude – 100 to 61,000 feet
Endurance – 90 minutes

Figure B-49. MQM-34D Firebee

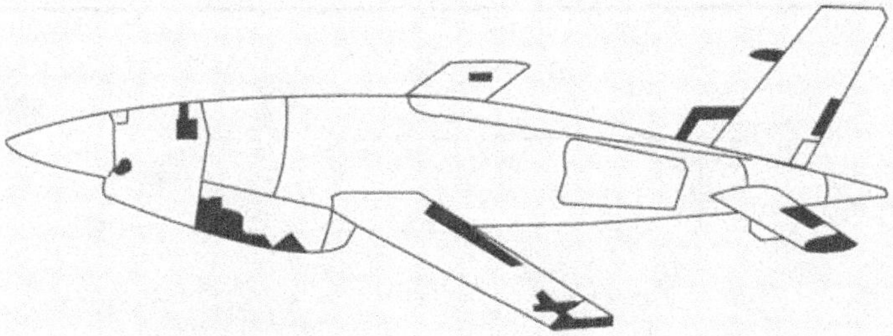

NOTE: The BQM-34 is the most powerful subscale target available.
Speed: 220 to 550 knots
Altitude: 100 to 55,000 feet
Flight time: 60 minutes
Payload: Internal 380 pounds, external 600 pounds each wing

Figure B-50. BQM-34 Firebee

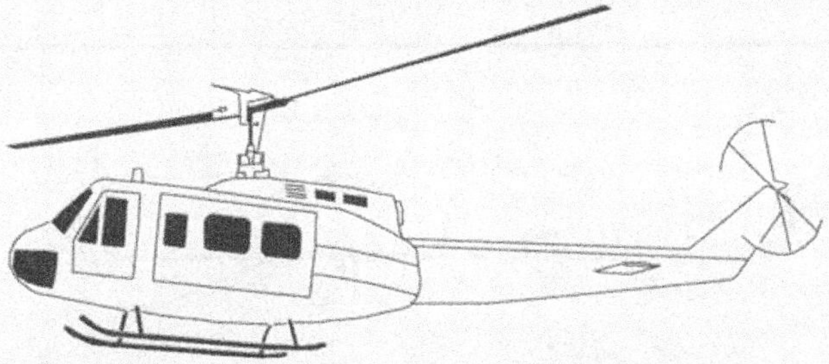

NOTE: The QUH-1 Huey is a full-scale aircraft that can perform numerous flight profiles from low altitude hover to normal flight altitudes.
Speed: Hover to 80 knots
Altitude: 10,000 feet
Flight time: 90 minutes
Payload: 1,000 pounds
Limited nap-of-the-earth

Figure B-51. QUH-1 Huey

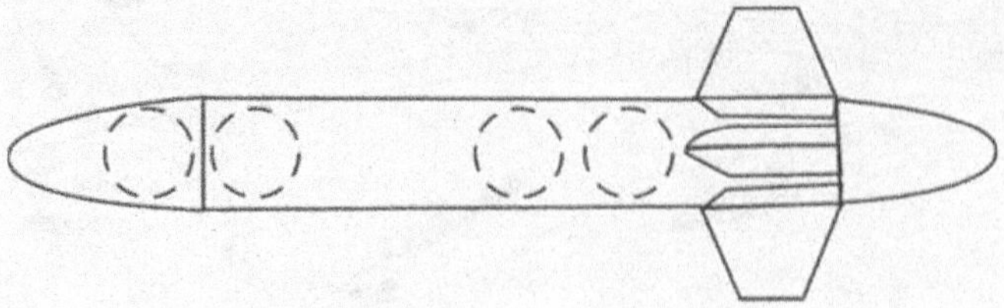

NOTE: The radar cross section (RCS) of the Patriot Omni directional Training Aerial (POTA)-Tow is designed to meet the requirements for Patriot live-fire training.

RCS: 3 to 5 square meters in C and S band
6,000 feet of 0.065-inch towline
Miss distance scoring available
Towed by MQM-107, T-38, or F-16

Figure B-52. POTA-Tow

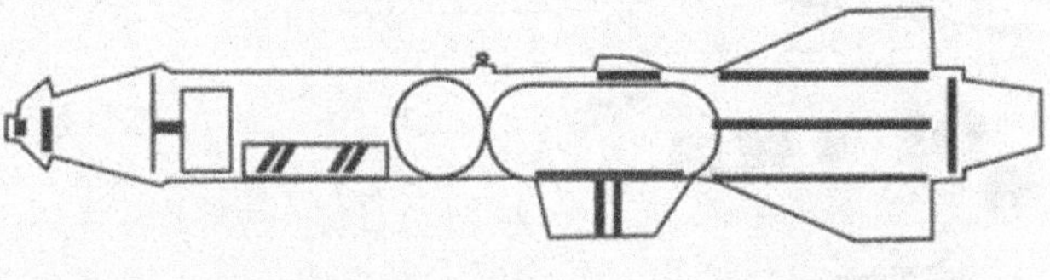

NOTE: This infrared threat target has both the visual and IR augmentation adequate for STINGER.

IR: +140 Watts/Sr in 3 to 5 micron band
600 feet of 0.064-inch steel towline
Miss distance scoring available
Towed by MQM-107 or T-38

Figure B-53. Infrared towed target

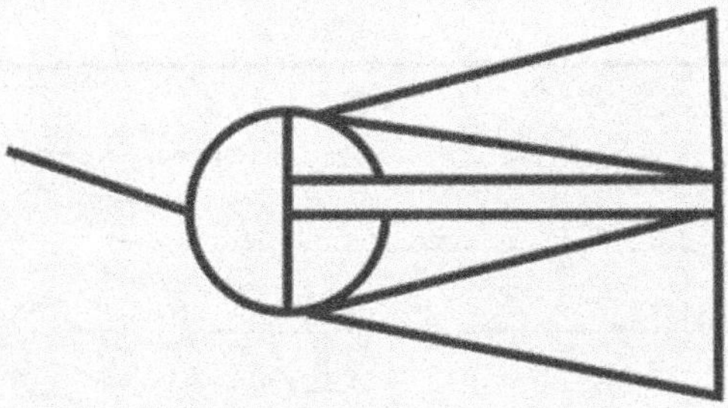

NOTES: The low RCS tow sphere target can be towed by a MQM-107 or a T-38 from an 8,000 foot kevlar cable.
RCS: Less than 0.1 square meter in the C-band.

Figure B-54. Low radar cross section tow sphere target

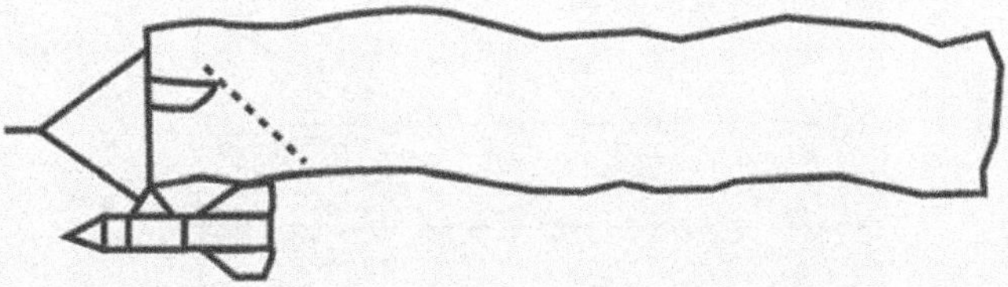

NOTES: The snatch banner consists of a 2 foot by 12 foot banner display with a corner reflector and bullet counter.
RCS: One square meter in X-band
600 feet of 3/8-inch diameter nylon tow rope.
Bullet counting scoring available.
Towed by MQM-107.

Figure B-55. Snatch banner target

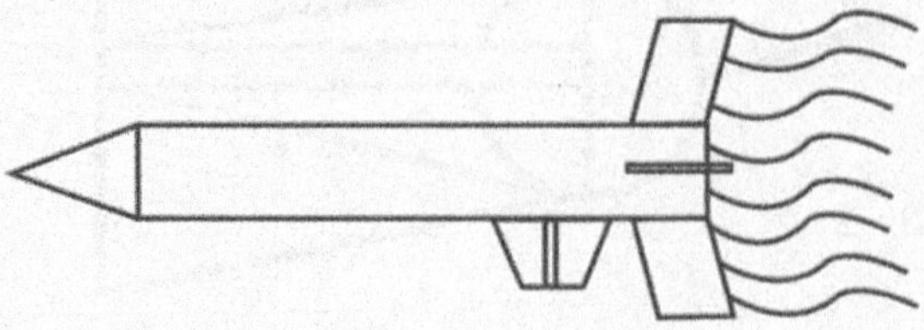

NOTE: RCS is one square meter in the X-band. It has 600 feet of 0.064-inch diameter steel towline, bullet counting scoring available and is towed by the MQM-107.

Figure B-56. Aerial gunnery tow target

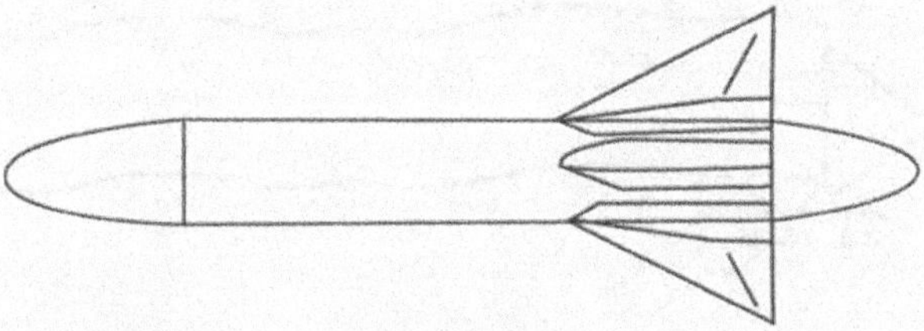

NOTE: An X-band radar augmented target towed on an 8,000 foot steel monofilament cable under the MQM-107. RCS: 8 square meters in X-band.

Figure B-57. TRX-4A Towbee

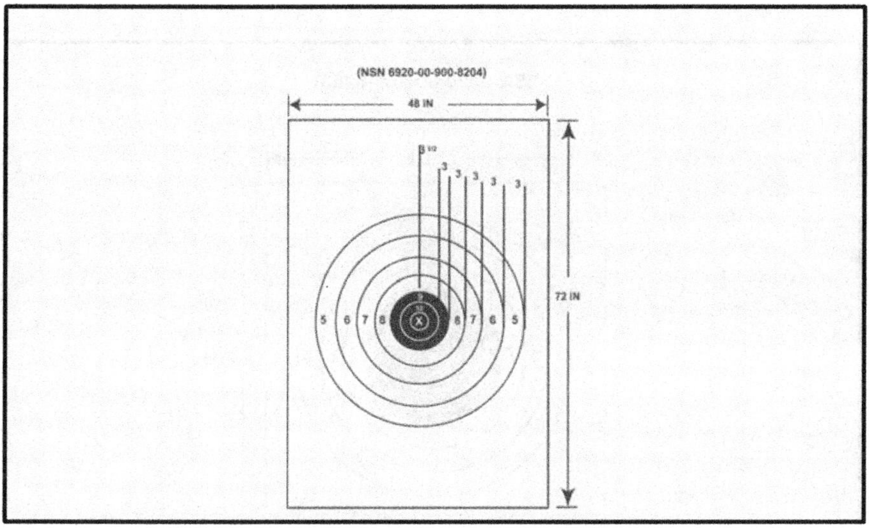

Figure B-58. Army rifle target A, 200-300 yards

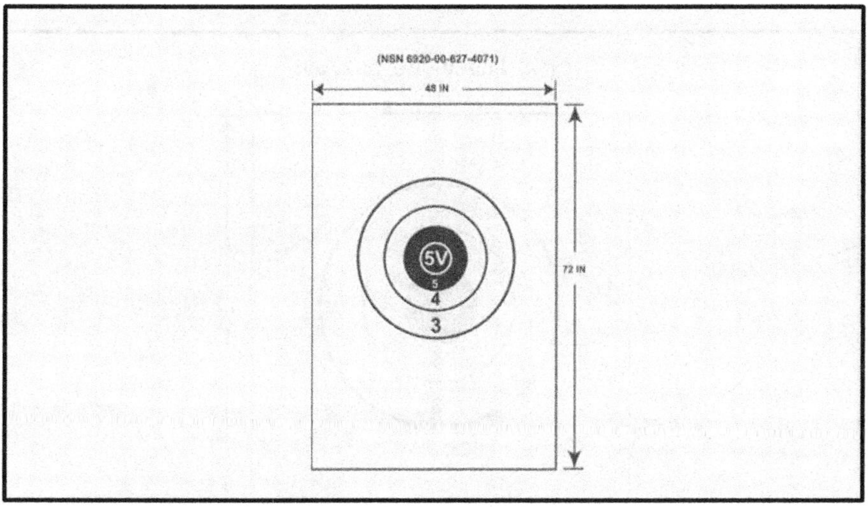

Figure B-59. Navy and Marine Corps rifle target A, 200-300 yards

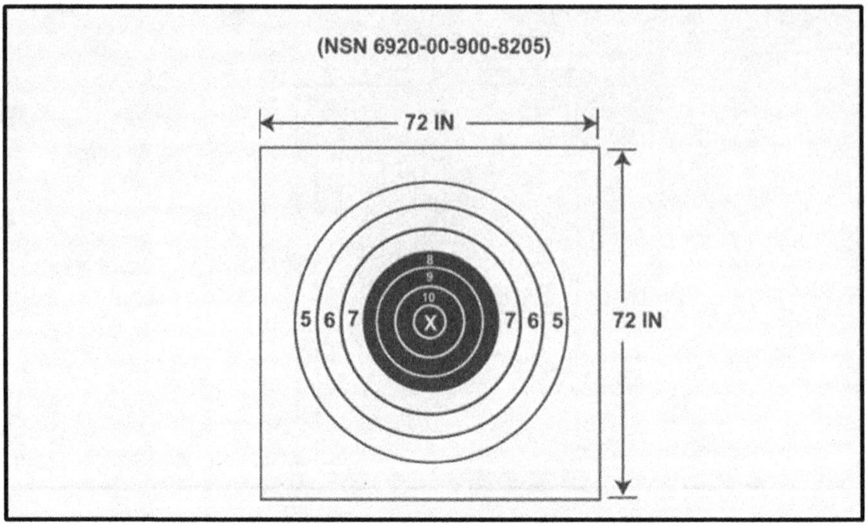

Figure B-60. Army rifle target B, 600 yards

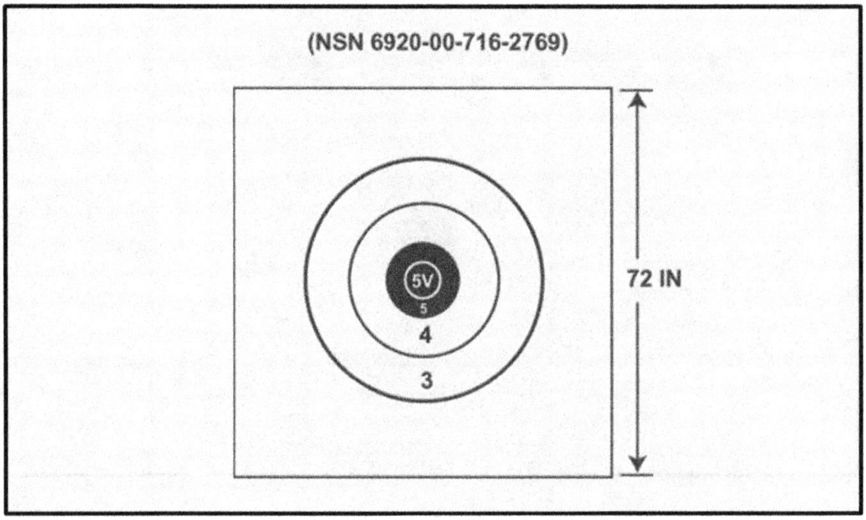

Figure B-61. Navy and Marine Corps rifle target B, 600 yards

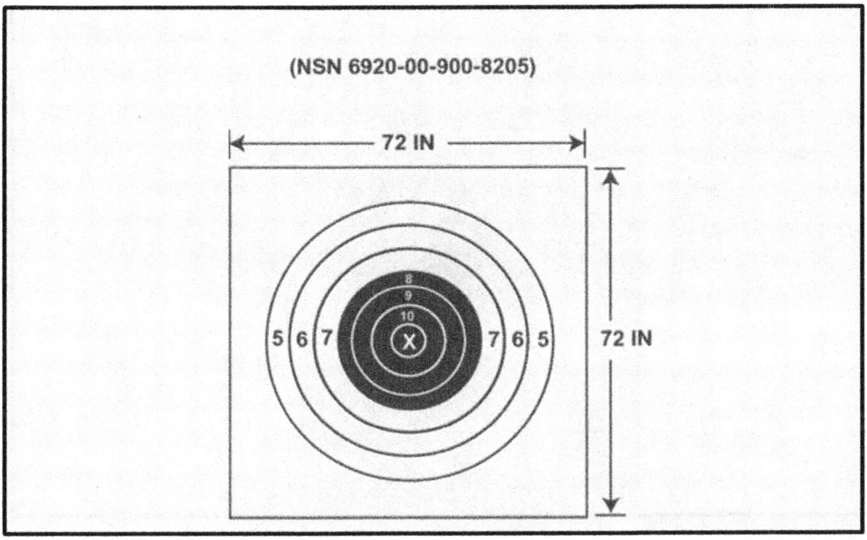

Figure B-62. Army rifle target C, 1000 yards

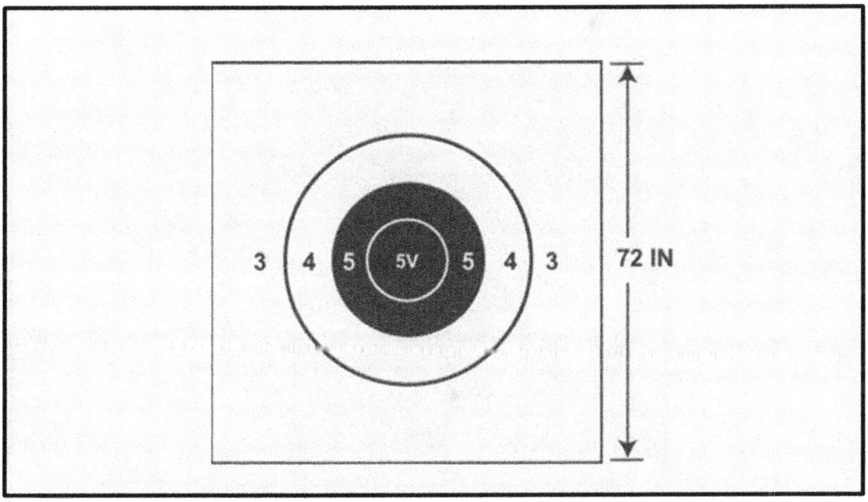

Figure B-63. Navy and Marine Corps rifle target C, 1000 yards

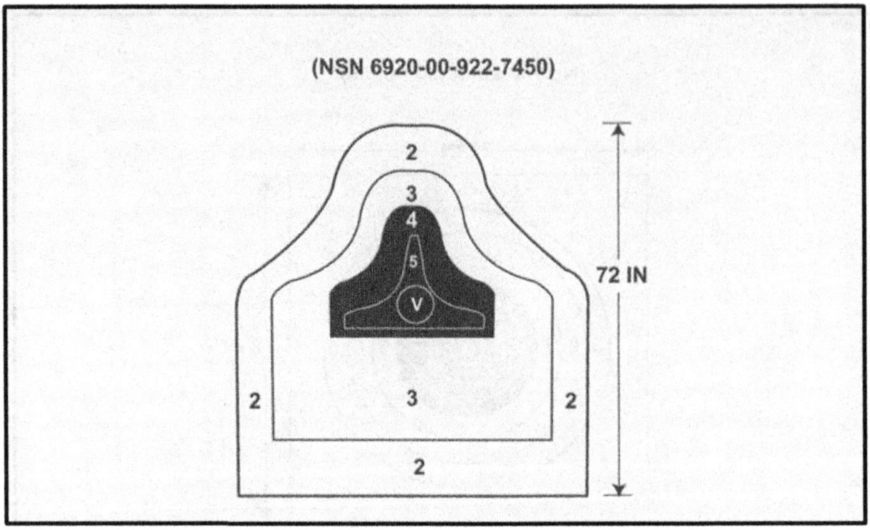

Figure B-64. Navy and Marine Corps target D, prone

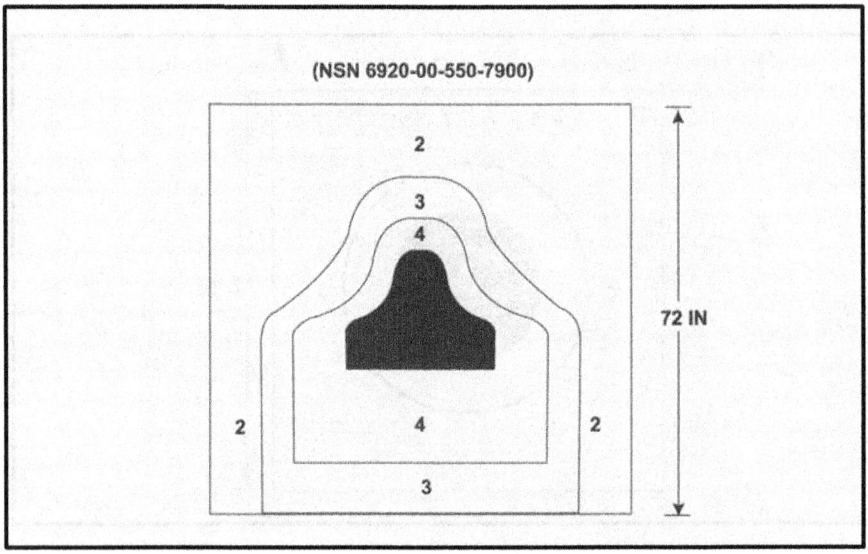

Figure B-65. Navy and Marine Corps target D, prone

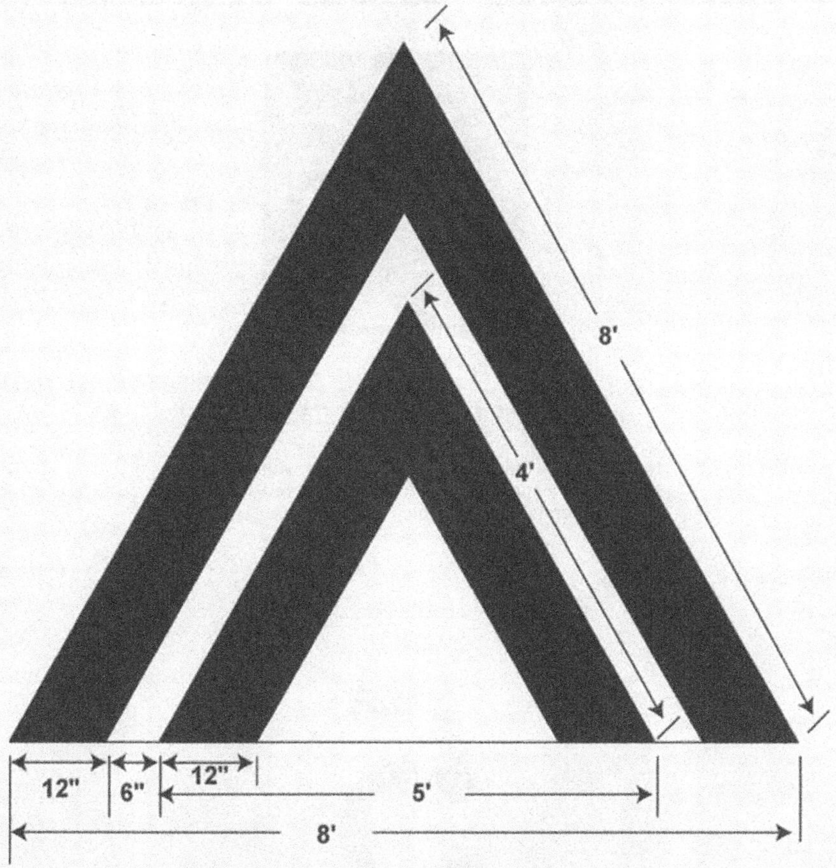

NOTE: Striping is painted international orange or fluorescent orange. Visual light and thermal signature are added for night firing.

Figure B-66. Range outer limit marker

NOTE: Mark even numbered lanes with white numerals on black background. Mark odd numbered lanes with black numerals on a white background.

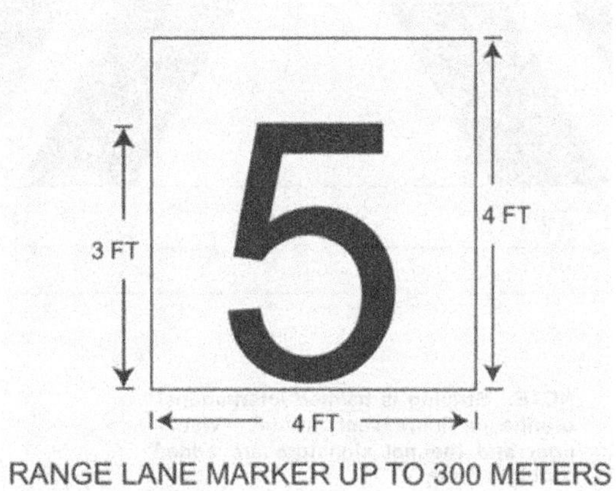

Figure B-67. Range lane markers

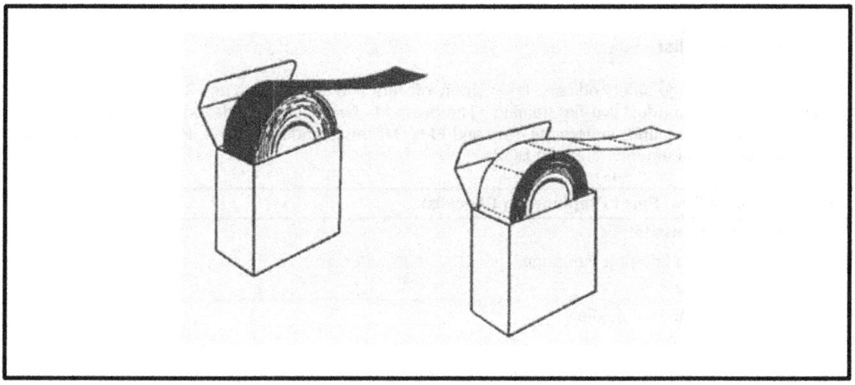

Figure B-68. Target pasters

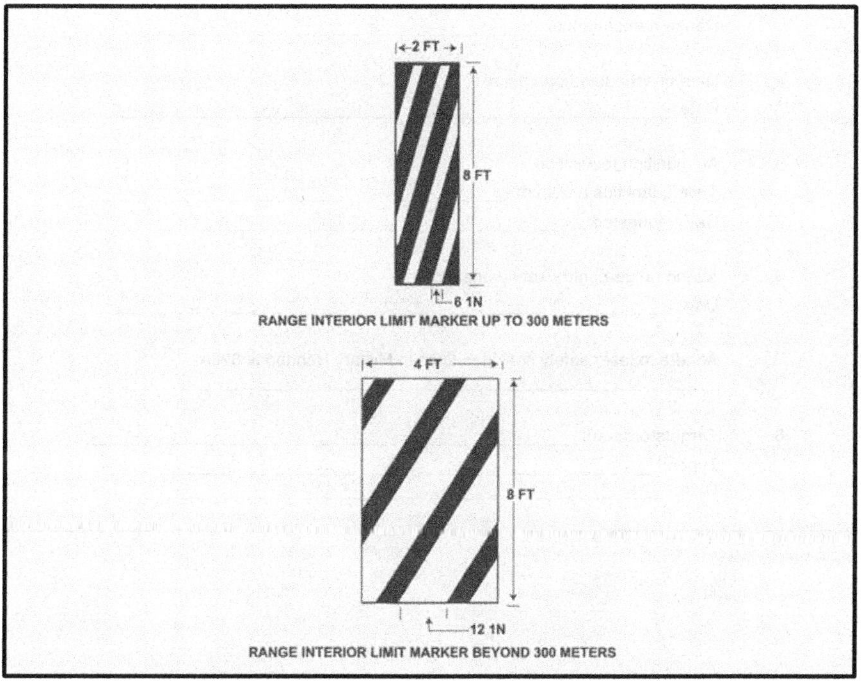

Figure B-69. Range interior limit markers

Appendix C

C-1 Live-Fire Checklist

Officers-in-charge, range safety officers, laser safety officers and unit trainers use Tables C-1 through C-5 checklist to plan and conduct live-fire training. The five-part checklist is a guide; it must be supplemented by AR/DA PAM 385-63, and appropriate TMs and FMs, Military Handbook 828A, local range regulations, special weapons requirements, and unit SOPs.

Table C-1 . Live Fire Part I Preplanning Checklist		
A.	**Mission analysis**	
	1.	Units firing on the range Units: ____________________ Number of soldiers: ____________________
	2.	Weapons and course/table to be fired Weapons: ____________________ Course/table: ____________________
	3.	Range requested Range name/number: ____________________
	4.	Date of scheduled operation Date: ____________________
	5.	Ammunition requisition Type/quantities required: ____________________ Date requested: ____________________
	6.	Attend range-control safety briefing Date: ____________________
	7.	Adhere to laser safety considerations in Military Handbook 828A ____________________
	8.	Targets ordered: ____________________ Types: ____________________ Date: ____________________

Table C-1 . Live Fire Part I Preplanning Checklist (cont)

			Y	N	NA
B.	**Administrative requirements**		**Y**	**N**	**NA**
	1.	Sufficient ammunition has been requested			
	2.	Overhead fire ammunition required			
	3.	Sufficient time has been scheduled to complete training			
	4.	Firing periods coordinated with Range Control			
	5.	Range-scheduling conflicts have been resolved			
C.	**Personnel requirements**		**Y**	**N**	**NA**
	1.	OIC			
		Attended safety course briefing			
		Completed unit certification			
	2.	RSO			
		Attended safety course briefing			
		Completed unit certification			
	3.	LRSO			
		Attended safety course briefing			
		Completed unit certification			
	4.	Assistant safety officers			
	5.	Medical support			
	6.	Ammunition detail			
	7.	Assistant instructors			
	8.	Range guards			
	9.	Range/target operator			
	10.	Maintenance personnel			
	11.	Target detail			
D.	**Equipment requirements**		**Y**	**N**	**NA**
	1.	Range packet/clearance form received			
	2.	Safety fan diagram/range overlay on hand			
		Firing points			
		Firing lanes			
		Firing boxes			
		Target locations			
	3.	Radios			
	4.	Range flag and light (night firing)			
	5.	Vehicle flag sets/lights			
	6.	Safety paddles			
	7.	Ambulance or designated vehicle			

Table C-1 . Live Fire Part I Preplanning Checklist (cont)

		Litter			
		Backboard			
		First-aid bag			
		Other equipment prescribed in local guidance			
	8.	Earplugs			
	9.	Master score sheet			
	10.	Scorecards			
	11.	Armorer's tool kit			
	12.	Weapons-cleaning equipment			
	13.	Fire extinguishers			
	14.	Repair parts/spare weapons			
	15.	Training publications, range and safety publications			
	16.	Report folder			
	17.	Lifesaving procedures			
	18.	Sanitary and cleanup supplies			

Table C-2. Live Fire Part II Range Occupation					
A.	**Establish communications with Range Control**		**Y**	**N**	**NA**
	1.	Establish FM radio range control firing net			
	2.	Establish FM radio range control administrative net			
	3.	Establish Radio or telephone range control back-up firing net			
B.	**Medical support is present**		**Y**	**N**	**NA**
	1.	Medics have communications with treatment facility			
	2.	Medics have strip map from range to treatment facility			
	3.	Medics (off-site) have communications with range			
	4.	Medics (off-site) have strip map to range			
	5.	Medics, in coordination with range OIC, select and clear air evacuation site near the range			
	6.	Target locations			
C.	**Prepare designated areas**		**Y**	**N**	**NA**
	1.	Ammunition point			
	2.	Medical station			
	3.	Concurrent training			
	4.	Parking			
	5.	Armorer			
	6.	Water point			
	7.	Mess			

Table C-2. Live Fire Part II Range Occupation (cont)

D. Range OIC brief	Y	N	NA
1. RSO			
2. LRSO (if required)			
3. Safety assistants			
4. Medical NCOIC			
5. Assistant instructors			
6. Vehicle commanders			
E. Range guards briefed and posted			
F. Inspect range to ensure targetry is present and operational			
G. Verify impact area clear of unauthorized personnel			
H. Raise range flag			
I. Check ammunition (No live-fire ammunition on non-firing range)			
J. Receive firing units			
K. Conduct safety checks on weapons			
1. Dud orientation			
2. Noise hazard briefing			
3. Misfire procedures			
M. Organize personnel into firing orders (keep unit integrity, if possible)			
N. For tanks and BFVs, inspect DA Form 2408-4 (Weapon Record Data) for each main gun to be fired			
O. Request clearance from Range Control to commence firing			

Table C-3. Live Fire Part III Firing Operations	Y	N	NA
A. Communications maintained with Range Control			
B. Ammunition accountability maintained			
C. Personnel accountability maintained			
D. Guards on duty/alert			
E. Earplugs in use			
F. Weapons cleared before departing firing line (rod on small-arms ranges)			
G. Surface danger zone is monitored to ensure it remains clear			
H. Laser danger zones are monitored for unsafe acts or improper safety equipment			
I. Weapon systems are cleared and checked during temporary suspensions of firing			
J. Cease-fire is called when unsafe act is observed or reported or when communications with Range Control is lost			
K. Handle misfires, hangfires, or cookoffs in accordance with technical manuals			

Table C-4. Post-firing Operations		Y	N	NA
A.	**Weapons cleared before leaving range**			
B.	**Conduct brass/ammunition check**			
C.	**Close range down in accordance with SOP**			
D.	**Firing status of range or vehicles reported to Range Control**			
E.	**Conduct police of range**			
F.	**Perform maintenance tasks as required by range SOP**			
G.	**Request range clearance and shut-down from Range Control**			
H.	**Secure range**			

Laser devices may be used only on those ranges that the installation commander has approved and established for such use. The following tables and checklists are furnished as guides for planning and conducting laser operations. Cautions, warnings, and hazards of laser systems must be strictly followed and enforced. Military Handbook 828A should be used as a reference and guide.

Table C-5. Laser Operations		Y	N	NA
A.	**Verify survey of proposed lasing and target area**			
B.	**Right and left laser safety limit stakes designate right and left limits of lasing at local training areas**			
C.	**Warning signs and barricades posted to prevent unauthorized entry**			
D.	**Warning signs posted at entrance to the range**			
E.	**Verify impact area clear of unauthorized personnel**			
F.	**Sweep range area to remove reflective material before lasing**			
	1. Target materials are non-reflective surfaces – cardboard, wood, or lusterless metal			
	2. Recommended target areas are free of reflective surfaces (glossy foliage and raindrops are not considered reflective surfaces)			
	3. Laser devices are not lased at specular (mirror like surfaces)			
G.	**Unprotected personnel are not exposed to either the direct beam or the beam reflected from a specular surface**			
H.	**Personnel within the laser safety danger zone are wearing protective eyewear**			
J.	**Appropriate laser safety filters are placed in daylight optical devices used to observe targets during lasing**			
K.	**Recommended target areas are free of calm, smooth water, and clean ice**			
L.	**Laser safety orientation is provided to personnel who work with or use lasers**			

Appendix D
Range Diagrams

D-1. Instrumented Ranges Functional Description

The Instrumented Range program is composed of the Digital Multi-Purpose Range Complex (DMPRC), the Digital Multi-Purpose Training Range (DMPTR), and the Battle Area Complex (BAX). Due to the commonality of the solutions among these various types of ranges, this consolidation represents the most economical and practical use of development assets. The individual functions of these three ranges are described below:

D-2. Digital Multi-Purpose Range Complex (DMPRC):

a. Supports platoon level collective training for all gunnery tables for Abrams, Bradley, (and all dismounted infantry) and aviation.
b. Supports individual and crew qualification training.
c. Uses thermal targets, night illumination devices, hostile-fire, target-kill, and visual flash simulators.

D-3. Digital Multi-Purpose Training Range (DMPTR):

a. Designed as one of two variants: Current Force or Stryker vehicle.
b. Supports crew qualification for up to table VIII.
c. Supports dismounted infantry squad training independently or with supporting vehicles.
d. Supports crew/squad level training and individual and crew qualifications.
e. Uses thermal targets, night illumination devices, hostile-fire, target-kill, and visual flash simulators.
f. Serves as feeder range for DMPRC, BAX and other collective training facilities.

D-4. Battle Area Complex (BAX):

a. Supports infantry-centric platoon/company CALFEX with MGS crew qualification overlay.
b. Incorporates complex terrain, bunkers, trenches, live fire villages, shoot houses and gunnery type facades.
c. Provides engineer area, add-on urban and mortar impact areas.
d. Uses thermal targets, night illumination devices, hostile-fire, target-kill, and visual flash simulators.

This "product line" of instrumented ranges provides enhanced training data collection for After Action Reviews (AARs), utilization of interfaces compliant with the Common Training Instrumentation Architecture (CTIA), use of advanced targetry to represent realistic threats, and the stimulation of digitized force systems. System and component commonality of the product line will offer a significant future payback in the development and fielding of the instrumented ranges.

D-5. Major Components:

The major components (common to all three instrumented range types) consist of the following:

a. Range Operations Center (ROC)
b. Player Unit (PU) Instrumentation
c. Player Instrumentation Radio (PIR)
d. Audio/Video Recording and Control
e. After Action Review (AAR)
f. Range Targetry
g. Battlefield Effects System (BES)

This drawing shows an example layout of an instrumented range:

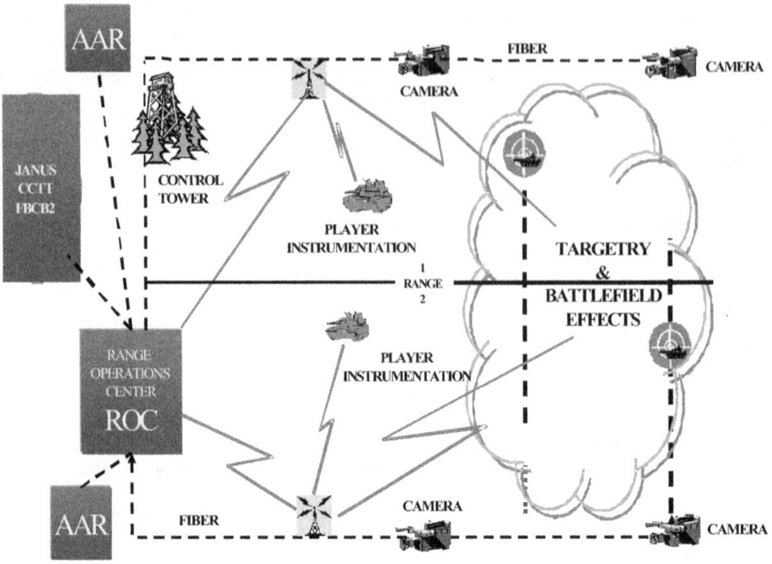

D-6. Range Operations Center (ROC)

a. The ROC is the heart of the instrumented range. It provides an Integrated Management System (IMS) to control all functions from a single workstation. As many workstations as required could be connected to the network to provide control or access to data. The ROC architecture is predicated on the philosophy that all data are stored in a relational database. This includes real-time range data, manually entered user data, data commands, user roles and simulation, targetry data, and overall results data. The ROC will use a platform-independent relational database.

b. The ROC will have multiple functions. It will serve as the central control point for all data collected and for control of all the components of the range facility. The ROC will provide the capability to control targetry, the interface for target data collection, and the capability to populate the range database with data collected from targets. The database will not only provide central processing of targetry information, but it will also collect information from the player instrumentation, video equipment, and the audio components. A High Level Architecture (HLA) interface will be provided at the ROC for external simulations (such as CCTT or Janus). A situational display will provide ROC operators with a view of the player unit (PU) status and range awareness for safety and control of the exercise. The ROC will also provide an overall mission scenario authoring capability to synchronize all Digital Range components.

D-7. Player Unit (PU) Instrumentation

The PU Instrumentation, comprised of all the communications, interfaces and other electronic packages mounted on the host platforms (target, vehicle, aircraft, or crew-served weapon) to be utilized in the training scenarios. The PU interfaces with the host weapons which records and forwards information to the ROC. Data that is forwarded includes Real Time Casualty Assessments (RTCA) results, Force XXI Battle Command Brigade and Below (FBCB2) traffic, through-sight video, position location, weapon system data bus parameters, and simulated live-fire pairing.

D-8. Player Instrumentation Radio (PIR)

The PIR is a bi-directional communication link between the ROC and PU instrumentation in support of near-real-time monitoring and controlling of the instrumented ranges live platform functions. Relayed data includes vehicle status, crew performance, AAR presentations, and range safety information. The PIR receives administrative commands from the ROC via a Network Control Processor (NCP) interface. The NCP ensures messages are received correctly, logged, and transmitted over the ROC LAN for near-real-time use.

D-9. Audio/Video Recording and Control

The audio/video recording system will enable users to easily tag, search, retrieve and playback key segments through a distributive system throughout the ROC. The audio recording system will record tactical radio transmissions, observer/controller (OC) radio systems and telephone systems. Through-sight video and range cameras will record activities in the exercise. The video and audio recording components will be integrated into a comprehensive system that provides an end-to-end digital asset management solution with the ability to acquire, retrieve, transmit, and display digital assets. Video feeds can be used for after-action reviews as well as for range safety monitoring.

D-10. After Action Review (AAR)

The AAR creation process starts with the execution of the exercise. Since there is a real time monitoring capability of the exercise, any operator can tag informational events, FBCB2 messages, or audio/video, for further analysis. The database allows data to be easily searched and retrieved based on time tags, or ad-hoc queries. Therefore, statistical and event data can be quickly formulated into a presentation, or slides that can be automatically created from saved templates. For situational awareness, the video and audio assets can be synchronized and played back in unison as part of an AAR. Users will be able to export situational displays along with multimedia assets into slide presentations, web pages, printed material, or CD/DVD. The ability to export to CD/DVD will make AAR products available for a take-home package to be provided to training units for at-home reviews.

D-11. Range Targetry

Targets will include moving and stationary vehicle and personnel targets. The range system will provide target controls and lifters; the actual targets will be procured separately. These targets will interface to the ROC to provide status, position and control. The targetry component will detect and score shots taken against each target. This data will be transmitted back and stored at the ROC database for use in the AAR. Targetry will be able to interact with the Tactical Engagement Simulation (TES). This will allow shootback capabilities.

D-12. Battlefield Effects System (BES)

Battlefield effects on the instrumented ranges will include, but are not limited to, smoke, sound, and pyrotechnics (explosions). These effects will be controlled from the ROC.

Figure D-1. FCC 17801 RIFLE/MACHINE GUN ZERO RANGE

This range is used to train individual soldiers on the skills necessary to align the sights and practice basic marksmanship techniques against stationary targets. The range is designed for training Shot-Grouping and Zeroing exercises with the M16 and M4 series rifles as well as crew served machine guns.

Primary features include:

- 32 Target frames at 25m
- 16 Target frames at 10m
- 32 Foxholes

This range requires no automation. All targets are fixed at 25 meters from the firing line for M16/M4 and fixed at 10 meters for machine gun.

Associated Range Operations and Control facilities:

- Standard Small Arms ROCA Facilities

Requirement Document: FM 3-22.9, FM 23-14, FM 23-65,
Additional Information: None

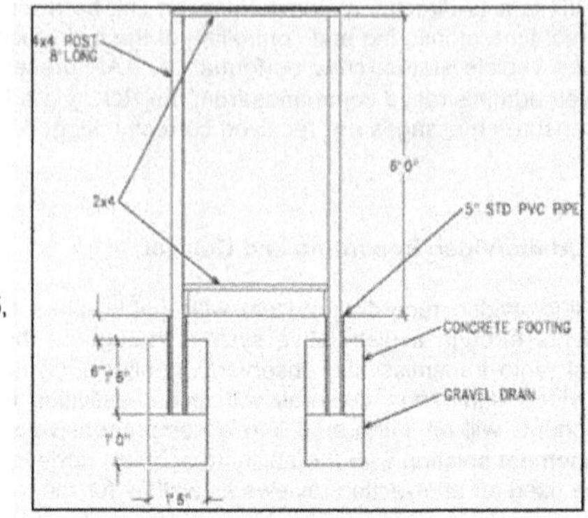

TARGET BOOT DETAIL

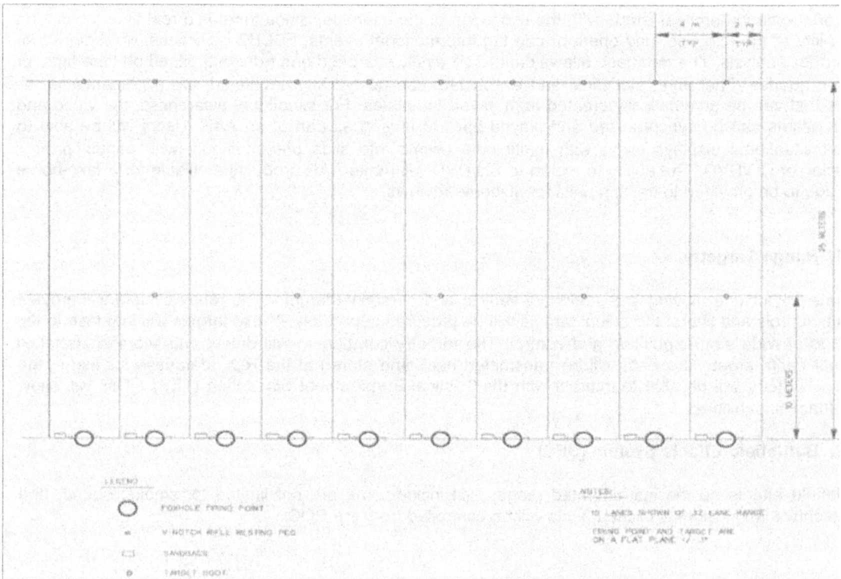

Figure D-2. FCC 17803 AUTOMATED FIELD FIRE (AFF) RANGE

This range is used to train and familiarize soldiers on the skills necessary to identify, engage and hit stationary infantry targets with the M16 & M4 rifles.

Primary features include:
- 96 Stationary infantry targets
- 32 Foxhole positions

All targets are fully automated and the event specific target scenario is computer driven and scored from the range operations center. The range operating system is fully capable of providing immediate performance feedback to the using participants.

Associated Range Operations and Control facilities:
- Standard Small Arms ROCA Facilities

Requirement Document: FM 3-22.9
Additional Information: This range can be used for automatic rifle practice.

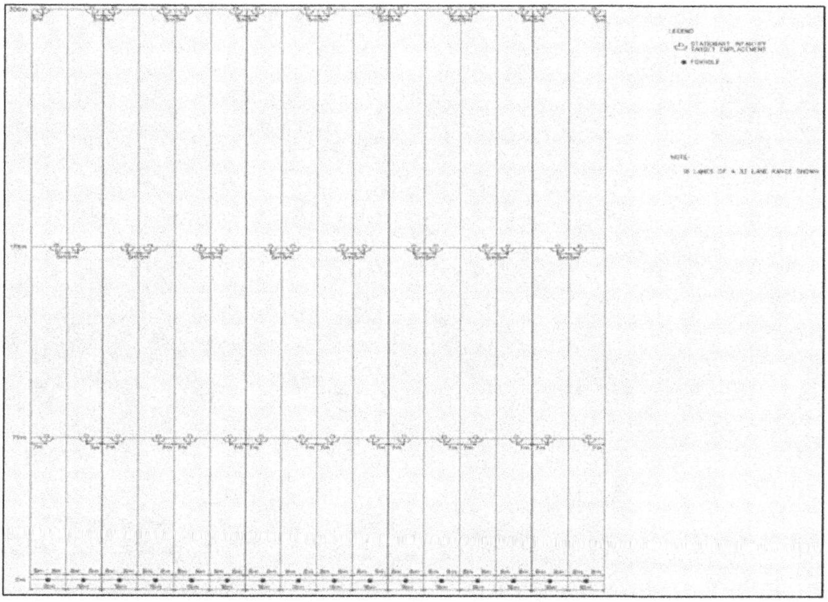

Figure D-3. FCC 17805 AUTOMATED RECORD FIRE (ARF) RANGE

This range is used to train and test individual soldiers on the skills necessary to identify, engage and defeat stationary infantry targets for day/night qualification requirements with the M16 & M4 rifles.

Primary features include:
- 112 Stationary infantry targets
- 16 Foxholes

All targets are fully automated and the event specific target scenario is computer driven and scored from the range operations center. The range operating system is fully capable of providing immediate performance feedback to the using participants.

Associated Range Operations and Control facilities:
- Standard Small Arms ROCA Facilities

Requirement Document: FM 3-22.9
Additional Information: Night firing is accomplished from the night-fire line and baseline, firing at the 50m targets. Replace one of the 50m F-type silhouettes with an E-type silhouette. Low light illumination capability is required in both of the 50m target emplacements. Engage the F-type from the 25m night-firing line, and the E-type from the baseline.

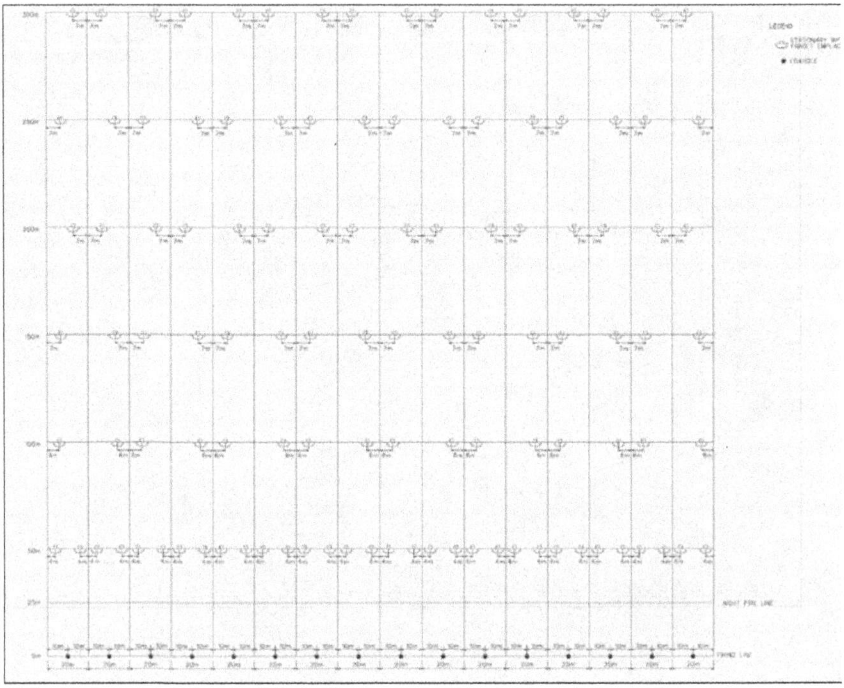

Figure D-4. FCC 17806 MODIFIED RECORD FIRE (MRF) RANGE

This range is used to train and test individual soldiers on the skills necessary to identify, engage and defeat stationary infantry targets for day/night qualification requirements with the M16 & M4 rifles. This range combines the capabilities of Automated Field Fire (17803), Automated Record Fire (17805), and the Automated Night Fire (17808) to reduce land and maintenance requirements and increase efficiencies.

Primary features include:

144 Stationary infantry targets
16 Foxholes

All targets are fully automated and the event specific target scenario is computer driven and scored from the range operations center. The range operating system is fully capable of providing immediate performance feedback to the using participants.

Associated Range Operations and Control facilities:

Standard Small Arms ROCA Facilities

Requirement Document: FM 3-22.9

Additional Information: Night firing accomplished from night-fire line and baseline, firing at the 50m targets. Replace one of the 50m F-type silhouettes with an E-type silhouette. Low light illumination capability is required in both of the 50m target emplacements. Engage the F-type from the 25m night-firing line, and the E-type from the baseline.

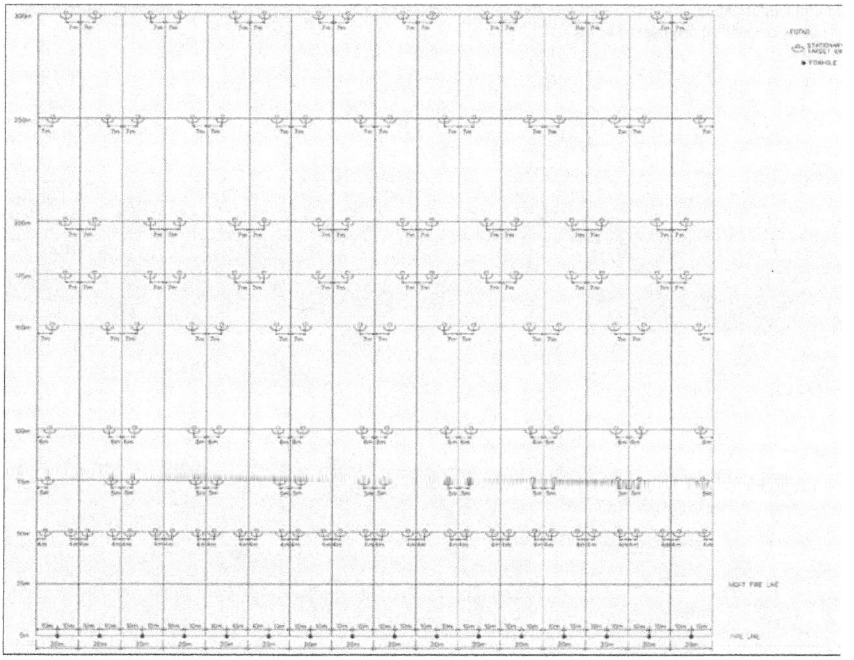

Figure D-5. FCC 17809 QUALIFICATION TRAINING RANGE (QTR)

This range is used to train and test soldiers on the skills necessary to detect, identify, engage and defeat stationary and moving infantry targets in a tactical array with their prescribed weapons. This range combines the capabilities of Modified Record Fire Range (17806), Automated Sniper Field Fire Range (17812), Combat Pistol/Military Police Firearms Qualification Course (17822), and the Multipurpose Machine Gun Range (17833) to centralize training and reduce land, maintenance, and unit overhead requirements.

Primary features include:
- 15 Lanes Combat Pistol Qualification
- 10 Lanes Sniper Field Fire (see notes)
- 16 Lanes Modified Record fire
- 10 Lanes Multipurpose Machine Gun
- 32 Lanes Rifle/Machine Gun Zero

429 Stationary infanty targets
20 Stationary armor targets
20 Moving infantry targets
10 Sationary infantry emplacements with
2 target mechanisms each (400 m and 700 m)

All targets are fully automated and the event specific target scenario is computer driven and scored from the range operations center. The range operating system is fully capable of providing immediate performance feedback to the using participants.

Associated Range Operations and Control facilities:
- Standard Small Arms ROCA Facilities
- Additional General Instruction Building (17120)
- Additional Ammo Breakdown Building (17129)

Requirement Document: FM 3-22.9, FM 3-22.10 (23-10), FM 3-22.35 (23-35), FM 19-10, FM 23-65, FM 3-22.68
Additional Information: This range enhances throughput capability for units with multiple weapons densities by consolidating unit efforts to operating one training facility.

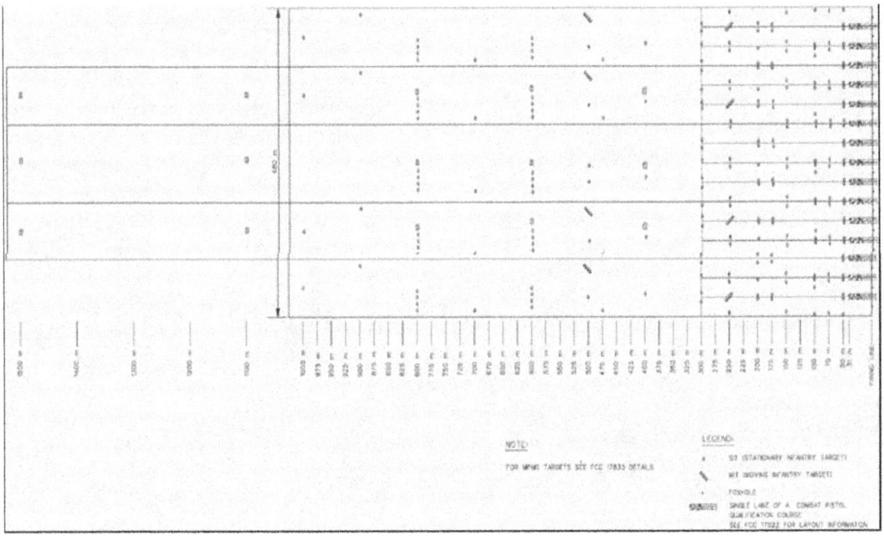

Figure D-6. FCC 17812 AUTOMATED SNIPER FIELD FIRE RANGE

This range is used to train and test soldiers on the skills necessary to detect, identify, engage and defeat stationary and moving infantry targets in a tactical array. This range is designed to satisfy the training and qualification requirements of the M24 Sniper rifle.

Primary features include:
- 40 stationary infantry targets
- 8 Moving infantry targets
- 4 firing positions

All targets are fully automated and the event specific target scenario is computer driven and scored from the range operations center. The range operating system is fully capable of providing immediate performance feedback to the using participants.

Associated Range Operations and Control facilities:
Standard Small Arms ROCA Facilities

Requirement Document: FM 3-22.10 (23-10)
Additional Information: Natural vegetation is required in the target area to provide realistic natural obstacles for the sniper to negotiate.

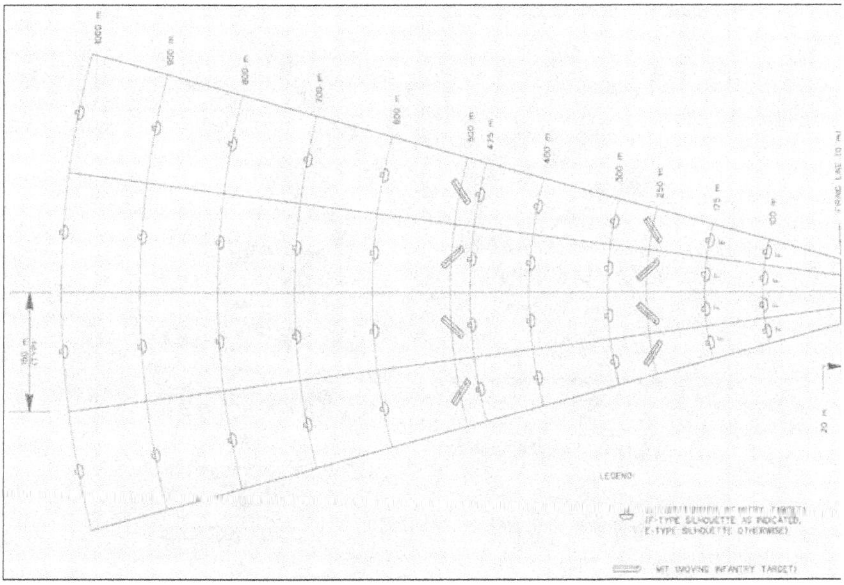

FCC 17812 AUTOMATED SNIPER FIELD FIRE RANGE

Figure D-7. FCC 17816 BAYONET ASSAULT COURSE (BAC)

This facility is used to train individual soldiers in assault techniques using a rifle and bayonet against dummy silhouettes as well as negotiating various obstacles.

Primary features include:
 9 Lanes
 72 Dummy silhouettes

This facility requires no automation.

Associated Range Operations and Control facilities:
 Operations/Storage Building (17122)
 Latrine (73075)

Requirement Document: FM 3-25.150
Additional Information: The number of lanes can vary due available terrain and the number of soldiers that must complete the course.

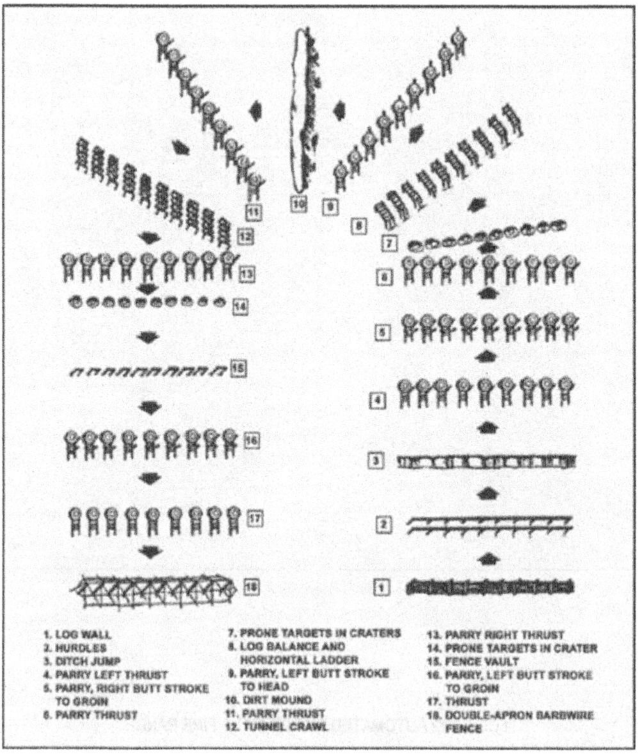

Figure D-8. FCC 17822 AUTOMATED COMBAT PISTOL/MILITARY POLICE FIREARMS QUALIFICATION COURSE (CP/MPQC)

This range is used to train and test soldiers on the skills necessary to detect, identify, engage and defeat stationary targets in a tactical array. The complex satisfies the training and qualification requirements of the 9mm, .38 caliber, and .45 caliber pistols.

Primary features include:

- 105 Stationary infantry targets
- 15 Firing lanes
- 15 Stationary silhouettes (booted, MP Qual)

All CPQC targets are fully automated and the event specific target scenario is computer driven and scored from the range operations center. The range operating system is fully capable of providing immediate performance feedback to the participants using the CPQC. MPQC silhouettes are non-automated and is manually scored.

Associated Range Operations and Control facilities:

Standard Small Arms ROCA Facilities

Requirement Document: CPQC: FM 3-22.35 (23-35)
MPFQC: FM1910.
Additional Information: None

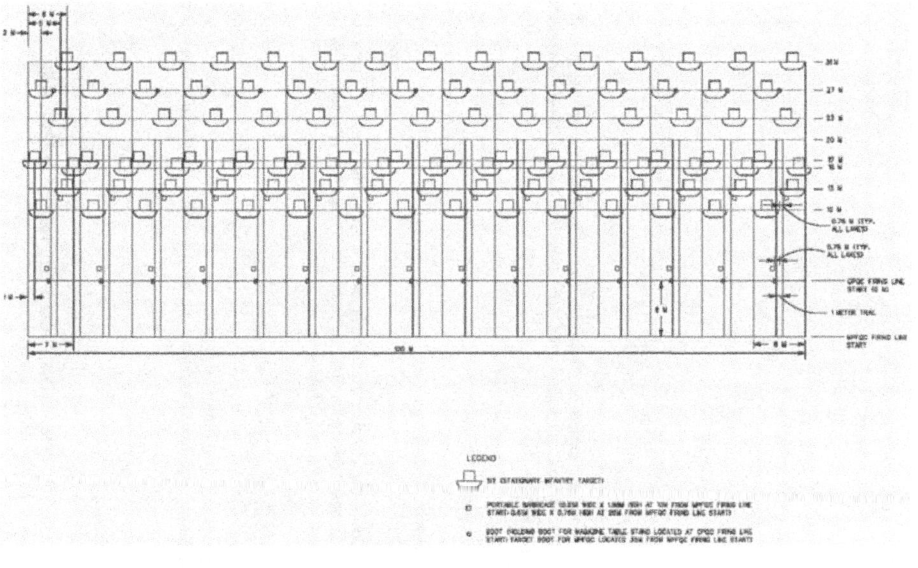

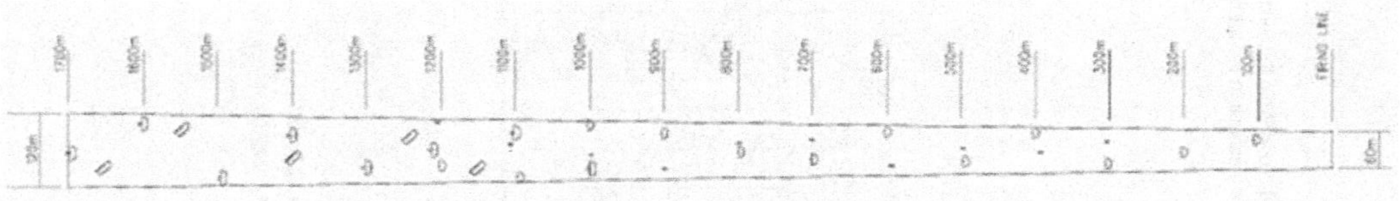

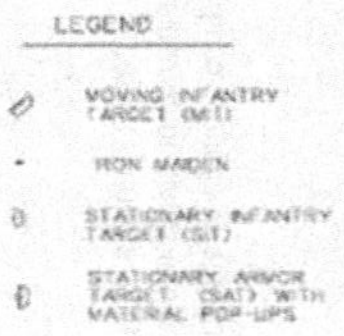

This range is used to train and test soldiers on the skills necessary to detect, identify, engage and defeat stationary and moving infantry targets in a tactical array. This range satisfies the training and qualification requirements of the heavy sniper rifle.

Primary features include:

12 Stationary infantry targets
10 Iron maidens
8 Stationary armor targets
5 Moving infantry targets

All targets, except iron maidens, are fully automated and the event specific target scenario is computer driven and scored from the range operations center. The range operating system is fully capable of providing immediate performance feedback to the using participants.

Associated Range Operations and Control facilities:
Standard Small Arms ROCA Facilities

Requirement Document: FM 3-22.10 (23-10).
Additional Information: Natural vegetation is required in the target area to provide realistic natural obstacles for the sniper to negotiate.

Figure D-9. FCC 17829 HEAVY SNIPER RANGE

Figure D-10. FCC 17833 MULTIPURPOSE MACHINE GUN (MPMG) RANGE

This range is used to train and test soldiers on the skills necessary to zero M249 SAW, M60 MG, M240B MG, MK19, Sniper Weapons and M2 MG weapon systems, detect, identify, engage and defeat stationary infantry targets in a tactical array.

Primary features include:

180 Stationary infantry targets (SIT)
20 Moving Infantry Targets emplacements
20 Stationary armor targets
10 Firing lanes
Note: Contains 20 emplacements with 2 single arm mechanisms

All targets are fully automated and the event specific target scenario is computer driven and scored from the range operations center. The range operating system is fully capable of providing immediate performance feedback to the using participants.

Associated Range Operations and Control facilities:

Standard Small Arms ROCA Facilities

Requirement Document: FM 23-14, FM 23-65, FM 3-22.68
Additional Information: Target beyond 800M will be battery powered/radio controlled.

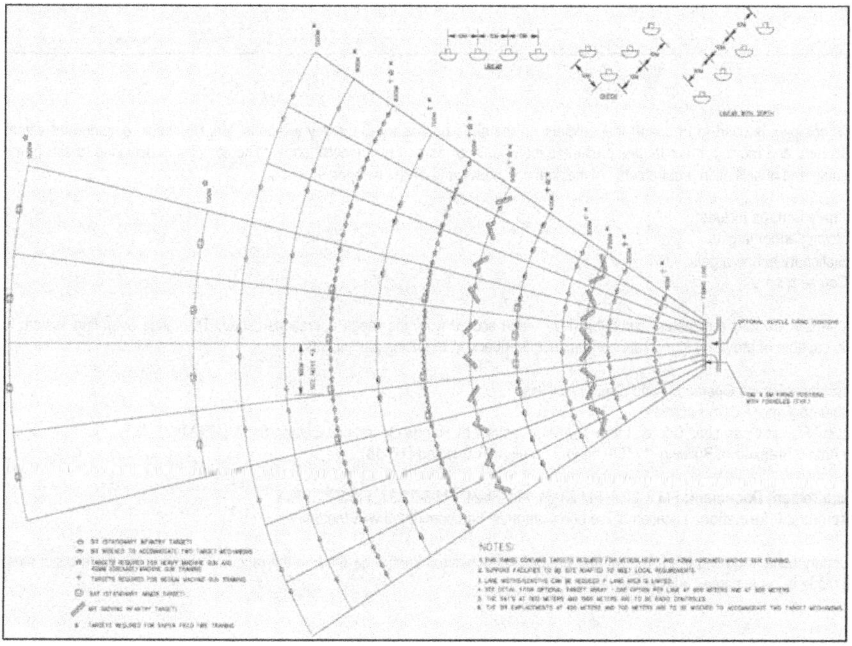

Figure D-11. FCC 17845 ANTIARMOR TRACKING AND LIVE-FIRE

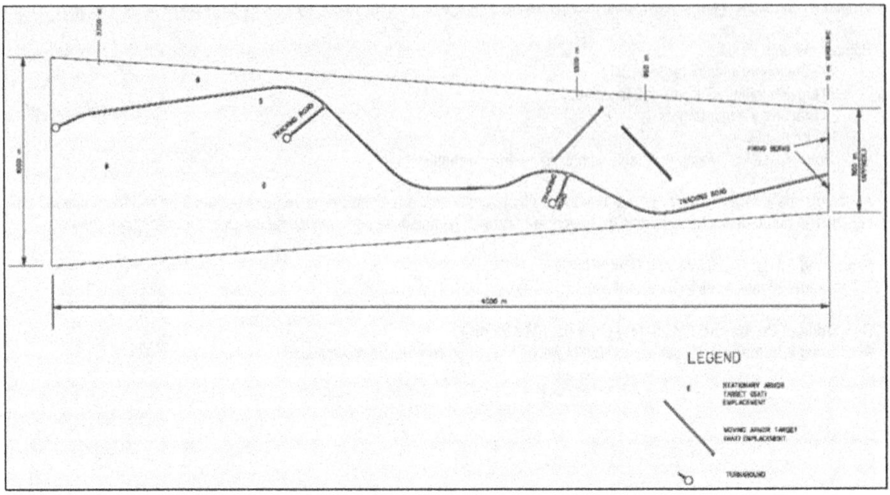

This complex is used to train and test soldiers on the skills necessary to employ weapons, identify, track, engage and defeat stationary and moving armor targets presented individually or as part of a tactical array. The complex is designed to satisfy the training and qualification requirements of medium and heavy anti-armor weapon systems.

Primary features include:
2 Moving armor targets
4 Stationary armor targets
1 Course road

All targets are fully automated, computer driven, and scored from the range operations center. The range operating system is fully capable of providing immediate performance feedback to the using participants.

Associated Range Operations and Control facilities:
Standard Armor ROCA Facilities
Except Range Operations Center, Large (17124) replaced by Range Operations Center, Small (17123);
no Ammo Breakdown Building (17129) and no Small AAR Building (17118)

Requirement Document: FM 3-20.8, FM 23-34, FM 23-24, FM 3-22.37, FM 3-22.999A
Additional Information: Location of the boresight must be coordinated with the trainer.

Gunnery tasks requiring the usage of dud producing ammuniton cannot be fired on the range. Provisions for these tasks must be made in impact areas adjacent to range.

Figure D-12. FCC 17852 MORTAR RANGE

This range is used to train mortar crews on the skills necessary to apply fire mission data, engage, and hit stationary targets in a tactical array.

Primary features include:
- Surveyed firing point(s)
- Dedicated impact area(s) with hard targets

No automation is required for this facility.

Associated Range Operations and Control facilities:
- Ammunition Loading Dock (14970)
- Latrine (73075)

Requirement Document: FM 7-90, FM 23-85, FM 23-90, FM 23-91, FM 23-92
Additional Information: This range may be next to a unit maneuver area. It can then support maneuver training with tactical smoke and illumination. Approximately 10 surveyed firing points are needed on each division post to calibrate aiming circles.

Natural terrain and vegetation should remain undisturbed to the maximum extent possible.

Based on safety, forward observation areas are located in front of firing areas and slightly to the flank of primary mortar to target lanes.

A common impact area is used for all types of mortars. It is at least 2,000 meters wide and 6,000 meters deep. Firing at maximum and minimum range is obtained by using different firing points.

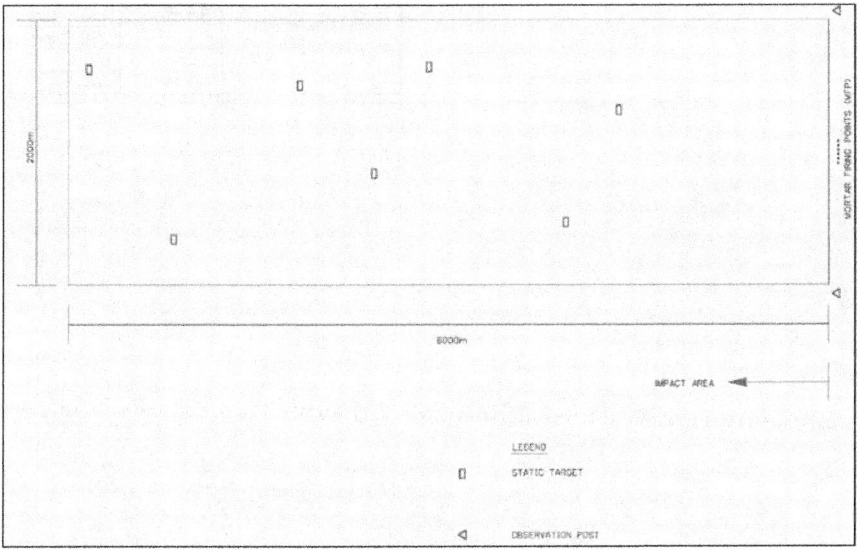

Figure D-13. FCC 17856 FIELD ARTILLERY INDIRECT RANGE

This range is used to train field artillery crews on the skills necessary to apply fire mission data, engage, and hit stationary targets in a tactical array.

Primary features include:
Surveyed firing point(s)
Dedicated impact area(s) with hard targets

No automation is required for this facility.

Associated Range Operations and Control facilities:
No standard facilities are associated with this range.

Requirement Document: FM 6-30, FM 6-40, FM 6-50,TC 25-1, FM 3-09.70
Additional Information: Ideally, this impact area is adjacent to direct fire ranges.
The alternate firing line should be located to fit the specific site. Natural terrain and vegetation should remain undisturbed to the maximum extent possible.
The battery firing position depends on the size of the weapon, formation and terrain at the firing point. Depending on the types of artillery planned, the following data should be uses.

Type	Dimensions (meters)	
M117/M102 (105mm Battery	Position	500 wide
	Range	1500 deep
		7500 wide
M109/M198 (155mm) Battery and M110 (203mm) Battery	Position	1000 wide
	Range	25,000 wide
		12,500 deep
MLRS Launcher (one weapon live fire)	Position	10 wide
	Range	15,000 wide
		30,000 deep
	Rear of launch blast	800 deep
		400 wide

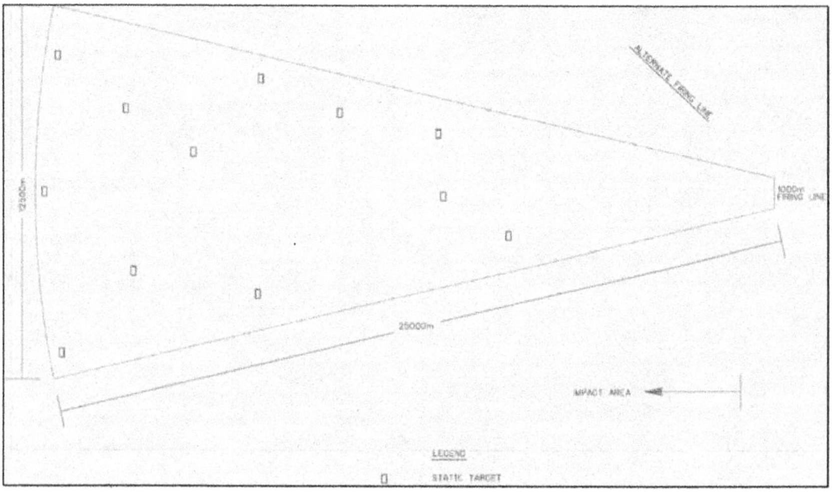

Figure D-14. FCC 17858 SCOUT/RECONNAISSANCE GUNNERY COMPLEX (SCOUT/RECCE)

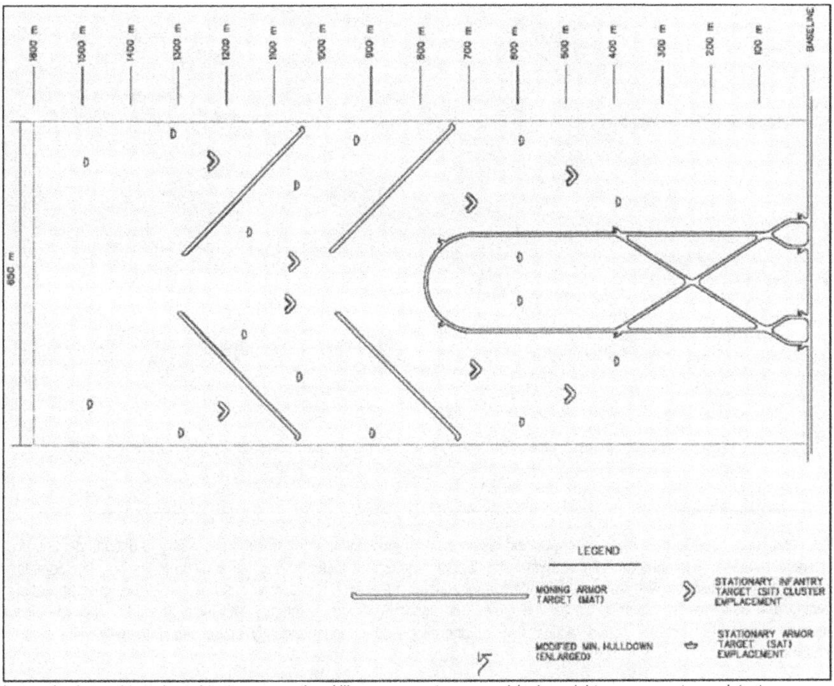

This complex is used to train and test crews on the skills necessary to zero and /or boresight weapon systems, detect, identify, engage and defeat stationary and moving infantry and armor targets in a tactical array.

Primary features include:

- 16 Stationary armor targets
- 4 Moving armor targets
- 56 Stationary infantry targets (8 clusters @ 7 SITs each)
- 1 Lane (2 course roads) with crossover capability
- 8 Battle positions per lane

All targets are fully automated, utilizing event-specific, computer-driven target scenarios and scoring. Targets will receive and transmit digital data from the range operations center. This captured data is compiled and available to the unit during the after action review.

Associated Range Operations and Control facilities:

- Standard Armor ROCA Facilities
- Except Range Operations Center, Large (17124) replaced by Range Operations Center, Small (17123)

Requirement Document: FM 3-20.8, FM 23-27
Additional Information: None

Figure D-15. FCC 17859 DIGITAL MULTIPURPOSE TRAINING RANGE (DMPTR)

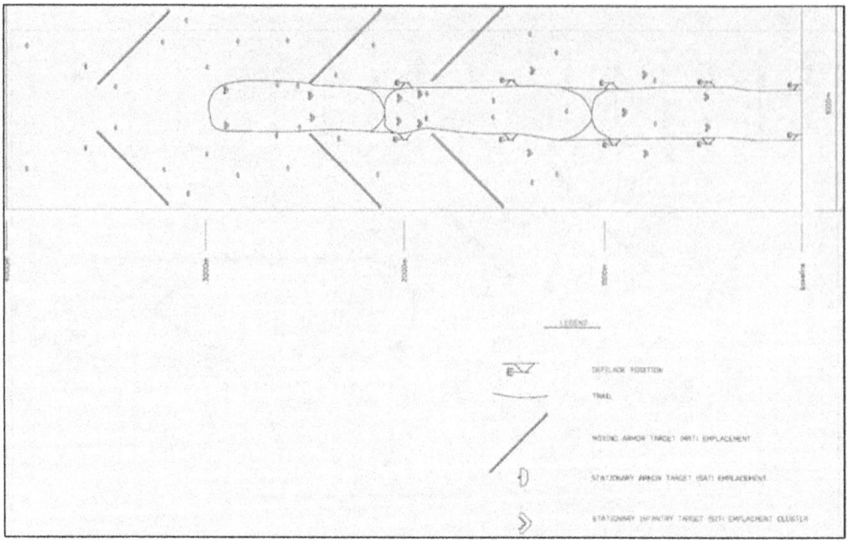

This complex is used to train and test crews and dismounted infantry squads on the skills necessary to detect, identify, engage and defeat stationary infantry and stationary/moving armor targets in a tactical array. In addition to live-fire, this complex can also be used for training with sub-caliber and/or laser training devices. The complex is specifically designed to satisfy the training and qualification requirements for the crews and sections of armor, infantry and aviation units. This complex also supports dismounted infantry squad tactical live-fire operations either independently of, or simultaneously with, supporting vehicles.

Primary features include:

- 35 Stationary armor targets
- 6 Moving armor targets
- 105 Stationary infantry targets (15 clusters @ 7 SITs each)
- 1 Lane (2 course roads) with midpoint crossover capability
- 10 Battle positions per lane

All targets are fully automated, utilizing event-specific, computer-driven target scenarios and scoring. Targets will receive and transmit digital data from the range operations center. The captured data is then compiled and available to the unit during the after action review.

Associated Range Operations and Control facilities:
Standard Armor ROCA Facilities

Requirement Document: FM 3-20.12, FM 23-1, FM 3-20.8, FM 3-04.140
Additional Information: This complex uses thermal targets, night illumination devices, and hostile-fire, target-kill, and visual flash simulators.
Location of the bore sight must be coordinated with the trainer. Gunnery tasks requiring the usage of dud producing ammunition cannot be fired on the complex. Provisions for these tasks must be made in impact areas adjacent to the complex.
Hard targets may be positioned in the impact area to facilitate RF/IR signatures for aviation gunnery.

Figure D-16. FCC 17860 DIGITAL MULTIPURPOSE RANGE COMPLEX (DMPRC)

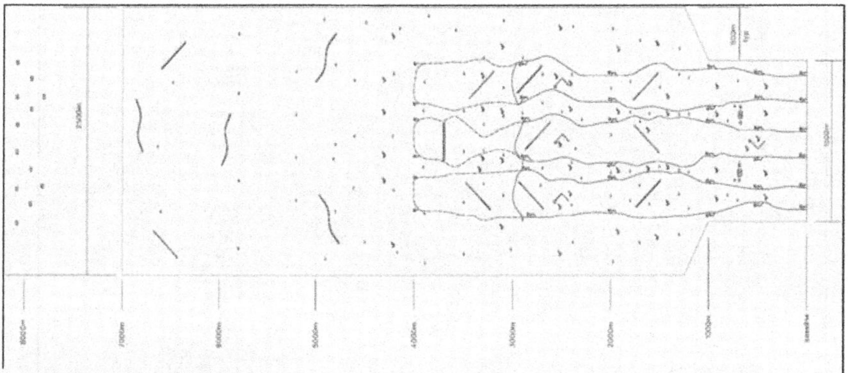

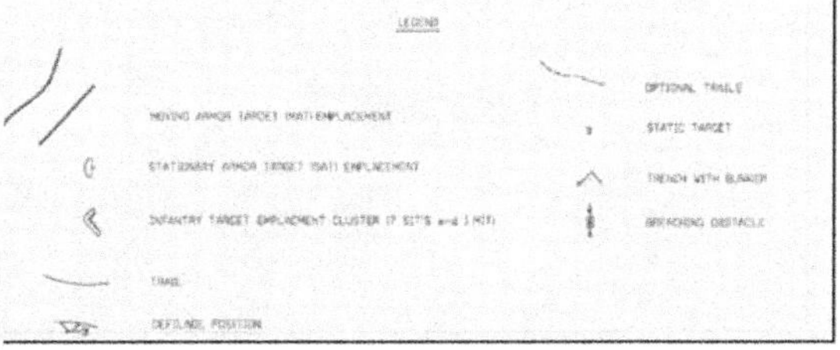

This complex is used to train and test armor, infantry and aviationcrews, sections, squads and platoons on skills necessary to detect, identify, engage and defeat stationary and moving infantry and armor targets in a tactical array. Company Combined Arms Live Fire Exercises (CALFEX) may also be conducted on this facility. This complex also accommodates training with sub-caliber and/or laser training devices. This complex supports dismounted infantry platoon tactical live-fire operations either independently of, or simultaneously with, supporting vehicles.

Primary features include:

DMPRC consists of a base area A and two optional areas B & C.

Area A - footprint is previous MPRC 1.5k x 5k with digital capability. (Meets minimum standards of armor/infantry/aviation units Tables I-XII & CALFEX)

- 69 Stationary armor targets
- 11 Moving armor targets
- 35 Infantry target emplacement clusters (245 SITs and 35 MITs total)
- 4 Trenches
- 2 Breaching obstacles
- 3 Lanes (6 course roads)
- 36 Defilade battle positions

Area B - 500m lateral extension on both sides of area A, extending to end of range (Allows greater dispersion of targets and separation of firing vehicles in order to maximize capabilities of digital platforms)

- 20 Stationary armor targets
- 10 Infantry target emplacement clusters (70 SITs and 10 MITs total)

Area C - 2k extension in depth, the width of range (Provides greater aviation capabilities and facilitates future long range direct-fire engagements from ground platforms)

- 11 Stationary armor targets
- 4 Moving armor targets
- Static targets

All targets are fully automated, utilizing event-specific, computer-driven target scenarios and scoring. Targets will receive and transmit digital data from the range operations center. The captured data is then compiled and available to the unit during the after action review.

Associated Range Operations and Control facilities:
Standard Armor ROCA Facilities
Except Small AAR (17118) replaced by Large AAR (17117);
and no General Instruction Building (17120)

Requirement Document: FM 3-20.12, FM 23-1, FM 3-20.8, FM 3-04.140, ARTEP 7-20 MTP, ARTEP 71-2
Additional Information: This complex uses thermal targets, night illumination devices, and hostile-fire, target-kill, and visual flash simulators. Location of the boresight must be coordinated with the trainer. Gunnery tasks requiring the usage of dud producing ammunition cannot be fired on the complex. Provisions for these tasks must be made in impact areas adjacent to the complex.

Figure D-17. FCC 17865 MULTIPURPOSE TRAINING RANGE (MPTR)

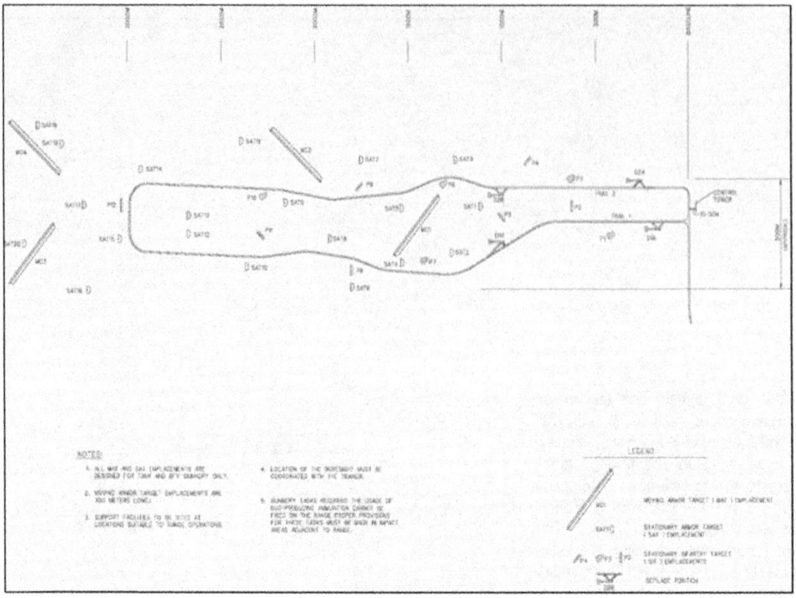

This complex is used to train and test crews and dismounted infantry squads on the skills necessary to detect, identify, and defeat stationary infantry and stationary/moving armor targets in a tactical array. In addition to live-fire, this complex can also be used for training with sub-caliber and/or laser training devices. The complex is specifically designed to satisfy the training and qualification requirements for the crews and sections of armor, infantry and aviation units. This complex also supports dismounted infantry squad tactical live-fire operations either independently of, or simultaneously with, supporting vehicles.

Primary features include:
- 4 Moving armor targets
- 10 clusters @ 7 SITs each
- 4 Battle positions
- 20 Stationary armor targets
- 70 Stationary infantry targets
- 1 Lane (2 course roads)

All targets are fully automated, utilizing event-specific, computer-driven target scenarios and scoring

Associated Range Operations and Control facilities:
Standard Armor ROCA Facilities except Range Operations Center, Large (17124) replaced by Range Operations Center, Small (17123)

Requirement Document: FM 3-20.12, FM 23-1, FM 3-20.8, FM 3-04.140
Additional Information: This complex uses thermal targets, night illumination devices, and hostile-fire, target-kill, and visual flash simulators.
Location of the boresight must be coordinated with the trainer. Gunnery tasks requiring the usage of dud producing ammunition cannot be fired on the complex. Provisions for these tasks must be made in impact areas adjacent to the complex.
Hard targets may be positioned in the impact area to facilitate RF/IR signatures for aviation gunnery.

Figure D-18. FCC 17867 MULTIPURPOSE RANGE COMPLEX (MPRC)

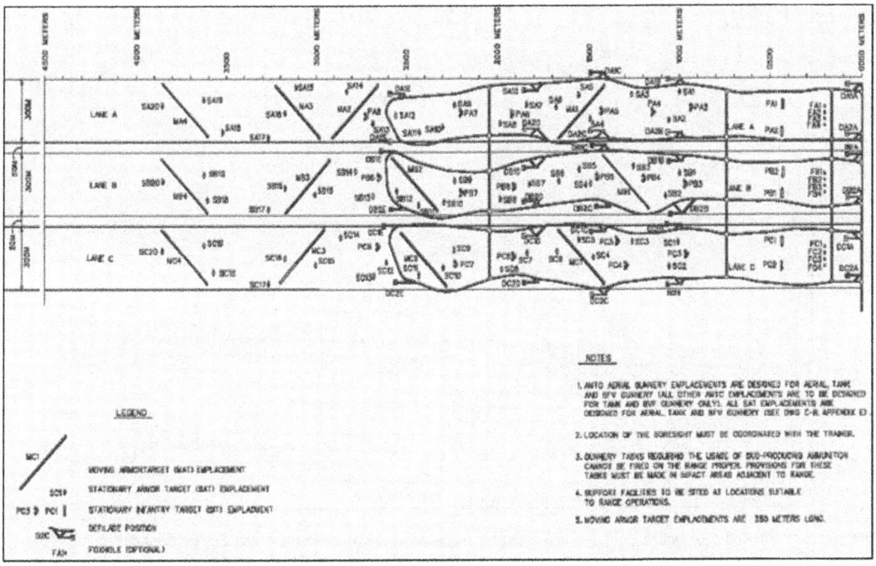

This complex is used to train and test armor, infantry and aviation crews, sections, squads and platoons on skills necessary to detect, identify, engage and defeat stationary and moving infantry and armor targets in a tactical array. This complex also accommodates training with sub-caliber and/or laser training devices.

Primary features include:

- 60 Stationary armor targets
- 12 Moving armor targets
- 154 Stationary infantry targets (22 clusters @ 7 SITs each)
- 22 Moving infantry targets
- 3 Lane (6 course roads)
- 36 Defilade battle positions

All targets are fully automated, utilizing event-specific, computer-driven target scenarios and scoring.

Associated Range Operations and Control facilities:

Standard Armor ROCA Facilities

Requirement Document: FM 3-20.12, FM 23-1, FM 3-20.8, FM 3-04.140, ARTEP 7-20 MTP, ARTEP 71-2

Additional Information: This complex uses thermal targets, night illumination devices, and hostile-fire, target-kill, and visual flash simulators. Location of the bore sight must be coordinated with the trainer. Gunnery tasks requiring the usage of dud producing ammunition cannot be fired on the complex. Provisions for these tasks must be made in impact areas adjacent to the complex.

Figure D-19. 17870 BATTLE AREA COMPLEX (BAX)

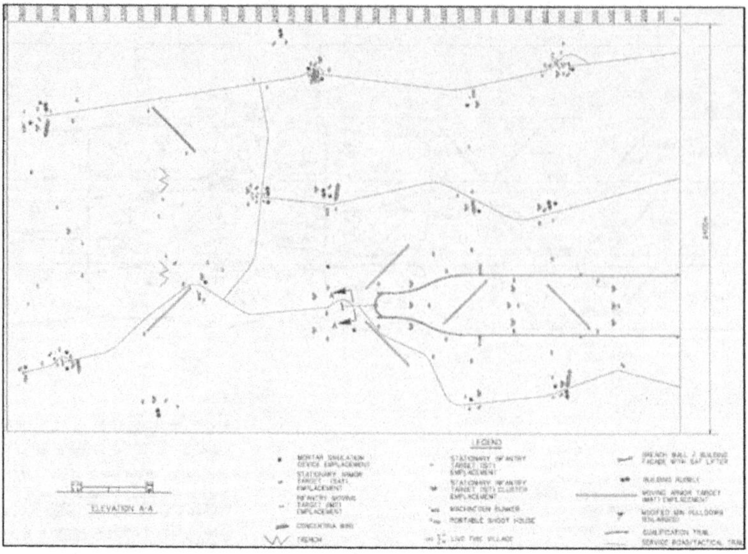

This complex provides a collective live fire home station training facility for all elements of the Stryker Brigade Combat Team (SBCT). SBCT crews and dismounted soldiers train and test their ability to detect, identify, engage and defeat stationary and moving combined arms targets in both open and urban terrain environments. This complex also supports tactical live fire operations independently of, or simultaneously with, supporting vehicles in free maneuver. This complex accommodates training with sub-caliber and/or Training devices.

Primary features include:

43 Stationary armor targets	6 Moving armor targets
35 Stationary infantry targets (SIT)	25 SIT clusters @ 7 emplacements each
14 Moving infantry targets	4 Machine gun bunkers w/sound effects simulator
2 Breach wall/Building facades	2 Live fire villages (1 with 7 buildings and 1 with 5 buildings)
2 Portable shoothouses	2 Trench lines
2 Lanes (2 course roads in 1 lane, no course roads in the free maneuver lane)	
8 Hasty battle positions	18 Mortar simulations devices (RGSS)
3 Landing zones (locations not shown)	

All targets are fully automated, utilizing event-specific, computer-driven target scenarios and scoring. Targets will receive and transmit digital data fom the range operations center. The captured data is then compiled and available to the unit during the after action review.

Associated Range Operations and Control facilities: Standard Armor ROCA Facilities
Except Small AAR (17118) replaced by Large AAR (17117); and no General Instruction Building (17120)

Requirement Document: FM 3-20.13, FM 3-20.8, FM 3-21.9, FM 3-21.11
Additional Information: This range uses thermal targets, night illumination devices, and hostile-fire, target-kill, and visual flash simulators.
Location of the bore sight must be coordinated with the trainer. Gunnery tasks requiring the usage of dud producing ammunition cannot be fired on the complex. Provisions for these tasks must be made in impact areas adjacent to complex.

Figure D-20. FCC 17872 AIR DEFENSE FIRING RANGE

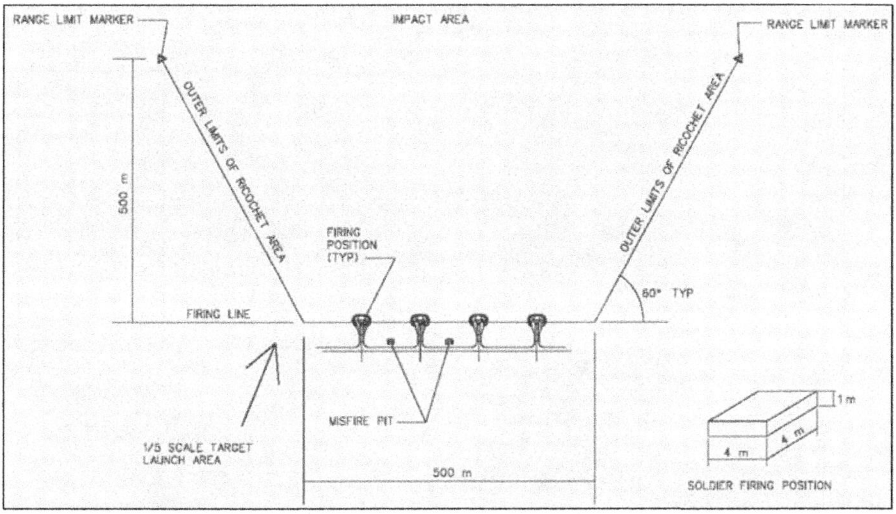

This complex is used to train and test Air Defense Artillery crews on the skills necessary to employ ground to air anti-aircraft missiles against ballistic aerial target systems (BATS) and drones. This complex supports shoulder-launched as well as vehicle-launched missiles.

Primary features include:
4 Shoulder/Vehicle launch firing positions

No automation is required for this facility.

Associated Range Operations and Control facilities:
Range Operations Center, Small (17123)
Bleacher Enclosure (75061)
Operations/Storage Building (17120)
Latrine (73075)

Requirement Document: FM 3-01.11, FM 44-43, FM 44-44, FM 44-18-1
Additional Information: For ground firing, large red arrows covered with highly visible paint are placed on right and left outer limits of the ricochet area. Red flags are placed on the firing line at start-fire and cease-fire points.

STINGER missiles have a 15 degree con-shaped danger zone centered directly to the rear of the launcher. To avoid hazards, the ground directly to the rear of the gunner is cleared of all obstacles and the launcher breech is placed one meter off the ground.

Figure D-21. FCC 17878 URBAN ASSAULT COURSE

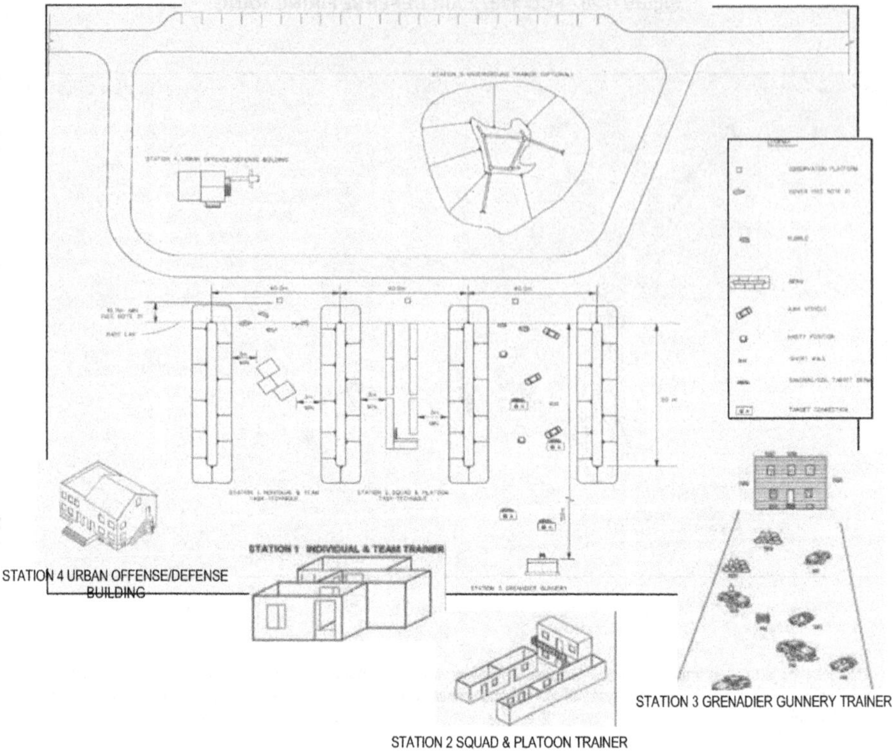

This facility is used to train individual soldiers, squads, and platoons on tasks necessary to operate within a built-up/urban area. The facility contains 5 stations:

Station 1 - Indvidual & Team Trainer
Station 2 - Squad & Platoon Trainer
Station 3 - Grenadier Gunnery Trainer
Station 4 - Urban Offense/Defense Trainer
Station 5 - Underground Trainer

Targetry Requirements:
Station 1 - 6 interior precision human urban targets (HUT)
Station 2 - 10 interior precision HUT
Station 3 –13 total targets: 6 non-precision exterior targets (SITs) and 7 non-precision fascade targets
Station 4 - 10 interior precision targets
Station 5 - no instrumentation required

All targets are fully automated and the event specific target scenario is computer driven and scored from the range operations center. Targetry in Stations 1, 2 and 4 areprecision human urban targets that have reconfigurable plug and play capability.

Associated Range Operations and Control facilities: Operations/Storage Building (17122) Latrine (73075)

Requirement Document: FM 3-0, FM 3-06, FM 3-06.11, ARTEP 7-8-MTP,7-3/4-1110, TC 90-1
Additional Information: Station 3 - 40mm HE grenades will not be used.
Station 4 - No live fire training.
Station 5 - Smoke will not be employed inside the underground trainer. Also, pyrotechnics are not authorized for station 5.

Figure D-22. FCC 17879 LIVE FIRE EXERCISE SHOOTHOUSE

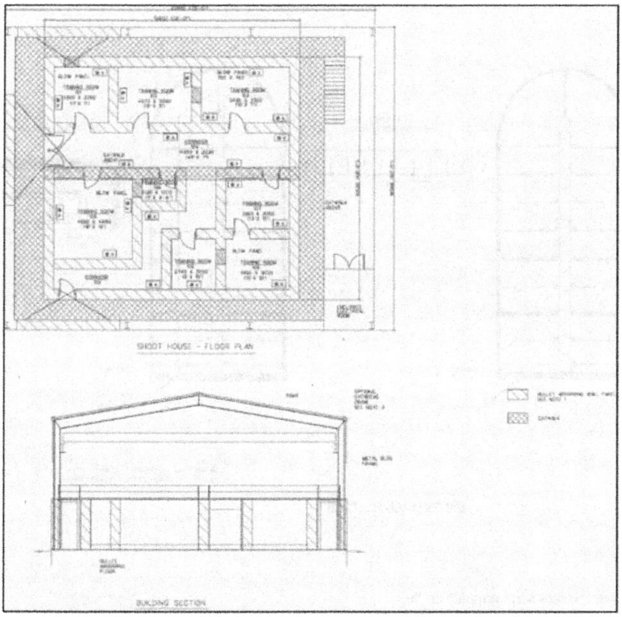

The Shoothouse provides the leader with a facility to train and evaluate the unit during a live fire exercise. Units are trained and evaluated on their ability to move tactically (enter and clear a room; enter and clear a building), engage targets, conduct breaches and practice target discrimination.

Primary features include:
Facility divided into 8 separate rooms & 2 corridors for a minimal net training capbiliby of 1400 square feet
20 Target/camera outlets
13 Precision human urban targets (HUT)

All targets are fully automated and the event specific target scenario is computer driven and scored from the range operations center. The range operating system is fully capable of providing immediate performance feedback to the using participants. All targetry are life-like precision targets that have reconfigurable plug and play capability.

Associated Range Operations and Control facilities:
Operations/Storage Building (17122)
Latrine (73075)
Small AAR Building (17118)

Requirement Document: TC 90-1
Additional Information: A roof does not reduce 360 degree SDZ, but is there to reduce light, weather protection, enhance realism and provide superstructure for an overhead crane

Figure D-23. FCC 17880 LIVE FIRE EXERCISE BREACH FACILITY

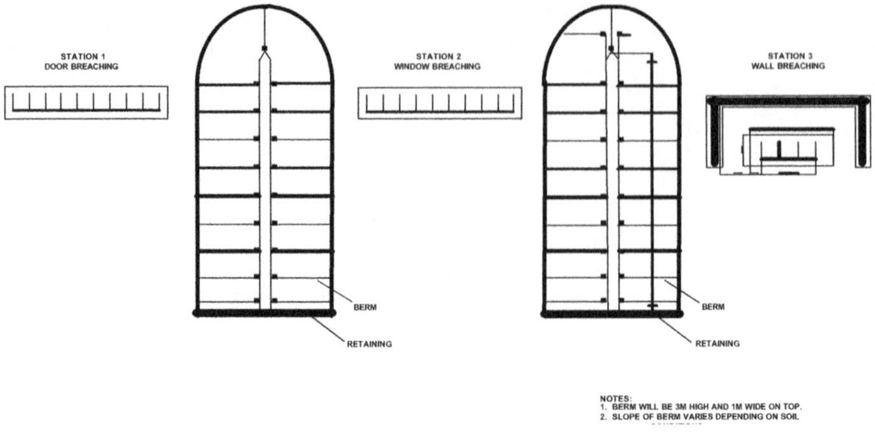

BREACH FACILTY – PLAN

The Breach Facility is used to train soldiers semi-annualy on the technical aspects of breaching techniques. It is also used to train TTP's and explosive techniques not trained on any other type of facility.

Primary features include:

- Station 1 - Door Breaching Structure
- Station 2 - Window Breaching Structure
- Station 3 - Wall Breaching Structure

No automation is required for this facility.

Associated Range Operations and Control facilities:
Latrine (73075) if not colocated with another range that has a ROCA

Requirement Document: TC 90-1
Additional Information: The trainees must retreat to a position behind the retaining wall before detonating the explosive charge.

Figure D-24. FCC 17882 HAND GRENADE QUALIFICATION COURSE

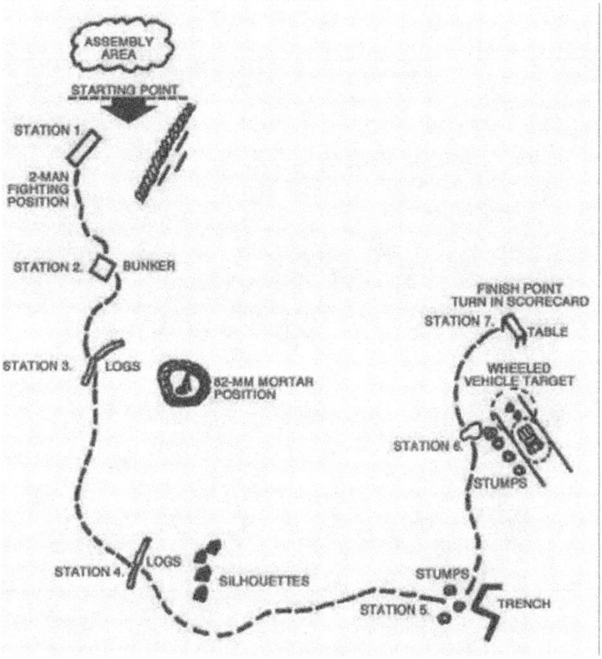

This range is used to train and test individual soldiers on the skills necessary to employ hand grenades against stationary target emplacements.

Primary features include:
7 Individual stations

No automation is required for this facility. All targets/facades are fixed at required distances.

Associated Range Operations and Control facilities:
Standard Small Arms ROCA Facilities
Except No Range Operations Center, Small (17123)

Requirement Document: FM 3-23.30
Additional Information: The qualification course allows soldiers to use fused practice hand grenades to engage targets in natural terrain under simulated combat conditions.
Station 1 - engage a group of F-type silhouettes in open from 2 man foxhole. Targets 35 meters to front beyond squad's protective wire.
Station 2 - engage a bunker with two firing portholes.
Station 3 - engage a fortified 82mm mortar position from 20 meters.
Station 4 - engage groups of targets behind cover 20 meters distance.
Station 5 - clear an entry point to a trench 25 meters distance.
Station 6 - engage troops in halted, open wheeled vehicle 25 meter distance.
Station 7 - identify hand grenades by shape, coloring, markings, and capabilities.

Figure D-25. FCC 17883 HAND GRENADE FAMILIARIZATION RANGE

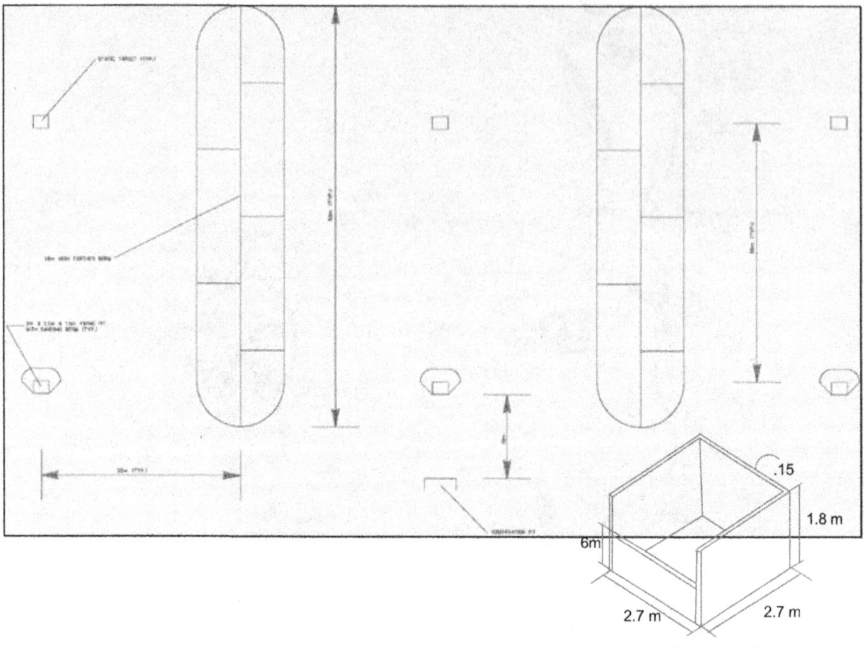

This range is used to train and test individual soldiers in the employment of live fragmentation hand grenades.

Primary features include:
4 Throwing bays

No automation is required for this facility. All targets are fixed at required distances

Associated Range Operations and Control facilities:
Standard Small Arms ROCA Facilities
Except No Range Operations Center, Small (17123)

Requirement Document: FM 3-23.30
Additional Information: Steel, concrete or wooden revetments or earthen berms 1.8m high should separate at least two of the bays from the other tow and extend out 50m. This permits practice to continue on half the range if a dud occurs. Ideally, dividers separate all of the bays.

Sand-filled drums provide an aiming point for the soldier as he throws the grenade.

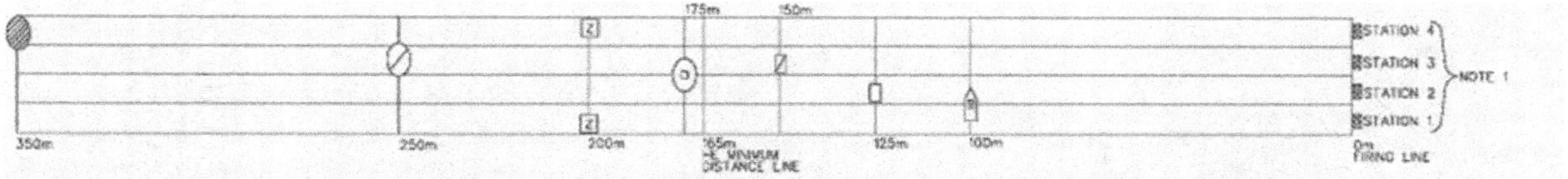

This facility is used to train and test individual solders on the skills necessary to engage and defeat stationary target emplacements with the 40mm Grenade Launcher.

Primary features include:
    4 Individual Firing Stations

No automation is required for this facility. All targets/facades are fixed at required distances.

Associated Range Operations and Control facilities:
    Standard Small Arms ROCA Facilities

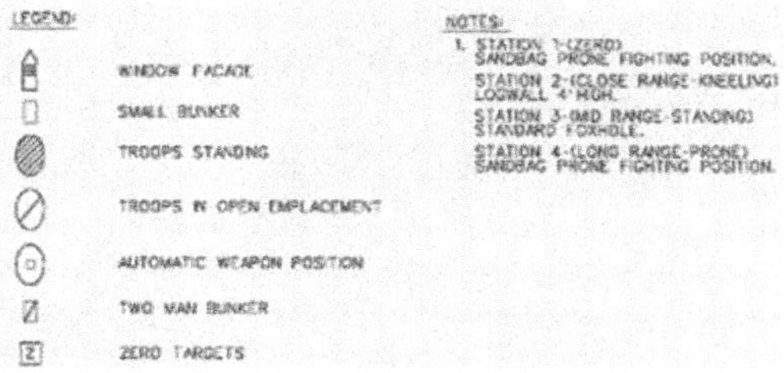

Requirement Document: FM 3-23.31, STP 21-1 SMCT
Additional Information: Zero targets are at least 2 meters high and 2 meters wide.
Zero target is clearly marked with a large contrasting "Z".
Targets are made of long-lasting, durable material that can withstand constant use with little maintenance. Salvaged oil drums filled with sand make excellent semi-permanent target material for this range.

Figure D-26. FCC 17884 GRENADE LAUNCHER RANGE

Figure D-27. FCC 17885 LIGHT DEMOLITION RANGE

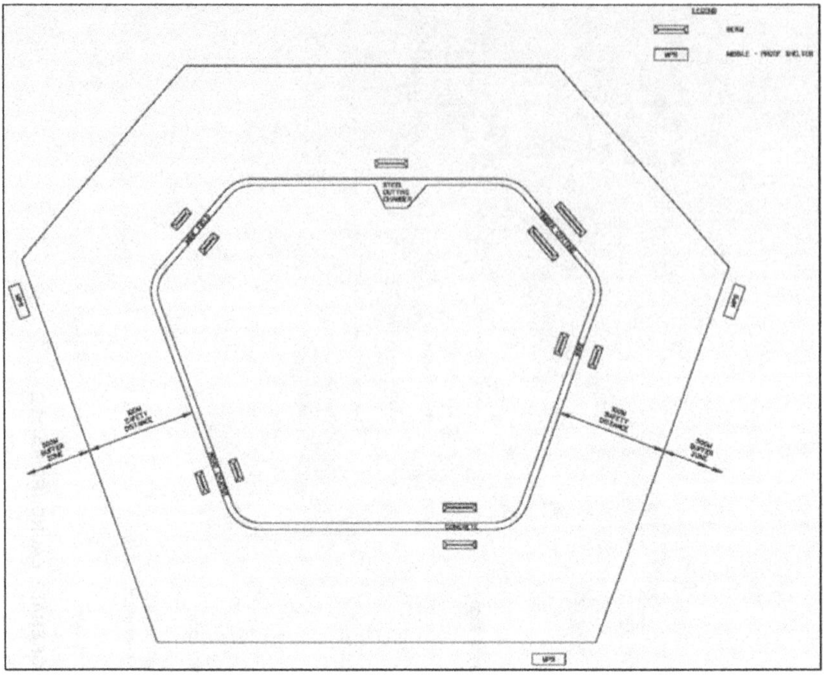

This range is used to train individual solders on the employment techniques of explosives and demolitions.

Primary features include:
- Safety berms (on each side of demolition points)
- 3 Missile-proof shelters
- Class V issue storage bunker

No automation is required for this facility.

Associated Range Operations and Control facilities:
- No ROCA facilities required.

Requirement Document: FM 5-250, FM 5-34, FM 5-102, FM 20-32
Additional Information: Wire Obstacle width - 7m; length - 20m. Use one Bangalore kit to breach the obstacle.
Minefield width - 10m; length - 20m. Mines may be placed on the surface or buried.
Steel-cutting Chamber has a ledge to mount steel or concrete beams. Steel and concrete beams vary in width and length.
Timber-cutting Site width - 10m; length - 35m. There are 8 concrete base supports (4 on each side of road) for placement of logs. Poles should be no larger than 36cm in dial nor taller than 10m. Charges should be placed to prevent damage to the concrete base.
Concrete Obstacle is constructed with concrete cubes or tetrahedrons. It is 10X30 m.
The road crater site must be refilled after each use.

Figure D-28. FCC 17891 INFILTRATION COURSE

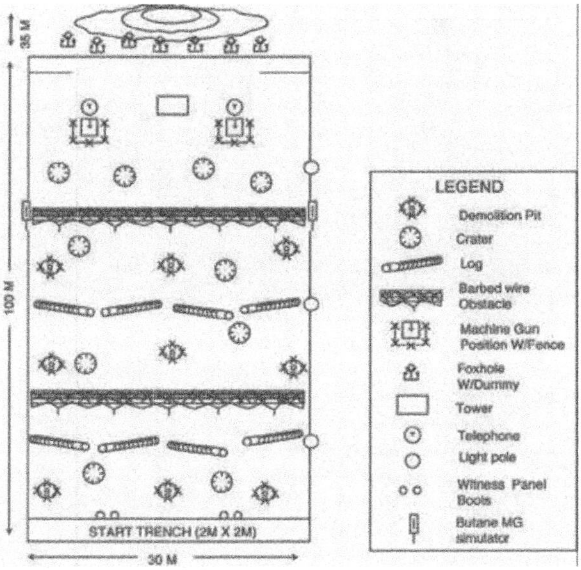

This range is used to train individual soldiers on combat movement techniques in a realistic live-fire environment.

Primary features include:

- 50 x 100 meter Movement Area
- 2 Fixed Machine Gun Positions (w/Safety Measures installed)
- 2 Barbed Wire Obstacles
- 9 Demolition Pits
- 2 Log Obstacles

No automation is required for this facility.

Associated Range Operations and Control facilities:

- Operations/Storage Building (17122)
- Bleacher Enclosure (75061)
- Latrine (73075)

Requirement Document: FM 7-93, FM 7-92
Additional Information: Construct surface danger fans for each machine gun; establish firing procedures and overhead clearance of personnel in accordance with AR/DA Pam 385-63.

When the range is opened, test fire each machine gun. Conduct test firings before each use of the range to determine dispersion pattern. Use machine gun traverse and elevating stops to prevent firing beyond the right and left limits of fire or below minimum elevations.

Figure D-29. FCC 17892 FIRE AND MOVEMENT RANGE

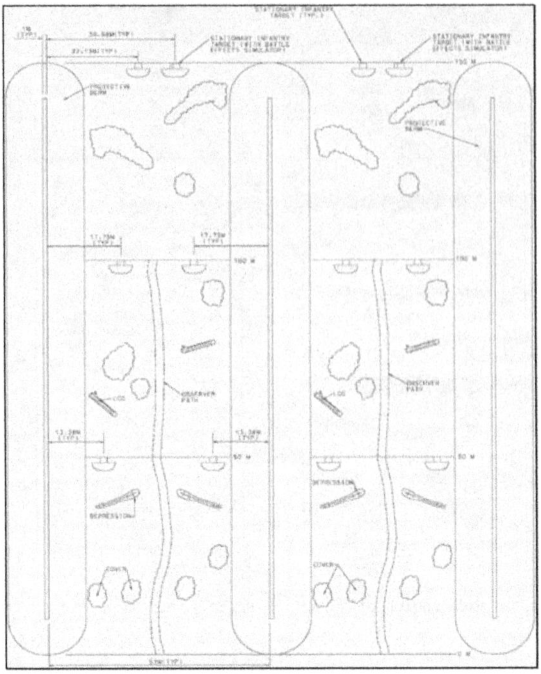

This range is used to train individual soldiers and buddy/teams on basic fire and movement techniques against stationary infantry targets. Soldiers show their ability to select covered and concealed positions, move while under fire, apply principles of teamwork, and use suppressive fire on enemy soldier targets.

Primary features include:
- 4 Lanes
- 6 Stationary infantry targets per lane
- 3 meter high berms along each side of lanes

Targets are not fully automated and/or the scenarios are not computer driven or scored.

Associated Range Operations and Control facilities:
- Range Operations Center, Small (17123)
- Operations/Storage Building (17122)
- Bleacher Enclosure (75061)
- Ammo Breakdown Building (17129)
- Latrine (73075)

Requirement Document: FM 7-8, TC 7-9
Additional Information: All lanes should have natural features that offer cover and concealment.

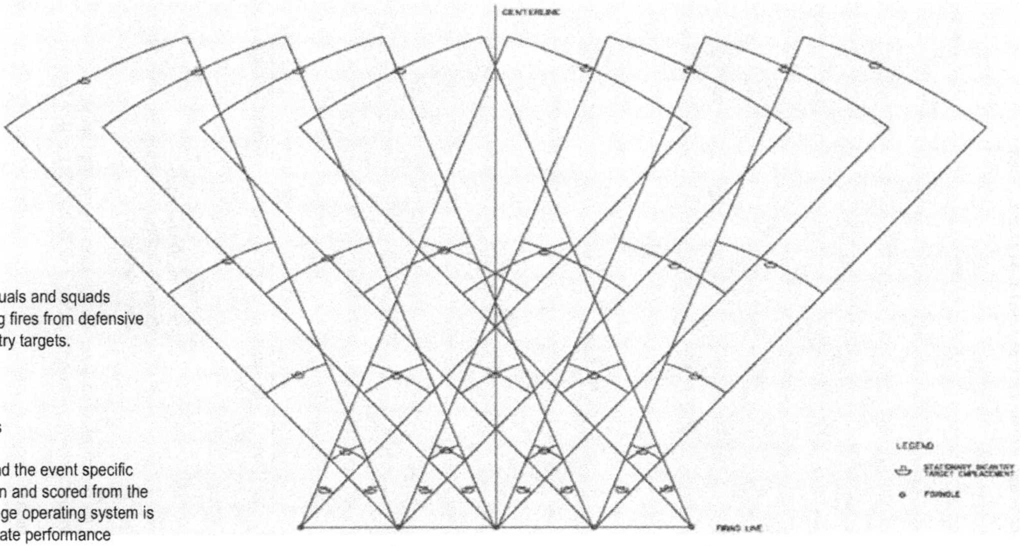

This range is used to train individuals and squads on employing mutually supporting fires from defensive positions against stationary infantry targets.

Primary features include:
- 5 Two-man foxholes
- 31 Stationary infantry targets

All targets are fully automated and the event specific target scenario is computer driven and scored from the range operations center. The range operating system is fully capable of providing immediate performance feedback to the using participants.

Associated Range Operations and Control facilities:
- Standard Small Arms ROCA Facilities

Requirement Document: FM 7-8, TC 7-9
Additional Information: Each defensive position will have a right and a left sector of fire.

Figure D-30. FCC 17893 SQUAD DEFENSE RANGE

Figure D-31. FCC 17895 INFANTRY SQUAD BATTLE COURSE

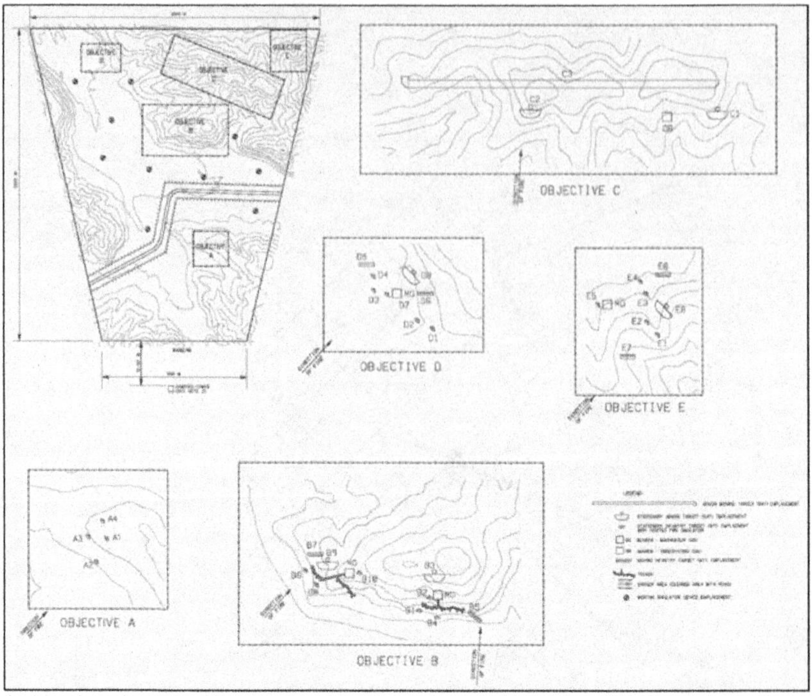

This complex is used to train and test infanty squads on the skills necessary to conduct tactical movement techniques, detect, identify, engage and defeat stationary and moving infantry and armor targets in a tactical array

Primary features include:

- 1 Moving armor targets
- 6 Moving infantry targets
- 2 Trench obstacles
- 6 Stationary armor targets
- 20 Stationary infantry targets
- 5 Machine gun/Observation bunkers (w/sound effects simulator)

All targets are fully automated and the event specific target scenario is computer driven and scored from the range operations center. The range operating system is fully capable of providing immediate performance feedback to the using participants.

Associated Range Operations and Control facilities:
Standard Small Arms ROCA Facilities

Requirement Document: FM 7-8, ARTEP 7-8-MTP, ARTEP 7-8 Drill, TC 7-9
Additional Information: Target locations are site adapted. All must be located in areas that support desired tactics and the user's training requirements.

All trenches, bunkers, and target emplacements must simulate typical threat scenarios.
Mortar simulation device emplacements are located in areas from which unfriendly mortar fire is to be simulated.
Helicopter landing areas, designed for heavy use, should be located to support aerial insertion and extraction of the squad.

Figure D-32. FCC 17897 INFANTRY PLATOON BATTLE COURSE

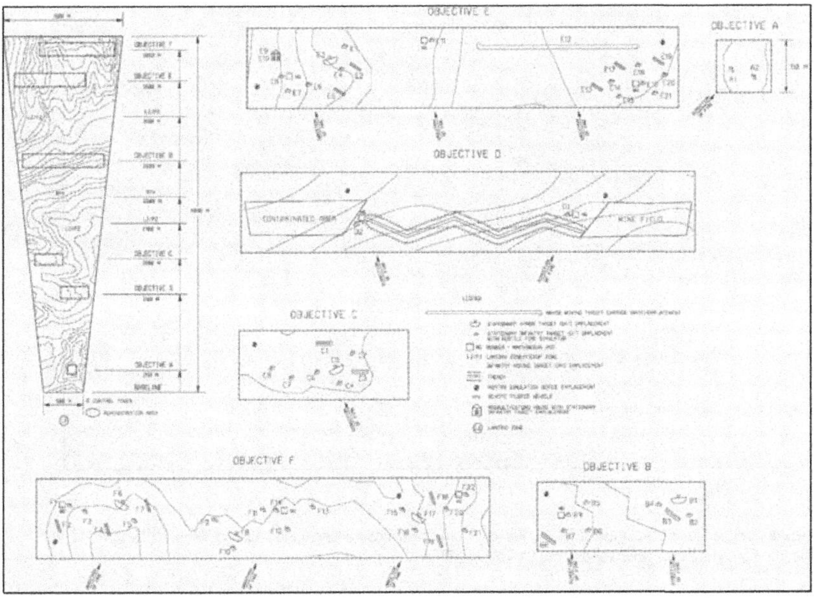

This complex is used to train and test infantry platoons, either mounted or dismounted, on the skills necessary to conduct tactical movement techniques, detect, identify, engage and defeat stationary and moving infantry and armor targets in a tactical array

Primary features include:
- 1 Moving armor targets
- 14 Moving infantry targets
- 1 Trench obstacles
- 2 Landing zones
- 6 Stationary armor targets
- 43 Stationary infantry targets
- 9 Machine gun bunkers (w/sound effects simulator)
- 1 Assault/defend house

All targets are fully automated and the event specific target scenario is computer driven and scored from the range operations center. The range operating system is fully capable of providing immediate performance feedback to the using participants.

Associated Range Operations and Control facilities:
Standard Small Arms ROCA Facilities

Requirement Document: FM 7-8, ARTEP 7-8 Drill, ARTEP 7-8 MTP, FM 23-1, FM 0 20.0, TC 7 0
Additional Information: Target locations are site adapted. All must be located in areas that support desired tactics and the user's training requirements.

All trenches, bunkers, and target emplacements must simulate typical threat scenarios.
Mortar

Figure D-33. FCC 17901 COMBINED ARMS COLLECTIVE TRAINING FACILITY (CACTF)

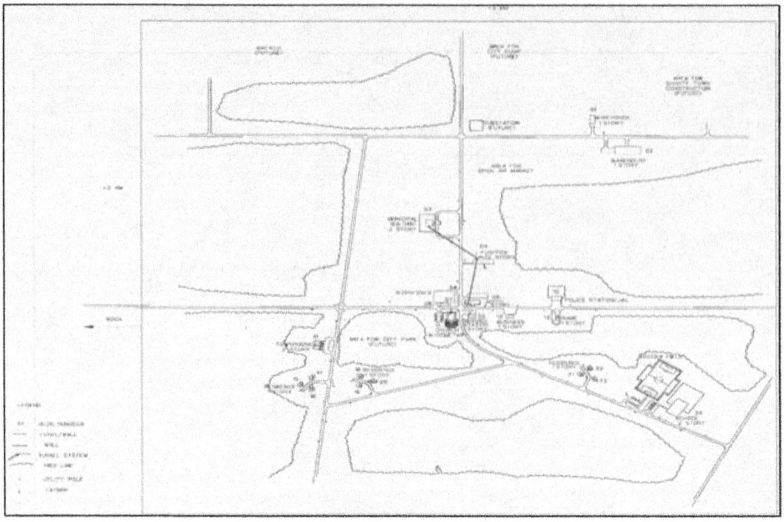

This facility is designed to conduct multi-echelon, full spectrum operations training up to battalion TF level. The CACTF will accommodate Force on Force (FOF) and Force on Targetry (FOT).

Primary features include:
1 School
1 Church/cemetery
1 Police station/Jail
1 Hotel
9 Residences
4 Businesses
1 Townhouse
1 Bank
2 Warehouses
1 Government Building
1 Office
1 Service Station

Recommended features Include:
Tunnel/Sewer System
Town Shanty
One, three-story building
Three, two-story buildings
Breachable walls
Loopholes
Mouse holes

Targets:
15 Precision/Human urban targets (HUT)
30 SITs (Recommend RF/Reconfigurable)
9 SATs (Recommend RF/Reconfigurable)

Associated Range Operations and Control facilities:
Range Operations Center, Large (17124)
Operations/Storage Building (17122)
Latrine (73075)
Covered Mess (17116)
Large AAR (17117)

All targets are fully automated, utilizing event-specific, computer-driven target scenarios and scoring. Targets will receive and transmit data from the range operations center. The captured data is then compiled and available to the unit during the after action review. All targetry are life-like precision targets that have reconfigurable plug and play capability.

Camera Coverage Requirement:
80% of Buildings with 80% room coverage; considerations should be given to hallways and stairwells.

Live-fire ballistic ammunition is prohibited on this facility. Pyrotechnics are not authorized for training in the tunnel system.

Requirement Document: TC 90-1, FM 3-22.9
Additional Information: The ROCA should not be located in the immediate vicinity of the urban training area.

Figure D-34. AROCA STANDARD ARMOR RANGE OPERATIONS AND CONTROL AREA FACILITIES

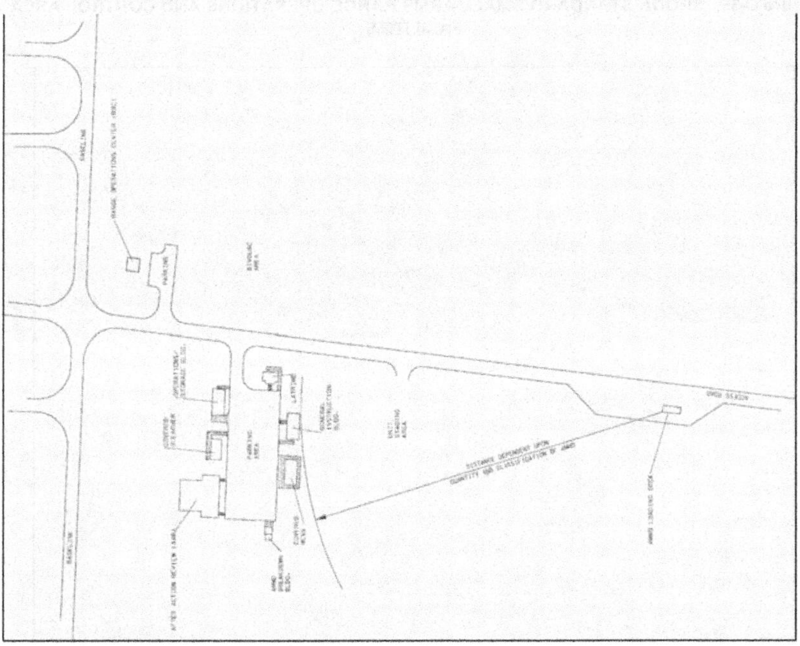

The Range Operation Center and Operations/Storage Building are used to operate and maintain the range. The Bleacher Enclosure, General Instruction and After Action Review buildings are used for pre and post event instruction. The remaining buildings are to support the training or the troops being trained.

Associated Range Operations and Control facilities:
Range Operations Center, Large (17124)
Operations/Storage Building (17122)
General Instruction Building (17120)
Latrine (73075)
Bleacher Enclosure (75061)
Covered Mess (17116)
Ammo Breakdown Building (17129)
After Action Review Building, Small (17118)
Ammunition Loading Dock (14970)
Bivouac Area
Unit Staging Area (85212)

Additional Information: The drawing depicts a typical layout and may be adjusted for project requirements. Minimum set back distances of building from the baseline are mandated in AR 385-63 and AR 385-64.

Figure D-35. SROCA STANDARD SMALL ARMS RANGE OPERATIONS AND CONTROL AREA FACILITIES

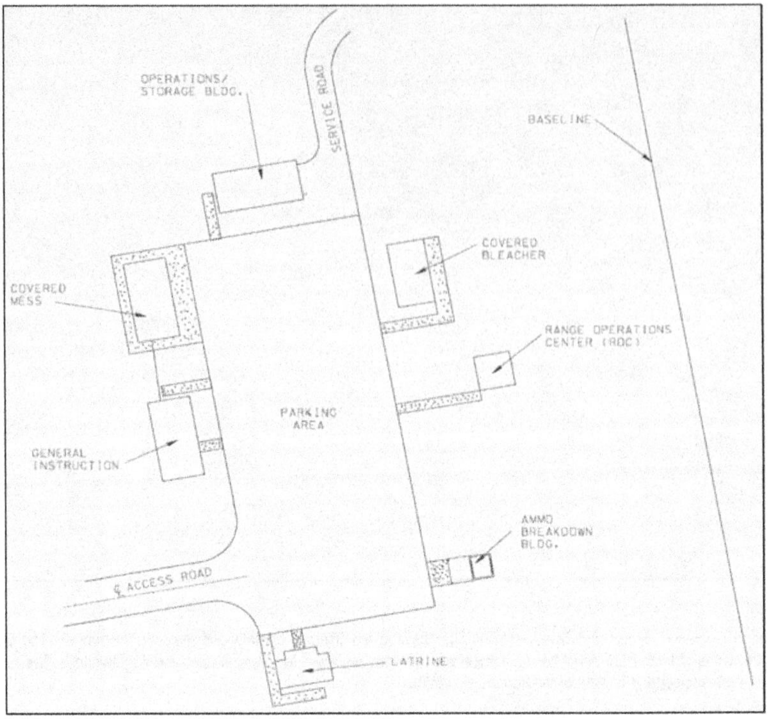

The Range Operation Center and Operations/Storage Building are used to operate and maintain the range. The Bleacher Enclosure and General Instruction Building are used for pre and post event instruction. The remaining buildings are to support the training or the troops being trained.

Associated Range Operations and Control facilities:
Range Operations Center, Small (17123)
Operations/Storage Building (17122)
General Instruction Building (17120)
Latrine (73075)
Bleacher Enclosure (75061)
Covered Mess (17116)
Ammo Breakdown Building (17129)

Additional Information: The drawing depicts a typical layout and may be adjusted for project requirements. Minimum set back distances of building from the baseline are mandted in AR 385-63 and AR 385-64.

Appendix E
RLTP Planning Process

E-1. Process

a. The RTLP planning process is a sequential six-step process:

(1) The "asset accounting" of existing training assets at a given installation or site.

(2) The "needs assessment" which includes METL requirements and an evaluation of range conditions to perform training missions.

(3) A "comparative analysis" to determine training asset shortfall(s) or adequacy.

(4) A "requirements definition" to develop specific capabilities, characteristics, and functionality.

(5) An "alternative analysis" identifies all viable or potential alternatives; defines mission support, environmental consequences and economic feasibility for each alternative identified; and site planning to determine any land use conflicts or environmental impacts.

(6) The "implementation plan" identifies preferred alternatives prioritized and an implementation timeline to establish critical milestones and resources.

(7) The requirement identified in the process culminating in the range development plan serves as the description of the mission requirement for land in the installation Integrated Natural Resources Management Plan (INRMP).

b. Planners (as identified in Chapter 1, paragraph 1-4) must comply with regulatory requirements, environmental matters, and examine a variety of factors when considering a site. Specific factors, procedures, and related policies are contained in TC 25-1 and this circular. Should the alternative require the development of a range, see the range project development process in paragraph 3-4, AR 210-21. Additional guidance on requirements identification and needs assessment follow.

E-2. Requirements Identification

a. Doctrine, force structure, weapon systems, and mission tasks are evaluated to determine the specific requirements necessary to train soldiers and units to meet operational readiness and Army training goals. Field Manuals provide the foundation for development of the unit METL. The following documents will assist in identifying requirements:

(1) Field Manuals 7-0 and appropriate type FMs assist in the development of METL that focus on those training tasks which are essential to accomplishing an organization's wartime mission.

(2) Unit ARTEPs and CATS provide unit-training requirements to be executed in maneuver areas and on live-fire ranges.

(3) Weapon system training strategies and standards in respective field manuals.

(4) Department of the Army Pamphlet 350-38, outlines recommended weapons training programs, qualification standards, training ammunition, operating tempo and suggested training strategies.

(5) Support Facility Annex (SFA) to the Integrated Logistics Support Plan published by HQUSACE identifies facility implications associated with material fielding, including training considerations and is continuously updated as the weapon system matures. These SFAs are available to

installations and MACOM via the Programming, Administration, and Execution (PAX) system under the Facility Planning System utility.

(6) Training Circular 25-1, this circular, and AR/DA Pam 385-63 provide guidance on site planning, range layouts, safety standards, SDZs, standard range facilities, training land, instrumentation, and specific weapon system requirements.

E-3. Needs Assessment

a. Planners must evaluate existing capabilities to meet standard training readiness requirements, alternative methods to maximize training effectiveness, and resources (for example, ammunition, fuel, training land and range facilities, manpower, life cycle O&M) to meet operational readiness. Inclusive is an assessment of range and training land effectiveness, conditions, utilization factors, and environmental influences to perform training missions. During the STRAC process, training strategies are developed incorporating combinations of devices, simulators and live fire requirements to satisfy training proficiency requirements in a holistic manner. Potential environmental restrictions related to range standardization or modernization, and training land requirements must be identified during the needs assessment and may mandate a tailored training strategy to accommodate existing training land restrictions. The following documents will assist in assessing potential range construction and training land needs:

(1) Army Regulations 200-1, 200-3, 200-4, and 210-20, and TC 25-1 outline planning considerations to be addressed during this assessment.

(2) Army Regulations 200-2, 415-15, 415-20, and the Master Planning Instructions outline technical procedures for planning, programming, estimating construction costs and milestones required to either construct or acquire facilities.

(3) United States Army Corps of Engineers (CEHNC 1110-1xx series) design manuals depict standardized facilities.

(4) AR 385-10 and DA PAM 385-63 outline range safety considerations and SDZ requirements.

b. Potential impacts of noise from a new or modified live-fire range on the surrounding civilian community can inhibit full utilization during the conduct of training. Environmental considerations as defined in AR 200-1 and applicable to Installation Compatible Use Zones will be evaluated when modifying existing ranges or siting new ranges.

c. The ITAM Training Requirements Integration (TRI) process is employed to identify optimum siting of training ranges and facilities and optimum modifications to training areas.

Glossary

Section I
Abbreviations

A	area target
AAR	after-action review
AATLFR	antiarmor tracking and live fire range
ADF	air defense firing
AE	architectural engineer
AFF	automated field-fire
AGES/AD	Air-ground engagement system/air defense
AGR	aerial gunnery range
AH	attack helicopter
AL	Alabama
AMC	US Army Materiel Command
AMRD	Army Modernization Reference Data
AP	armor-piercing
approx	approximately
AR	Army regulation
ARF	automated record-fire
ARTEP	Army Training and Evaluation Program
ATS	Army target system
ATTN	attention
BATS	ballistic aerial target system
BCTP	Battle Command Training Program
BES	Battlefield Effects Simulator
BFV	Bradley Fighting Vehicle
Btry	Battery
C	civilian; troops
CALFEX	combined arms live-fire exercise
cal	caliber
CATS	Combined Arms Training Strategy
CCTT	close combat tactical trainer
CD-ROM	compact disk – read only memory
CEHNC	Corps of Engineers, Huntsville Engineering and Support Center
CEHNDM	Corps of Engineers, Huntsville Division, manual
CEV	combat engineer vehicle
CFV	Cavalry Fighting Vehicle
CH	Cargo helicopter
cm	centimeter
CMTC	Combat Maneuver Training Center
Co	Company
COE	Corps of Engineers
cont	continued
CONUS	continental United States
CPQC	combat pistol qualification course
CPT	captain
CS	combat support
CSS	combat service support
CTC	combat training center
DA	Department of the Army
DCSOPS	Deputy Chief of Staff for Operations and Plans

def	defensive
DMPRC	digital multipurpose range complex
DPTM	directorate of plans, training, and mobilization
DPW	directorate of public works
DRM	directorate of resource management
DSN	Defense Switched Network
DTLOMS	doctrine, training, leader development, organization, materiel, and soldiers
EA	environmental assessment
EIS	environmental impact statement
EOD	explosive ordnance disposal
EPLRS	Enhanced Position Location Reporting System
FBCB2	Force XXI Battle Command Brigade and Below
FEE	functional estimating equations
FM	field manual; frequency modulated
FP	firing position
FTX	field training exercise
G3	Assistant Chief of Staff, G3 (Operations and Plans)
GIS	geographic information system
GLLD	Ground laser locator designator
GS	General Schedule
GTA	Grafenwoehr Training Area
GTL	gun-target line
HE	high explosive
HG	hand grenade
HMMWV	high-mobility multi-purpose wheeled vehicle
HQDA	Headquarters, Department of the Army
HTA	Hohenfels Training Area
IBCT	Interim Brigade Combat Team
IHFS	infantry hostile-fire simulator
IN	inch
INRMP	Integrated Natural Resources Management Plan
IPBC	infantry platoon battle course
IR	infrared
ISBC	infantry squad battle course
ITAM	Integrated Training Area Management
ITAS	Improved Target Acquisition System
ITV	Improved TOW Vehicle
JAAT	joint air-attack team
JRTC	Joint Readiness Training Center
kmh	kilometers per hour
LAV	light armored vehicle
LAW	light anti-tank weapon
lb	pound
LD	line of departure
LFX	live-fire exercise
LOS	line-of-sight
LSO	laser safety officer
LT	lieutenant

LTA	local training area
LURS	land use requirements study
LZ	landing zone
M	military/meter
MACOM	major Army command
MAJ	major
MAT	moving armor target
MD	Maryland
METL	mission-essential task list
MFS	Muzzle Flash Simulator
MG	machine gun
MILES	multiple integrated laser engagement system
MIT	moving infantry target
MLRS	multiple launch rocket system
mm	millimeter
MOS	military occupational specialty
MOUT	military operations on urbanized terrain
MPB	Master Planning Board
MPFQC	military police firearms qualification course
MPH	miles per hour
MPMG	multi-purpose machine gun
MPRC	multipurpose range complex
MPS	missile-proof shelter
MPTR	multipurpose training range
MRA	maneuver rights area
MRF	modified record fire
MSD	MILES Shoot back Device
MS-3	Manpower Staffing Standards System
MTA	major training area
MTP	mission training plan
MULE	modular universal laser equipment
NA	not applicable
NBC	nuclear, biological, chemical
NCO	noncommissioned officer
NCOIC	noncommissioned officer-in-charge
NdYAG	neodymium doped yttrium aluminum garnet
NGB	National Guard Bureau
no	number
NSN	national stock number
NTC	National Training Center
Obi	objective
OC	observer-controller
OE-CX	Ordnance and Explosives Center of Expertise
off	offensive
OIC	officer in charge
O&M	operation and maintenance
OPFOR	opposing force
P	point target
Pam	pamphlet
PEO	program executive officer
PGTS	precision gunnery target system
Plt	platoon

PM	project manager
POTA	Patriot omni-directional training, aerial
PPBES	planning, programming, budgeting, and execution system
PZ	pickup zone
QTR	qualification training range
RADS	range analysis and design system
RC	Reserve Component
RCMAT	radio-controlled miniature aerial target
RCS	radar cross section/Range Control System
RDP	range development plan
Recon	reconnaissance
RETS	remoted target system
RFMSS	Range Facility Management Support System
RLEM	rifle launched entry munition
RNG-MCX	Range-Mandatory Center of Expertise
RPV	remotely piloted vehicle
RSO	range safety officer
RTLP	Range and Training Land Program
RTLP-MCX	Range and Training Land Program – Mandatory Center of Expertise
S	Army wide standard
SAT	stationary armor target
SAW	squad automatic weapon
SDZ	surface danger zone
SES	Sound Effects Simulator
SESAMS	special effects small arms marking system
SFA	support facility annex
SGR	scaled gunnery range
SIT	stationary infantry target
SMR	scaled mortar range
SOP	standing operating procedure
SRTA	short range training ammunition
STRAC	Standards in Training Commission
STX	situational training exercise
T	tank
T&EO	training and evaluation outline
TADS	target acquisition detection system
TADSS	training aids, devices, simulators, and simulations
TBD	to be determined
TC	training circular; throughput capacity
TCPC	tank crew proficiency course
TES	tactical engagement simulation
TGT	target
THP	take-home package
TOW	tube-launched, optically-tracked, wire-guided
TP	target practice
TR	throughput requirement
TRADOC	US Army Training and Doctrine Command
TRC	training readiness condition
TRI	training requirements integration
TSM	TRADOC system manager
TTM	tank target mechanism
TTP	tactics, techniques, and procedures

TWGSS/PGS	tank weapons gunnery simulation system/precision gunnery system
UCOFT	Unit Conduct of Fire Trainer
UH	utility helicopter
US	United States
USACE	US Army Corps of Engineers
USAESCH	US Army Engineering Support Center, Huntsville
USAEDH	US Army Engineer Division, Huntsville
USAF	US Air Force
USAREUR	US Army, Europe
USATSC	US Army Training Support Center
VA	Virginia
VIP	very important person
WB	wage board
WG	wage grade

Section II
Terms

Active Army
The active Army consists of (1) members of the Regular Army on active duty; (2) members of the Army National Guard of the United States and Army Reserves on active duty (other than for training); (3) members of the Army National Guard in the service of the United States pursuant to a call; and (4) persons appointed, enlisted, or inducted into the Army without component.

After-action review (AAR)
A professional discussion that focuses on the objectives of ongoing or completed training. It reviews a training activity to permit participants to discover for themselves what happened and why.

Army National Guard
The Army portion of the organized militia of the states, Commonwealth of Puerto Rico, and the District of Columbia whose units and members are federally recognized.

Army National Guard of the United States
A reserve component of the Army, all of whose members are members of the Army National Guard.

Category code
A numerical code used to classify and categorize Army real property. The Army generally uses a five-digit code to plan, program, budget, design, construct, inventory, and maintain its facilities.

Collective training
Training either in institutions or units that prepares cohesive teams and units to accomplish their missions on the battlefield and in operations other than war.

Combined arms live-fire exercise (CALFEX)
A combat exercise in which a combined-arms team, in combat formation, conducts a coordinated combat firing and maneuver practice to assault, seize, and defend appropriate objectives. The CALFEX may include tactical air support.

Cook off
A functioning of any of the ammunition explosive components because of high temperatures within the weapon.

Crew-served weapon
A weapon system for which a crew of at least two members is required.

Force XXI Battle Command Brigade and Below
A digital command-and-control system that provides battle command and situational awareness information from brigade down to the soldier and platform level.

Hang fire
A delay in functioning of a weapon.

Impact area
The land area and associated airspace within a training complex intended to capture or contain ammunition, munitions, or explosives to include resulting debris, fragments, and components from weapon system employment.

Installation
An aggregation of contiguous or near-contiguous, common mission-supporting real property holdings under the jurisdiction of the Department of Defense or a state, the District of Columbia, territory, commonwealth, or possession, controlled by and at which a Department of Defense unit or activity (active or Reserve Component) is permanently assigned. A fixed location together with its land, buildings, structures, utilities, and improvements.

Misfire
A failure to fire, not necessarily hazardous. Since it cannot readily be distinguished from a delay in functioning (hang fire), it must be handled as worst case in accordance with procedures for the weapon system.

Mission Training Plan (MTP)
A guide for units on what and how to train to achieve mission proficiency. It identifies unit wartime missions and the critical tasks that support them, provides comprehensive training and evaluation outlines, and provides exercises and other management aids to assist field commanders in planning and executing unit training.

Range
An area that is reserved and normally equipped for practice in weapons delivery or shooting at targets.

Reserve Components (RC)
The US Army Reserve (USAR) and the US Army National Guard (ARNG).

Stryker Brigade Combat Team (SBCT)
The Army's evolving structure – called the "transformation force." The restructured Army will have improved deployment characteristics while maintaining the capability to provide humanitarian assistance and prosecute military operations in low to high intensity conflicts.

Sub caliber ammunition
Practice ammunition of a caliber smaller than standard for the gun on which practice is being given. Sub caliber ammunition is economical and may be fired in relatively crowded areas. It is used with special sub caliber equipment to simulate firing conditions with standard ammunition.

Surface danger zone (SDZ)
The area designated on the ground of a training complex (to include associated safety areas) for the vertical and lateral containment of projectiles, fragments, debris, and components resulting from the firing or detonation of weapon systems. Army Regulation and Department of the Army Pamphlet 385-63 describes and illustrates weapon-system SDZs.

Sustainment training

Individual and collective training conducted in the unit or resident school to ensure continued expertise on the operations, employment, and logistics support of fielded systems or equipment.

Tactical engagement simulation training system
An advanced collective training methodology supported by a family of TADSS used in free-play force-on-force, field training exercises. The TES training system consists of three subsystems – the simulator subsystem which includes TADSS and supporting procedures that simulate casualty-producing effects of weapons in real time; the control subsystem which includes a staff of trained observer-controllers who referee, ensure realism, record events, and report actions observed through AARs and unit take-home packages; and the management subsystem which includes activities and computers to plan, schedule, conduct, and evaluate the training of battle-focused METL tasks to standard.

Take-home package (THP)
A compilation of a unit's experience at a combat training center (CTC) provided to the unit after completing training at the CTC. The observer-controllers compile the THP using engagement and other data collected by the CTC instrumentation system. The THP lets the unit replay a training exercise at an LTA or MTA, focusing on areas that need improvement to achieve and sustain readiness.

Throughput
The number of individuals, crews, or units that use a range. The number of individuals, crews, or units required to train is the throughput requirement. The number of individuals, crews, or units that can accomplish all required iterations of training on a given range during a single year is the annual throughput capacity of the range. When comparing the annual throughput requirement and capacity, if the throughput capacity exceeds the throughput requirement of a given range, there exists excess capacity. If the throughput requirement exceeds the throughput capacity, a need exists for additional training capability.

Training aids, devices, simulators, and simulations (TADSS)
A term that includes training range and combat training center (CTC) instrumentation, tactical engagement simulation (TES), battle simulation, targets, training-unique ammunition, and dummy, drill, and inert munitions. TADSS are subject to the public laws and regulatory guidance governing the acquisition of materiel.

Training land
The term "training land" encompasses facilities (e.g., ranges, maneuver land, proficiency courses, and direct-support facilities) dedicated to preparing and sustaining personnel and units to meet mission roles and standards. Examples include facilities to support training personnel in weapon systems use and proficiency, occupational skills, and standards developed.

References

Section I
Sources

These are the sources quoted or paraphrased in this publication.

AR 210-20. *Master Planning for Army Installations.* 30 Jul 93.
AR 210-21. *Army Ranges and Training Land Program.* 1 May 97.
AR 350-38. *Training Device Policies and Management.* 15 October 1993.
AR 385-40. *Accident Reporting and Records.* 1 Nov 94.
AR 385-63/MCO P3570.1B. *Range Safety.* 19 May 03.
AR 415-28. *Real Property Category Codes*. 10 Oct 96.
DA Pam 350-9. *Index and Description of Army Training Devices.* 3 Sep 2002
DA Pam 350-38. *Standards in Weapon Training.* 3 Jul 97.
DA Pam 385-63. *Range Safety.* 10 Apr 03
DA Pam 415-28. *Guide to Army Real Property Category Codes.* 11 Feb 00.
FM 7-0. *Training the Force.* 10 Oct 02.
FM 17-12-7. *Tank Gunnery Training Devices and Usage Strategies.* 1 May 00.
FM 17-12-8. *Light Cavalry Gunnery.* 16 Feb 99.
FM 100-14. *Risk Management,* Apr 98.
TC 25-1. *Training Land.* 30 Sep 91.
ARTEP 7-7J-MTP. *Mission Training Plan for the Mechanized Infantry Platoon and Squad,* 05 Oct 2001
ARTEP 7-8-DRILL. *Battle Drills for the Infantry Rifle Platoon and Squad.* 3 Mar 00.
ARTEP 7-8-MTP. *Mission Training Plan for the Infantry Rifle Platoon and Squad.* 05 Oct 01.
ARTEP 7-20-MTP. *Mission Training Plan for the Infantry Battalion.* 27 Nov 2001.
ARTEP 71-2-MTP. *Mission Training Plan for the Tank and Mechanized Infantry Battalion Task Force.* 27 Nov 2001.

Section II
Required Documents

These documents must be available to users of this publication.

AR 350-1. *Army Training and Education.* 9 Apr 03.
FM 7-1. *Battle Focused Training.* 6 Jun 03.
DA Form 2028. *Recommended Changes to Publications and Blank Forms.* Feb 74.
DA Form 2408-4. *Weapon Record Data.* Jan 79.

Section III
Recommended Readings

These publications contain relevant supplemental information.

AR 1-1. *Planning Programming and Budgeting and Execution System.* 13Aug 96.
AR 40-5. *Preventive Medicine.* 15 Oct 90.
AR 75-1. *Malfunctions Involving Ammunition and Explosives (RCS CSGLD-1961(MIN)),* 23 Apr, 01.
AR 95-2. *Air Traffic Control, Airspace, Airfields, Flight Activities and Navigation Aids.* 10 Aug 90.
AR 140-483. Army Reserve Land and Facilities Management. 30 Jul 94.
AR 200-1. *Environmental Protection and Enhancement.* 21 Feb 97.
AR 200-2. *Environmental Effects of Army Actions.* 23 Dec 88.
AR 200-3. *National Resources—Land, Forest, and Wildlife Management.* 28 Feb 95.
AR 200-4. *Cultural Resources Management.* 1 Oct 98

AR 350-2. *Opposing Force Program.* 15 Jun 83.
AR 385-10. *Army Safety Program.* 29 Feb 00.
AR 385-64. *Ammunition and Explosives Safety Standards.* 1 Mar 00.
AR 415-15. *Army Military Construction Program Development and Execution.* 9 Apr 98.
AR 415-35. *Minor Construction.* 15 Sep 83.
AR 420-74. *Natural Resources.* 10 Oct 97.
AR 700-127. *Integrated Logistical Support (ILS) Manager's Guide,* 11 Oct 99.
DA Pam 200-4. *Cultural Resources Management.* 1 Oct 98.
DA Pam 350-39. *Standards in Weapons Training (Special Operation Forces).* 3 Jul 97.
FM 3-04.140. *Helicopter Gunnery.* 14 Jul 03.
FM 3-20.12. *Tank Gunnery (Abrams),* 1 Oct 01.
FM 5-34. *Engineer Field Data.* 30 Aug 99.
FM 5-102. *Countermobility.* 14 Mar 85.
FM 5-250. *Explosives and Demolitions.* 10 Mar 86.
FM 6-30. *Tactics, Techniques, and Procedures for Observed Fire.* 16 Jul 91.
FM 6-71. *Tactics, Techniques and Procedures for Fire Support for the Combined Arms Commander.* 20 Sep 94.
FM 7-90. *Tactical Employment of Mortars.* 9 Oct 92.
FM 9-15. *Explosive Ordnance Disposal Service and Unit Operations.* 8 May 96.
FM 20-32. *Mine/Countermine Operations.* 29 May 98.
FM 23-1. *Bradley Gunnery.* 13 Mar 96.
FM 23-8. *M14 and M14A1 Rifles and Rifle Marksmanship.* 15 Apr 74.
FM 23-9. *M16A1 Rifle and M16A2 Rifle Marksmanship.* 3 Jul 89.
FM 23-10. *Sniper Training.* 17 Aug 94.
FM 23-11. *90MM Recoiless Rifle, M67.* 6 Jul 65
FM 23-14. *M249 Light Machine Gun in the Automatic Rifle Role.* 26 Jul 94.
FM 23-24 *DRAGON Medium Antitank/Assault Weapon System.* 3 Apr 90.
FM 23-25. *Light Antiarmor Weapons.* 17 Aug 94.
FM 23-27. *MK 19, 40mm Grenade Machine Gun, MOD 3.* 27 Dec 88.
FM 23-30. *Grenades and Pyrotechnic Signals.* 27 Dec 88.
FM 23-31. *40mm Grenade Launcher M203.* 20 Sep 94.
FM 23-34. *TOW Weapon Systems.* 17 Aug 94.
FM 23-35. *Combat Training with Pistols and Revolvers.* 3 Oct 88.
FM 23-41. *Submachineguns, Caliber .45, M3 and M3A1.* 28 Jun 74.
FM 23-65. *Browning Machinegun, Caliber .50 HB, M2.* 23 Dec 02.
FM 23-67. *Machinegun, 7.62-mm, M60.* 29 Feb 84.
FM 23-90. *Mortars.* 9 Dec 02.
FM 23-91. *Mortar Gunnery.* 1 Mar 00.
FM 44-1-2. *Air Defense Artillery Reference Handbook.* 15 Jun 84.
FM 44-62. *Air Defense Artillery Automatic Weapon Gunnery.* 6 Aug 69.
FM 44-102. *Procedures for Ballistic Aerial Target System.* 30 May 74.
FM 90-10. *Military Operations on Urbanized Terrain.* 15 Aug 79.
FM 90-10-1. *An Infantryman's Guide to Combat in Built-up Areas.* 12 May 93.
MIL-HDBK-828A. *Laser Safety on Ranges and in Other Outdoors Areas.* 15 Apr 96
NG Pam 25-1. *Training Site General Information Summary.* 28 Apr 95.
NG Pam 415-10. *Army National Guard Facilities Construction.* 15 Aug 94.
NGR 415-5. *Military Construction, Army National Guard Facilities Construction. 30 Nov 90*
NGR 415-10. *Army National Guard Facilities Construction.* 15 Aug 94.
NGR 420-10. *Real Property Operations, Maintenance, and OMARNG Minor Construction, Army National Guard.* 15 Feb 90.

STP 7-11BCHM14-SM-TG. *Soldier's Manual, MOS 11B, 11C, 11H, and 11M, Infantry Skill Levels 1/2/3/4.* 30 Sep 88.
STP 7-11BCHM24-SM-TG. *Soldier's Manual and Trainer's Guide, MOS 11B, 11C, 11H, and 11M, Infantry Skill Levels 1/2/3/4.* 19 Jan 00.
STP 7-11C14-SM-TG. *Soldier's Manual and Trainer's Guide MOS 11C Indirect Fire Infantryman, Skill Levels 1/2/3/4.* 16 Aug 94.
STP 21-1-SMCT. *Soldier's Manual of Common Tasks Skill Level 1.* Aug 03.
TC 6-40. *Field Artillery, Manual Cannon Gunnery.* 27 Dec 88.
TC 23-14. *Sniper Training and Employment.* 14 Jun 89.
CEHNC 1110-1-xxx series.

Index

S

T

U

V

W

X-Y-Z

***Field Manual**
No. 3-22.9

Headquarters
Department of the Army
Washington, DC, 12 August 2008

Rifle Marksmanship M16-/M4-Series Weapons

Contents

DISTRIBUTION RESTRICTION: Approved for public release; distribution is unlimited.

***This publication supersedes FM 3-22.9, 24 April 2003.**

Figures

Tables

Preface

This manual provides guidance for planning and executing training on the 5.56-millimeter M16-series rifle (M16A1/A2/A3/A4) and M4 carbine. It is a guide for commanders, leaders, and instructors to develop training programs, plans, and lessons that meet the objectives or intent of the United States Army rifle marksmanship program and FM 7-0.

This manual is organized to lead the trainer through the material needed to conduct training during initial entry training (IET) and unit sustainment training. Preliminary subjects include discussion on the weapon's capabilities, mechanical training, and the fundamentals and principles of rifle marksmanship. Live-fire applications are scheduled after the Soldier has demonstrated preliminary skills.

This manual was revised to include references to new materiel and systems. This revision includes—

- The new Army total marksmanship training strategy, to include specific strategies for the United States Army Reserve (USAR) and the Army National Guard (ARNG).
- Information about the advanced combat optical gunsight (ACOG), the AN/PEQ-15 advanced target pointer/illuminator aiming light (ATPIAL), various thermal sights, and the MK 262 round.
- Information about the alternate qualification record fire courses (known distance [KD] record fire, 25-meter scaled target alternate course, 15-meter scaled target alternate course).
- Information about the rapid magazine change and barricade transition fire for short-range marksmanship (SRM).
- Changes to all of the scorecards.
- Updated terminology.

This publication prescribes DA Form 3595-R (Record Fire Scorecard), DA Form 3601-R (Single Target—Field Fire I Scorecard), DA Form 5239-R (100-, 200-, and 300-Meter Downrange Feedback Scorecard), DA Form 5241-R (Single and Multiple Targets—Field Fire II Scorecard), DA Form 5789-R (Record Fire Scorecard—Known-Distance Course), DA Form 5790-R (Record Fire Scorecard—Scaled Target Alternate Course), DA Form 7489-R (Record Night Fire Scorecard), DA Form 7649-R (Squad Designated Marksman—Record Fire I and II), and DA Form 7650-R (Squad Designated Marksman—Position Evaluation).

This publication applies to the Active Army, the Army National Guard (ARNG)/National Guard of the United States (ARNGUS), and the US Army Reserve (USAR).

Terms that have joint or Army definitions are identified in both the glossary and the text. Terms for which FM 3-22.9 is the proponent FM are indicated with an asterisk in the glossary.

Uniforms depicted in this manual were drawn without camouflage for clarity of the illustration. Unless this publication states otherwise, masculine nouns and pronouns refer to both men and women.

The proponent for this publication is the US Army Training and Doctrine Command. The preparing agency is the United States Army Infantry School (USAIS). You may send comments and recommendations by any means (US mail, e-mail, fax, or telephone) as long as you use DA Form 2028 (Recommended Changes to Publications and Blank Forms) or follow its format. Point of contact information is as follows:

E-mail:	benn.29IN.229-S3-DOC-LIT@conus.army.mil
Phone:	Commercial: 706-545-8623 DSN: 835-8623
Fax:	Commercial: 706-545-8600 DSN: 835-8600
US Mail:	Commandant, USAIS ATTN: ATSH-INB 6650 Wilkin Drive, Bldg 74, Rm 102 Fort Benning, GA 31905-5593

Chapter 1

Marksmanship Training

An effective marksmanship program can be measured by the unit's ability to put effective fire on a target. This chapter provides a proven strategy for establishing and conducting an effective rifle marksmanship training program. The strategy begins with the progressive individual training periods taught during initial entry training (IET) and culminates with advanced rifle marksmanship (ARM) skills. Refresher training is conducted only when necessary.

A Soldier's marksmanship proficiency depends on proper training and application of the basic marksmanship fundamentals. During initial marksmanship training, emphasis is placed on learning the firing fundamentals, which are taught in four phases—preliminary marksmanship instruction (PMI), downrange feedback, field firing, and advanced firing exercises. This prepares Soldiers for advanced optic and laser training for combat-type collective exercises and real-world deployments.

SECTION I. TRAINING STRATEGY

The total Army marksmanship training strategy is the overall concept for integrating resources into a program designed to train, sustain, and improve the individual and collective skills needed to achieve proficiency in individual and collective gunnery tasks. Training strategies for rifle marksmanship are implemented in TRADOC institutions (IET, Noncommissioned Officers Education System [NCOES], Basic Officer Leaders Course [BOLC]) and in units. The overall training strategy is multifaceted and includes supporting strategies that use resources such as publications; ranges; ammunition; and training aids, devices, simulators, and simulations (TADSS). These strategies focus on developing the Soldier and leader skills required for success in combat.

NOTE: See AR 350-1 for specific requirements pertaining to marksmanship training and DA Pam 350-38 for live-fire frequency requirements.

OBJECTIVES

1-1. The procedures and techniques for implementing the total Army rifle marksmanship training strategy are based on the concept that Soldiers must become skilled marksmen. FM 7-0 stresses marksmanship as the paramount Soldier skill. Further, Soldiers should understand common firing principles and be confident in applying their firing skills in combat. Unit leaders accomplish proficiency through practice supervised by qualified instructors/trainers and thorough objective performance assessments.

1-2. The basic firing skills and exercises outlined in this manual must be part of every unit's marksmanship training program. Unit commanders must focus their basic and advanced marksmanship training programs to support their mission-essential task lists (METLs).

MARKSMANSHIP TRAINING STRATEGY

1-3. The following marksmanship training strategy guide contains the tasks that are currently trained in basic rifle marksmanship (BRM) programs, during basic combat training at Army training centers (ATCs), and during infantry one-station unit training (OSUT). It also provides a basis for structuring unit sustainment programs for active Army, Army National Guard, and United States Army Reserve units. Units normally perform diagnostic tests of the tasks and only conduct training on specific periods for Soldiers who must improve their basic firing skills. Unit training is usually conducted in less time than at IET.

1-4. There are two primary components of a marksmanship training strategy: initial training and sustainment training. Both may include individual and collective tasks and skills. Initial training must be taught correctly the first time. A task taught correctly and learned well is retained longer, and skills can be more easily sustained. However, an individual or unit eventually loses skill proficiency if the training is not reinforced. This learning decay depends on many factors, such as the difficulty and complexity of the task. Personnel turnover is a main factor in the decay of collective skills, since the loss of critical team members requires retraining to regain proficiency. If a long period elapses between initial and sustainment sessions or training doctrine is altered, retraining may be required.

INITIAL ENTRY TRAINING

1-5. The training strategy for BRM begins in IET and continues in the unit. Figure 1-1 shows the IET training strategy.

1-6. In IET, Soldiers learn how to maintain a rifle, hit a point target, and apply the four marksmanship fundamentals and other skills needed to engage a target. Once Soldiers understand the weapon and have demonstrated skill in zeroing, additional simulations and live-fire training exercises are conducted to prepare Soldiers for qualification. During these exercises, Soldiers master target types and scenarios of increasing difficulty to develop their proficiency.

1-7. IET culminates in the Soldier's proficiency assessment (which is conducted on the standard record fire range or approved alternates), followed by instruction on advanced firing techniques (a night-fire exercise with iron sights [unassisted] or night vision goggles [assisted]). This evaluation allows leaders to determine the effectiveness of the training.

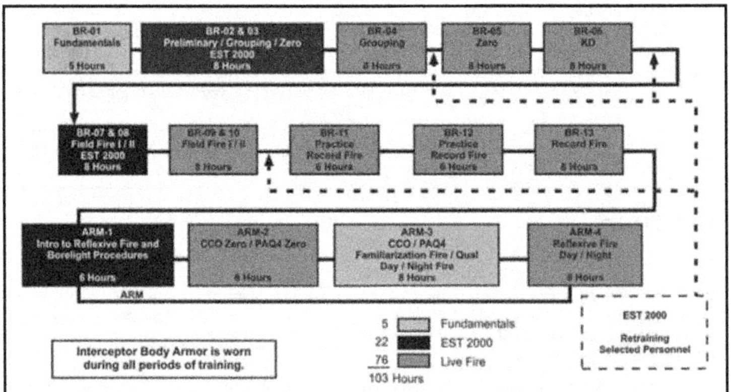

Figure 1-1. Initial entry training marksmanship training strategy.

SUSTAINMENT TRAINING

1-8. Training continues in active Army, National Guard, and Army Reserve units using the same basic skills taught in IET. Units must set up a year-round program to sustain skills and have a plan not only for when they are at their home station, but for when they are deployed as well.

1-9. To sustain the basic marksmanship skills taught in IET, periodic PMI is conducted, followed by instructional and qualification range firing. Key elements include—

- The training of trainers.
- Refresher training of nonfiring skills.
- The use of the Engagement Skills Trainer (EST) 2000, Laser Marksmanship Training System (LMTS), or other devices.
- Sustainment training.
- Remedial training.

NOTE: See Appendix A for more information about TADSS.

1-10. Additional skills trained in the unit include—

- Semiautomatic and automatic fires.
- Night fire.
- Mission-oriented protective posture (MOPP) firing.
- Firing using aiming devices.
- Moving target training techniques.
- Squad designated marksman (SDM) training.
- Short-range marksmanship (SRM).

1-11. These skills are trained and integrated into collective training exercises, such as platoon and squad live-fire situation training exercises (STXs). Figure 1-2 shows a year-round training strategy guide.

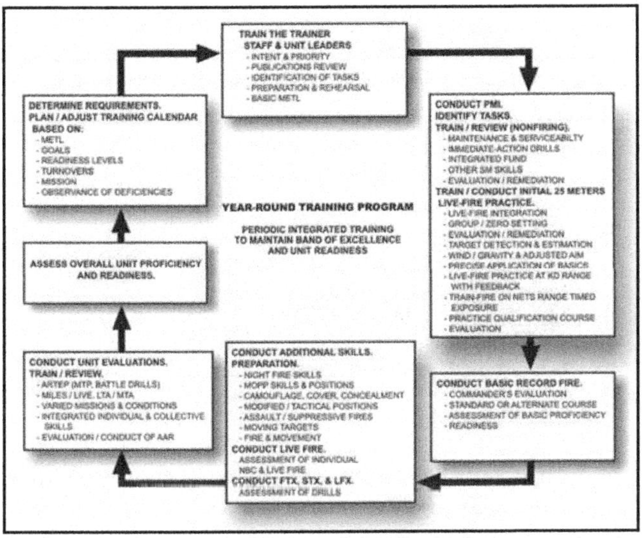

Figure 1-2. Unit marksmanship sustainment strategy.

1-12. General marksmanship knowledge and weapon proficiency are perishable skills. Most units have a readiness requirement that all Soldiers must zero their rifles within a certain time after unit assignment.

Soldiers must confirm the battlesight zeros of their assigned rifles before qualifying. Units should conduct PMI and practice firing throughout the year due to personnel turnover. A year-round marksmanship sustainment program is needed for the unit to maintain the individual and collective firing proficiency requirements to accomplish its mission. The following figures depict marksmanship training programs for active Army home station units (Figure 1-3), National Guard home station units (Figure 1-4), Army Reserve home station units (Figure 1-5), and deployed units (Figure 1-6).

NOTE: While the training strategy depicted in Figures 1-3 through 1-6 is not compulsory, it is strongly recommended for Army-wide application. The strategy has proven success during IET, producing more proficient Soldiers while using less ammunition. Commanders are urged to follow this strategy closely.

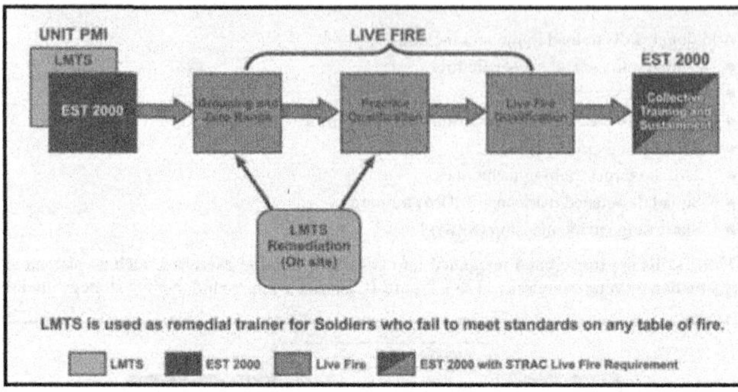

Figure 1-3. Active Army home station marksmanship training strategy.

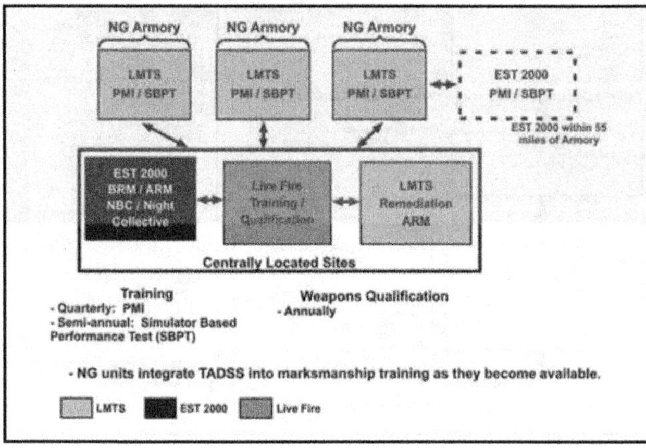

Figure 1-4. National Guard home station marksmanship training strategy.

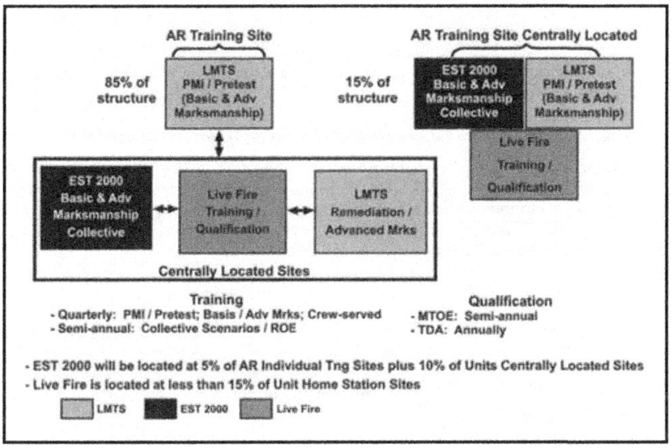

Figure 1-5. Army Reserve home station marksmanship training strategy.

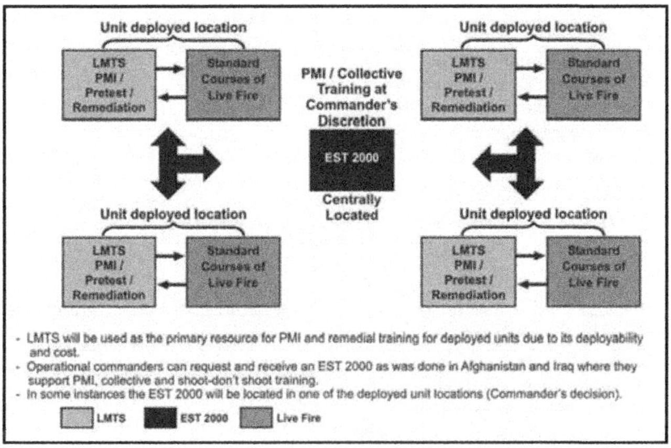

Figure 1-6. Deployed unit marksmanship training strategy.

TRAINING PHASES

1-13. Soldiers progress through five phases of rifle marksmanship training:

- Phase I—Basic Rifle Marksmanship Preliminary Marksmanship Instruction.
- Phase II—Basic Rifle Marksmanship Downrange Feedback Range Firing.
- Phase III—Basic Rifle Marksmanship Field Firing.

- Phase IV—Advanced Rifle Marksmanship.
- Phase V—Advanced Optics, Lasers, and Iron Sights.

1-14. When Soldiers are trained in all phases of rifle marksmanship, a solid sustainment program is key to mission readiness.

PHASE I—BASIC RIFLE MARKSMANSHIP PRELIMINARY MARKSMANSHIP INSTRUCTION

1-15. Understanding the operation and functions of any machine is vital to becoming an expert with that machine. The same theory applies to rifle marksmanship. Soldiers must master weapon maintenance, function checks, and firing fundamentals before progressing to advanced skills and firing exercises under tactical conditions. Armed with this knowledge, a Soldier is able to assess and correct any malfunction to keep the weapon operating properly.

Introduction to Basic Rifle Marksmanship and Mechanical Training

1-16. During this period of instruction, Soldiers are trained to understand the operation and functions of the weapon. This period of instruction includes the following topics:

- General safety rules and weapon clearing procedures.
- Characteristics, components, accessories, and ammunition.
- Disassembly.
- Inspection.
- Lubrication.
- Assembly.
- Function check.
- Loading/unloading the magazine.
- Loading/unloading the weapon.
- Immediate and remedial action.
- Adjusting the front and rear sights.
- Peer coaching.
- Eight cycles of function.
- Troubleshooting.

NOTE: The IET program of instruction (POI) allots four hours for this phase of training.

Marksmanship Fundamentals I

1-17. During this period of instruction, Soldiers are trained to demonstrate the integrated act of shooting using the EST 2000 or LMTS.

NOTE: EST 2000 will not be used in lieu of live-fire qualification except for those outlined in DA Pam 350-38.

1-18. This period of instruction includes the following topics:

- The four fundamentals.
- Basic firing positions.
- Range and safety procedures.
- Dominant eye training.
- Demonstrating the integrated act of shooting during dry-fire exercises.

NOTE: The IET POI allots six hours for this phase of training.

Marksmanship Fundamentals II

1-19. During this period of instruction, Soldiers are trained to demonstrate the integrated act of shooting using the EST 2000 or LMTS.

NOTES: 1. The IET POI allots six hours for this phase of training.

2. See paragraphs 4-76 through 4-80 and Appendix A for more information about EST 2000 and LMTS training.

PHASE II—BASIC RIFLE MARKSMANSHIP DOWNRANGE FEEDBACK RANGE FIRING

1-20. During Phase II, instructors/trainers outline grouping and zeroing procedures and demonstrate how to conduct the three types of known distance (KD) ranges. Downrange feedback provides precise knowledge of what happens to bullets at a given range and allows Soldiers to transition between 25-meter firing and firing on the field fire range. Knowing precisely where bullets hit allows poor firers (with the assistance of instructors/trainers) to improve their performance and good firers to bring their shots to the center of the target. Firers develop the knowledge and skills required to perform with confidence on the field fire range, where only hit-or-miss information is available.

Grouping Procedures

1-21. Grouping is a form of practice firing with two primary objectives: firing tight shot groups and consistently placing those groups in the same location. Frequent use of the EST 2000 or LMTS greatly reduces live-fire grouping time.

Zeroing Procedures

1-22. Zeroing allows firers to use standard issue ammunition to align the sights with the weapon's barrel. When this is accomplished correctly, the aimpoint and the impact point are the same at a given range. For most combat targets, this sight setting provides the greatest probability of a hit with minimum adjustment to the aimpoint. When followed, a properly zeroed rifle for one Soldier is close to zero for another Soldier.

Downrange Feedback

1-23. On KD ranges, Soldiers fire tight shot groups at a known distance and make sight adjustments at a given range while experiencing the effects of wind, gravity, and other environmental factors. The advantage of a KD range is the ability to see precisely where each bullet hits. KD firing is conducted with a single, clearly visible target at a known distance, and the Soldier can establish a position that provides a natural aimpoint on that single target.

NOTES: 1. See Chapter 5 for more information about downrange feedback.

2. See paragraphs 4-76 through 4-80 and Appendix A for more information about EST 2000 and LMTS training.

PHASE III—BASIC RIFLE MARKSMANSHIP FIELD FIRING

1-24. Field fire begins a critical transition from unstressed firing at single targets at known distances to requiring the Soldier to refine techniques for scanning the range for targets, estimating range, and firing quickly and accurately. Phase III includes the following:

- Field Fire I (single timed targets at 75 meters, 175 meters, and 300 meters).
- Field Fire II (single and multiple targets at 75 meters, 175 meters, and 300 meters).
- Practice Record Fire.
- Practice Record Fire II.
- Record Fire.

NOTES: 1. See Chapter 6 for more information about BRM field fire.

2. See paragraphs 4-76 through 4-80 and Appendix A for more information about EST 2000 and LMTS training.

PHASE IV—ADVANCED RIFLE MARKSMANSHIP

1-25. ARM focuses on the techniques and procedures that the Soldier will need to participate in collective training. This phase addresses the following topics:

- Alternate firing positions.
- Burst fire.
- Quick fire.
- Chemical, biological, radiological, and nuclear (CBRN) fire.
- Moving targets.
- SDM.
- Unassisted night record fire.
- M68 close combat optic (CCO).
- Advanced combat optical gunsight (ACOG).
- AN/PAQ-4C infrared (IR) aiming laser.
- Assisted night record fire IR.
- Assisted night record fire (thermal).
- SRM.

NOTE: See Chapter 7 for more information about ARM.

PHASE V—ADVANCED OPTICS, LASERS, AND IRON SIGHTS

1-26. BRM teaches Soldiers to effectively engage the enemy with the basic rifle using the iron sights (primarily during the day). ARM adds additional situations for the firer. To enhance the lethality of night firing, Soldiers participate in training with optics and lasers. This phase addresses the following topics:

- Backup iron sight (BUIS).
- M68 CCO.
- ACOG.
- AN/PAQ-4C IR aiming laser.
- AN/PEQ-2A/B target pointer illuminator/aiming light (TPIAL).
- Thermal weapon sight (TWS).
- AN/PVS-4 night vision sight.

NOTE: See Chapter 8 for more information about advanced optics, lasers, and iron sights.

SECTION II. UNIT MARKSMANSHIP TRAINING PROGRAM

An effective unit marksmanship program reflects the priority, emphasis, and interest of commanders and trainers. This section outlines a rifle marksmanship training program strategy as guidance in establishing and conducting an effective unit training program. The strategy consists of the individual and leader refresher training for maintaining the basic skills learned during IET. It progresses to training advanced and collective skills under near-combat conditions during live-fire STXs.

MISSION-ESSENTIAL TASKS

1-27. Marksmanship proficiency is critical to soldiering and is required for any unit deployed to a wartime theater. All commanders should develop a METL and organize a training program that devotes adequate time to marksmanship. The unit's combat mission must be considered when establishing training priorities. This not only applies to the tasks selected for the unit's METL, but also to the conditions under which the tasks are to be performed. If a unit may be employed in an urban environment, the effects of range, gravity, and wind may not be as important as automatic or burst fire, quick fire, or assault fire. The reverse may be true of a unit that expects to engage the enemy at long ranges.

TRAINING ASSESSMENT

1-28. To conduct an effective marksmanship program, the unit commander must determine the current marksmanship proficiency of all assigned personnel. Constant evaluation provides commanders understanding of where training emphasis is needed. All results are reviewed to determine any areas that need strengthening, along with any individuals that require special attention. Based on this evaluation, marksmanship training programs are developed and executed. Commanders continually assess the program and modify it as required. To develop a training plan and assess the marksmanship program, commanders should use the following tools:

- Direct observation of training.
- Spot checks.
- Review of past training.

1-29. Based on the commander's evaluation, goals, and missions, quarterly, semiannual, or annual training events are identified. Marksmanship programs must be continuous, and to sustain an effective marksmanship program, resources are required. While the unit may only qualify its Soldiers annually or semiannually, test results show that sustainment training is required at least quarterly to maintain marksmanship skills.

Direct Observation of Training

1-30. Observing and accurately recording performance reveals the status of weapon maintenance, Soldier zero and qualification results, and each Soldier's ability to hit targets. This also allows the commander to identify Soldiers who need special assistance to reach required standards and those who exceed these standards.

Spot Checks

1-31. Spot checks of individual marksmanship performance, such as interviews and evaluations of Soldiers, provide commanders with valuable information about Soldier proficiency and knowledge of the marksmanship tasks.

Review of Past Training

1-32. Commanders review past training to gain valuable information for developing a training plan. The assessment should include—

- The frequency and results of training.
- The basic and advanced record fire results.
- The frequency of unit-conducted collective CBRN or night fire training.

Commander's Evaluation Guide

1-33. The commander's evaluation guide contains three sections:

- Commander's priorities and intent.
- Soldier assessment.
- Trainer assessment.

1-34. The following is an example of a commander's evaluation guide. Commanders can use this guide not only to assess their unit's marksmanship proficiency, but to assess the unit leaders and their ability to effectively implement a marksmanship program. They can also use it to develop NCOs into subject matter experts.

Commander's Priorities and Intent

1-35. When considering their priorities and intent, commanders answer the following questions:

- Have you clearly stated the priority of rifle (small arms) proficiency in your unit? What is it? Do the staff and subordinates support this priority? Is it based on your METL and an understanding of FM 7-0 and FM 7-1?
- Have you clearly stated the intent of record fire? Are leaders evaluating firing performance based on accurately recorded data and results?
- Have you clearly stated that weapon qualification or record fire is one of the commander's opportunities to assess several skills relating to small arms readiness?
- What qualification course will be used to evaluate your unit's marksmanship readiness (small arms)?
  - Is the standard 300-meter, 300-yard KD, or 25-meter scaled target alternate qualification course used?
  - How will it be conducted? Will the prescribed procedures be followed?
  - Who will collect the data?
- Have you clearly stated the purpose and intent of PMI?
  - What skills will PMI address?
  - Will PMI be performance-oriented? Are tasks integrated?

Soldier Assessment

1-36. During Soldier assessment, commanders answer the following questions:

- Do Soldiers maintain their assigned weapons and magazines IAW the TM? Do they have a manual?
- Do Soldiers conduct serviceability checks of weapons and magazines before training? Were maintenance deficiencies corrected?
- Do Soldiers demonstrate an understanding of the weapon's operation, functioning, and capabilities?
- Can Soldiers correctly apply immediate action procedures to reduce weapon stoppages and then continue to fire? Have they demonstrated this during dry-fire exercises?
- Are Soldiers firing their assigned weapons?
  - How often are weapons reassigned between individuals?
  - What is the value of a recorded zero?
- Can Soldiers precisely and consistently apply the four fundamentals of rifle marksmanship? To what standard have they demonstrated their mastery?
  - During a dry-fire exercise?
  - During a live-fire exercise (LFX)?
  - When firing on the 25-meter course?
  - During KD firing?
- Can Soldiers accurately battlesight zero their assigned rifle to standard?
  - Do they understand sight adjustment procedures?
  - Do they record their rifle's zero? How is it done? Why?
  - Do they record the date the Soldier last zeroed his rifle? What is the sight setting? Are these linked? How do you check this?
- Do Soldiers demonstrate their knowledge of the effects of wind and gravity while firing out to 300 meters? What feedback was provided? How?

- Can Soldiers scan a designated area or sector of fire and detect all targets out to 300 meters? If not, why?
- Can Soldiers quickly engage timed single and multiple targets from both supported and unsupported firing positions out to 300 meters? If not, which targets were not engaged? Which were missed? Why?
- During individual and collective training, do Soldiers demonstrate their ability to manage allocated ammunition and to engage all targets? Do they fire several rounds at one target? Which targets? Why?
- Based on an analysis of individual qualification scores, what is the distribution?
  - Are most Soldiers just meeting the minimum acceptable performance (marksman)?
  - Are most Soldiers distributed in the upper half of the performance spectrum (sharpshooter, expert)?
- What is the hit distribution during collective LFXs?
- Do Soldiers demonstrate proficiency during night-fire, target detection and acquisition, and night fire engagement techniques? When using night vision devices (NVDs)?
- Do Soldiers demonstrate individual marksmanship proficiency during MOPP firing conditions? During collective exercises?
- Do Soldiers demonstrate proficiency during moving target engagements? Do they demonstrate proficiency collectively by hitting moving targets at the multipurpose range complex? If not, is moving target training conducted?
- Are marksmanship skills integrated into tactical exercises and unit LFXs? If so, is suppressive fire, rapid semiautomatic fire, and automatic or burst fire conducted? What tasks in the mission training plan are evaluated?
- Based on onsite observations and analysis of training and firing performance, what skills or tasks show a readiness deficiency?
  - What skills need training emphasis? Individual emphasis? Leader emphasis?
  - What are the performance goals?

Trainer Assessment

1-37. During trainer assessment, commanders answer the following questions:

- Who has trained or will train the trainers?
  - What is the subject matter expertise of the cadre?
  - Are they actually training the critical skills?
  - Have they addressed the nonfiring skills first?
  - What aids and devices are being used? Are EST 2000 and LMTS being properly used?
- What administrative constraints or training distracters can you overcome for the junior officer and NCO? At what level are the resources necessary to train marksmanship controlled (time, training aids, weapons, ammunition, ranges)? Do the sergeants do the job they are charged with?

TRAINERS

1-38. Knowledgeable instructors or cadre are the key to marksmanship performance. All commanders must be aware of maintaining expertise in marksmanship instruction/training.

Instructor/Trainer Selection

1-39. Institutional and unit instructors/trainers are selected and assigned from the most highly qualified Soldiers. These Soldiers must demonstrate proficiency in all aspects of rifle marksmanship, be proficient in applying these fundamentals, know the importance of marksmanship training, and have a competent and professional attitude. The commander must ensure that selected instructors/trainers can effectively train other Soldiers. Local instructor/trainer training courses and marksmanship certification programs must be established to ensure that instructor/trainer skills are developed.

CADRE/TRAINER

1-40. Cadre/trainer refers to a marksmanship instructor/trainer that has more experience and expertise than the firer does. He trains Soldiers in the effective use of the rifle by maintaining strict discipline on the firing line, insisting on compliance with range procedures and program objectives, and enforcing safety regulations. A good cadre/trainer must understand the training phases and techniques for developing marksmanship skills. Each cadre/trainer must have the following attributes:

- Knowledge.
- Patience.
- Understanding.
- Consideration.
- Respect.
- Encouragement.

Knowledge

1-41. Effective cadre/trainers must possess a thorough knowledge of the rifle, proficiency in firing, and a complete understanding of this manual and supporting manuals.

Patience

1-42. Effective cadre/trainers relate to the Soldier calmly, persistently, and patiently.

Understanding

1-43. Cadre/trainers enhance success and understanding by emphasizing close observance of rules and instructions.

Consideration

1-44. Most Soldiers, even those who do not fire well, enjoy firing and begin their training with great enthusiasm. A cadre/trainer who is considerate of Soldiers' feelings and who encourages them throughout their training will find training a pleasant and rewarding duty.

Respect

1-45. An experienced cadre is assigned the duties of instructor/trainer, which classifies him as a technical expert and authority. A good cadre/trainer watches for mistakes and patiently makes needed corrections.

Encouragement

1-46. A cadre/trainer can encourage Soldiers by convincing them to achieve good firing performance through practice. He imparts his knowledge and helps Soldiers gain the practical experience necessary.

TRAINING THE TRAINER

1-47. Knowledgeable small-unit leaders are key to marksmanship training. This manual and other training publications provide the unit instructor with the required information for developing a good train-the-trainer program.

1-48. The goal of a progressive train-the-trainer program is to achieve a high state of combat readiness. Through the active and aggressive leadership of the chain of command, a perpetual base of expertise is established and maintained.

1-49. The commander should identify unit personnel who have had assignments as marksmanship instructors. These individuals should be used to train other unit cadre by conducting PMI and LFXs for their Soldiers.

1-50. Assistance and expertise from outside the unit may also be available, such as the Army Marksmanship Unit (AMU) at Fort Benning, Georgia. A suggested train-the-trainer program is outlined below:

- Conduct marksmanship diagnostic test.
- Review operation and function of the rifle and ammunition, immediate action, and safety.
- Conduct PMI; review the four fundamentals.
- Review coaching techniques and device usage.
- Establish grouping and zeroing procedures.
- Review effects of wind and gravity when firing as far away as 300 meters (600 meters for ARM).
- Conduct range operations.
- Conduct qualification/record firing.
- Diagnose firing problems.

DUTIES OF THE INSTRUCTOR/TRAINER

1-51. The instructor/trainer helps the firer master the fundamentals of rifle marksmanship. He ensures that the firer consistently applies what he has learned. When training the beginner, the instructor/trainer could confront problems, such as fear, nervousness, forgetfulness, failure to understand, and a lack of coordination or determination, which may be compounded by arrogance or carelessness. With all types of firers, the instructor/trainer must ensure that firers are aware of their firing errors, understand the causes, and apply remedies. To perform these duties, he—

- Observes the firer.
- Questions the firer.
- Analyzes the shot group.

Observing the Firer

1-52. To pinpoint errors, the instructor/trainer observes the firer during drills and when firing. If there is no indication of probable error, the firer's position, breath control, shot anticipation, and trigger squeeze are closely observed.

Questioning the Firer

1-53. The firer is asked to detect his errors and to explain his firing procedure (position, aiming, breath control, and trigger squeeze).

Analyzing the Shot Group

1-54. Analyzing the shot group is an important step in detecting and correcting errors. When analyzing a target, the instructor/trainer correlates observations of the firer to probable errors in performance, according to the shape and size of shot groups. A poor shot group is usually caused by more than one observable error.

TRAINER CERTIFICATION PROGRAM

1-55. The certification program sustains the trainers' expertise and develops methods of training. The program standardizes procedures for certifying marksmanship trainers. Trainers' technical expertise must be continuously refreshed, updated, and closely managed.

TRAINING BASE

1-56. The training base can expect the same personnel changes as any other organization. Soldiers assigned as marksmanship trainers will have varying degrees of experience and knowledge of training procedures and methods. Therefore, the trainer certification program must be an ongoing process that is tailored to address these variables. At a minimum, formal records should document program progression for each trainer. All marksmanship trainers must complete the four phases of training using the progression steps, and the records should be updated on a quarterly basis.

Certification Program Outline

1-57. Before certification, all trainers must attend all phases of the train-the-trainer program in the following order:

- Phase I—Program Orientation.
- Phase II—Preliminary Marksmanship Training.
- Phase III—Basic Marksmanship Training.
- Phase IV—Advanced Marksmanship Training.

1-58. Then, they conduct all phases to demonstrate their ability to train Soldiers and to diagnose and correct problems. Trainers who fail to attend or do not pass any phase of the diagnostic examination will be assigned to subsequent training.

Phase I—Program Orientation

1-59. During this phase, the trainer must accomplish the following tasks and be certified by the chain of command:

- Be briefed on the concept of the certification program.
- Be briefed on the unit marksmanship training strategy.
- Review the unit marksmanship training outlines.
- Review issued reference material.
- Visit training sites and firing ranges.

Phase II—Preliminary Marksmanship Training

1-60. Phase II should be completed no more than two weeks following the conclusion of Phase I. During Phase II, the trainer demonstrates his ability to master the marksmanship fundamentals, and his performance is reviewed by the chain of command. The results of this review are recorded and maintained on the trainer's progression sheet, which is, in turn, designed IAW the unit SOP.

- Safety, clearing, and unloading procedures.
- Characteristics.
- Capabilities.
- Accessories.
- Ammunition.
- Disassembly.
- Cleaning, inspection, and lubrication procedures.
- Assembly.
- Function check.
- Loading.
- Immediate and remedial actions.
- Four fundamentals.
- Use of EST 2000, LMTS, and other TADSS.
- Sight manipulation.
- Boresighting procedures.
- Zeroing procedures.
- Effect of wind, gravity, and other environmental factors.
- Range determination and estimation.
- Classes of fire.

- Application of fire.
- Range operations, courses of fire, conduct of fire, and fire commands.
- Optics, NVDs, night aiming devices, and thermal devices.
- Scanning techniques.

Phase III—Basic Marksmanship Training

1-61. During this phase, the trainer demonstrates and reinforces what he has learned during Phase II. The trainer sets up and conducts firing on the various ranges. He explains the targets and the zeroing and scoring procedures. The trainer explains the purpose of transition firing, field zeroing procedures, range layout, and the conduct of training on the transition range. This validates that the trainer has gained the knowledge necessary to conduct training. The results of this review are recorded and maintained on the trainer's progression sheet.

Phase IV—Advanced Marksmanship Training

1-62. The final phase of the train-the-trainer program tests the trainer. During this phase, the trainer sets up a range and conducts training for at least one person. If ammunition is available, the trainer conducts a firing exercise. If ammunition is not available, the evaluation is based on the quality of training given.

QUALIFICATION TRAINING

1-63. Although marksmanship is a continuous training requirement, units normally conduct a refresher program before qualification. Soldiers must be well-versed in marksmanship fundamentals and have preparatory marksmanship training before qualification. This applies to qualification for the entire unit or for newly assigned personnel. Trainers must understand that unit rifle marksmanship is not a series of exercises to be trained in a planned sequence as is done during IET, but trainers can use the exercises and POI events covered during IET to identify events that the unit can use for a sustainable and effective unit marksmanship program. The unit must prepare for training by—

- Issuing Soldiers a serviceable weapon.
- Maintaining and replacing bad magazines.
- Issuing and assigning each Soldier his own rifle. Only he will zero and fire the weapon assigned to him.
- Considering available or required resources (targets, ranges, ammunition, training aids, devices, and publications) early in the process.

1-64. Before the Soldier can fire, he must know how to adjust the rifle's sights and should understand ballistics (for example, the effects of wind and gravity on a bullet strike). A refresher training program prevents Soldiers from becoming frustrated and losing confidence, and conserves ammunition and training time. All Soldiers attend this program so they can meet the standards outlined in this manual and supporting manuals.

NOTE: Many individual marksmanship tasks, such as operation and function checks, immediate action, target detection, and dry-fire, do not require live-firing. Live and virtual simulators can be used to reinforce PMI, grouping, zeroing, practice record fire, record fire, CBRN fire, and assisted and unassisted night fire by simulating the LFXs. Building marksmanship confidence by repetition can bring consistency to the unit marksmanship training program.

FEEDBACK

1-65. Feedback must be included in all live-fire training. Soldiers must have precise knowledge of a bullet strike; feedback is not adequate when bullets from previous firings cannot be identified. To provide accurate feedback, trainers ensure that Soldiers triangulate and clearly mark previous shot groups on a zeroing target.

Grouping and Zeroing

1-66. The initial LFX should be a grouping exercise, during which Soldiers apply marksmanship fundamentals to obtain tight, consistent shot groups. Following a successful grouping exercise, Soldiers zero their weapons quickly using only a few rounds.

Downrange Feedback

1-67. After zeroing, downrange feedback should be conducted. If modified field fire or KD ranges are not available, a series of scaled silhouette targets can be used for training on the 25-meter range. A timed-fire scaled silhouette target can add to successful record fire performance, since it represents targets at six different ranges, requires quick response, and allows precise feedback. It also serves as another way to confirm zero and requires the application of the four fundamentals.

Field Fire

1-68. Field fire training is a transitional phase that stresses focusing on a certain area. Soldiers must detect the target as soon as it appears and quickly fire; this is an important combat skill. Soldiers who are exposed to the field fire range before they have refined their basic firing skills cannot benefit from the exercise, and feedback from this exercise is hit-or-miss. If most 175- and 300-meter targets are missed, additional feedback or PMI training should be conducted.

Record Fire

1-69. The intent of record fire is to facilitate the commander's evaluation of several individual tasks and integrated marksmanship skill performances, and to provide unit readiness indicators. The qualification standards are specifically related to a prescribed procedure for the conduct of record fire. Individual performance must be evaluated IAW three components:

- What test was used (standard, known distance, or scaled)?
- How was the test administered?
- How were individual and unit performances distributed (marksman, sharpshooter, and expert) and at what target ranges?

Remedial Training

1-70. For firers who need additional training to meet the requirements, remedial training is conducted using the EST 2000 or LMTS. Soldiers proficient in marksmanship skills can assist in the remedial training effort.

NOTE: Table 1-1 shows the training devices that a commander may use to assist in remedial training and sustain weapon proficiency. See Appendix A for more information about these training devices. The devices replicate, but are not intended to replace, LFXs or qualifications. Active and reserve component units should consult DA Pam 350-38 for regulatory guidance on mandatory live-fire training and qualification events.

Table 1-1. Training simulators, devices, and exercises.

TRAINING DEVICE / EXERCISE	Engagement Skills Trainer (EST 2000)	Laser Marksmanship Training System (LMTS)	Location of Misses and Hits (LOMAH)	Short-Range Training Ammunition (SRTA) and M2 Bolt	Weaponeer
Zero	X	X	X	X	X
Practice Fire	X	X	X	25 m	X
Simulated Record Fire	X	X	X	25 m	X
CBRN Practice	X	X	X	X	X
Simulated CBRN Record	X	X	X	X	X
Unassisted Night Practice	X	X	X	50 m	
Unassisted Simulated Night Record	X	X	X	50 m	
NVD Zero	X	X	X		
NVD Practice	X	X	X	50 m	
NVD Simulated Record	X	X	X	50 m	
Advanced Skills	X	X	X	X	

UNIT LIVE-FIRE EXERCISES

1-71. Unit LFXs are planned, prepared, and performed as outlined in the mission training plan for the infantry platoon and squad. Within the framework of these exercises, Soldiers perform marksmanship tasks under realistic combat conditions.

FUNDAMENTALS

1-72. During training, the fundamentals must apply to combat, as well as to the range. Too often Soldiers disregard the fundamentals while under the pressure of combat, so it is imperative that the Soldier receives feedback regarding his firing results and his use of the fundamentals during collective LFXs. This training should also discuss target acquisition, area fire, quick fire, assuming firing positions, responding to oral fire commands, and safety. To learn SOPs and proper procedures, Soldiers must participate in dry-fire or MILES rehearsals at crawl, walk, and run paces.

EVALUATORS

1-73. During training, enough evaluators must be present to observe each Soldier and provide performance feedback. Evaluators must know the scenario, the location of targets, the friendly plan, and SOPs. They must watch each Soldier to determine if he identifies targets in his sector and successfully engages them. Evaluators must also understand the fundamentals of marksmanship to detect Soldiers' mistakes and review them during the after-action review (AAR).

Chapter 2

Weapon Characteristics, Accessories, and Ammunition

This chapter describes the general components, characteristics, accessories, and ammunition for M16- and M4-series weapons, and includes a brief explanation of how to mount the various accessories.

SECTION I. RIFLES AND CARBINES

All M16-/M4-series weapons are magazine-fed, gas-operated, air-cooled, shoulder-fired 5.56-millimeter weapons. This section describes the general characteristics and components of M16-/M4-series weapons.

CHARACTERISTICS OF M16-/M4-SERIES WEAPONS

2-1. Table 2-1 describes the general characteristics of M16-/M4-series weapons.

Table 2-1. Characteristics of M16-/M4-series weapons.

CHARACTERISTICS	M4-SERIES	M16A2/A3	M16A4	M16A1
WEIGHT (lb)				
Without magazine and sling	6.49	7.78	9.08	6.35
With sling and loaded:				
20-round magazine	7.19	8.48	9.78	6.75
30-round magazine	7.50	8.79	10.09	8.06
Bayonet knife, M9	1.50	1.50	1.50	1.50
Scabbard	0.30	0.30	0.30	0.30
Sling, M1	0.40	0.40	0.40	0.40
LENGTH (in)				
Rifle w/bayonet knife	N/A	44.88	44.88	44.25
Overall rifle length	N/A	39.63	39.63	39.00
Buttstock closed	29.75	N/A	N/A	N/A
Buttstock open	33.0	N/A	N/A	N/A
OPERATIONAL CHARACTERISTICS				
Barrel rifling-righthand 1 twist (in)	7	7	7	12
Muzzle velocity (fps)	2,970	3,100	3,100	3,250
Cyclic rate of fire (rounds per min)	700-900	700-900	800	700-800
MAXIMUM EFFECTIVE RATE OF FIRE (rounds per min)				
Semiautomatic	45	45	45	45-65
3-round burst	90	90 (A2)	90	N/A
Automatic	150-200 A1	150-200 A3	N/A	150-200
Sustained	12-15	12-15	12-15	12-15
RANGE (m)				
Maximum range	3,600	3,600	3,600	2,653
Maximum effective range:				
Point target	500	550	550	460
Area target	600	800	600	N/A

NOTE: For further technical information, refer to TM 9-1005-319-10 and TM 9-1005-249-10.

M4-SERIES CARBINE

2-2. The M4-series carbine consists of the M4, the M4A1, and the M4 modular weapon system (MWS) (Figures 2-1 and 2-2) and features several modifications to equip it for close combat operations:

- The M4-series carbine buttstock has four positions:
  - Closed.
  - ½ open.
  - ¾ open.
  - Fully open.
- The M4 carbine becomes the M4 MWS when the M4 adapter rail system (ARS) is installed (Figure 2-3).
- When operating an M4 or M4 MWS, the firer can move the selector lever (SAFE, SEMI, and BURST) to fire a semiautomatic or a three-round burst.
- The M4A1 is fully automatic.

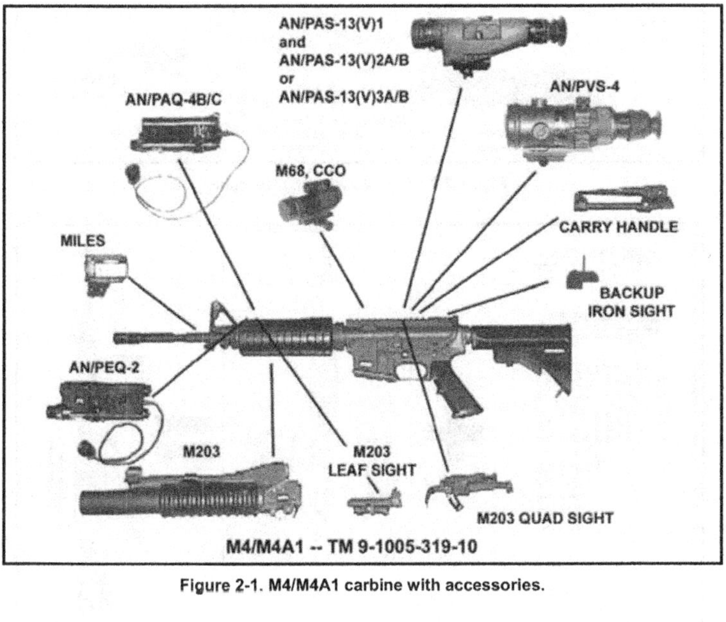

Figure 2-1. M4/M4A1 carbine with accessories.

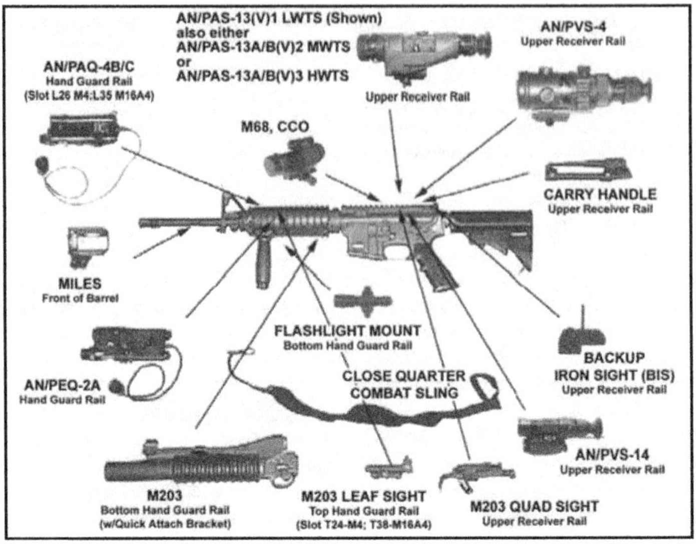

Figure 2-2. M4 MWS with accessories.

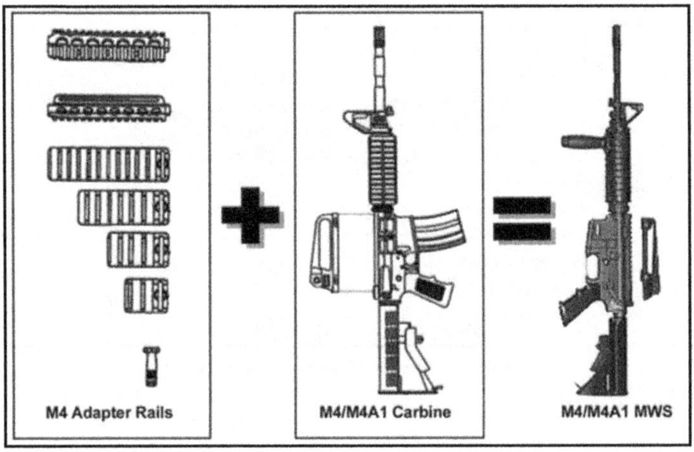

Figure 2-3. M4/M4A1 and M4/M4A1 MWS.

Mechanically Zeroing the M4/M4A1 Carbine or M4 MWS

NOTE: Mechanically zeroing the weapon is only necessary when the weapon zero is questionable, the weapon is newly assigned to the unit, or the weapon sights have been serviced.

2-3. To mechanically zero an M4/M4A1 or M4 MWS—

NOTE: Reference the weapon components using the numbers listed in Figure 2-4.

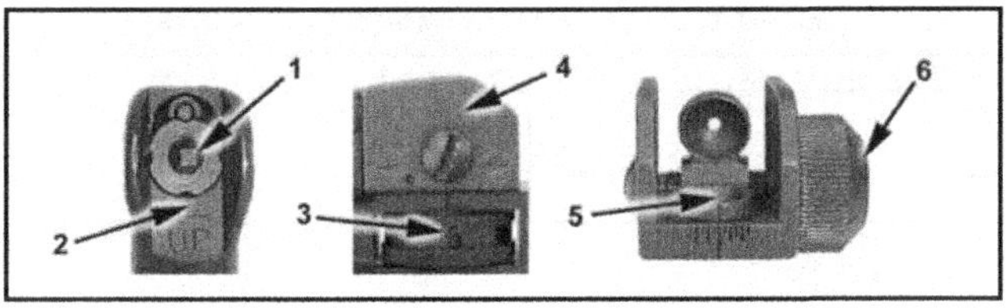

Figure 2-4. M4/M4A1 or M4 MWS mechanical zero.

(1) Adjust the front sightpost (1) until the base of the front sightpost is flush with the front sightpost housing (2).

(2) Turn the elevation knob (3, shown as viewed from above) counterclockwise until the rear sight assembly (4) rests flush with the detachable carrying handle and the 6/3 marking is aligned with the index line (5) on the left side of the carrying handle.

(3) Position the apertures (6) so the unmarked aperture is up and the 0-200 meter aperture is down.

(4) Turn the windage knob (7) to align the index mark (8) on the 0-200 meter aperture with the long center index line on the rear sight assembly.

Battlesight Zeroing the M4/M4A1 Carbine or M4 MWS

2-4. To battlesight zero an M4/M4A1 or M4 MWS (Figure 2-5)—

NOTE: Reference the weapon components using the numbers listed in Figure 2-5.

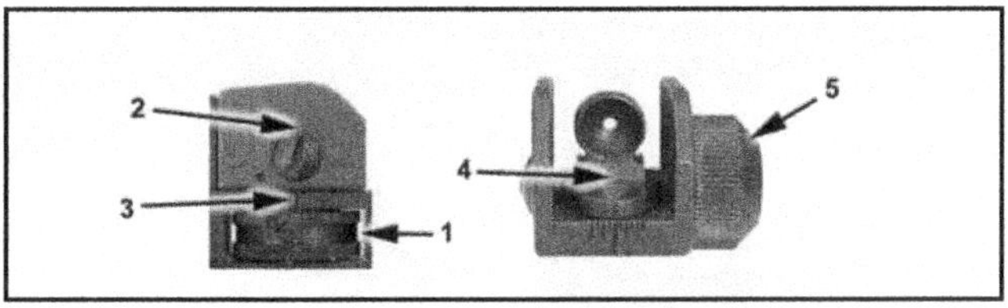

Figure 2-5. M4/M4A1 or M4 MWS battlesight zero.

(1) Turn the elevation knob (1, shown as viewed from above) counterclockwise until the rear sight assembly (2) rests flush with the detachable carrying handle and the 6/3 marking is aligned with the index line (3) on the left side of the detachable carrying handle.

NOTE: The elevation knob remains flush.

(2) Position the apertures (4) so the unmarked aperture is up and the 0-200 meter aperture is down.

(3) Turn the windage knob (5) to align the index mark (6) on the 0-200 meter aperture with the long center index line on the rear sight assembly.

NOTE: The "Z" marking on the elevation knob (employed when using the M4-series weapon's detachable carrying handle) should be ignored. The "Z" marking is only used when the M16A4 is being zeroed.

2-5. Table 2-2 shows how much one click of elevation or windage will move the strike of the round at ranges from 25 to 500 meters.

Table 2-2. Point of impact for the M4/M4A1 and M4 MWS.

RANGE (m)	25	100	200	300	400	500
Elevation	3/4 in 0.9 cm	1 3/8 in 3.5 cm	2 3/4 in 7 cm	4 1/8 in 10.5 cm	5 1/2 in 14 cm	6 7/8 in 17.5 cm
Windage	1/8 in 0.3 cm	1/2 in 1.25 cm	1 in 2.5 cm	1 1/2 in 3.8 cm	2 in 5 cm	2 1/2 in 6.3 cm

M16A2/A3 RIFLE

2-6. Figure 2-6 shows the M16A2/A3 rifle. When operating an M16A2 rifle, the firer can move the selector lever (SAFE, SEMI, and BURST) to fire a semiautomatic or a three-round burst. The M16A3 has the same characteristics as the M16A2, with the exception of the selector lever (SAFE, SEMI, and AUTO) and the addition of the automatic mode.

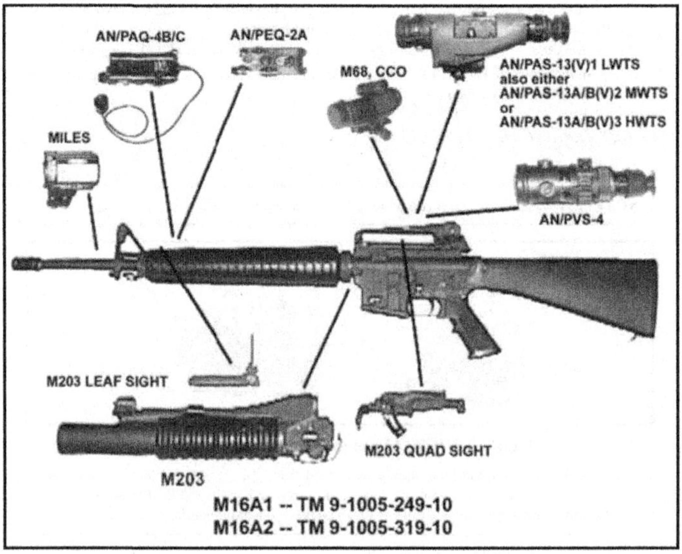

Figure 2-6. M16A2/A3 rifle with accessories.

MECHANICALLY ZEROING THE M16A2/A3 RIFLE

NOTE: Mechanically zeroing the weapon is only necessary when the weapon zero is questionable, the weapon is newly assigned to the unit, or the weapon sights have been serviced.

2-7. To mechanically zero an M16A2/A3 rifle—

NOTE: Reference the weapon components using the numbers listed in Figure 2-7.

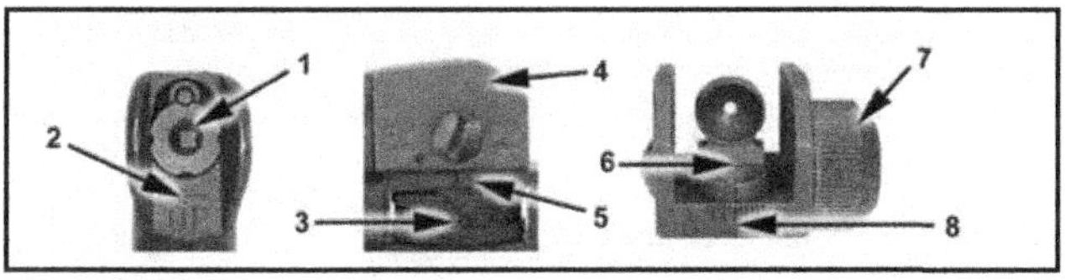

Figure 2-7. M16A2/A3 mechanical zero.

(1) Adjust the front sightpost (1) until the base of the front sightpost is flush with the front sightpost housing (2).

(2) Turn the elevation knob (3, shown as viewed from above) counterclockwise until the rear sight assembly (4) rests flush with the carrying handle and the 8/3 marking is aligned with the index line on the left side of the carrying handle.

(3) Position the apertures (5) so the unmarked aperture is up and the 0-200 meter aperture is down.

(4) Turn the windage knob (6) to align the index mark on the 0-200 meter aperture with the long center index line on the rear sight assembly.

BATTLESIGHT ZEROING THE M16A2/A3 RIFLE

2-8. To battlesight zero an M16A2/A3 rifle—

NOTE: Reference the weapon components using the numbers listed in Figure 2-8.

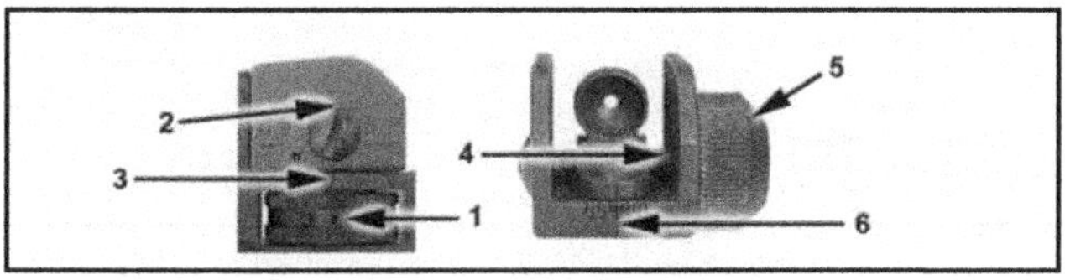

Figure 2-8. M16A2/A3 battlesight zero.

(1) Turn the elevation knob (1, shown as viewed from above) counterclockwise until the rear sight assembly (2) rests flush with the carrying handle and the 8/3 marking is aligned with the index line (3) on the left side of the carrying handle.

(2) Turn the elevation knob one more click clockwise.

(3) Position the apertures (4) so the unmarked aperture is up and the 0-200 meter aperture is down.

(4) Turn the windage knob (5) to align the index mark on the 0-200 meter aperture with the long center index line on the rear sight assembly.

2-9. Table 2-3 shows how much one click of elevation or windage will move the strike of the round at ranges from 25 to 600 meters.

Table 2-3. Point of impact for the M16A2/A3 rifle.

RANGE (m)	25	100	200	300	400	500	600
Elevation	3/4 in 0.9 cm	1 3/8 in 3.5 cm	2 3/4 in 7 cm	4 1/8 in 10.5 cm	5 1/2 in 14 cm	6 7/8 in 17.5 cm	8 1/4 in 20.9 cm
Windage	1/8 in 0.3 cm	1/2 in 1.25 cm	1 in 2.5 cm	1 1/2 in 3.8 cm	2 in 5 cm	2 1/2 in 6.3 cm	3 in 7.6 cm

M16A4 RIFLE

2-10. The M16A4 rifle (Figure 2-9) is the same as the M16A2 rifle, with the addition of a flat top upper receiver (with an integral rail and detachable carrying handle) and the M5 ARS.

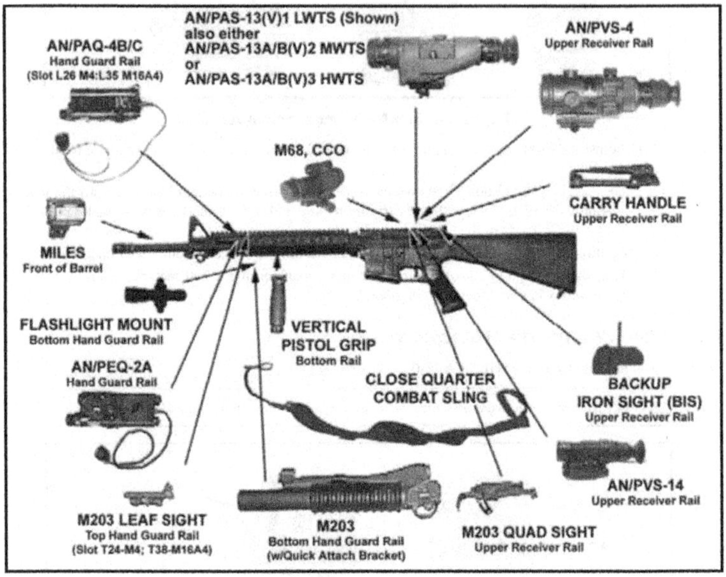

Figure 2-9. M16A4 rifle with accessories.

Mechanically Zeroing the M16A4 Rifle

NOTE: Mechanically zeroing the weapon is only necessary when the weapon zero is questionable, the weapon is newly assigned to the unit, or the weapon sights have been serviced.

2-11. To mechanically zero an M16A4 rifle—

NOTE: Reference the weapon components using the numbers listed in Figure 2-10.

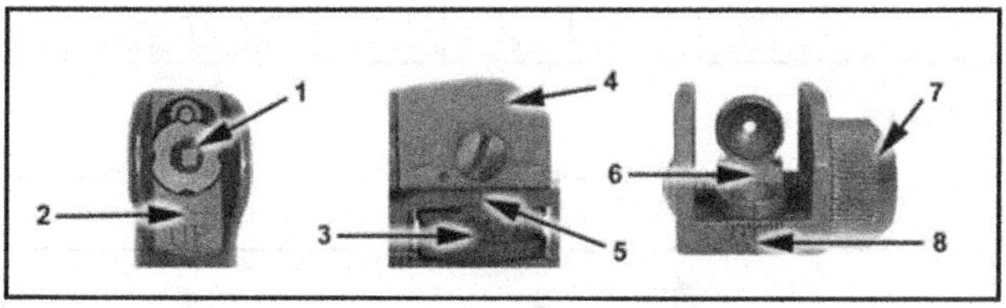

Figure 2-10. M16A4 mechanical zero.

(1) Adjust the front sightpost (1) until the base of the front sightpost is flush with the front sightpost housing (2).

(2) Turn the elevation knob (3, shown as viewed from above) counterclockwise until the rear sight assembly (4) rests flush with the carrying handle and the 6/3 marking is aligned with the index line (5) on the left side of the carrying handle.

(3) Position the apertures (6) so the unmarked aperture is up and the 0-200 meter aperture is down.

(4) Turn the windage knob (7) to align the index mark on the 0-200 meter aperture with the long center index line (8) on the rear sight assembly.

Battlesight Zeroing the M16A4 Rifle

2-12. To battlesight zero an M16A4 rifle—

NOTE: Reference the weapon components using the numbers listed in Figure 2-11.

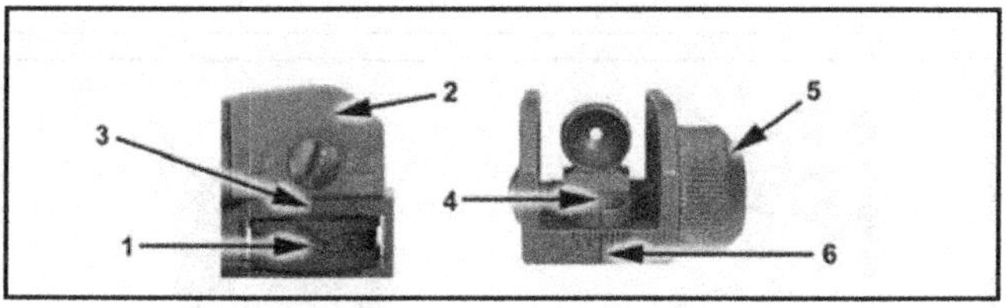

Figure 2-11. M16A4 battlesight zero.

(1) Turn the elevation knob (1, shown as viewed from above) counterclockwise until the rear sight assembly (2) rests flush with the detachable carrying handle and the 6/3 marking is aligned with the index line (3) on the left side of the detachable carrying handle.

(2) Turn the elevation knob two clicks clockwise so the index line on the left side of the detachable carrying handle is aligned with the "Z" on the elevation knob.

(3) Position the apertures (4) so the unmarked aperture is up and the 0-200 meter aperture is down.

(4) Turn the windage knob (5) to align the index mark on the 0-200 meter aperture with the long center index line (6) on the rear sight assembly.

2-13. Table 2-4 shows how much one click of elevation or windage will move the strike of the round at ranges of 25 to 600 meters.

Table 2-4. Point of impact for the M16A4 rifle.

RANGE (m)	25	100	200	300	400	500	600
Elevation	3/4 in 0.9 cm	1 3/8 in 3.5 cm	2 3/4 in 7 cm	4 1/8 in 10.5 cm	5 1/2 in .14 cm	6 7/8 in 17.5 cm	8 1/4 in 20.9 cm
Windage	1/8 in 0.3 cm	1/2 in 1.25 cm	1 in 2.5 cm	1 1/2 in 3.8 cm	2 in 5 cm	2 1/2 in 6.3 cm	3 in 7.6 cm

M16A1 RIFLE

2-14. When operating an M16A1 rifle (Figure 2-12), the firer can move the selector lever (SAFE, SEMI, and AUTO) to fire in the semiautomatic or automatic mode.

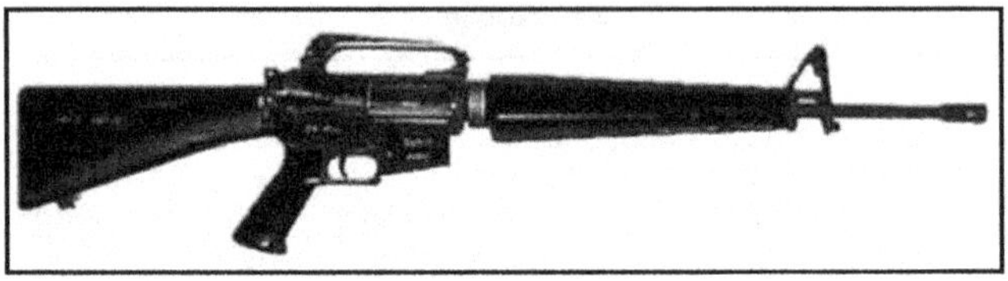

Figure 2-12. M16A1 rifle.

MECHANICALLY ZEROING THE M16A1 RIFLE

NOTE: Mechanically zeroing the weapon is only necessary when the weapon zero is questionable, the weapon is newly assigned to the unit, or the weapon sights have been serviced.

2-15. To mechanically zero an M16A1 rifle—

NOTE: Reference the weapon components using the numbers listed in Figure 2-13.

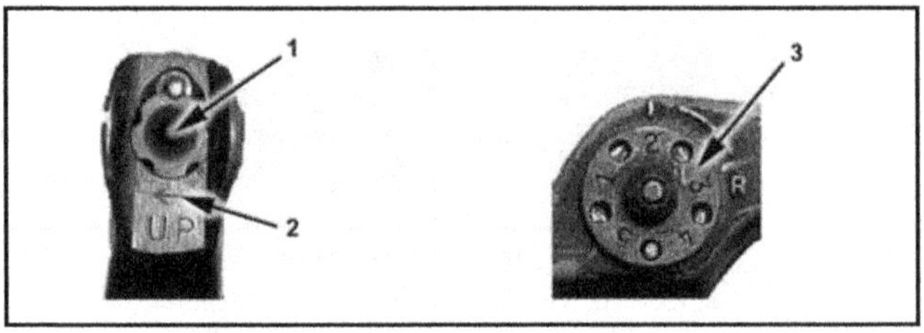

Figure 2-13. M16A1 mechanical zero.

(1) Adjust the front sightpost (1) until the base of the front sightpost is flush with the front sightpost housing (2).
(2) Adjust the front sightpost 11 clicks in the direction of UP.
(3) Turn the rear sight windage drum (3) left until it stops.
(4) Turn the windage drum right 17 clicks so the rear sight is approximately centered.

BATTLESIGHT ZEROING THE M16A1 RIFLE

2-16. Soldiers should use the aperture marked "L" (Figure 2-14) to battlesight zero their weapons.

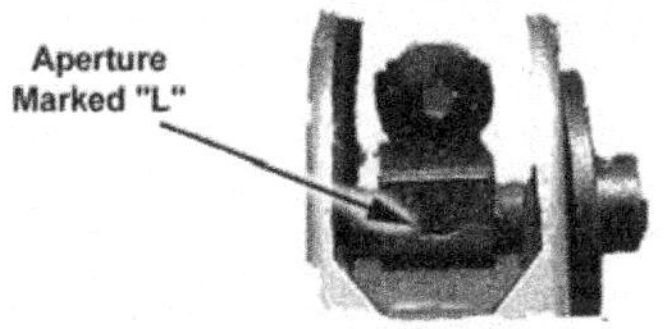

Figure 2-14. M16A1 battlesight zero.

2-17. Table 2-5 shows how much one click of elevation or windage will move the strike of the round at ranges of 25 to 500 meters using standard sights or the low light level sight system.

Table 2-5. Point of impact for the M16A1 rifle.

STANDARD SIGHTS						
Range (m)	25	100	200	300	400	500
Elevation	17/64 in 0.7 cm	1 3/32 in 2.8 cm	2 13/64 in 5.6 cm	3 9/32 in 8.4 cm	4 3/8 in 11.2 cm	5 15/32 in 14 cm
Windage	17/64 in 0.7 cm	1 3/32 in 2.8 cm	2 13/64 in 5.6 cm	3 9/32 in 8.4 cm	4 3/8 in 11.2 cm	5 15/32 in 14 cm
LOW LIGHT LEVEL SIGHT SYSTEM						
Elevation	23/64 in 0.9 cm	1 3/4 in 3.5 cm	2 3/4 in 7 cm	5 1/4 in 10.5 cm	7 in 17.7 cm	8 3/4 in 22.2 cm
Windage	17/64 in 0.7 cm	1 3/32 in 2.8 cm	2 13/64 in 5.6 cm	3 9/32 in 8.4 cm	4 3/8 in 11.2 cm	5 15/32 in 14 cm

SECTION II. ACCESSORY MOUNTING

The M4/M5 ARS and rail grabbers are designed to mount accessories on M16- and M4-series weapons.

M4/M5 ADAPTER RAIL SYSTEM

2-18. The ARS provides a secure mounting point for various accessories that may be mounted on the weapon's top, left, and right. The M4 ARS (Figure 2-15) consists of lightweight sections that replace the standard handguards on the M4 carbine. The M5 ARS replaces the handguards on the M16A4 rifle.

NOTES:

1. Only the armorer is authorized to remove the ARS from a weapon. However, the user may remove only the lower assembly to perform preventive maintenance checks and services (PMCS). If the top section is removed, the accessories mounted on the rail will lose their zero.
2. To prevent the screw from vibrating loose during firing, apply locking compound to the original screw or use a new screw when reinstalling the ARS.
3. For optics and lasers to retain zero, the ARS must be properly installed. Mounted accessories will not retain zero unless proper procedures are followed, as outlined in TM 9-1005-319-23& P.
4. Accessories may be mounted on the right side of the ARS, but are not currently supported with zeroing procedures. Further, the bottom rail of the ARS will not retain zero. Only accessories that do not require zero retention, such as a flashlight or vertical pistol grip, can be mounted on the bottom rail.

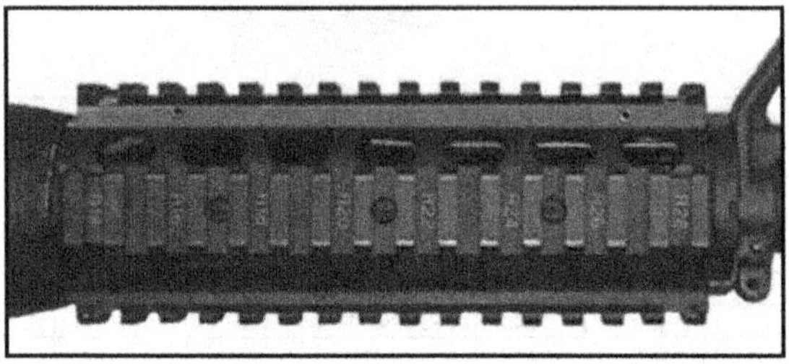

Figure 2-15. Adapter rail system.

RAIL COVERS

2-19. Rail covers protect the firer's hands from direct contact with the metal parts of the ARS and protect the ARS surfaces from excess wear and damage. The rail covers are available in 11-, 9-, 6-, 5-, and 4-rib sections. For ease of reference, the shorter lengths can be referred to by the number of ribs along their outer surfaces. 11-, 9-, and 5-rib sections are shown in Figure 2-16.

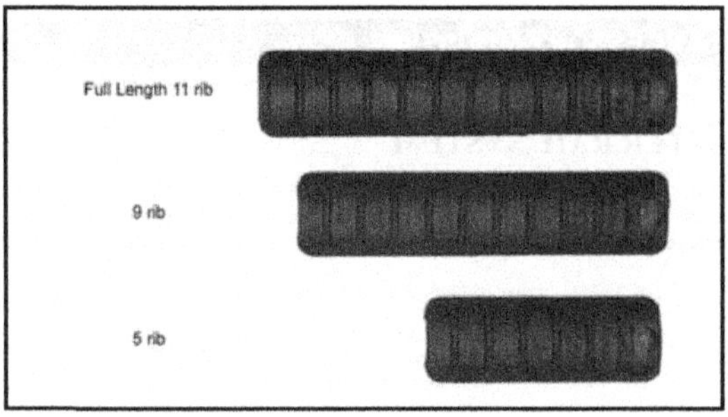

Figure 2-16. M5 rail covers/heat shields.

2-20. ARS rail covers can be quickly attached and detached from the ARS. A spring latch at one end of each rail cover automatically engages cutouts at the end of each rail. To slide the rail cover beyond a cutout or to remove it, apply thumb pressure to the center of the spring latch, and slide it in the desired direction.

NOTES: 1. Keep unused bottom, left, and right rail sections covered with 11-rib rail cover sections.

2. If any accessories are mounted on a rail, cover the remaining rail surface with an appropriately-sized rail cover.

3. If a BUIS is installed, permanently remove the top full length rail cover, and replace it with a shorter rail cover to protect the nonfiring hand when the barrel is hot.

WARNING

When firing at high rates of sustained fire, the barrel and metal components of the ARS can become hot enough to inflict serious burns. Cover exposed metal portions with plastic rail covers. Use the vertical pistol grip (Figure 2-18) during heavy, sustained fire.

NUMBERED RECOIL GROOVES

2-21. The even numbered recoil grooves on each rail of the ARS are sequentially numbered within the recoil grooves themselves (Figure 2-17). Each number is preceded by a letter prefix indicating a specific slot on the ARS. The numbers on the top rail have a "T" prefix, while those on the bottom have a "B" prefix. Additionally, the numbers on the rail to the firer's left have an "L" prefix, while those to the firer's right have an "R" prefix. These addresses—

- Assist the user in remounting an accessory in the same position.
- Allow standardization in the location of mounted accessories.
- Identify reference points for discussions on accessory mounting locations.

Figure 2-17. Address markings on the adapter rail system.

NOTE: Within the notches, each ARS also contains holes that are threaded ¼ inch deep, with 20 threads per inch (Figure 2-17). This is the standard thread size for a camera tripod adapter, which is used to attach standard camera or video accessories. For example, an M4 MWS with an NVD mounted may be attached to a standard camera tripod for hands-free support during long periods of surveillance.

VERTICAL PISTOL GRIP

2-22. Each ARS comes with a vertical pistol grip. To install the pistol grip—

NOTE: Reference the weapon components using the numbers listed in Figure 2-18.

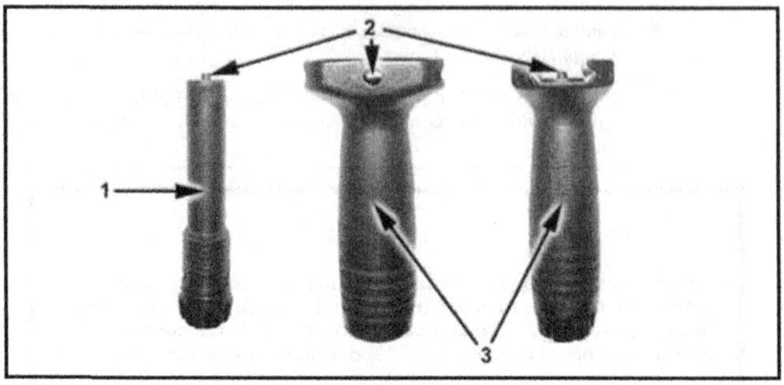

Figure 2-18. Vertical pistol grip.

(1) Remove the rail cover.

(2) Unscrew the pistol grip lock (1) until the tip (2) is no longer visible through the hole in the pistol grip.

(3) Slide the pistol grip (3) onto the ARS.

NOTE: The pistol grip will cover five notches on the ARS.

(4) Align the tip on the top of the pistol grip lock (1) with a notch.

(5) Hand-tighten.

NOTE: For further information about these accessories, refer to TM 9-1005-319-10.

RAIL GRABBERS

2-23. The Insight (Figure 2-19) and Picatinny (Figure 2-21) rail grabbers are designed to mount accessories onto the M16A4 and M4-series weapons. These rail grabbers enable the user to attach accessories on the upper receiver and on all four sides of the ARS, while retaining zero (when installed and tightened properly). Once zeroed, they will retain zero—even when removed from the weapon—as long as the rail grabber is not separated from the accessory and is remounted on the same notch it was zeroed on.

NOTE:
1. Retighten the rail grabber and accessory after firing the first three rounds to fully seat both.
2. The bottom rail of the ARS will not retain zero.

INSIGHT RAIL GRABBER

2-24. The Insight rail grabber (Figure 2-19) can be mounted where the tightening screw (3, Figure 2-19) is on either the left or right side (when top-mounted) or top or bottom (when mounted on the left side) so it does not interfere with weapon operation.

Figure 2-19. Insight rail grabber.

2-25. Unless command-directed, all devices in a unit do not have to be mounted in the same location as long as individual users record or mark (with paint markers or grease pencils) the mounting location on their weapons to avoid unnecessary rezeroing.

CAUTION

Both of the holes (2, Figure 2-19) located in the top of the rail grabber can be used to mount accessories, but the hole closest to the muzzle must be used. This ensures that the majority of the rail grabber is supporting the accessory to prevent damage to the accessory.

NOTES: 1. The Insight rail grabber must fully rest on the ARS in order to retain zero. The locking clamp (1, Figure 2-19) must grasp the ARS, and the screw that tightens the rail grabber must be tightened with a field tool, such as a multipurpose tool.

2. Even if the rail grabber is resting entirely on the ARS, accessories should not make contact with the front sight assembly or the collar of the barrel. The vibrations that occur during firing will interfere with the rail grabber's and accessory's zero retention capabilities.

MILES Training Extender for the Insight Rail Grabber

2-26. The training extender (Figure 2-20) is used to elevate an accessory above the MILES laser during force-on-force training. The training extender is only used when the Insight rail grabber is top-mounted.

2-27. To install the extender and accessories—

NOTE: Reference the weapon components using the numbers listed in Figure 2-20.

Figure 2-20. Insight rail grabber MILES training extender.

(1) Use the thumbscrew (1) to hand-tighten the extender into the mounting hole closest to the muzzle on the Insight rail grabber.

(2) Install the accessory on top of the extender and tighten.

NOTE: The AN/PEQ-2A/B and AN/PAQ-4B/C must be zeroed before and after using the MILES training extender.

Picatinny Rail Grabber

2-28. The Picatinny rail grabber (Figure 2-21) must fully rest on the ARS in order to retain zero. The locking clamp must grasp the ARS, and the torque-limiting knob (2, Figure 2-21) that tightens the rail grabber must be hand-tightened until it clicks twice.

Figure 2-21. Picatinny rail grabber.

CAUTION

Both of the holes located in the top of the rail grabber (1, Figure 2-21) can be used to mount accessories, but the hole closest to the muzzle must be used for the Picatinny rail grabber. This ensures that the majority of the rail grabber is supporting the accessory to prevent damage to the accessory. It also allows the torque-limiting knob to be mounted on either the left or right side (when top-mounted) or top or bottom (when side-mounted) so that it will not interfere with weapon operation.

2-29. Unless command-directed, all devices in a unit do not have to be mounted in the same location as long as individual users record or mark (with paint markers or grease pencils) the mounting location on their weapons to avoid unnecessary rezeroing.

NOTE: Even if the rail grabber is resting entirely on the ARS, accessories should not make contact with the front sight assembly or the collar of the barrel. The vibrations that occur during firing will interfere with the rail grabber's and accessory's zero retention capabilities.

SECTION III. ACCESSORIES

Table 2-6 shows the weapons with which the various accessories are compatible, along with the mounting device and TM number. Table 2-7 shows select characteristics of the various accessories.

Table 2-6. Accessory compatibility and mounting.

ACCESSORY	TECHNICAL MANUAL	M4 MWS , M16A4	M4/M4A1	M16A1/A2/A3
M68, CCO	TM 9-1240-413-13&P	* Upper receiver	* Upper receiver	M16 mount
AN/PVS-14	TM 11-5855-306-10	***Upper receiver	N/A	N/A
AN/PAC-4C	TM 11-5855-301-12&P	** Rail grabber	Bracket assembly	Bracket assembly
AN/PEQ-2A/B	TM 11-5855-308-12&P	** Rail grabber	Bracket assembly	Bracket assembly
AN/PAQ-4	TM 11-5855-261-10	** Rail grabber	Bracket assembly	Bracket assembly
AN/PAS-13B(V1), LWTS	TM 11-5855-312-10	Upper receiver	Upper receiver	M16 mount
AN/PAS-13B(V3), HWTS	TM 11-5855-312-10	Upper receiver	Upper receiver	M16 mount
AN/PAS-13C(V1), LWTS	TM 11-5855-316-10	Upper receiver	Upper receiver	M16 mount
AN/PAS-13C(V3), HWTS	TM 11-5855-316-10	Upper receiver	Upper receiver	M16 mount
AN/PAS-13D(V1), LWTS	TM 11-5855-317-10	Upper receiver	Upper receiver	M16 mount
AN/PAS-13D(V3), HWTS	TM 11-5855-317-10	Upper receiver	Upper receiver	M16 mount
AN/PVS-4(A)	TM 11-5855-213-10	Upper receiver	Upper receiver	M16 mount
BUIS		Upper receiver	Upper receiver	N/A
ACOG		Upper receiver	Upper receiver	M16 mount

* With a half-moon spacer installed.
** Picatinny or Insight rail grabbers may be used.
*** If used in conjunction with the CCO, the CCO will mount on the top rail of the ARS.

Table 2-7. Characteristics of various accessories.

ACCESSORY	WEIGHT	LENGTH	HEIGHT	RANGE
M68, CCO	6.2 oz	4.9 in	2.5 in	300 m
AN/PVS-14	14.8 oz	4.5 in	2.25 in	
AN/PAC-4C	5.78 oz	5.5 in	1.2 in	*600 m +
AN/PEQ-2A/B	7.5 oz	6.4 in	1.2 in	*600 m +
LWTS	4.1 lb	15.5 in	6.25 in	1,600 m +
HWTS	4.5 lb	18 in	6.25 in	1,600 m +
AN/PVS-4	3.8 lb	9.45 in	4.7 in	*600 m
ACOG	15.1 oz	5.8 in	2.17 in	800 m

*Actual range is dependent upon ambient light, NVGs, and background contrast.

M68 CLOSE COMBAT OPTIC

2-30. The M68 CCO (Figure 2-22) is a reflex (nontelescopic) sight that is designed for the "two eyes open" method of sighting, but can be shot with only one eye open. The red dot aiming point follows the horizontal and vertical movement of the firer's eye, while remaining fixed on the target.

NOTES: 1. Retighten the torque-limiting knob after firing the first three rounds to fully seat the M68.

2. No centering or focusing is required beyond 50 meters.

Figure 2-22. M68 close combat optic.

MOUNTING ON THE M16A4 RIFLE OR M4-SERIES CARBINE

2-31. The M68 mounts directly on the integrated rail on top of M16A4 rifles and M4-series carbines (Figure 2-23). The half-moon spacer (1, Figure 23) must be installed to raise the M68 above the front sightpost. The Soldier's preference dictates the notch that the M68 is mounted on. Although any notch is acceptable, testing has shown that the farther away the M68 is from the Soldier's eyes, the better his field of view.

2-32. To mount the M68 on an M16A4 rifle or M4-series carbine (Figure 2-23)—

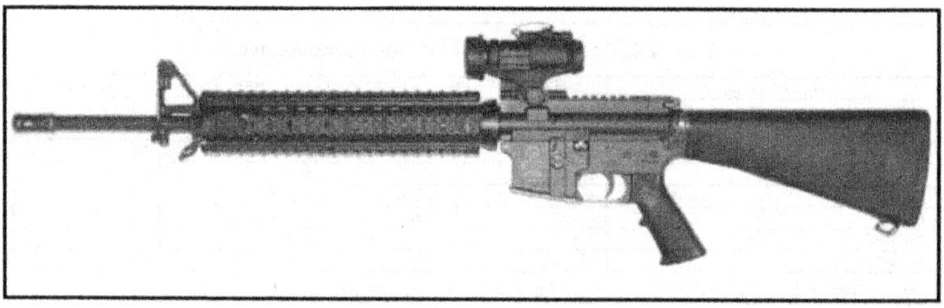

Figure 2-23. Mounting the M68 on an M16A4 rifle or M4-series carbine.

(1) Remove the carrying handle.

(2) Align the locking bar with a notch.

(3) Tighten the torque-limiting knob until it clicks twice.

NOTE: If the M68 is remounted on the same notch, it will retain zero.

MOUNTING ON THE M16A1/A2/A3 RIFLE

2-33. The M68 mounts on the M16A1 mounting bracket (1, Figure 2-24) that attaches to the carrying handle on the M16A1/A2/A3 rifle. The half-moon spacer should not be installed, but will not hinder firing performance.

2-34. To mount the M68 on an M16A1/A2/A3 rifle—

NOTE: Reference the weapon components using the numbers listed in Figure 2-24.

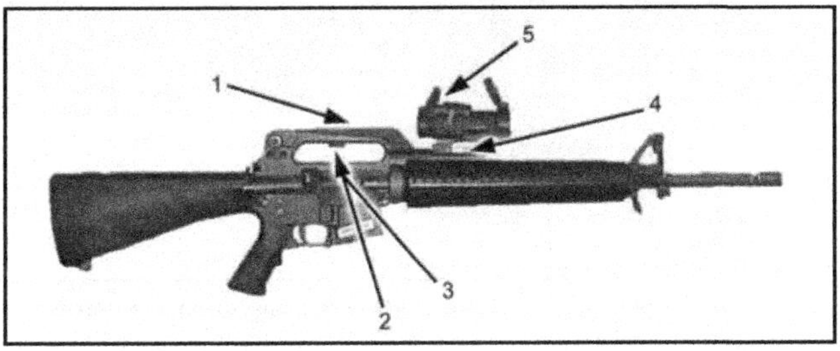

Figure 2-24. Mounting the M68 on an M16A1/A2/A3 rifle.

(1) Firmly hand-tighten the bracket (1), O-ring (2), and machine screw (3).

(2) Align the locking bar (4) under the M68 with the notch in the rail, ensuring that the rotary switch (5) is facing the firer.

(3) Tighten the torque-limiting knob (not shown here) until it clicks twice.

MOUNTING ON THE M16A4 RIFLE OR M4 MWS

2-35. This combination (Figure 2-25) is an effective means of engaging targets during hours of limited visibility. The brightness knob on the M68 should be on the lowest setting (2 or 3) that presents the red dot clearly when viewed through the AN/PVS-14. When employing the AN/PVS-14, Soldiers must consider the following factors:

- The AN/PVS-14 should be mounted where the firer can acquire a good sight picture while performing the integrated act of shooting.
- The M68 can be mounted and zeroed on any slot forward of the AN/PVS-14 as long as the rail grabber fully rests on the ARS and the M68 lens does not rest on the front sightpost.
- The closer the AN/PVS-14 is mounted to the M68, the larger the field of view will be.
- In order to get a clear sight picture with this configuration, fine adjustments must be made to the AN/PVS-14's range focus, gain-control, and diopter.
- The eyecup should be exchanged with the eye guard that is shipped with the AN/PVS-14 to reduce the light signature from the display when not viewing.
- The red dots on settings 2 and 3 project a negligible light signature at night, which can only be seen through an NVD. Settings 4 through 10 project a noticeable signature detectable by opposing forces using NVDs.
- The brighter the dot, the larger the blooming effect becomes in the AN/PVS-14. The blooming effect reduces the Soldier's field of view and prevents him from seeing targets behind the blooming.

2-36. To mount the M68 and AN/PVS-14 on an M16A4 rifle or M4 MWS—

NOTE: Reference the weapon components using the numbers listed in Figure 2-25.

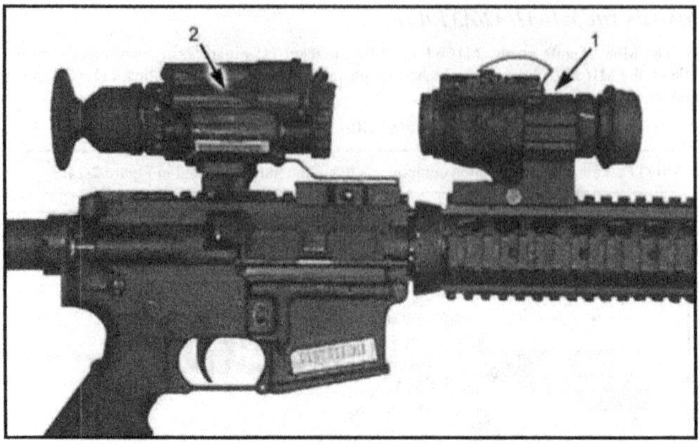

Figure 2-25. Mounting the M68/AN/PVS-14 combination on an M4 MWS.

(1) Remove the carrying handle.
(2) Mount the M68 (1) by tightening the torque-limiting knob.
(3) Mount the AN/PVS-14 (2) where the best field of view is achieved.
(4) Once the preferred location for the M68 is located, zero the M68 to that notch (if different from the notch the M68 was previously zeroed on).

NOTE: For more information about the M68, refer to TM 9-1240-413-13&P. For more information about the AN/PVS-14, refer to TM 11-5855-306-10.

AN/PAQ-4B/C INFRARED AIMING LIGHT

2-37. The AN/PAQ-4B/C IR aiming light (Figure 2-26) projects an IR laser beam that is invisible to the naked eye, but can be seen with NVDs. This aiming light works with the AN/PVS-7-series goggles and the AN/PVS-14. The AN/PAQ-4B/C mounts on various M16-/M4-series weapons with mounting brackets or rail grabbers.

Figure 2-26. AN/PAQ-4B/C infrared aiming light.

MOUNTING ON THE M16A4 RIFLE OR M4 MWS

2-38. The Picatinny (1, Figure 2-27) or Insight (5, Figure 2-27) rail grabber is used to mount the AN/PAQ-4B/C on the ARS.

2-39. To mount the AN/PAQ-4B/C on the ARS—

NOTE: Reference the weapon components using the numbers listed in Figure 2-27.

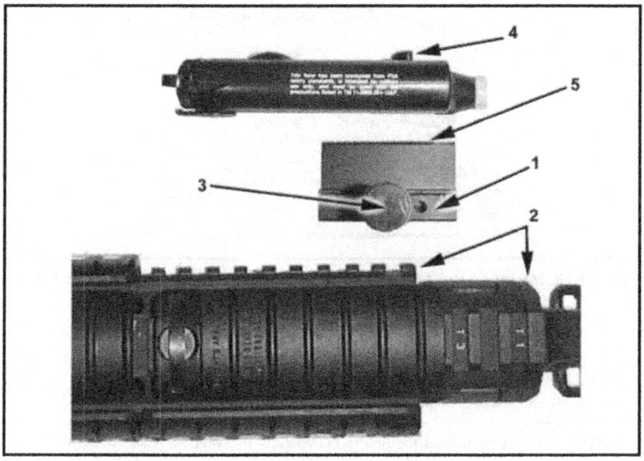

Figure 2-27. Mounting the AN/PAQ-4B/C on the M4 MWS top or left.

(1) Mount the rail grabber all the way forward on the top or either side of the ARS (2), ensuring that it does not extend beyond the end of the ARS.

NOTE: The AN/PAQ-4B/C will not retain zero if the rail grabber extends beyond the end of the integrated rail when mounted.

(2) Tighten the torque-limiting knob (3) until it clicks twice.

(3) Align the thumbscrew (4) on the AN/PAQ-4B/C with the thumbscrew hole in the rail grabber nearest the muzzle.

2-40. The mounting procedures are identical for the M16A4 rifle and M4 MWS. The remote switch should be attached to the weapon where it will not interfere with the functioning of the weapon or hinder the firer's ability to fire the weapon. If the aiming light and rail grabber are removed as a whole unit and mounted onto the same rail, the system will retain zero.

NOTE:
1. If the rail grabber and AN/PAQ-4B/C are separated, the AN/PAQ-4B/C must be rezeroed to the weapon.
2. For further information, refer to TM 11-5855-301-12&P.

Mounting on the M16A1/A2/A3 Rifle or M4 Carbine

2-41. To mount the AN/PAQ-4B/C on an M16A1/A2/A3 rifle or M4 carbine—

NOTE: Reference the weapon components using the numbers listed in Figure 2-28.

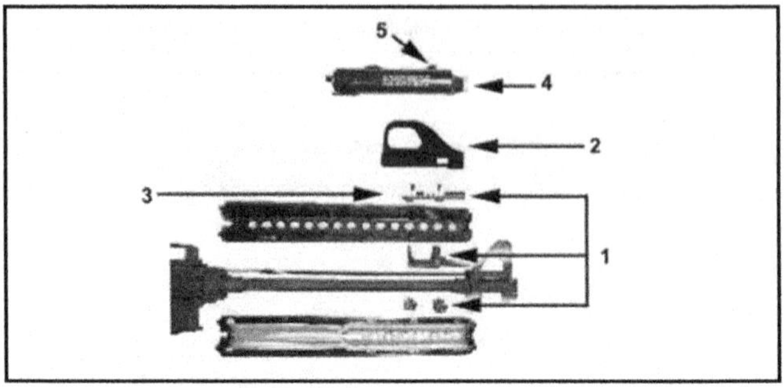

Figure 2-28. Mounting the AN/PAQ-4B/C on the M16A1/A2/A3 rifle and M4 carbine.

(1) Install the bracket assembly (1).
(2) Align the switch lever shroud (2) with the notches on the mounting rail (3).
(3) Lower the ON/OFF switch.
(4) Align the AN/PAQ-4B/C (4) with the notches on the switch lever shroud.
(5) Hand-tighten using the thumbscrew (5).

NOTE: 1. Hand-tighten the plastic thumbscrew to avoid breakage. If the thumbscrew is metal, tool-tightening is recommended to ensure zero retention.

2. To ensure zero retention, retighten the thumbscrew after firing a few rounds.

2-42. The remote switch should be attached to the weapon where it will not interfere with the functioning of the weapon or hinder the firer's ability to fire the weapon.

AN/PEQ-2A/B TARGET POINTER/ILLUMINATOR/AIMING LIGHT

2-43. AN/PEQ-2A and AN/PEQ-2B aiming lights (Figure 2-29) are Class IIIb laser devices that emit a collimated beam of IR light for precise aiming and a separate IR beam for illumination of the target or target area. Both beams can be independently zeroed to the weapon and to each other. The beams can be operated individually or in combination in both high and low power settings.

NOTES: 1. The IR illuminator is equipped with an adjustable bezel to vary the size of the illumination beam based on the size and distance of the target.

2. A safety block is provided for training purposes to limit the operator from selecting high power modes of operation.

2-44. The aiming lights are used with NVDs and can be used as handheld illuminators/ pointers or mounted on the weapon with the included brackets and accessory mounts. In the weapon-mounted mode, the aiming lights can be used to direct fire and to illuminate and designate targets.

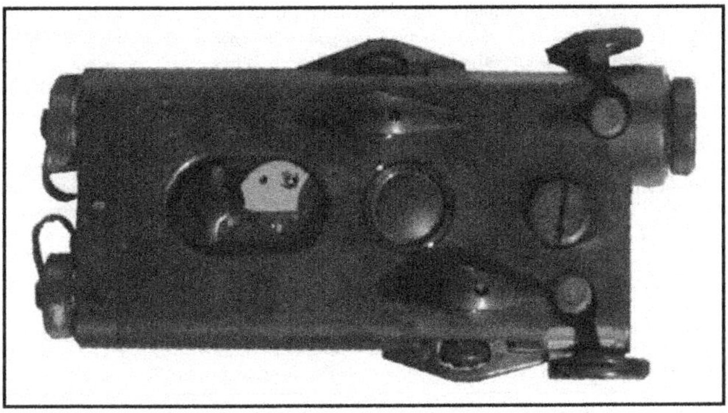

Figure 2-29. AN/PEQ-2A/B target pointer/illuminator/aiming light.

MOUNTING ON THE M16A4 RIFLE OR M4 MWS

2-45. The Picatinny (1, Figure 2-30) or Insight (2, Figure 2-30) rail grabber may be used to mount the AN/PEQ-2A/B on the ARS.

2-46. To mount the AN/PEQ-2A/B on the ARS—

NOTE: Reference the weapon components using the numbers listed in Figure 2-30.

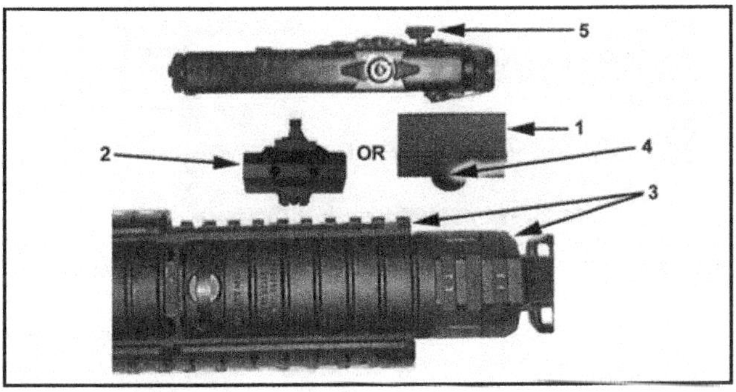

Figure 2-30. Mounting the AN/PEQ-2A/B on the M16A4 rifle and M4 MWS.

(1) Mount the rail grabber all the way forward on the top or either side of the ARS, ensuring that it does not extend beyond the end of the ARS.

NOTE: The AN/PEQ-2A/B will not retain zero if the rail grabber extends beyond the end of the integrated rail when mounted.

(2) Tighten the torque-limiting knob (4) until it clicks twice.

(3) If installing the AN/PEQ-2A/B with the Insight rail grabber, tool-tighten the AN/PEQ-2A/B and rail grabber so that it will not come loose.

(4) Align the thumbscrew (5) on the AN/PEQ-2A/B with the hole that is closest to the front sight assembly located on the top of the rail grabber.

2-47. The mounting procedures are identical for the M16A4 rifle and M4 MWS.

NOTE: If the aiming light and rail grabber are removed as a whole unit and mounted onto the same rail, the system will retain zero. If the rail grabber and AN/PEQ-2A/B are separated, the AN/PEQ-2A/B must be rezeroed to the weapon.

MOUNTING ON THE M16A1/A2/A3 RIFLE OR M4 CARBINE

2-48. To mount the AN/PEQ-2A/B on an M16A1/A2/A3 rifle or M4 carbine—

NOTE: Reference the weapon components using the numbers listed in Figure 2-31.

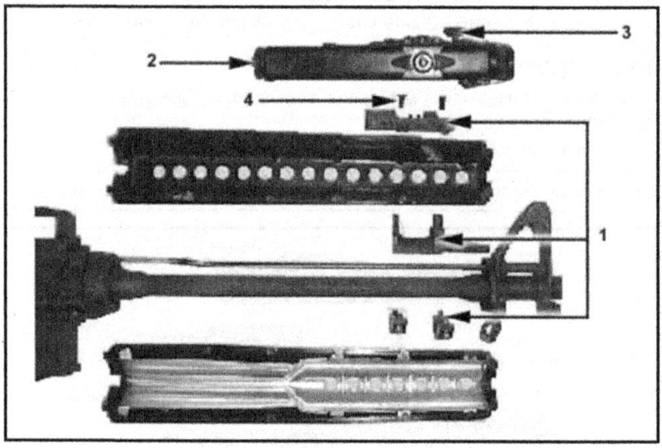

Figure 2-31. Mounting the AN/PEQ-2A/B on M16A1/A2/A3 rifles and M4 carbines.

(1) Install the bracket assembly (1).

(2) Align the AN/PEQ-2A/B (2) thumbscrew (3) with the hole in the mounting rail (4) nearest to the muzzle.

(3) Tool-tighten.

2-49. The remote switch should be attached to the weapon where it will not interfere with the functioning of the weapon or hinder the firer's ability to fire the weapon.

NOTE: To ensure zero retention, retighten the rail grabber and thumbscrew after firing a few rounds.

USE IN CONJUNCTION WITH MILES

2-50. When conducting MILES training with the Insight rail grabber (1, Figure 2-32) or bracket assembly (4, Figure 2-32), the AN/PEQ-2A/B is attached to the M16-/M4-series weapon using the training extender bracket (2, Figure 2-32). The training extender is not required when—

- Mounting the AN/PEQ-2A/B onto the side of the MWS.
- Using the Picatinny rail grabber.

2-51. To attach the training extender, hand-tighten it by turning the thumb wheel (3, Figure 2-32) on the training extender clockwise.

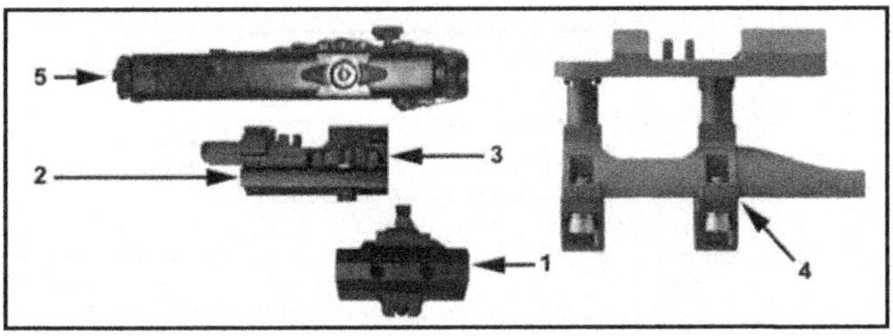

Figure 2-32. MILES training extender bracket installation on M16-/M4-series weapons.

NOTES: 1. The AN/PEQ-2A/B and AN/PAQ-4B/C must be zeroed before and after using the MILES training extender.

2. For further information, refer to TM 11-5855-308-12&P.

AN/PEQ-15 ADVANCED TARGET POINTER/ILLUMINATOR AIMING LIGHT

2-52. The AN/PEQ-15 advanced target pointer/illuminator aiming light (ATPIAL, Figure 2-33) is a multifunctional laser that emits visible or IR light for precise weapon aiming and target/area illumination. This ruggedized system can be used as a handheld illuminator/pointer or can be mounted to weapons equipped with a MIL-STD-1913 rail. Unlike the AN/PAQ-4 and AN/PEQ-2A/2B, the AN/PEQ-15 has an integrated rail grabber molded into the body.

NOTE: The AN/PEQ-15 can be used during force-on-force training in the low power modes only. High power modes can be used on live-fire ranges exceeding 220 meters only. Refer to TM 9-5855-1914-13&P for more information.

2-53. The AN/PEQ-15 ATPIAL's visible aiming laser provides for active target acquisition in low light conditions and close quarters combat situations, and allows users to zero using the borelight without using NVDs. When used in conjunction with NVDs, its IR aiming and illumination lasers provide for active, covert target acquisition in low light or complete darkness.

NOTES: 1. The ATPIAL visible and IR aiming lasers are aligned. A single set of adjusters moves both aiming beams, and the user can boresight/zero using either aiming laser.

2. In addition to momentary and continuous modes of operation, the IR illuminator may be programmed to operate at 1, 2, 4, or 8 pulses per second to eliminate confusion on the battlefield.

Figure 2-33. AN/PEQ-15 advanced target pointer illuminator aiming light.

Mounting on the M16A4 Rifle or M4 MWS

2-54. To mount the AN/PEQ-15 on an M16A4 rifle or M4 MWS (Figure 2-34)—

(1) Loosen the clamping knob on the integral rail grabber bracket until the jaws sufficiently fit over the MIL-STD-1913 rail.

(2) Position the integral rail grabber bracket on the rail, ensuring that the recoil lug is seated in the desired recoil groove of the rail.

NOTE: The ATPIAL may be positioned anywhere on the rail where it is most convenient for the operator; however, ensure that it does not extend beyond the end of the ARS or touch the collar of the barrel. The AN/PEQ-15 will not retain zero if the rail grabber extends beyond the end of the integrated rail when mounted.

---

(3) Turn the clamping knob clockwise to tighten.

(4) Tool-tighten the AN/PEQ-15 onto the MIL-STD-1913 rail to ensure that it does not come loose.

NOTE: Failure to properly secure the AN/PEQ-15 to the rail may lead to zero retention issues.

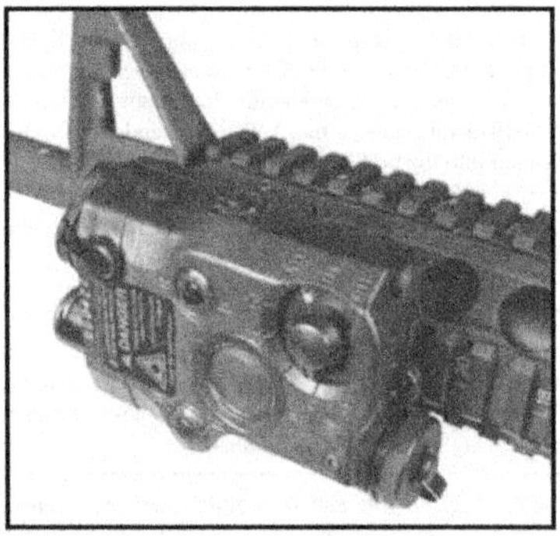

Figure 2-34. AN/PEQ-15 mounted on M4 carbine.

NOTE: If the ATPIAL is removed from the rail, the operator must note the notch that it was zeroed on, and return it to the same position to retain zero.

AN/PAS-13B/C/D (V1) LIGHT WEAPON THERMAL SIGHT AND AN/PAS-13B/C/D (V3) HEAVY WEAPON THERMAL SIGHT

2-55. The AN/PAS-13B/C/D (V1) light weapon thermal sight (LWTS) and the AN/PAS-13B/C/D (V3) heavy weapon thermal sight (HWTS) (Figure 2-35) are silent, lightweight, compact, and durable battery-powered IR imaging sensors that operate with low battery consumption. The AN/PAS-13B/C/D (V1) is used on riflemen's M4s and M16A4s. The AN/PAS-13B/C/D (V3) is used on squad leaders' M4s and M16A4s only.

NOTE: Both the LWTS and the HWTS are referred to henceforth as a singular TWS. For more information, refer to TM 11-5855-312-10, TM 11-5855-316-10, and TM 11-5855-317-10.

2-56. The TWS is capable of target acquisition under conditions of limited visibility, such as darkness, smoke, fog, dust, and haze. The TWS operates effectively both at night and during the day. The TWS is composed of two functional groups: the telescope and the basic sensor.

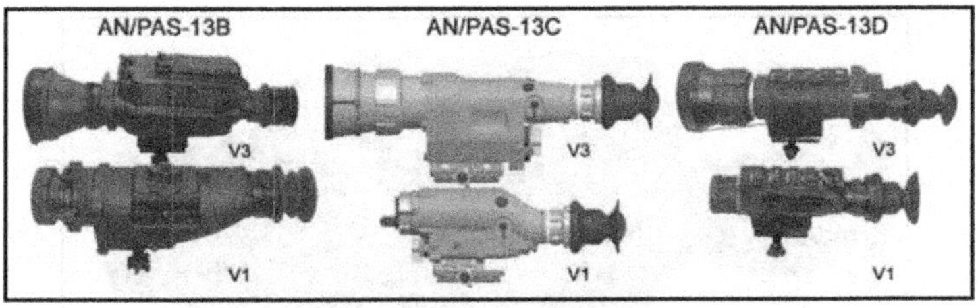

Figure 2-35. AN/PAS-13B/C/D (V1) light weapon thermal sight and AN/PAS-13B/C/D (V3) heavy weapon thermal sight.

Mounting the TWS on an M16A4 Rifle or M4 Carbine

2-57. To mount the TWS on an M16A4 rifle or M4 carbine—

NOTE: Reference the weapon components using the numbers listed in Figure 2-36.

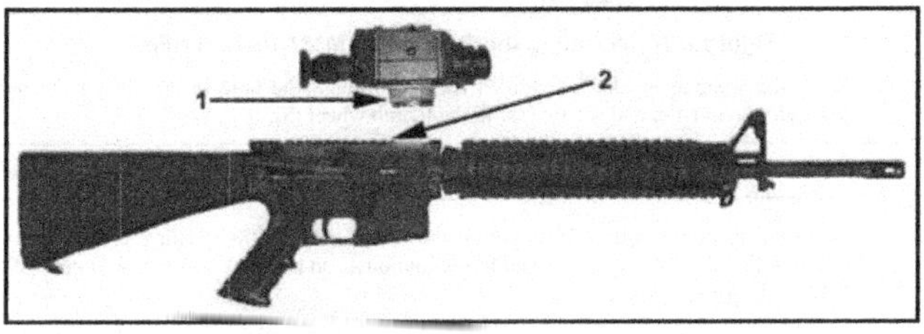

Figure 2-36. Mounting the TWS on M16A4 rifle or M4 carbine.

(1) Align the Picatinny rail grabber with the spacer on the bottom of the TWS with a notch on the integrated rail of the M16A4 rifle or M4 carbine, ensuring that the TWS is positioned to accommodate an effective firing position once the eyecup is depressed.

NOTE: The TWS will not retain zero if the rail grabber extends beyond the end of the integrated rail when mounted.

(2) Tighten the torque-limiting knob clockwise until it clicks twice.

NOTE: Retighten the rail grabber after firing a few rounds to ensure that the sight is fully seated.

2-58. The mounting procedures are identical for the M16A4 rifle and M4 MWS.

MOUNTING THE TWS ON AN M16A1/A2/A3 RIFLE

2-59. To mount the TWS on an M16A1/A2/A3 rifle—

NOTE: Reference the weapon components using the numbers listed in Figure 2-37.

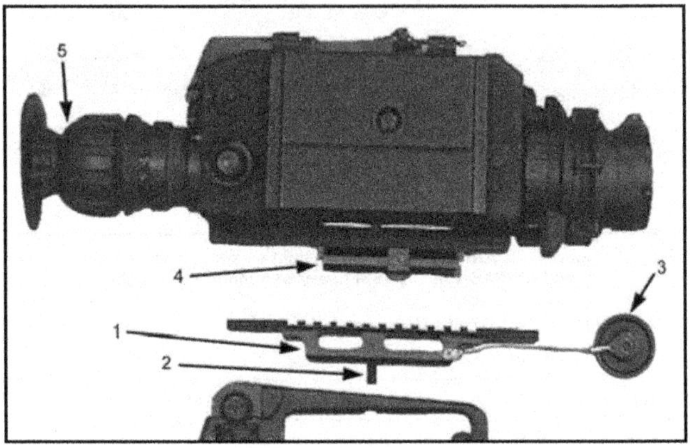

Figure 2-37. Mounting the TWS on an M16A1/A2/A3 rifle.

(1) Insert the weapon bracket's threaded rod (2) through the hole in the carrying handle of the M16A1/A2/A3 rifle, and secure it with the thumb wheel (3).

NOTE: The M16A1/A2/A3 weapon bracket (1) is a standard item in the TWS carrying case.

(2) Align the Picatinny rail grabber (4) on the bottom of the TWS with a notch on the bracket, ensuring that the TWS is positioned to accommodate an effective firing position once the eyecup (5) is depressed.

NOTES:
1. Ensure that the rail grabber fully rests on the bracket when mounting the TWS or the sight will not retain zero.
2. It is not necessary to use a spacer when mounting the HWTS on the M16A1/A2/A3.

AN/PVS-4 NIGHT VISION SIGHT

2-60. The AN/PVS-4 (Figure 2-38) is a passive, battery-operated night sight used for observation and accurate firing of individual-served weapons during nighttime conditions (for example, moonlight, starlight, skyglow) against targets as far away as 400 meters.

NOTE: The sight is less effective when viewing into rain, fog, sleet, snow, smoke, shadows, and other obscurants.

2-61. The sight is portable and mounts to the following weapons:

- M16A1/A2 and M14 rifles.
- M203 and M79 grenade launchers.
- M249 squad automatic weapon (SAW).
- M60 machine gun.
- M67 recoilless rifle.
- M72A1 rocket launcher.

Figure 2-38. AN/PVS-4 night vision sight.

Mounting on the M16A4 Rifle or M4 Carbine

2-62. To mount the AN/PVS-4 night vision sight on an M16A4 rifle or M4 carbine—

NOTE: Reference the weapon components using the numbers listed in Figure 2-39.

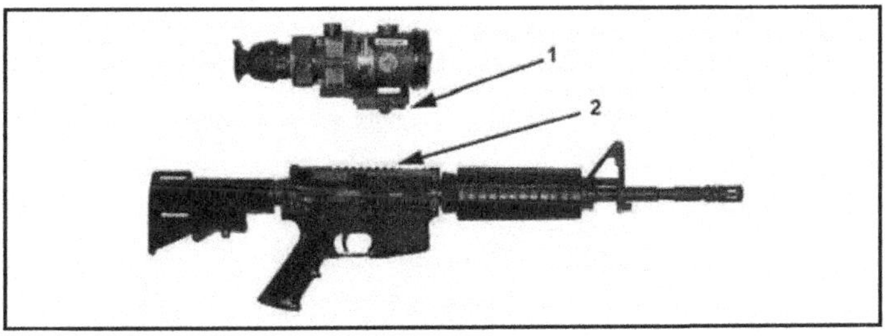

Figure 2-39. Mounting the AN/PVS-4 on an M4 carbine or M4 MWS.

(1) Align the Picatinny rail grabber with a mounting adapter (1) on the bottom of the AN/PVS-4 with a notch on the integrated rail (2) of an M16A4 rifle, M4 carbine, or M4 MWS, ensuring that the AN/PVS-4 is positioned to accommodate an effective firing position once the eyecup is depressed.

NOTE: The AN/PVS-4 will not retain zero if the rail grabber extends beyond the end of the integrated rail when mounted.

(2) Tighten the torque-limiting knob clockwise until it clicks twice. .

NOTE: Retighten the rail grabber after firing a few rounds to ensure that the sight is fully seated.

2-63. The mounting procedures are identical for the M4 carbine and M4 MWS.

MOUNTING ON THE M16A1/A2/A3 RIFLE

2-64. The AN/PVS-4 is mounted on the carrying handle of the M16A1/A2/A3 rifle (Figure 2-40). To mount the AN/PVS-4 night vision sight on an M16A1/A2/A3 rifle—

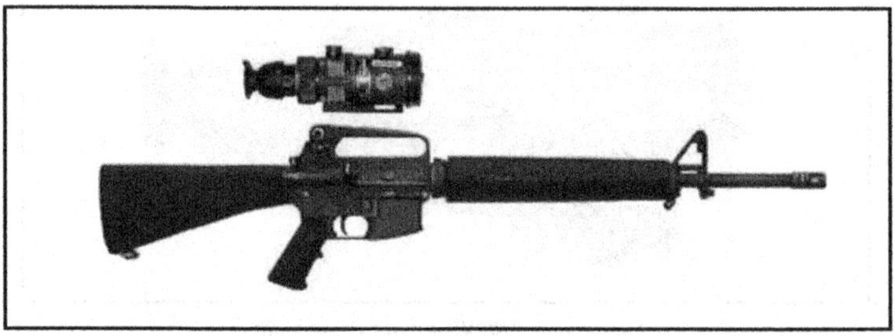

Figure 2-40. Mounting the AN/PVS-4 on an M16A1/A2/A3 rifle.

(1) Position the sight in the groove on the top of the carrying handle.

(2) Align the threaded hole in the base of the sight-mounting adapter over the hole in the handle.

(3) Insert the mounting knob assembly through the hole in the carrying handle, and screw it firmly clockwise into the sight-mounting adapter.

2-65. If difficulty is encountered—

(1) Turn the sight and the rifle upside down.

(2) Place the rifle handle onto the sight-mounting adapter, lining up the hole in the carrying handle with the hole in the sight-mounting adapter.

(3) Place the mounting knob assembly through the hole in the carrying handle, and screw it clockwise.

BACKUP IRON SIGHT

2-66. The BUIS (Figure 2-41) is a semi-permanent flip-up iron sight equipped with a rail-grabbing base. It is intended to remain on the M4 MWS while the M68 CCO is used as the primary means of day fire control. If the M68 fails, the prezeroed BUIS can be flipped up and used to continue the mission. The BUIS provides a backup capability effective when firing 600 meters away and farther and can be installed on the M16A4 rifle and M4 carbine. It provides a sighting capability when all other accessories have been removed, and it can be used to establish approximate zeros for other sighting components without requiring live-fire. Zeros established using this method are only effective to approximately 20 meters and should be refined by a live-fire zero.

2-67. The armorer installs the BUIS on the first notch of the integrated rail, nearest to the charging handle. Before the armorer installs the BUIS, he removes all rail covers/heat shields from the top, except one 4-, 5-, or 6-rib shield. The remaining rail cover/heat shield can be positioned to accommodate accessories and protect the nonfiring hand when the barrel is hot. Once installed and zeroed, the BUIS should be left in the stowed position (collapsing toward the firer, as shown in Figure 2-42) for best durability and minimal interference unless its use is imminent. The BUIS should only be removed by the armorer and remains on the MWS unless the carrying handle/sight is installed.

NOTE: Failure to install the BUIS on the first notch will lead to an improper zero and inaccuracy at longer ranges when the range lever is used.

Figure 2-41. Backup iron sight.

Figure 2-42. Backup iron sight in the stowed position

ADVANCED COMBAT OPTICAL GUNSIGHT

2-68. The ACOG (Figure 2-43) is designed to provide enhanced target identification and hit probability for the M4A1 or M16-series weapon when firing as far as 800 meters (approximately 870 yards) away. It is designed with dual illuminated technology, using fiber optics for daytime employment and tritium for nighttime and low-light use. The ACOG is a lightweight, rugged, fast, and accurate 3.5 power optic scope. The body is machined from aluminum forgings; both the material and finish are identical to the M4A1. It is internally adjustable to allow the shock of rough handling to be carried by the scope body instead of the adjustment mechanism.

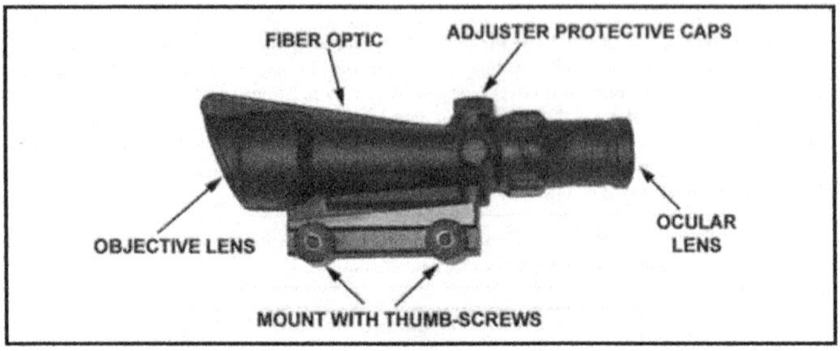

Figure 2-43. Advanced combat optical gunsight.

MOUNTING ON THE M16A4 RIFLE OR M4 CARBINE

2-69. The ACOG can be attached to the flattop easily using the adapter that comes from the factory (Figure 2-44). To attach the ACOG to the flattop with the adapter—

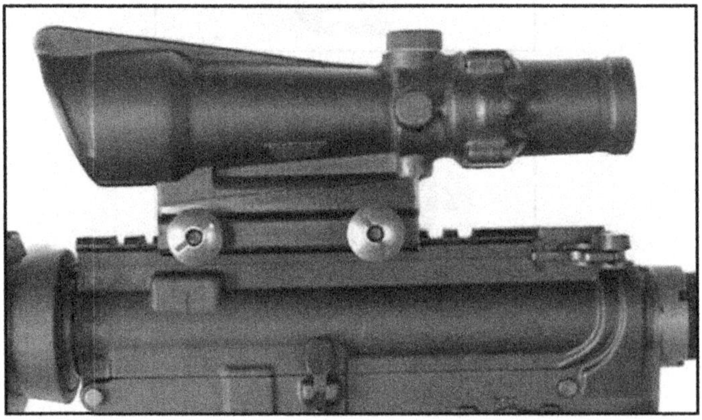

Figure 2-44. Mounting the ACOG on M16A4 rifle and M4 carbine.

(1) Loosen the thumb screws.
(2) Pull the interface clamp bar back against the knobs.
(3) Place the ACOG onto the flattop upper receiver rail surface. Be sure to engage the interface studs on the bottom of the adapter with the grooves on the top mounting surface of the upper receiver.

NOTE: 1. The ACOG can be placed in any of the slots on top of the weapon to allow for eye relief adjustments.

2. To retain zero, place the ACOG into the same slots during each installation.

(4) Using a finger, tighten the knobs firmly.
(5) Add another ¼ of a turn using a coin or a blade screwdriver.

NOTE: The slots on the knobs are only for removing the ACOG from the weapon if the knobs have been attached too tightly for hand-removal. If the slots are used to tighten the knobs, the ACOG will be difficult to remove.

MOUNTING ON THE M16A1/A2/A3 RIFLE

2-70. Although the ACOG is designed for use on the flattop rail of M16A4 rifles and M4 carbines, it can be attached to the carrying handle (Figure 2-45). To attach the ACOG to the carrying handle—

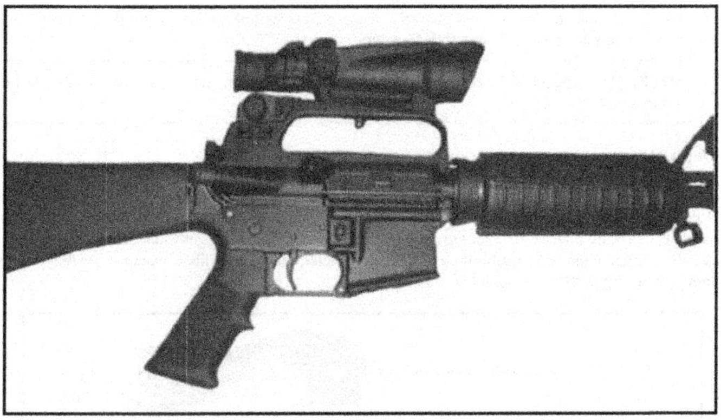

Figure 2-45. Mounting the ACOG on an M16A1/A2/A3 rifle.

(1) Align the forward-most threaded mounting hole on the bottom of the ACOG with the existing hole in the M4/M16 carrying handle.
(2) Press down firmly on the optic until it is seated inside the carrying handle rail.
(3) Insert the locking screw and washer set from the bottom of the carrying handle. Ensure that the locking screw and washer are inserted in the proper order (screw, locking nut, washer), or the screw will eventually shake itself loose. Figure 2-46 shows the proper assembly of the screw set.

Figure 2-46. ACOG locking screw.

(4) Tighten the locking screw with the factory-supplied ⅛" Allen wrench, and then tighten another ¼ turn.

NOTE: When the ACOG is attached to the carrying handle, the bullet drop compensated (BDC) reticle will be off slightly at extended distances. For example, at 500 meters, the impact will be 5 inches high; at 600 meters, the impact will be 6 inches high. To compensate for this, aim slightly low once the target is ranged beyond 400 meters.

SECTION IV. 10-METER BORESIGHT AND 25-METER ZERO OFFSET

Boresighting is conducted at 10 meters with the borelight, weapon, aiming device, and a 10-meter offset. Each aiming device and weapon combination has a unique 10-meter offset. The 25-meter zero offset target is used when live-firing at 25 meters. When used properly, these offsets will align the aiming device on the selected weapon to engage a target's center of mass at 300 meters.

NOTE: See Appendix F for more information about 10-meter offsets for each weapon combination.

BORELIGHT

2-71. The borelight (Figure 2-47) is an eye-safe laser that is used to zero aiming lasers (such as the AN/PAQ-4 or AN/PEQ-2) without a 25-meter confirmation. The borelight will also boresight optics and iron sights to ensure that the first shot group hits the 25-meter zeroing target when zeroing the weapon. The borelight comes with a 5.56-millimeter, 7.62-millimeter, and .50-caliber mandrel. M203 and MK19 mandrels can be purchased separately.

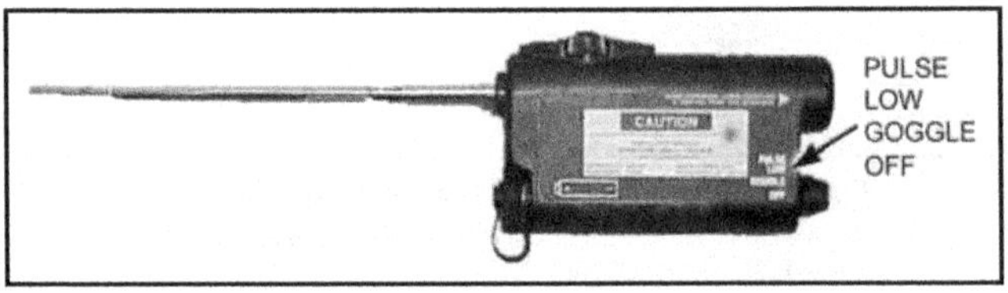

Figure 2-47. Borelight with a 5.56-millimeter mandrel.

2-72. The borelight has four settings:

- OFF.
- GOGGLE.
- LOW.
- PULSE.

OFF SETTING

2-73. This setting is used when the borelight is not in use.

GOGGLE SETTING

2-74. This setting is used when the boresight is being used with NVGs in a tactical environment.

LOW SETTING

2-75. This setting is used during normal operations.

PULSE SETTING

2-76. This setting is used during dry-fire training.

10-METER BORESIGHT

2-77. The 10-meter boresighting target is used in conjunction with the borelight. The 10-meter boresighting target is a 1-centimeter grid system with a crosshair and a circle (Figure 2-48A). The crosshair is the point of aim for the aiming device, and the circle is the point of impact for the borelight.

NOTE: Refer to Chapter 8 for a detailed explanation of boresighting procedures.

25-METER ZERO OFFSET

2-78. The M16A2 25-meter zeroing target is used when live-firing all optics and IR aiming lights at 25 meters, with the appropriate strike zone marked on the target (Figure 2-48B). The M4 zeroing target is only used when zeroing the iron sights on the M4. The M4 25-meter zeroing target will not allow the placement of the correct offset due to the size of the squares, which are 13 millimeters high and wide. The point of aim is always the 300-meter scaled silhouette's center of mass. The designated strike zone is a 4x4 square designating where the rounds should impact when the firer aims at the center of mass.

NOTE: Refer to Chapter 8 for a detailed explanation of the 25-meter offset zeroing procedures.

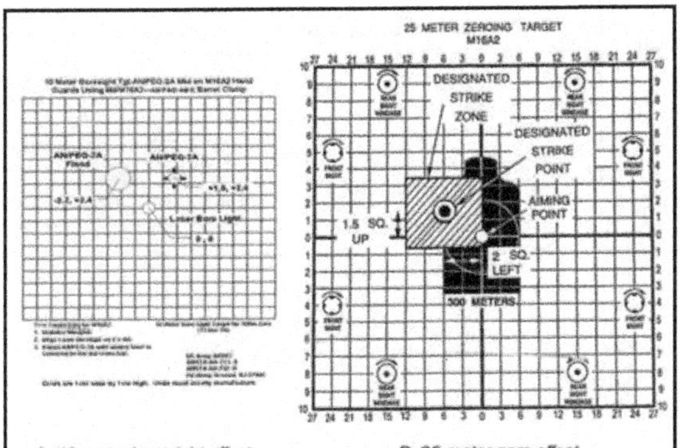

A. 10-meter boresight offset B. 25-meter zero offset

Figure 2-48. 10-meter boresighting target and 25-meter zero offset.

SECTION V. AMMUNITION

This section provides information about different types of standard military ammunition used in M16- and M4-series weapons.

AUTHORIZED AMMUNITION

2-79. Use only authorized ammunition manufactured to U.S. and NATO specifications (Table 2-8).

Table 2-8. Authorized ammunition.

CARTRIDGE/ROUND	IDENTIFICATION	USE	ADDITIONAL INFORMATION
M193 cartridge – 5.56-mm, ball 	Plain tip	The M193 is the standard cartridge for field use with the M16A1 rifle.	The M193 cartridge is a center-fire cartridge with a 55-grain, gilded metal-jacketed, lead alloy core bullet.
M196 cartridge – 5.56-mm, tracer 	Red or orange tip	The M196 cartridge is used only in the M16A1 rifle. Its main uses are for observation of fire, incendiary effect, and signaling.	Soldiers should avoid long-term use of 100 percent tracer rounds, which could cause deposits of incendiary material or chemical compounds that could damage the barrel. When tracer rounds are fired, they are mixed with ball ammunition in a ratio of no greater than one-to-one with a preferred ratio of three or four ball rounds to one tracer round.
M199 cartridge – 5.56-mm, dummy 	Six grooves along the sides of the case beginning about 1/2 inch from its tip	The M199 dummy cartridge is used in all M16-/M4-series weapons during dry-firing and other training.	This cartridge contains no propellant or primer. The primer well is open to prevent damage to the firing pin.
M200 cartridge – 5.56-mm, blank (no projectile) 	Case mouth is closed with a seven-petal rosette crimp, violet tip	The M200 blank cartridge is used in all M16-/M4-series weapons.	N/A
M855 cartridge – 5.56-mm, ball 	Green tip	The M855 cartridge is used in the M16A2/3/4 and in M4-series weapons.	The M855 cartridge has a 62-grain, gilded metal-jacketed, lead alloy core bullet with a steel penetrator. The primer and case are waterproof. This round is also linked and used in the M249. NOTE: This ammunition should not be used in the M16A1 except under emergency conditions, and only at targets less than 90 meters away. The twist of the M16A1 rifling is not sufficient to stabilize the length of the round's projectile.

Table 2-8. Authorized ammunition (continued).

CARTRIDGE/ROUND	IDENTIFICATION	USE	ADDITIONAL INFORMATION
M856 cartridge – 5.56-mm, tracer 	Red tip (orange when linked 4 to 1 for the M249)	The M856 tracer cartridge is used in the M16A2/3/4 and M4-series weapons.	The M856 tracer cartridge has characteristics similar to the M196 tracer, with a slightly longer tracer burnout distance. This cartridge has a 63.7-grain bullet. The M856 does not have a steel penetrator. NOTE: This ammunition should not be used in the M16A1 except under emergency conditions, and only at targets less than 90 meters away. The twist of the M16A1 rifling is not sufficient to stabilize the length of the round's projectile.
M862 cartridge – 5.56-mm, short-range training ammunition (SRTA) 	N/A	The M862 SRTA is used in all rifles and is designed exclusively for training.	The M862 SRTA can be used in lieu of service ammunition on indoor ranges and by units who have a limited range fan that does not allow the firing of service ammunition. If adequate range facilities are not available for sustainment training, SRTA can be used for any firing exercise of 25 meters or less. This includes the 25-meter scaled silhouette, 25-meter alternate qualification course, and quick-fire training. SRTA can also be used for urban operations (UO) training. NOTES: 1. See Appendix A for use of SRTA in training. 2. Although SRTA closely replicates the trajectory and characteristics of service ammunition out to 25 meters, the settings placed on the sights for SRTA could be different for service ammunition. SRTA should not be used to battlesight zero weapons that will fire service ammunition. 3. SRTA ammunition must be used with the M2 training bolt.
M995 cartridge – 5.56-mm, armor-piercing (AP) 	Conventional brass cartridge case Aluminum cup sits at the rear of the projectile (for the purpose of properly locating the penetrator within the projectile)	The M995 cartridge is used by the M249 (SAW), M16/A2/A3/A4, and M4-series weapons. It is intended for use against light armored targets.	The M995 offers the capability to defeat light armored targets at ranges two to three times that of currently available 5.56-mm ammunition. The M995 cartridge consists of a projectile and a propelling charge contained in a brass cartridge case. The projectile is a dense metal penetrator (tungsten carbide) enclosed by a standard gilded metal jacket. The cartridge utilizes a double base propellant. A standard rifle cartridge primer is used in the case to initiate the propelling charge.

TRAJECTORY

2-80. Figures 2-49 through 2-53 show trajectory data for M855 cartridges when fired from M16A2 rifles and M4 carbines.

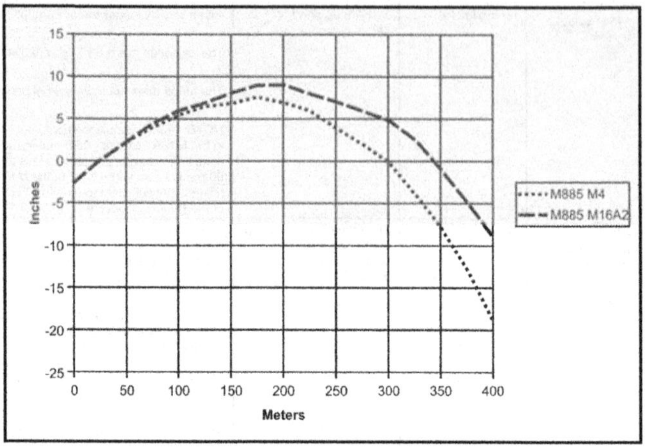

Figure 2-49. M855 drop during 25-meter zeroing (M16A2 at 8/3+1, M4 at 6/3).

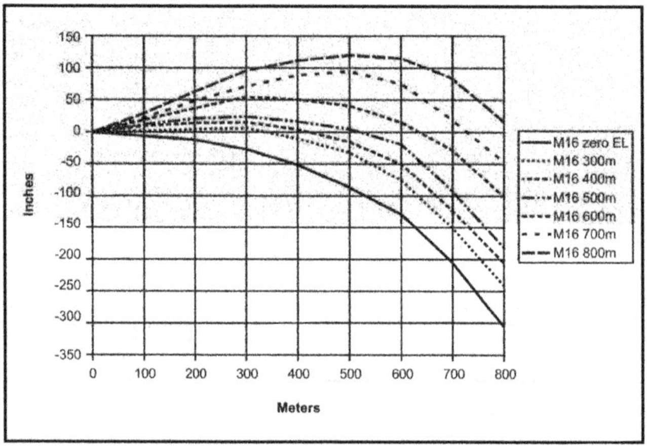

Figure 2-50. Bullet drop of M855 ammunition with M16A2 (8/3).

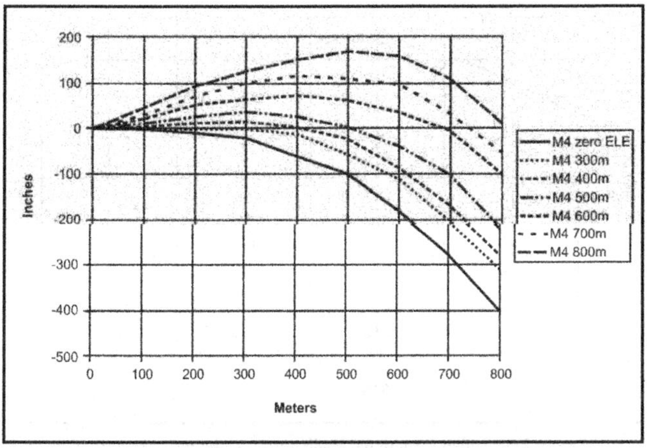

Figure 2-51. Bullet drop of M855 ammunition with M4 (6/3).

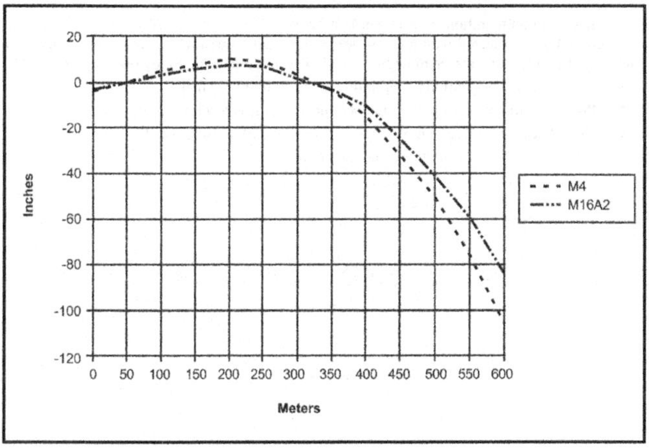

Figure 2-52. M4 carbine and M16A2 rifle bullet trajectory comparison.

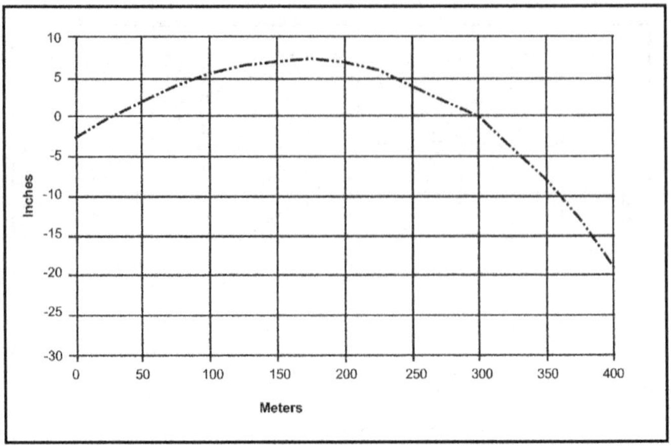

Figure 2-53. Bullet drop of M4/M855 during 25-meter zeroing on 6/3.

STORAGE

2-81. When storing ammunition in the open is necessary, it must be raised on dunnage at least 6 inches from the ground and protected with a cover, leaving enough space for air circulation. Since moisture and high temperatures adversely affect ammunition and explosives, take the following precautions:

- Do not open ammunition boxes until you are ready to use them.
- Protect ammunition from high temperatures and the direct rays of the sun.
- Do not attempt to disassemble ammunition or any of its components.
- Never use lubricants or grease on ammunition.

Chapter 3

Troubleshooting and Destruction

Commanders and unit armorers are responsible for the field level maintenance of weapons and for the destruction of weapons, when necessary. Soldiers are responsible for keeping their weapons clean and operational at all times—in training and in combat—therefore, they should be issued an operator's TM and cleaning equipment for their assigned weapons.

STOPPAGES

3-1. A stoppage is a failure of an automatic or semiautomatic firearm to complete the cycle of operation. The firer can apply immediate or remedial action to clear the stoppage. Some stoppages cannot be cleared by immediate or remedial action and may require weapon repair to correct the problem. A complete understanding of how the weapon functions is an integral part of applying immediate action procedures.

IMMEDIATE ACTION

3-2. Immediate action involves quickly applying a possible correction to reduce a stoppage without performing troubleshooting procedures to determine the actual cause. Apply immediate action only once for a stoppage. If the rifle fails to fire a second time for the same malfunction, inspect the weapon to determine the cause of the stoppage or malfunction, and take the appropriate remedial action. The key word **SPORTS** will help the firer remember the steps for immediate action:

(1) **S**lap gently upward on the magazine to ensure that it is fully seated and that the magazine follower is not jammed.

NOTE: When slapping up on the magazine, be careful not to knock a round out of the magazine into the line of the bolt carrier, causing more problems. Slap only hard enough to ensure that the magazine is fully seated. Ensure that the magazine is locked into place by quickly pulling down on the magazine.

---

(2) **P**ull the charging handle fully to the rear.

(3) **O**bserve the ejection of a live round or expended cartridge.

NOTE: If the weapon fails to eject a cartridge, perform remedial action.

---

(4) **R**elease the charging handle; do not ride it forward.

(5) **T**ap the forward assist assembly to ensure that the bolt is closed.

(6) **S**queeze the trigger and try to fire the rifle.

REMEDIAL ACTION

3-3. Remedial action is the continuing effort to determine the cause of a stoppage or malfunction and attempt to clear the stoppage once it has been identified. To apply the corrective steps for remedial action—

(1) Try to place the weapon on SAFE.

NOTE: A bolt override may not allow the weapon to be placed on SAFE.

---

(2) Remove the magazine.

(3) Lock the bolt to the rear.

(4) Place the weapon on SAFE (if not already done).

MALFUNCTIONS

3-4. Malfunctions are caused by procedural or mechanical failures of the rifle, magazine, or ammunition. Prefiring checks and serviceability inspections identify potential problems before they become malfunctions. This section describes the primary categories of malfunctions.

NOTE: In training, Soldiers must alert other Soldiers and range personnel when experiencing weapon malfunctions.

Failure to Feed, Chamber, or Lock

3-5. This malfunction can occur when loading the rifle or during the cycle of operation. Once the magazine has been loaded into the rifle, the forward movement of the bolt carrier group could lack enough force (generated by the expansion of the action spring) to feed, chamber, or lock the bolt (Figure 3-1).

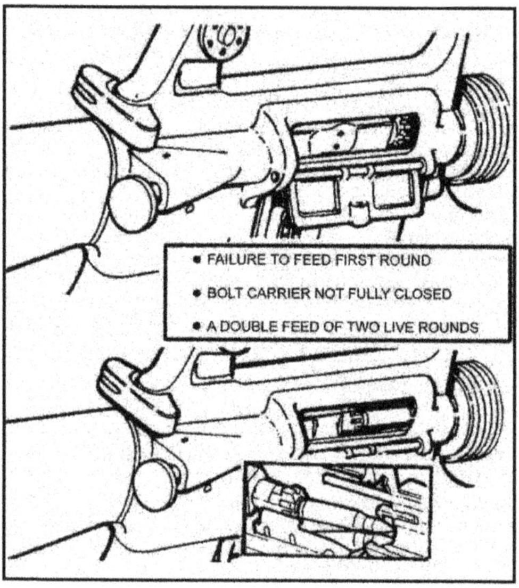

Figure 3-1. Failure to feed, chamber, or lock.

Probable Causes

3-6. The malfunction could be the result of one or more of the following:

- Excess accumulation of dirt or fouling in and around the bolt and bolt carrier.
- Defective magazine (dented, bulged, or a weak magazine spring).
- Improperly loaded magazine.
- Defective round (projectile forced back into the cartridge case, which could result in a stubbed round, or the base of the previous cartridge could be separated, leaving the remainder in the chamber).
- Damaged or broken action spring.
- Exterior accumulation of dirt in the lower receiver extension.
- Fouled gas tube (resulting in short recoil).
- A magazine resting on the ground or pushed forward (causing an improper lock).

Corrective Action

3-7. Applying immediate action usually corrects the malfunction. To avoid the risk of further jamming, watch for ejection of a cartridge and ensure that the upper receiver is free of loose rounds.

3-8. If immediate action fails to clear the malfunction, take remedial action.

NOTE: Do not force the bolt carrier.

3-9. If resistance is encountered (which can occur with an unserviceable round)—

(1) Lock the bolt to the rear.

(2) Remove the magazine.

(3) Clear the malfunction.

3-10. For example, to correct a bolt override (a cartridge has wedged itself between the bolt and charging handle)—

(1) Ensure that the charging handle is pushed forward and locked into place.

(2) Attempt to place the weapon on SAFE.

(3) Secure the rifle, and pull the bolt to the rear until the bolt seats completely into the buffer well.

(4) Turn the rifle upright and allow the overridden cartridge to fall out.

Failure to Fire Cartridge

3-11. Despite the fact that a round has been chambered, the trigger has been pulled, and the sear has released the hammer, a cartridge may fail to fire. This occurs when the firing pin fails to strike the primer with enough force or when the ammunition is defective. The firer must follow unit safety guidelines until the determining factors of the misfire have been identified and corrected.

Probable Causes

3-12. Excessive carbon buildup on the firing pin (Figure 3-2, A) is often the cause, because the full forward travel of the firing pin is restricted. A defective or worn firing pin can give the same results. Inspection of the ammunition could reveal a shallow indentation or no mark on the primer, indicating a firing pin malfunction (Figure 3-2, B). Cartridges that show a normal indentation on the primer (but did not fire) indicate faulty ammunition or failure of the cartridge to fully seat in the chamber.

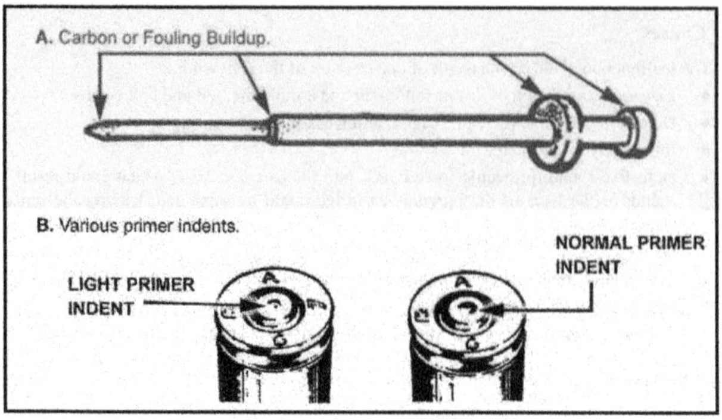

Figure 3-2. Failure to fire.

Corrective Action

3-13. If the malfunction continues—

(1) Inspect the firing pin, bolt, bolt carrier, and locking lug recesses of the barrel extension.

(2) Remove any accumulation of excessive carbon or fouling.

(3) Inspect the firing pin for damage.

3-14. If the round is suspected to be faulty, dispose of it IAW the unit SOP, and ensure that it is reported and returned to the agency responsible for issuing ammunition.

WARNING

If an audible pop or reduced recoil occurs during firing, immediately cease fire. This could be the result of a round being fired without enough force to send the projectile out of the barrel. Do not apply immediate action. Instead, perform the following actions:

1. **Lock the bolt to the rear.**
2. **Place the selector lever in the SAFE position.**
3. **Remove the magazine.**
4. **Visually inspect the bore to ensure that a projectile is not lodged in the barrel.**
5. **If a projectile is lodged in the barrel, do not try to remove it. Turn the rifle in to the unit armorer.**

FAILURE TO EXTRACT

3-15. A failure to extract results when the cartridge case remains in the chamber of the rifle. The bolt and bolt carrier might move rearward only a short distance, but more commonly, the bolt and bolt carrier recoil fully to the rear, leaving the cartridge case in the chamber. A live round is then forced into the base of the cartridge case as the bolt returns in the next feed cycle.

WARNING

A failure to extract is an extremely serious malfunction, requiring the use of tools to clear. A live round could be left in the chamber and accidentally discharged. If a second live round is fed into the primer of the chambered live round, the rifle could explode and cause personal injury. This malfunction must be properly identified and reported.

Ejection failures should not be reported as extraction failures.

Probable Cause

3-16. Short recoil cycles and fouled or corroded rifle chambers are the most common causes of failures to extract. A damaged extractor or a weak or broken extractor spring can also cause this malfunction.

Corrective Action

3-17. This malfunction is one of the hardest to clear; the severity of the failure determines the corrective action procedures.

NOTE: If the bolt has moved rearward far enough to strip a live round from the magazine in its forward motion, the bolt and bolt carrier must be locked to the rear.

3-18. With the bolt locked to the rear and the weapon on SAFE, perform the following actions:

(1) Remove the magazine and all loose rounds.

(2) Tap the weapon's buttstock on a hard surface to cause the cartridge to fall out of the chamber.

3-19. However, if the cartridge case is ruptured, it can be seized. When this occurs, perform the following actions:

(1) Insert a cleaning rod into the bore from the muzzle end.

(2) Force the cartridge case from the chamber by tapping the cleaning rod against the inside base of the fired cartridge.

3-20. If cleaning and inspecting the mechanism and chamber reveals no defects but failures to extract persist, the extractor and extractor spring should be replaced. If the chamber surface is damaged, the entire barrel must be replaced.

FAILURE TO EJECT

3-21. Ejection of a cartridge is an element in the rifle's cycle of functioning, regardless of the mode of fire. This malfunction occurs when the cartridge is not ejected through the ejection port and either remains partly in the chamber or becomes jammed in the upper receiver as the bolt closes. When the firer initially clears the rifle, the cartridge could strike an inside surface of the receiver and bounce back into the path of the bolt.

Probable Cause

3-22. The cartridge must be extracted before it can eject. Failures to eject can also be caused by a buildup of carbon or fouling on the extractor or by short recoil. Short recoil is usually due to a buildup of fouling in the bolt carrier mechanism or gas tube. Resistance caused by a carbon-coated or corroded chamber can impede the extraction and ejection of a cartridge.

Corrective Action

3-23. While retraction of the charging handle usually frees the cartridge and permits removal, the charging handle must not be released until the position of the next live round is determined.

3-24. If another live round has been sufficiently stripped from the magazine or remains in the chamber, remove the magazine and all live rounds before releasing the charging handle.

3-25. If several malfunctions occur and are not corrected by cleaning and lubricating, replace the ejector spring, extractor spring, and extractor.

OTHER MALFUNCTIONS

3-26. Table 3-1 describes other malfunctions that can occur and the appropriate corrective actions.

Table 3-1. Other malfunctions.

MALFUNCTION	CORRECTIVE ACTION
The bolt fails to lock in the rearward position after the last round in the magazine is fired.	Check for a bad magazine or short recoil.
The bolt fails to lock in the rearward position when the bolt catch has been engaged.	Check the bolt catch; turn in the weapon to the unit armorer.
The weapon fires two or more rounds when the trigger is pulled and the selection lever is in the SEMI position.	This indicates a worn sear, cam, or disconnector. Turn in the weapon to the unit armorer so that he can repair and replace the trigger group parts.
The trigger fails to pull or return after release with the selector set in a firing position.	This indicates that the trigger pin has backed out of the receiver or the hammer spring is broken (A, Figure 3-3). Turn in the weapon to the unit armorer so that he can repair and replace it.
The magazine fails to lock into the magazine well.	Check the magazine and magazine catch for damage (B, Figure 3-3). Turn in to the unit armorer to adjust or replace the catch.
Any part of the bolt carrier group fails to function.	Check for incorrect assembly of components (C, Figure 3-3). Correctly clean and assemble the bolt carrier group, or replace damaged parts.
The ammunition fails to feed from the magazine.	Check for a damaged magazine (D, Figure 3-3). A damaged magazine could cause repeated feeding failures and should be turned in to the unit armorer or exchanged.

NOTE: For more information about troubleshooting malfunctions and replacing components, see organizational and direct support maintenance publications and manuals.

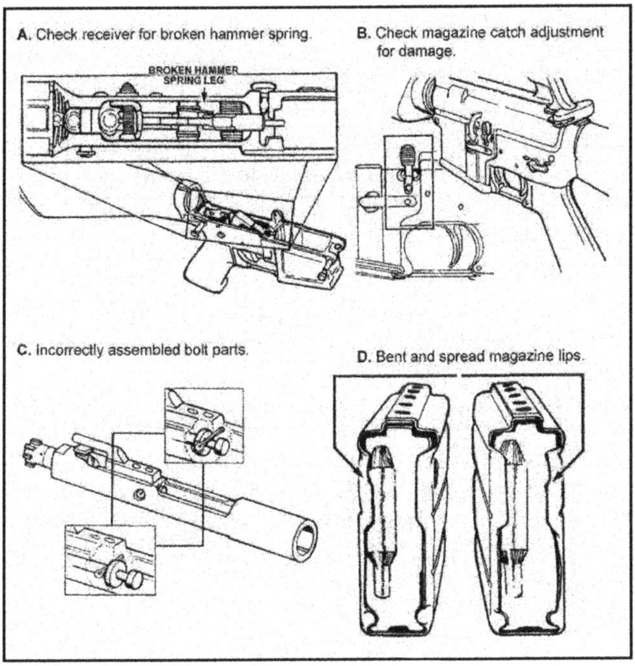

Figure 3-3. Other possible malfunctions.

DESTRUCTION PROCEDURES

3-27. Destruction of any military weapon is authorized only as a last resort to prevent the enemy from capturing or using it. In combat situations, the commander has the authority to destroy weapons, but he must report doing so through the proper channels.

NOTE: Certain procedures outlined require the use of explosives and incendiary grenades. Related principles and the specific conditions under which destruction occurs are command decisions.

METHODS OF DESTRUCTION

3-28. Equipment may be destroyed using several methods. The commander must use his imagination and resourcefulness to select the best method of destruction based on the facilities available. Time is usually critical. Table 3-2 outlines the methods of destruction.

NOTE: If destruction is directed, appropriate safety precautions must be observed.

Table 3-2. Methods of destruction and their applications.

METHOD OF DESTRUCTION	APPLICATION
Mechanical	Use an axe, pick, mattock, sledgehammer, crowbar, or other heavy implement.
Burning	Use gasoline, oil, incendiary grenades, other flammables, or a welding/cutting torch.
Demolition	Use suitable explosives, ammunition or, as a last resort, hand grenades.
Disposal	Bury essential parts, dump them in streams or marshes, or scatter them so widely that recovering them would be impossible.

DEGREE OF DAMAGE

3-29. The method of destruction used must damage equipment and essential spare parts to the extent that they cannot be restored to usable condition (by repair or by cannibalization) in the combat zone.

PRIORITIES OF DESTRUCTION

3-30. When lack of time prevents completely destroying equipment, Soldiers must destroy the same essential parts on all like equipment. The order in which the parts should be destroyed (priority of destruction) is as follows:

(1) Bolt carrier group.
(2) Upper receiver group.
(3) Lower receiver group.

Chapter 4

Preliminary Marksmanship Instruction

An Infantryman's basic battlefield tool is his weapon. To effectively employ his weapon, he must master marksmanship—from the basics of rifle marksmanship to the advanced stages of target engagement.

Understanding the operation and functions of any machine is vital to becoming an expert with that machine. The same theory applies to rifle marksmanship. Commanders must keep this in mind when setting up a training program.

SECTION I. INTRODUCTION TO BASIC RIFLE MARKSMANSHIP AND MECHANICAL TRAINING

This section covers the mechanical training of M16- and M4-series weapons. With this knowledge, a Soldier is able to assess and correct any malfunction to keep the weapon operating properly. This training program (Table 4-1) introduces Soldiers to BRM and teaches them how to maintain, operate, and correct malfunctions on M16- and M4-series weapons. It also teaches peer coaching responsibilities and sight manipulation, while emphasizing safety.

Table 4-1. Introduction to basic rifle marksmanship and mechanical training.

INTRODUCTION TO BASIC RIFLE MARKSMANSHIP AND MECHANICAL TRAINING
Period 1 (4 hours)
Instructional Intent
• Introduce the Soldiers to BRM and teach them how to maintain, operate, and correct malfunctions on a M16-/M4-series weapon. • Teach peer coaching responsibilities and sight manipulation, while emphasizing safety.
Observables
Soldiers— • Emphasize safety throughout training IAW TM 9-1005-319-10 (refer to this TM for more information). • Clear their weapons IAW this manual and TM 9-1005-319-10. • Identify all components of their weapon IAW TM 9-1005-319-10 (refer to this TM for more information). • Handle and identify 5.56-mm ammunition IAW TM 9-1005-319-10 (refer to this TM for more information). • Understand the eight cycles of function IAW this manual. • Understand the modes of fire IAW this manual. • Disassemble and assemble their weapon IAW TM 9-1005-319-10 (refer to this TM for more information). • Perform a function check on their weapon IAW TM 9-1005-319-10 (refer to this TM for more information). • Maintain, load, and unload their magazines IAW TM 9-1005-319-10 (refer to this TM for more information). • Maintain, load, unload, and clear their weapons IAW TM 9-1005-319-10 (refer to this TM for more information). • Perform SPORTS on their weapon within five seconds IAW TM 9-1005-319-10 (refer to this TM for more information). • Correctly manipulate their sights without assistance IAW TM 9-1005-319-10 (refer to this TM for more information). • Are taught peer-coaching techniques and responsibilities IAW this manual.

NOTE: Soldiers who do not meet the standard will receive remedial training before continuing with subsequent instruction.

CLEARING

WARNING

To maintain safety, the weapon must be cleared before disassembling, cleaning, inspecting, transporting, or storing.

NOTE: Additional mechanical training, including disassembly, maintenance, assembly, loading, and sight manipulation, is available in TM 9-1005-319-10.

4-1. To clear an M16-/M4-series weapon (Figure 4-1), perform the following actions:

(1) Point the muzzle in a designated safe direction or in a clearing barrel.

(2) Attempt to place the selector lever on SAFE. If the selector lever cannot be placed on SAFE because the weapon is not cocked, proceed to Step 3.

(3) Remove the magazine. To perform this procedure—

- Depress the magazine catch button.
- Pull the magazine down.

(4) Lock the bolt open. To perform this procedure—

- Pull the charging handle rearward.
- Press the bottom of the bolt catch.
- Allow the bolt to move forward until it engages the bolt catch.
- Return the charging handle to the full forward position.
- If you have not done so already, place the selector lever on SAFE.

(5) Visually inspect the receiver and chamber to ensure that these areas contain no ammunition.

(6) Allow the bolt to move forward by pressing the upper portion of the bolt catch.

(7) Place the selector lever on SEMI.

(8) Squeeze the trigger.

(9) Pull the charging handle fully rearward and release it, allowing the bolt to return to the full forward position.

(10) Place the selector lever on SAFE.

NOTE: The weapon is now clear.

(11) Close the ejection port cover.

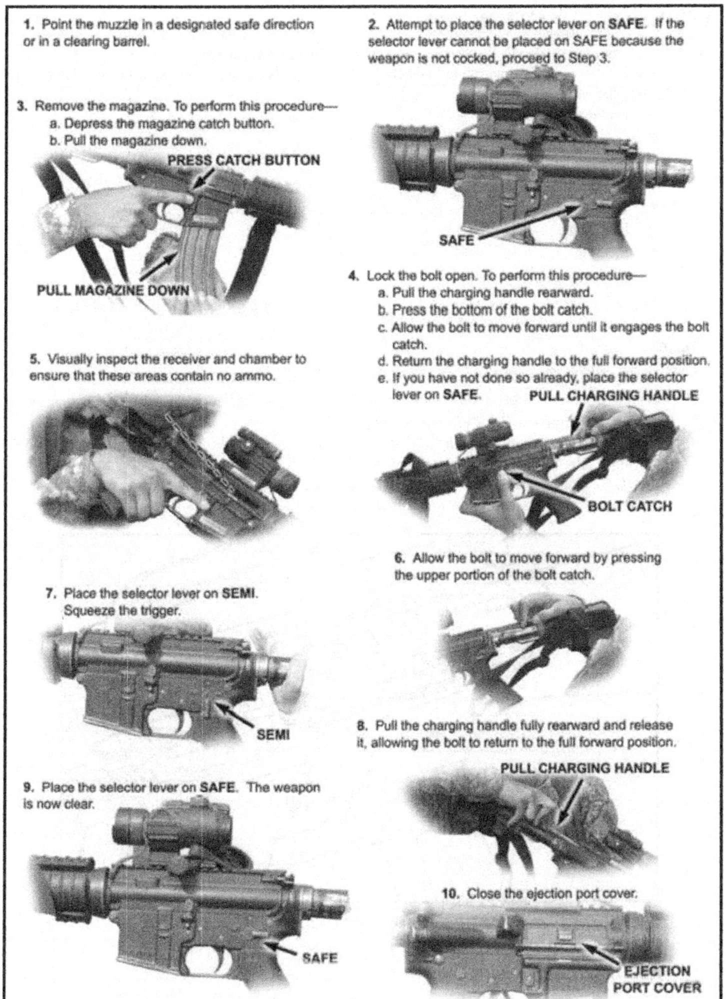

Figure 4-1. Clearing.

CYCLES OF FUNCTIONING

4-2. Soldiers must understand the rifle components and the mechanical sequence of events during the firing cycle. After a loaded magazine has been inserted into the weapon, the eight cycles of functioning begin. They include—

- Feeding.
- Chambering.
- Locking.
- Firing.
- Unlocking.
- Extracting.
- Ejecting.
- Cocking.

4-3. The following paragraphs describe the actions that occur during each cycle of functioning.

FEEDING

4-4. During feeding, the following actions occur. This process is shown in Figure 4-2.

(1) As the bolt carrier group moves rearward, it engages the buffer assembly and compresses the action spring into the lower receiver extension.

(2) When the bolt carrier group clears the top of the magazine, the expansion of the magazine spring forces the follower and a new round into the path of the forward movement of the bolt.

(3) The expansion of the action spring sends the buffer assembly and bolt carrier group forward with enough force to strip a new round from the magazine.

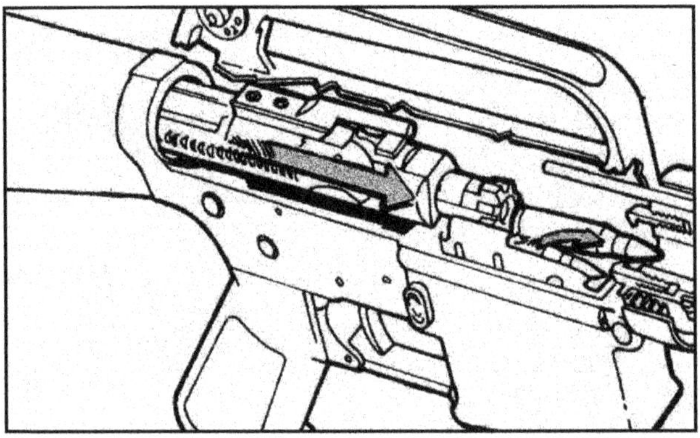

Figure 4-2. Feeding.

CHAMBERING

4-5. During chambering, the following actions occur. This process is shown in Figure 4-3.

(1) As the bolt carrier group continues to move forward, the face of the bolt thrusts the new round into the chamber.

(2) At the same time, the extractor claw grips the rim of the cartridge and the ejector is compressed.

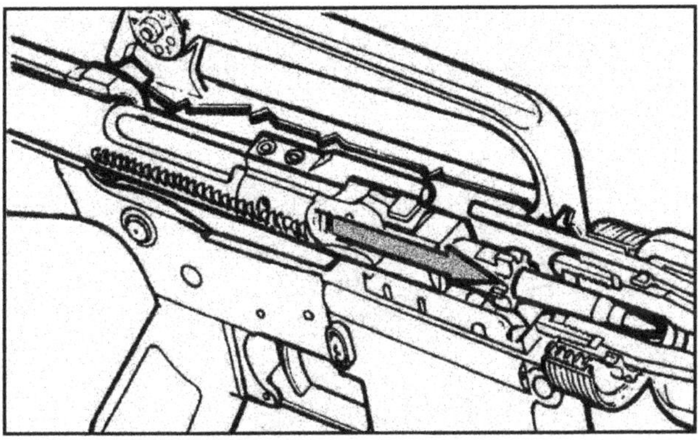

Figure 4-3. Chambering.

LOCKING

4-6. During locking, the following actions occur. This process is shown in Figure 4-4.

(1) As the bolt carrier group moves forward, the bolt cam pin riding in the guide channel in the upper receiver keeps the bolt in its most forward position.

(2) Just before the bolt locking lugs make contact with the barrel extension, the bolt cam pin emerges from the guide channel.

(3) The pressure exerted by the contact of the bolt locking lugs and barrel extension causes the bolt cam pin to move along the cam track (located in the bolt carrier) in a counterclockwise direction, rotating the bolt locking lugs in line behind the barrel extension locking lugs.

NOTE: The rifle is ready to fire.

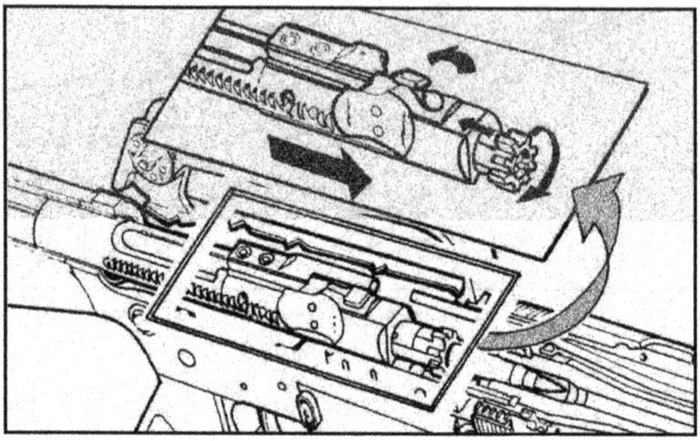

Figure 4-4. Locking.

FIRING

4-7. During firing, the following actions occur. This process is shown in Figure 4-5.

(1) With a round in the chamber, the hammer cocked, and the selector on SEMI, the firer squeezes the trigger.

(2) The trigger rotates on the trigger pin, depressing the nose of the trigger and disengaging the notch on the bottom of the hammer.

(3) The hammer spring drives the hammer forward.

(4) The hammer strikes the head of the firing pin, driving the firing pin through the bolt and into the primer of the round.

(5) The primer ignites, causing the powder in the cartridge to ignite.

(6) The gas generated by the rapid burning of the powder forces the projectile from the cartridge and propels it through the barrel.

(7) After the projectile has passed the gas port (located on the upper surface of the barrel under the front sight, as shown in Figure 4-5) and before it leaves the barrel, gas enters the gas port and moves into the gas tube.

(8) The gas tube directs the gas into the bolt carrier. It passes down through the key and into a space between the rear of the carrier's bolt cavity and the rear of the bolt itself.

(9) Then, the gas expands.

(10) The bolt is locked into the barrel extension, unable to move forward; the carrier is forced to the rear by the expanding gas.

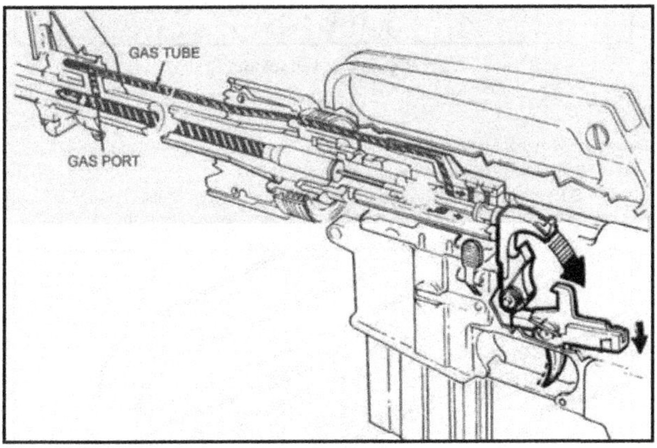

Figure 4-5. Firing.

Unlocking

4-8. During unlocking, the following actions occur. This process is shown in Figure 4-6.

(1) As the bolt carrier moves to the rear, the bolt cam pin follows the path of the cam track (located in the bolt carrier).

(2) The cam pin and bolt assembly rotate simultaneously until the locking lugs of the bolt are no longer in line behind the locking lugs of the barrel extension.

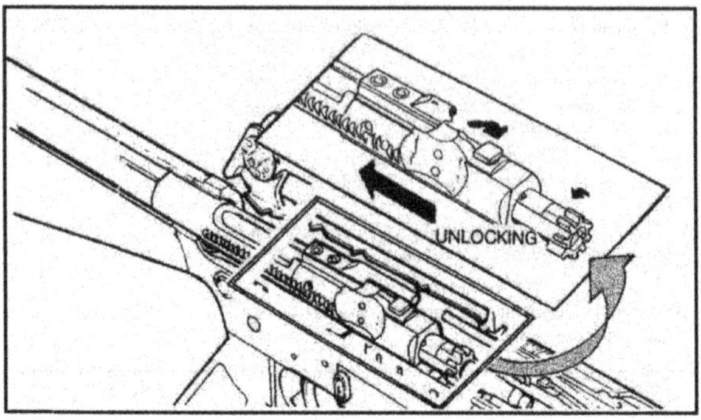

Figure 4-6. Unlocking.

Extracting

4-9. During extracting, the following actions occur. This process is shown in Figure 4-7.

(1) The bolt carrier group continues to move to the rear.

(2) The extractor (which is attached to the bolt) grips the rim of the cartridge case, holds it firmly against the face of the bolt, and withdraws the cartridge case from the chamber.

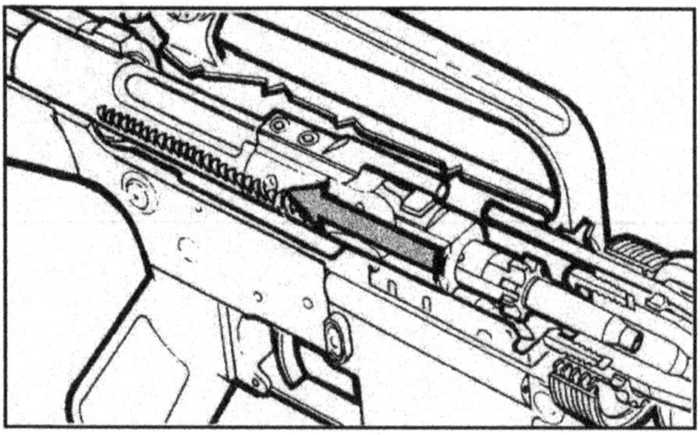

Figure 4-7. Extracting.

EJECTING

4-10. During ejecting, the following actions occur. This process is shown in Figure 4-8.

(1) With the base of a cartridge case firmly against the face of the bolt, the ejector and ejector spring are compressed into the bolt body.

(2) As the rearward movement of the bolt carrier group allows the nose of the cartridge case to clear the front of the ejection port, the cartridge is pushed out by the action of the ejector and ejector spring.

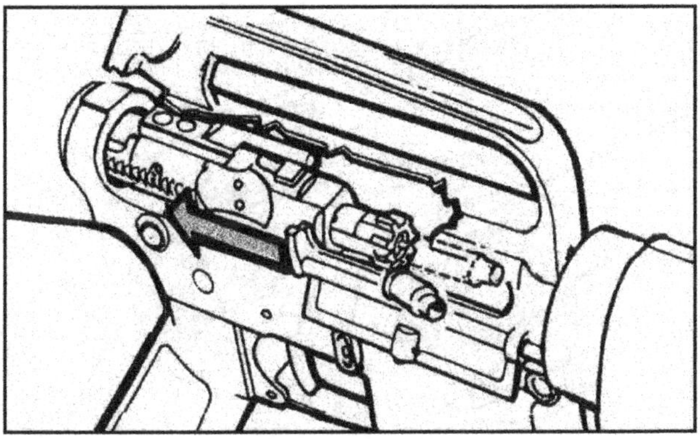

Figure 4-8. Ejecting.

COCKING

4-11. During cocking, the following actions occur. This process is shown in Figure 4-9.

(1) The rearward movement of the bolt carrier overrides the hammer.

(2) The hammer is forced down into the receiver, and the hammer spring is compressed.

NOTE: This action cocks the hammer in the firing position.

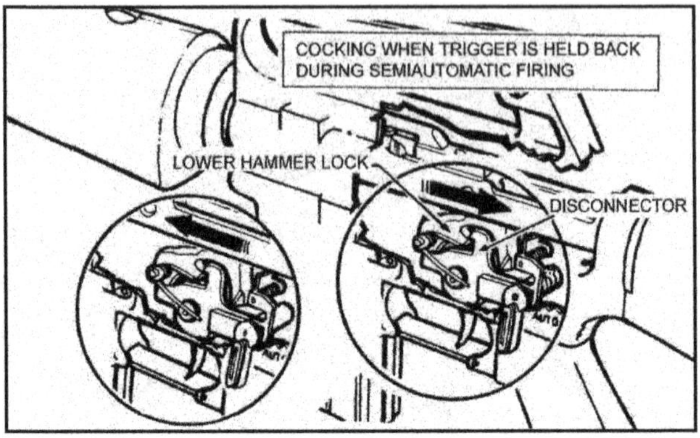

Figure 4-9. Cocking.

MODES OF FIRE

4-12. Weapons may fire using the three modes of fire:

- Semiautomatic (Figure 4-10).
- Automatic (Figure 4-10).
- Burst (Figure 4-11).

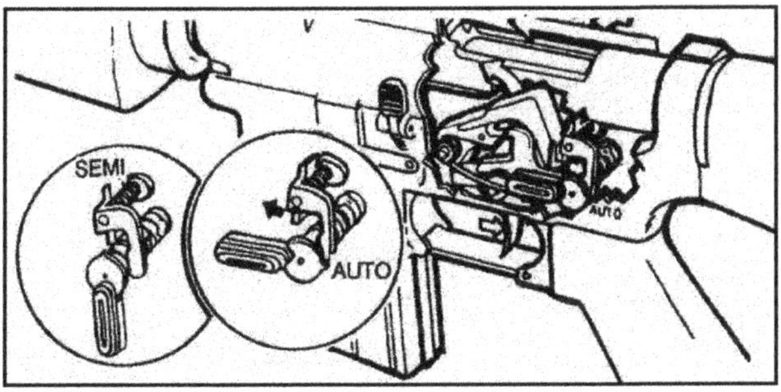

Figure 4-10. Semiautomatic and automatic fire mode selector positions.

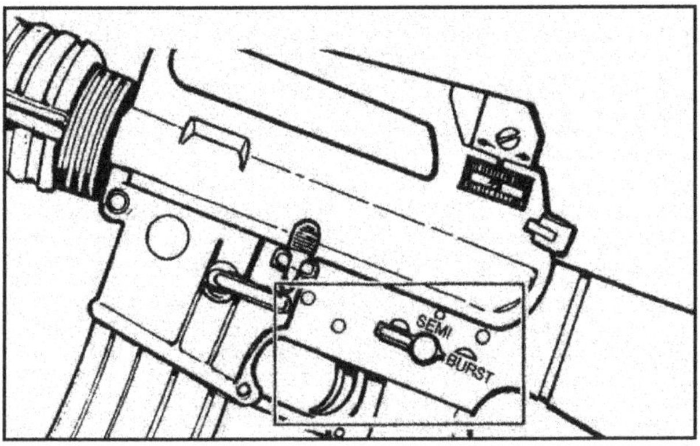

Figure 4-11. Burst fire mode selector position.

SEMIAUTOMATIC MODE OF FIRE

4-13. Weapons that function in the semiautomatic mode of fire are—

- M16A1/A2/A3/A4 rifles.
- M4/M4A1carbines.

4-14. Before a Soldier fires a weapon using the semiautomatic mode of fire, a disconnector (mechanism installed so the firer can fire single rounds) is attached to the trigger and rotated forward by the action of the disconnector spring. When the recoil of the bolt carrier cocks the hammer—

(1) The disconnector engages the lower hook of the hammer and holds it until the trigger is released.

(2) The disconnector rotates to the rear and down, disengaging the hammer and allowing it to rotate forward until caught by the nose of the trigger (this prevents the hammer from following the bolt carrier forward and causing multiple firings).

(3) The trigger must be squeezed again before the next round will fire.

AUTOMATIC MODE OF FIRE

4-15. Weapons that function in the automatic mode of fire are—

- M16A1/A3 rifles.
- M4A1 carbines.

4-16. When the selector lever is in the AUTO position, the rifle continues to fire as long as the trigger is held back and ammunition is in the magazine. The functioning of certain parts of the rifle changes when firing automatically. Once the trigger is squeezed and the round is fired—

(1) The bolt carrier group moves to the rear, and the hammer is cocked.

(2) The center cam of the selector depresses the rear of the disconnector and prevents the nose of the disconnector from engaging the lower hammer hook.

(3) The bottom part of the automatic sear catches the upper hammer hook and holds it until the bolt carrier group moves forward.

(4) The bottom part strikes the top of the sear and releases the hammer, causing the rifle to fire automatically.

(5) If the trigger is released, the hammer moves forward and is caught by the nose of the trigger.

NOTE: This ends the automatic cycle of fire until the trigger is squeezed again.

BURST MODE OF FIRE

4-17. Weapons that function in the burst mode of fire are—

- M16A2/A4 rifles.
- M4 carbines.

4-18. When the selector lever is in the BURST position, the rifle fires a three-round burst if the trigger is held to the rear during the complete cycle. The weapon continues to fire three-round bursts with each separate trigger pull as long as ammunition is in the magazine.

NOTE: Releasing the trigger or exhausting ammunition at any point in the three-round cycle interrupts fire, producing one or two shots. Reapplying the trigger only completes the interrupted cycle; it does not begin a new one. This is not a malfunction.

4-19. The M16A2/4 and M4 disconnectors have a three-cam mechanism that continuously rotates with each firing cycle. Based on the position of the disconnector cam, the first trigger pull (after initial selection of the BURST position) can produce one, two, or three firing cycles before the trigger must be pulled again. The burst cam rotates until it reaches the stop notch.

PEER COACHING

4-20. Peer coaching involves using two Soldiers of equal firing proficiency and experience to assist each other during marksmanship training.

ADVANTAGES AND DISADVANTAGES

4-21. Some problems exist with peer coaching. If the new Soldier does not have adequate guidance, a "blind leading the blind" situation may result, leading to negative training and safety violations. However, when adequate instruction is provided, peer coaching can be helpful even in the IET environment. Since all Soldiers in units have completed BRM, peer coaching within units should yield better results.

BENEFITS

4-22. The pairing of Soldiers can enhance learning for both participants. The coach learns what to look for as he provides guidance to the firer, and the firer has a chance to ask simple questions and to discuss areas that are not understood. Pairing Soldiers who have demonstrated good firing proficiency with those who have firing problems can improve the performance of problem firers.

DUTIES

4-23. The peer coach—

- Constantly checks factors that the firer is unable to observe for himself.
- Prevents the firer from repeating errors.
- Assists the firer in applying marksmanship fundamentals during firing.
- Assists the firer in obtaining a good position and in adjusting sandbags.
- Adds to range safety procedures by helping safety personnel with preliminary rifle checks.
- Watches the firer—not the target—to ensure that the firer maintains proper position, breath control, trigger pressure, and trigger squeeze.

NOTE: There are times when the peer coach may be required to observe the target area and not the firer for example, when field fire targets are being engaged and the firer cannot see where he is hitting or missing targets).

CHECKLIST FOR THE COACH

4-24. Coaches follow these procedures to determine and eliminate rifle and firer deficiencies:

- The coach checks to see that the—
  - Rifle is cleared, and defective parts have been replaced.
  - Ammunition is clean, and the magazine is properly placed in the pouch.
  - Sights are blackened and set correctly for long- or short-range firing.
- The coach observes the firer to see if he—
  - Uses the correct position and properly applies the steady position elements.
  - Properly loads the rifle.
  - Obtains the correct sight alignment (with the aid of an M16 sighting device).
  - Holds his breath correctly (by watching his back).
  - Applies proper trigger squeeze (determines whether he flinches or jerks by watching his head, shoulders, trigger finger, and firing hand and arm).
  - Is tense and nervous. If the firer is nervous, the coach has the firer breathe deeply several times to relax.
- Supervisory personnel and peer coaches correct errors as they are detected. If many common errors are observed, it is appropriate to call the group together for discussion, demonstration of proper procedures, and feedback.

POSITION OF THE COACH

4-25. During an exercise, the coach should be positioned where he can best observe the firer as he assumes position. Then, he moves to various points around the firer (sides and rear) to check the correctness of the firer's position. The coach requires the firer to make adjustments until the firer obtains a correct position.

4-26. When the coach is satisfied with the firing position, he assumes a coaching position alongside the firer. The coach usually assumes a position like that of the firer on the firing side of the firer (Figure 4-12).

Figure 4-12. Prone position of coach (right-handed firer).

SECTION II. MARKSMANSHIP FUNDAMENTALS I

This training program (Table 4-2) reinforces BRM and trains the four fundamentals to standard through dry-firing and simulation circuit training. It also teaches and reinforces range and safety procedures.

Table 4-2. Marksmanship Fundamentals I training program.

MARKSMANSHIP FUNDAMENTALS I
Period 2 (8 hours)
Instructional Intent
• Reinforce BRM I, and train the four fundamentals to standard with hands-on training, simulation, and dry-firing during circuit training with an M16-/M4-series weapon. • Teach and reinforce range and safety procedures.
Observables
Ensure that— • All equipment (helmet, IBA) is fitted properly to maximize training IAW the local SOP. • Live-fire range procedures are replicated and enforced IAW the local SOP. • The four fundamentals are being integrated into all exercises IAW this manual. • All dry-firing is well-aimed using 25-meter zeroing targets, EST, and LMTS. • Peer coaching is being emphasized IAW this manual.
Tasks
• The four fundamentals (IAW with this manual). • Dominant eye training (IAW with this manual). • Basic firing positions (IAW with this manual). • Range and safety procedures (IAW with the local SOP). • Demonstrate the integrated act of shooting during dry-fire exercises utilizing simulators and training devices (IAW this manual).

NOTES: 1. Simulators and training devices are listed in Appendix A of this manual.

2. Soldiers who do not meet the standard will receive remedial training before continuing with subsequent instruction.

INTERCEPTOR BODY ARMOR

4-27. BRM strategy includes the wearing of interceptor body armor (IBA, shown in Figure 4-13), if it is available—minus the throat, collar, and groin attachments—during all BRM periods and concurrent training. Whether or not IBA is worn, marksmanship fundamentals remain the same.

4-28. Prior to BRM training, use an IBA immersion approach so the Soldier can adapt to weight and movement restrictions. For an easier weight transition, incrementally introduce the outer tactical vest (OTV) and front and back small arms protective insert (SAPI) plates.

Figure 4-13. Interceptor body armor.

PROPER WEAR AND FIT

4-29. When using IBA, adhere to the following guidelines:

- Have adequate IBA quantities on hand for all Soldiers.
- Properly size IBA to the Soldier by conducting deliberate fit procedures to reduce or eliminate fit and size problems.

NOTE: Improper wear and fit of IBA impedes a Soldier's marksmanship ability.

- Ensure that the SAPI plate size corresponds to the OTV.
- Make sure that nothing else is in the OTV/SAPI compartment.

WEAR OF HELMETS WITH INTERCEPTOR BODY ARMOR

4-30. When using helmets with IBA, adhere to the following guidelines:

- When in the prone position, the IBA's back plate tends to shove the personnel armor system for ground troops (PASGT) helmet over Soldiers' eyes. To minimize the PASGT helmet positioning problem, make sure that the helmet is properly sized and fitted. Female shooters with long hair will find that wearing their hair in a bun adds material between the IBA and helmet, further forcing the helmet down over their eyes. Encourage female Soldiers to wear a short (chin length) haircut or cornrow hairstyle. If the female Soldier chooses not to wear short hair, allow her to wear her hair down when firing. Tightening the suspension harness and sweat band (raising the helmet higher on the head) can lessen interference with the IBA, hair, and helmet.
- The Army combat helmet (ACH) is lighter than the PASGT helmet, has better weight distribution, and contains less material that can impede a Soldier's firing vision when in the prone position. The ACH does not interfere with the IBA or block a Soldier's vision while in the prone position.

ADJUSTMENTS TO FIRING POSITIONS

4-31. When using IBA, adhere to the following guidelines:

- To increase comfort and stability while wearing IBA in the prone position, scoop sand or dirt underneath the chest while preparing to fire.
- To alleviate the pain and pressure on elbows and knees that the added weight of IBA causes, use elbow and knee pads. If used in the kneeling position, do not rest the elbow pad on the knee pad; hard plastic on hard plastic is not conducive to a steady position. To help with stability while firing in the kneeling position, squeeze the rifle buttstock between the SAPI plate and bicep. Loosen the firing-side straps and tighten the nonfiring-side straps to shift the SAPI plate away from the firing side.
- Instead of using load-bearing equipment (LBE), attach canteens, ammunition pouches, or first aid pouches directly to the IBA to minimize interference with LBE shoulder straps, IBA, and helmet.
- Reserve IBA firing with throat, collar, or groin protectors for ARM.
- To position themselves more comfortably and be able to reach the handguards, Soldiers of shorter stature may have to increase their body-line-to-rifle axis angle to more of an "L" shape.
- Soldiers should be in a comfortable firing position to leverage the natural point of aim. The more the target and rifle are naturally in line (as when in a relaxed position), the less movement is needed to acquire a proper sight picture.

FOUR FNDAMENTALS

4-32. Before the Soldier approaches the firing line, he must understand and apply the four fundamentals:

- Steady position.
- Aiming.
- Breath control.
- Trigger squeeze.

4-33. Soldiers apply these four fundamentals rapidly and consistently to perform the integrated act of firing. These fundamentals should be practiced while the Soldier is wearing all of his equipment, including his helmet and IBA (if available).

STEADY POSITION

4-34. When the Soldier approaches the firing line, he assumes a comfortable, steady firing position (Figure 4-14). The firer is the best judge of the quality of his position. If he can hold the front sightpost steady through the fall of the hammer, he has a good position. Steady position incorporates the following elements:

- Nonfiring hand grip.
- Rifle's buttstock position.
- Firing hand grip.
- Firing elbow placement.
- Nonfiring-side elbow placement.
- Cheek-to-stock weld.
- Support and muscle relaxation.
- Natural point of aim.

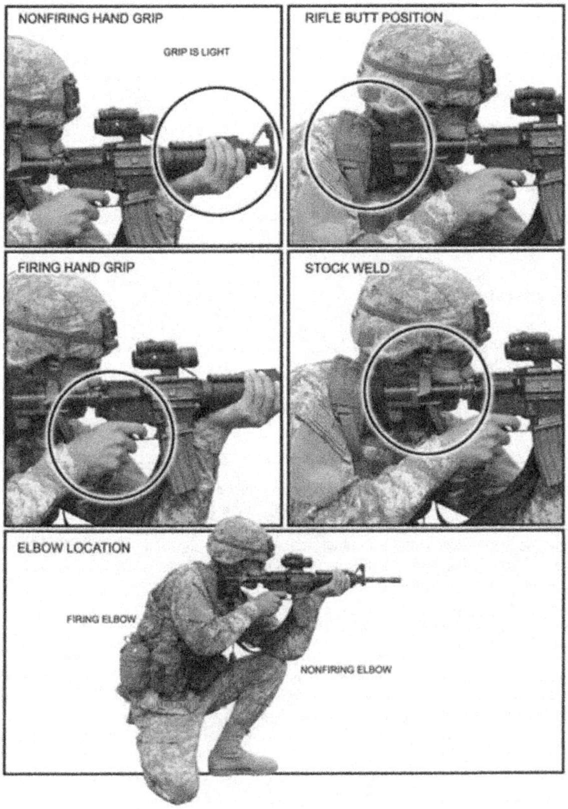

Figure 4-14. Steady position.

Nonfiring Hand Grip

4-35. The weapon's handguard rests lightly on the heel of the nonfiring hand, in the "V" formed by the thumb and fingers.

Rifle's Buttstock Position

4-36. Place the weapon's buttstock into the pocket of the firing shoulder. When wearing IBA, place the weapon's buttstock where the pocket should be; this reduces the effect of recoil and ensures a steady position.

Firing Hand Grip

4-37. The firing hand grasps the pistol grip so that it fits in the "V" formed by the thumb and forefinger. The forefinger is placed on the trigger so that the lay of the weapon is not disturbed when the trigger is squeezed. The remaining three fingers exert a slight rearward pressure to ensure that the buttstock remains in the pocket of the shoulder.

Firing Elbow Placement

4-38. The firing elbow is important in providing balance. Its exact location depends on the firing or fighting position used. Placement of the firing elbow should allow the firer's shoulders to remain level.

Nonfiring-Side Elbow Placement

4-39. The nonfiring-side elbow is positioned firmly under the weapon to allow a comfortable and stable position. When the Soldier engages a wide sector of fire, moving targets, and targets at various elevations, his nonfiring-side elbow should remain free from support.

Cheek-to-Stock Weld

4-40. The cheek-to-stock weld should provide a natural line of sight through the center of the rear sight aperture to the front sightpost and onto the target. The firer's neck should be relaxed, allowing his cheek to fall naturally onto the stock.

NOTE: Proper eye relief is obtained when a Soldier establishes a good cheek-to-stock weld. A small change in eye relief normally occurs each time that the firer assumes a different firing position.

4-41. Through dry-fire training, the Soldier practices this position until he assumes the same cheek-to-stock weld each time he assumes a given position, which provides consistency in aiming. To learn to maintain the same cheek-to-stock weld each time the weapon is aimed, the Soldier should begin by trying to touch the charging handle with his nose when assuming a firing position. The Soldier should be mindful of how the nose touches the charging handle and should be consistent when doing so. This position should be critiqued and reinforced during dry-fire training.

Support and Muscle Relaxation

4-42. When artificial support (for example, sandbags, logs, or stumps) is available, it should be used to steady the position and support the weapon. If support is used properly, the Soldier should be able to relax most of his muscles. If artificial support is not available, the bones—not the muscles—in the firer's upper body must support the weapon. Using muscles to support the rifle can cause muscle fatigue, which in turn, causes the weapon to move.

Natural Point of Aim

4-43. When the Soldier first assumes his firing position, he orients his weapon in the general direction of his target. Then, he adjusts his body to align the weapon and sights with the desired point of aim. When using proper support and consistent cheek-to-stock weld, the Soldier should have his weapon and sights naturally aligned on the target.

4-44. If correct body-rifle-target alignment cannot be achieved, the front sightpost must be held on the target using muscular support and effort. As the weapon fires, muscles tend to relax, causing the front sight to move away from the target, toward the natural point of aim. Adjusting this natural point of aim to the target eliminates this movement. When multiple target exposures are expected or a sector of fire must be covered, the Soldier adjusts his natural point of aim to the center of the expected target exposure area or sector.

AIMING

4-45. Having mastered the task of holding the rifle steady, the Soldier must align the rifle with the target in exactly the same way for each firing. The firer is the final judge as to where his eye is focused. The instructor or trainer emphasizes this point by having the firer focus on the target and then on the front sightpost. He checks the position of the firing eye to ensure that it is in line with the rear sight aperture. The elements of aiming training are as follows:

- Sight alignment.
- Focus of the eye.
- Sight picture.
- Front sightpost.
- Aiming practice.

Sight Alignment

4-46. The weapon must be aligned with the target; to do so, Soldiers place the tip of the front sightpost in the center of the rear sight aperture (Figure 4-15). Any alignment error between the front and rear sights repeats itself for every ½ meter the bullet travels. For example, at the 25-meter line, any error in rifle alignment is multiplied 50 times. If the bullet is misaligned by 1/10 of an inch, it causes a target 300 meters away to be missed by 5 feet.

Figure 4-15. Correct sight alignment.

Focus of the Eye

4-47. A proper firing position aligns the eye with the center of the rear sight aperture. When the eye is focused on the front sightpost, the eye's natural ability to center objects in a circle and to seek the point of greatest light (center of the aperture) aid in providing correct sight alignment. For the average Soldier firing at combat-type targets, the eye's natural ability can accurately align the sights. Therefore, the firer can place the tip of the front sightpost on the point of aim, but the eye must be focused on the tip of the front sightpost. This causes the target to appear blurry, while the front sightpost is seen clearly. Two reasons for focusing on the front sightpost are:

(1) Only a minor aiming error should occur, since the error reflects only as much as the Soldier fails to determine the target's center. A greater aiming error can result if the front sightpost is blurry due to focusing on the target or other objects.

(2) Focusing on the tip of the front sightpost aids the firer in maintaining proper sight alignment.

Sight Picture

4-48. Once the Soldier can correctly align his sights, he can obtain a correct sight picture. A correct sight picture has the target, front sightpost, and rear sightpost aligned. The sight picture includes two basic elements: sight alignment and placement of the point of aim. Placement of the point of aim varies, depending on the engagement range. For example, Figure 4-16 shows a silhouette at 300 meters where the point of aim is the center of mass and the sights are aligned for a correct sight picture.

Figure 4-16. Focus of the eye and correct sight picture.

4-49. The side aiming technique can be used to obtain a correct sight picture (Figure 4-17). It involves positioning the front sightpost to the side of the target in line with the vertical center of mass, keeping the sights aligned. The front sightpost is moved horizontally until the target is directly centered on the front sightpost.

Figure 4-17. Side aiming technique.

Front Sightpost

4-50. The front sightpost is vital to proper firing and should be replaced when damaged. The post should be blackened; when it is shiny, the firer cannot focus precisely on the tip of the front sightpost.

Aiming Practice

4-51. Aiming practice is conducted before firing live rounds. During dry-firing, the Soldier should practice sight alignment and placement of the point of aim. Training aids, such as the M15A1 aiming card, can be used to do this.

BREATH CONTROL

4-52. While sighted on a target, the firer must be aware of the rifle's movement as a result of breathing. Two breath control techniques are practiced during dry-fire:

- Breath control for engaging single targets.
- Breath control for engaging short-exposure targets.

4-53. As the firer's skills improve and as timed or multiple targets are presented, he must learn to control his breath at any part of the breathing cycle. The coach/trainer ensures that the firer uses both breathing techniques and understands them by instructing him to exaggerate his breathing.

Breath Control for Engaging Single Targets

4-54. When zeroing or when time is available to fire a shot, Soldiers fire when there is a natural respiratory pause, when most of the air has been exhaled from the lungs and before inhaling (Figure 4-18). The shot must be fired before the Soldier feels any discomfort.

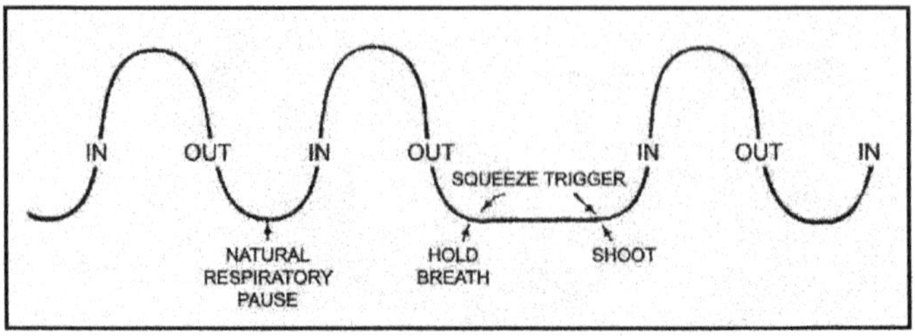

Figure 4-18. Breath control for engaging single targets.

Breath Control for Engaging Short-Exposure Targets

4-55. When employing rapid fire (engaging short-exposure targets), Soldiers stop their breath when they are about to squeeze the trigger (Figure 4-19).

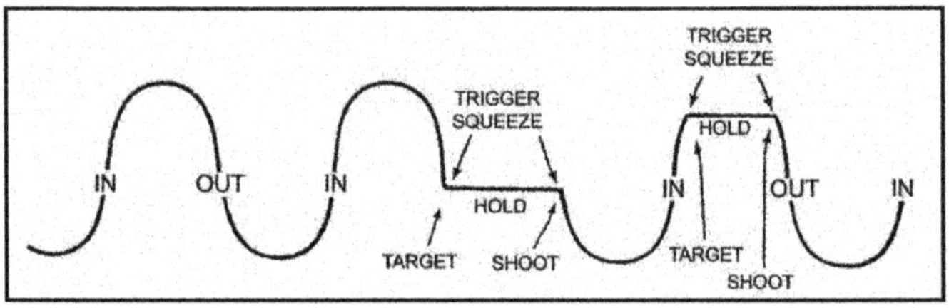

Figure 4-19. Breath control for engaging short-exposure targets.

TRIGGER SQUEEZE

4-56. A steady position reduces disturbance of the rifle during trigger squeeze. If the trigger is not properly squeezed, the rifle will be misaligned with the target at the moment of firing. The elements of trigger squeeze training are as follows:

- Rifle movement.
- Trigger finger.
- Trigger squeeze time.
- Coaching trigger squeeze.
- Wobble area.

Rifle Movement

4-57. Trigger squeeze is important for two reasons:

- Any sudden movement of the finger on the trigger can disturb the lay of the rifle and cause the shot to miss the target.
- The precise instant of firing should be a surprise to the Soldier. If a Soldier knows the exact instant that the rifle will fire, the Soldier will naturally compensate for the weapon's noise and recoil, causing him to miss the target. Soldiers usually tense their shoulders when expecting the rifle to fire; it is difficult to detect since the Soldier does not realize that he is flinching.

Trigger Finger

4-58. The Soldier places his trigger finger (index finger on the firing hand) on the trigger between the first joint and the tip of the finger—not the very end of the finger—and adjusts depending on his hand size and grip. The trigger finger must squeeze the trigger to the rear so the hammer falls without disturbing the lay of the rifle.

4-59. When a live round is fired, it is difficult to see the effect that the trigger pull had on the lay of the rifle. It is important to experiment with many finger positions during dry-fire training to ensure that the hammer is falling with little disturbance to the aiming process.

Trigger Squeeze Time

4-60. The proper trigger squeeze should start with slight pressure on the trigger during the initial aiming process. The firer applies more pressure after the front sightpost is steady on the target and he is holding his breath.

4-61. As the firer's skills increase with practice, he needs less time spent on trigger squeeze. A novice firer can take five seconds to perform an adequate trigger squeeze, but as skills improve, he can squeeze the trigger in a second or less.

Coaching Trigger Squeeze

4-62. The coach/trainer—

- Observes the trigger squeeze, emphasizes the correct procedure, and checks the firer's applied pressure.
- Places his finger on the trigger and has the firer squeeze the trigger by applying pressure to his finger.
- Ensures that the firer squeezes straight to the rear on the trigger, avoiding a left or right twisting movement.
- Observes that the firer follows through and holds the trigger to the rear for approximately one second after the round has been fired.

Wobble Area

4-63. Wobble area is the movement of the front sight around the point of aim when the rifle is in the steadiest position.

4-64. The position must provide for the smallest possible wobble area.

- From a supported position, there should be minimal wobble area and little reason to detect movement. If movement of the rifle causes the front sight to leave the target, more practice is needed.
- From an unsupported position, the firer experiences a greater wobble area than from a supported position. If the front sight strays from the target during the firing process, the firer should hold constant pressure on the trigger and resume as soon as he corrects the sighting.

NOTE: The firer should never try to quickly squeeze the trigger while the sight is on the target. The best firing performance results when the trigger is squeezed continuously and the rifle is fired without disturbing its lay.

DOMINANT EYE TRAINING

4-65. This exercise assists the coach and the firer in determining which eye the firer should use when engaging targets. The firer's dominant eye should be identified early in the training process to prevent unnecessary problems, such as a blurred sight picture or the inability to acquire a tight shot group during the grouping exercise. To perform dominant eye training—

(1) The trainer cuts a 1-inch circular hole in the center of an 8- by 10-inch piece of material (can be anything from paper to plywood).

(2) The trainer positions himself approximately 5 feet in front of the Soldier. The trainer closes his nondominant eye and holds his finger in front of and just below his dominant eye to provide the Soldier with a point of aim.

(3) The Soldier holds the training aid with both hands at waist level and looks with both eyes open at the trainer's open eye. With both eyes focused on the trainer's open eye and arms fully extended, the Soldier raises the training aid between himself and the trainer while continuing to look at the trainer's eye through the hole in the training aid. The Soldier's eye that the trainer sees through the hole in the training aid is the Soldier's dominant eye.

BASIC FIRING POSITIONS

4-66. Basic firing positions are taught during PMI. Other advanced firing positions are added later in training to support tactical conditions. The firing positions used during initial training are—

- Individual foxhole supported.

NOTE: Once the individual foxhole supported fighting position has been mastered, the firer should practice various unsupported positions to obtain the smallest possible wobble area during final aiming and hammer fall. The coach/trainer can check the steadiness of the position by observing movement at the forward part of the rifle, by looking through the Ml6 sighting device, or by checking to see if support is being used. The objective is to establish a steady position under various conditions.

- Prone unsupported firing.
- Prone supported firing.
- Kneeling unsupported.

4-67. These firing positions offer a stable platform for firing the rifle. These positions are used during basic record fire.

INDIVIDUAL FOXHOLE SUPPORTED FIRING POSITION

4-68. This position provides the most stable platform for engaging targets. To assume the individual foxhole supported firing position (Figure 4-20)—

(1) Add or remove dirt, sandbags, or other supports to adjust for your height.

(2) Face the target.

(3) Execute a half-face to the firing side.

(4) Lean forward until the chest is against the firing hand corner of the position.

(5) Place the rifle handguard in the "V" formed by the thumb and fingers of the nonfiring hand.

(6) Rest the nonfiring hand on the material (sandbags or berm) to the front of the position.

(7) Place the weapon's buttstock into the pocket of the firing shoulder.

(8) Rest the firing elbow on the ground outside of the position.

NOTES: 1. When prepared positions are not available, the prone supported position can be substituted.

2. The objective is to establish a steady position under various conditions. The ultimate performance of this task is combat. Although the firer must be positioned high enough to observe all targets, he must remain as low as possible to provide added protection from enemy fire.

Figure 4-20. Individual foxhole supported firing position.

PRONE UNSUPPORTED FIRING POSITION

4-69. This firing position offers another stable firing platform for engaging targets. To assume the prone unsupported firing position—

(1) Face the target.

(2) Spread the feet a comfortable distance apart.

(3) Drop to the knees, breaking the fall with the weapon's buttstock.

(4) Using the rifle's buttstock as a pivot, roll onto the nonfiring side, placing the nonfiring-side elbow close to the side of the magazine.

(5) For the basic prone unsupported position (Figure 4-21), spread the legs apart, with the inside of the feet flat on the ground. For the alternate prone unsupported firing position (Figure 4-22), bend the firing leg to relieve pressure on the lower back.

(6) Place the weapon's buttstock between the SAPI plate and bicep to stabilize the weapon and absorb recoil.

(7) Grasp the pistol grip with the firing hand.

(8) Lower the firing elbow to the ground.

(9) Place both elbows on the ground to support the upper body.

(10) Rest the rifle in the "V" formed by the thumb and fingers of the nonfiring hand.

(11) Adjust the position of the firing elbow until the shoulders are approximately level.

(12) Pull back firmly on the rifle with both hands.

(13) Obtain a stock weld and relax, keeping the heels close to the ground.

NOTE: To increase comfort and stability while wearing IBA in the prone position, sand or dirt should be scooped underneath the chest while preparing to fire. Elbow and knee pads can be worn to relieve IBA-induced pain and pressure.

Figure 4-21. Basic prone unsupported firing position.

Figure 4-22. Alternate prone unsupported firing position.

PRONE SUPPORTED FIRING POSITION

4-70. To assume the prone supported firing position—

(1) Face the target and drop to the ground, breaking the fall with the weapon's buttstock.

(2) For the basic prone supported firing position (Figure 4-23), spread the legs apart, with the inside of the feet flat on the ground. For the alternate prone supported firing position (Figure 4-24), bend the firing leg to relieve pressure on the lower back.

(3) Use sandbags or any other suitable object to support the handguard. Keep the nonfiring hand free for use on any part of the rifle.

(4) Place both elbows on the ground to support the upper body.

(5) Place the firing hand on the pistol grip.

(6) Place the nonfiring hand on the upper handguard.

(7) Place the weapon's buttstock between the SAPI plate and bicep to stabilize the weapon and absorb recoil.

NOTE: Elbow and knee pads can be worn to relieve IBA-induced pain and pressure.

Figure 4-23. Basic prone supported firing position.

Figure 4-24. Alternate prone supported firing position.

KNEELING UNSUPPORTED FIRING POSITION

4-71. To assume the kneeling unsupported firing position (Figure 4-25)—

(1) Keep the left foot in place.

(2) Step back with the right foot.

(3) Drop to the right knee.

(4) Place the left nonfiring hand on the upper handguard with the upper arm (triceps) on the left knee for support.

(5) Place the right firing hand on the pistol grip, with the weapon's buttstock between the SAPI plate and bicep to stabilize the weapon and absorb recoil.

(6) Rest the ball of the right foot firmly on the ground.

(7) Rest the buttock on the heel.

(8) Relax and lean forward into the position to help absorb recoil.

Figure 4-25. Kneeling unsupported firing position.

TRAINING DEVICES AND EXERCISES

4-72. When used alone or in combinations with the appropriate training strategies, training devices and aids can be used to help individuals or squads sustain or practice basic marksmanship skills. They are beneficial when ammunition is limited for training or practice exercises. Some training devices are complex, costly, and in limited supply, while others are relatively simple, cheap, and in large supply.

M15A1 AIMING CARD

4-73. This exercise measures the firer's ability to acquire the same sight picture each time he places his iron sights on a target.

NOTE: Refer to Appendix A for a detailed explanation of training aids and devices.

TARGET BOX AND PADDLE EXERCISE

4-74. This exercise checks the consistency of aiming and placement of three-round shot groups in a dry-fire environment. It incorporates the Soldier's position, breathing, and sight picture to simulate a live-fire 25-meter engagement. Further, it reinforces the basic fundamentals, while refining the Soldier's muscle memory during the integrated act of dry-firing.

NOTE: Refer to Appendix A for a detailed explanation of the target box and paddle exercise and training standards.

DIME/WASHER EXERCISE

4-75. The primary purpose of this exercise is to practice trigger control; it is also used to reinforce good body position and breath control. The Soldier must successfully dry-fire his weapon six consecutive times without the dime or washer falling to the ground.

NOTE: Refer to Appendix A for a detailed explanation of the dime/washer exercise and training standards.

SECTION III. MARKSMANSHIP FUNDAMENTALS II

This training program (Table 4-3) reinforces BRM and the four fundamentals, while demonstrating the integrated act of shooting on the EST 2000 or LMTS.

Table 4-3. Marksmanship Fundamentals II training program.

MARKSMANSHIP FUNDAMENTALS II
Period 3 (8 hours)
Instructional Intent
• Reinforce BRM I and II and the four fundamentals, while demonstrating the integrated act of shooting on the EST or LMTS.
Observables
Ensure that—
• All fundamentals are emphasized and applied on the EST or LMTS.
• Weapon safety is reinforced on the EST or LMTS.
• Peer coaching is emphasized during EST or LMTS firing.
• All Soldiers who fail to hit six out of nine shots at the 300-meter EST target receive remedial training.
Tasks
• Demonstrate the integrated act of firing while using the EST.

NOTE: Soldiers who do not meet the standard will receive remedial training before continuing with subsequent instruction.

ENGAGEMENT SKILLS TRAINER 2000

4-76. The EST 2000 is a home station, indoor, multipurpose, multilane, small arms, crew-served, shoulder-launched munitions gunnery simulator with superior accuracy and state-of-the-art graphics.

4-77. The EST 2000 is used to provide small arms weapon training on—

- Marksmanship.
- Squad tactical procedures.
- Close-range shoot/don't shoot techniques and skills.

4-78. Commanders should review DA Pam 350-38 for live-fire events that can be executed using the EST 2000.

NOTE: Refer to Appendix A for a detailed explanation of the EST 2000 training simulator.

LASER MARKSMANSHIP TRAINING SYSTEM

4-79. The LMTS helps bridge the existing gap in individual marksmanship training by providing a more widely available, flexible, transportable, and lightweight means of maintaining marksmanship skills. The LMTS large-company suite with the warrior kit is a home station or deployed, multipurpose, modular, scaleable, individual and unit small arms marksmanship sustainment simulator that accommodates the M16-series rifle and M4 carbine.

4-80. The LMTS supports both IET and unit sustainment training for static mounted and dismounted small units. When incorporated into BRM and Marksmanship Fundamentals I and II, it greatly improves Soldier feedback.

NOTES:

1. The LMTS is not intended to replace live-firing, qualification, or the EST 2000.
2. Refer to Appendix A for a detailed explanation of the LMTS.
3. Soldiers failing to maintain a three-centimeter shot group on the LMTS 25-meter target should receive remedial feedback.

Chapter 5

Downrange Feedback

Downrange feedback provides precise knowledge of what happens to bullets at range. It provides for an effective transition between 25-meter firing and firing on the field fire range. Knowing precisely where all bullets are hitting or missing the target, a poor firer (with instructor/trainer assistance) can improve his performance, and a good firer can bring his shots to the target's center.

This chapter contains guidelines for the instructor/trainer to conduct training on ranges that provide detailed feedback from the targets downrange. This chapter contains grouping and zeroing procedures and procedures for conducting the three types of KD ranges.

SECTION I. GROUPING PROCEDURES

This section provides guidelines for conducting a grouping range (Table 5-1). It includes concept, organization, shot group marking, shot group analysis, multiple shot group analysis, and troubleshooting of the fundamentals.

Table 5-1. Grouping procedures.

GROUPING PROCEDURES
Instructional Intent
• Reinforce PMI by performing the integrated act of shooting. • Shoot two consecutive 3-round shot groups within a 4-centimeter circle at 25 meters.
Special Instructions
Ensure that— • The rear sight is on the proper setting (zero; M16A2/3=8/3+1; M16A4=6/3+2; M4=6/3). • The rear sight aperture is set on 300+1, not 800+1. • The M16A1's rear sight is set on the aperture marked L. • The small aperture is being used. • Proper and accurate shot group marking is enforced. • The firer's name is clearly marked on the target.
Observables
• Coaches analyze the firer's fundamentals. • Majority of the round must be inside the circle to be counted. • Two consecutive 3-round groups are shot with 6 of 6 rounds impacting inside of the 4- centimeter circle.

CONCEPT

5-1. Shot grouping is a form of practice firing with two primary objectives: firing tight shot groups and consistently placing those groups in the same location. Grouping exercises can be conducted anywhere that provides precise location of bullet hits and misses, such as a 25-meter live-fire zeroing range, KD range, EST 2000, LMTS, or location of misses and hits (LOMAH) system.

NOTES: 1. Shot grouping should be conducted between dry-fire training and zeroing.

2. The initial live-fire training should be a grouping exercise with the purpose of practicing and refining marksmanship fundamentals.

ORGANIZATION OF A 25-METER GROUPING RANGE

5-2. The organization and conduct of a grouping range are based on the availability of ammunition, number of personnel, and the firing ability of personnel in training.

5-3. To properly conduct a 25-meter grouping range, perform the following actions:

- Divide the unit into firing orders. The first order fires, while the second order coaches.
- Reserve ten firing points to conduct corrective instruction.
- Provide sandbags at each firing point to accommodate supported firing positions.
- Set up the 25-meter grouping range as depicted in Figure 5-1.

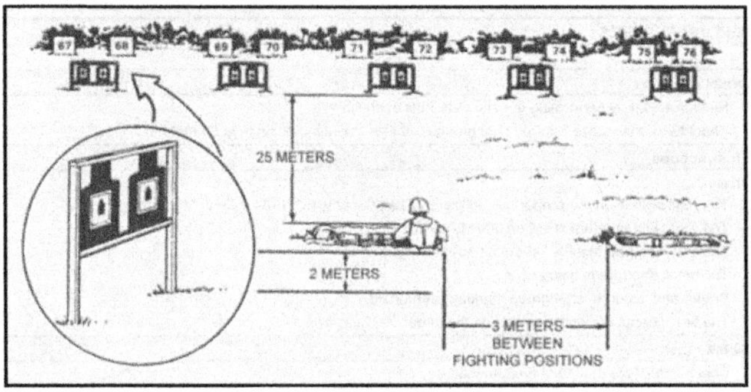

Figure 5-1. 25-meter range.

CONDUCT OF A 25-METER GROUPING FIRING

5-4. Each shot is fired from a supported firing position using the same point of aim (25-meter zeroing target's center of mass). The objective is to fire tight shot groups and to place those shot groups inside a 4-centimeter circle (the actual location of groups on the target is not important).

NOTES: 1. Since this is not a zeroing exercise, few sight adjustments are made unless the shot group is off of or barely on the 25-meter zeroing target.

2. No sight adjustments should be made until the firer can shoot six consecutive shots (two shot groups) inside a 4-centimeter circle. Once this is accomplished, the Soldier is ready to conduct zeroing procedures.

5-5. To conduct a 25-meter grouping firing—

NOTE: Before beginning the 25-meter grouping firing, each Soldier ensures that his sights are set for 25-meter firing.

(1) The Soldier fires a three-round shot group at the 25-meter zeroing target.

(2) The firing line is cleared, and the Soldier and coach move downrange to examine the shot group for fundamental errors, triangulate the shot group, and put the number 1 in the center of the shot group (Figures 5-2 and 5-3).

NOTE: If the shot group is off of the 25-meter zeroing target, the Soldier should mechanically zero the weapon. If the shot group is barely on the target, the Soldier should make a bold adjustment.

(3) The Soldier returns to the firing line and fires a second three-round shot group.

(4) The firing line is cleared, and the Soldier moves downrange to examine the second shot group, triangulate, and mark the center of the shot group with the number 2.

(5) The Soldier groups the two shot groups and marks the center.

5-6. The Soldier repeats Steps 1 through 5 until he places six out of six consecutive rounds inside a 4-centimeter circle. If the Soldier has not grouped in 18 rounds, he should be removed from the firing line and given remedial training before attempting to group again.

NOTE: To be counted, the majority of the round must be inside of the circle.

5-7. Once the Soldier has demonstrated firing proficiency from the supported firing position, grouping exercises can be conducted from the unsupported firing position. For example, 27 rounds are allocated for the grouping exercise (IET only). If the Soldier groups in 18 rounds, he can fire the remaining 9 rounds from the unsupported firing position.

SHOT GROUP MARKING

5-8. If the Soldier is to benefit from this exercise and if the instructor/trainer (or coach) is to provide useful guidance, the Soldier must mark each shot group for a clear record of his firing practice. The instructor/trainer must understand how to analyze shot groups correctly.

5-9. To properly mark the shot groups (Figure 5-2)—

(1) Connect the three bullet holes on the target with a straight line.

(2) Place a number inside of the shot group.

NOTES: 1. The number represents the center of the three shots

2. When two shots are near one end of the group and the third shot is toward the other end, the number is placed closer to the two near shots (Figure 5-3).

3. This is not a precise marking that requires a measurement, but a procedure to help with shot group analysis.

5-10. The three-round shot group allows the firer's performance to be evaluated.

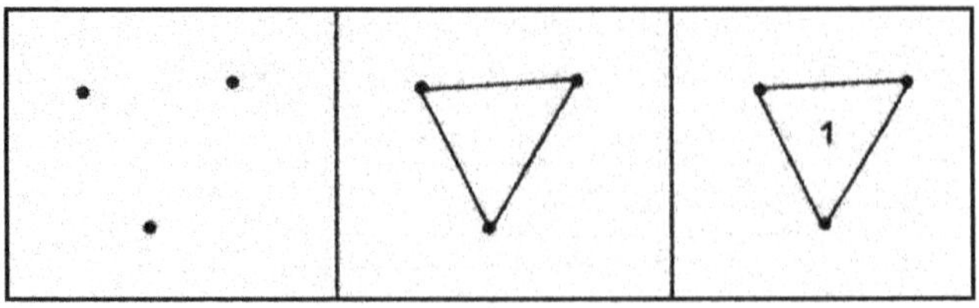

Figure 5-2. Shot group marking.

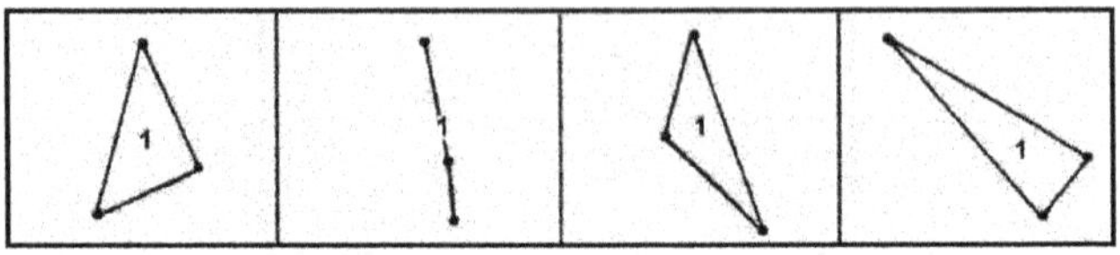

Figure 5-3. Central point of an odd-shaped group.

SINGLE SHOT GROUP ANALYSIS

5-11. The purpose of single shot group analysis is to identify firer errors on the single shots of a shot group so the Soldier can correct these errors while firing the next shot group.

5-12. Shot group analysis begins with the instructor/trainer observing the Soldier while he fires, looking for proper position, aim, trigger squeeze, and breathing. Then, the instructor/trainer analyzes the shot group to confirm problem areas.

NOTE: Coaches should not use shot group analysis without observing the firer.

5-13. The ideal shot group will have all three rounds within a 2-centimeter circle. Three rounds within a 4-centimeter circle is the minimum standard.

NOTE: M16A2 zeroing target squares are .96 centimeter in size, while M4 zeroing target squares are 1.3 centimeters in size.

MATCH-GRADE PERFORMANCE

5-14. The target shown in Figure 5-4 illustrates a match-grade quality weapon/ammunition combination. This combination places all bullets in almost the same hole and helps detect the firer's slightest errors.

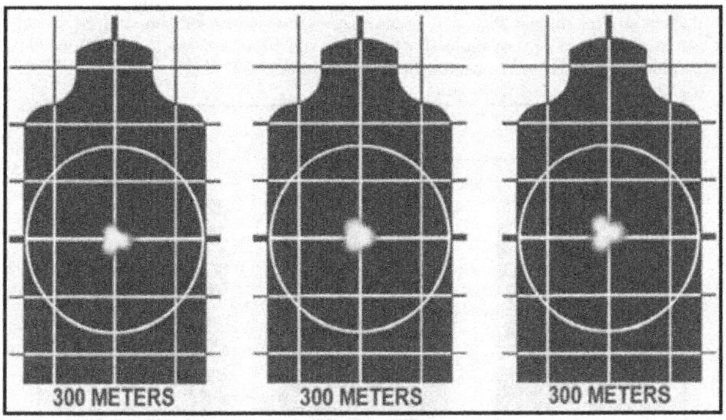

Figure 5-4. 25-meter match grade performance.

2-CENTIMETER SHOT GROUPS

5-15. The variances of standard weapons and ammunition must be considered during shot group analysis. When firing a standard service weapon/ammunition combination, the dispersion pattern may be up to 2 centimeters apart without human error. The instructor/trainer must ensure the Soldier understands that his weapon or ammunition may not be capable of placing three rounds within a 1-centimeter square.

NOTE: The dispersion pattern is not considered a firer error.

5-16. The targets shown in Figure 5-5 reflect proper 25-meter shot group performances using standard weapon/ammunition combinations.

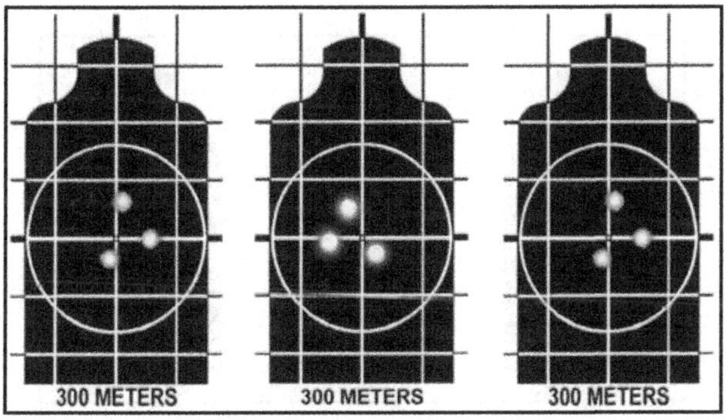

Figure 5-5. Shot groups with no firer error.

3- TO 4-CENTIMETER SHOT GROUPS

5-17. The targets shown in Figure 5-6 represent minimum acceptable firing performances; a better firing performance should be expected.

5-18. The instructor/trainer should ensure that the Soldier is properly applying the four marksmanship fundamentals and explain that this shot group size is due to minor shooting error, not weapon or ammunition performance. Any of these shot groups could have resulted from a minor change in sight picture, breathing, trigger squeeze, position, or an erratic round.

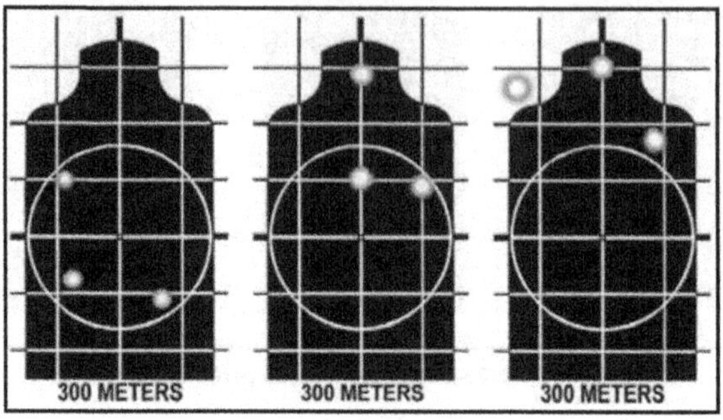

Figure 5-6. Shot groups with minor shooting error.

OVER 4- TO 5-CENTIMETER SHOT GROUPS

5-19. The targets shown in Figure 5-7 represent unacceptable firing performance.

5-20. The instructor/trainer should ensure that the Soldier is properly applying the four marksmanship fundamentals and explain that this shot group size is due to considerable shooting error, not weapon or ammunition performance. Any of these shot groups could have resulted from a change in position, sight picture, breathing, trigger squeeze, position, or an erratic round.

5-21. Soldiers who fire these shot groups should receive dry-fire training or remedial training on the EST 2000 or LMTS to help correct firing problems.

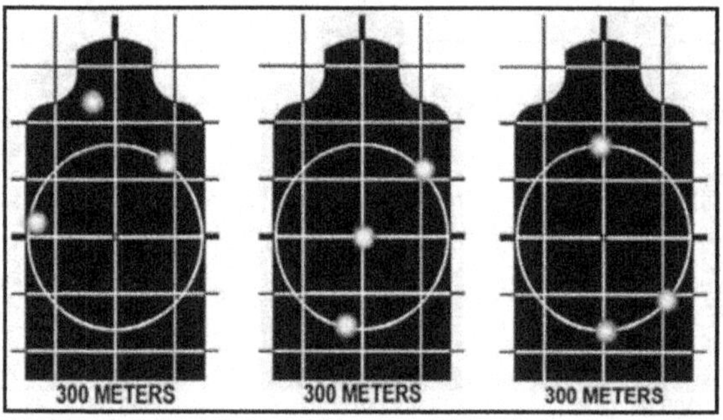

Figure 5-7. Shot groups with considerable shooting error.

Shot Groups Larger than 5 Centimeters

5-22. The targets shown in Figure 5-8 represent unacceptable firing performance; a better firing performance should be expected.

5-23. The instructor/trainer should ensure that the Soldier is properly applying the four marksmanship fundamentals and explain that this shot group size is due to major shooting error, not weapon or ammunition performance. Any of these shot groups could have resulted from a change in position, sight picture, breathing, or trigger squeeze, or the firer may have anticipated the shot.

5-24. Soldiers who fire these shot groups should receive extensive dry-fire training or remedial training on the EST 2000 or LMTS to help correct firing problems.

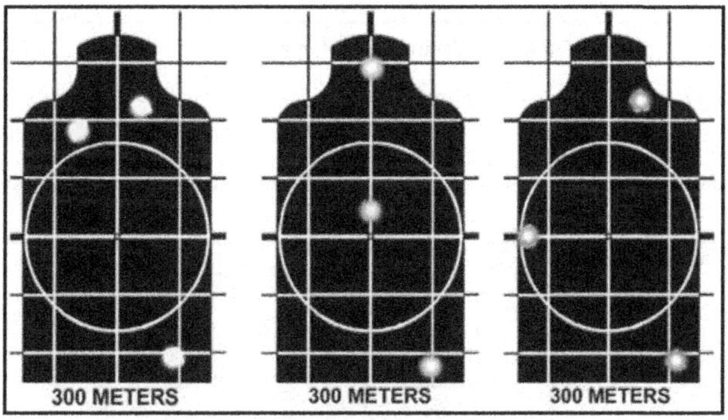

Figure 5-8. Shot groups with major shooting error.

MULTIPLE SHOT GROUP ANALYSIS

5-25. Multiple shot group analysis involves performing ongoing analysis of individual shot groups, while comparing them to each other for consistent aiming.

NOTE: If the Soldier is to benefit from this exercise, and if the instructor/trainer (or coach) is to provide useful guidance, the Soldier must mark each shot group individually and locate the center of more than one shot group.

5-26. To properly mark the shot groups (Figure 5-9)—

(1) Connect each individual shot group on the target with a straight line.

(2) Place a number inside the shot group.

NOTE: The number represents the center of the three shots.

(3) Connect the numbers, and place an X in the center.

NOTE: The X represents the center of the shot groups.

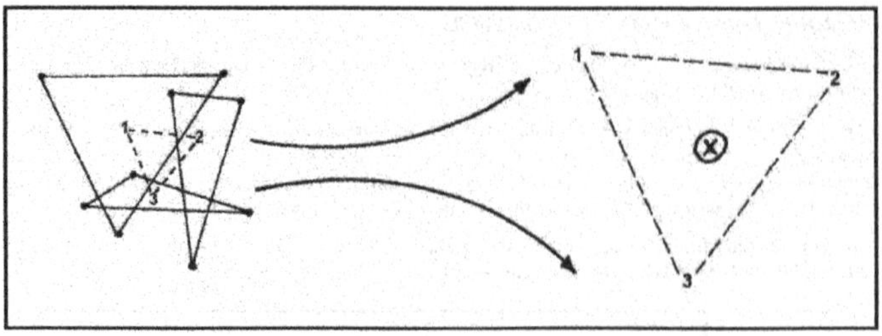

Figure 5-9. Central point of three shot groups.

Acceptable Shot Grouping Performance

5-27. The shot groups in Figure 5-10 represent acceptable shot groups (4 centimeters or less) in the same location. The Soldier firing this shot grouping should make a sight change of left 10 and down 4. Any change should be clearly marked on the target and saved for reference. The Soldier is then ready to zero his weapon.

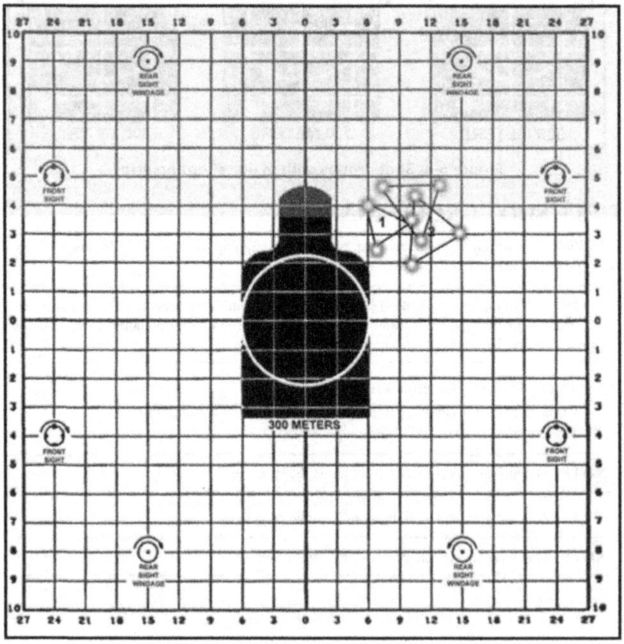

Figure 5-10. Acceptable shot grouping performance.

NOTES: 1. Location of the shot group on the 25-meter target is not important when conducting a grouping exercise. The size and dispersion of the shot groups are the main focus of this exercise.

2. Before the Soldier should be allowed to make any adjustments or start zeroing procedures, two consecutive shot groups must fall within a 4-centimeter circle when fired at 25 meters.

SHOT GROUPS WITH INCONSISTENT AIMING

5-28. The groups in Figure 5-11 indicate that the Soldier firing the shot groups is applying proper firing fundamentals, but is using a different point of aim each time a shot group is fired.

5-29. The instructor/trainer should question the Soldier's understanding of the aiming process and check his position for consistency. The instructor/trainer cannot determine which shot group best represents the firer's zero.

Figure 5-11. Shot groups with inconsistent aiming.

Shot Groups with Consistent Aiming and Major Shooting Error

5-30. The groups in Figure 5-12 indicate consistent aiming, but the Soldier is not applying the four fundamentals properly while firing each shot group.

5-31. The firer should be assigned a coach to troubleshoot his application of the four fundamentals in an attempt to isolate his firing errors.

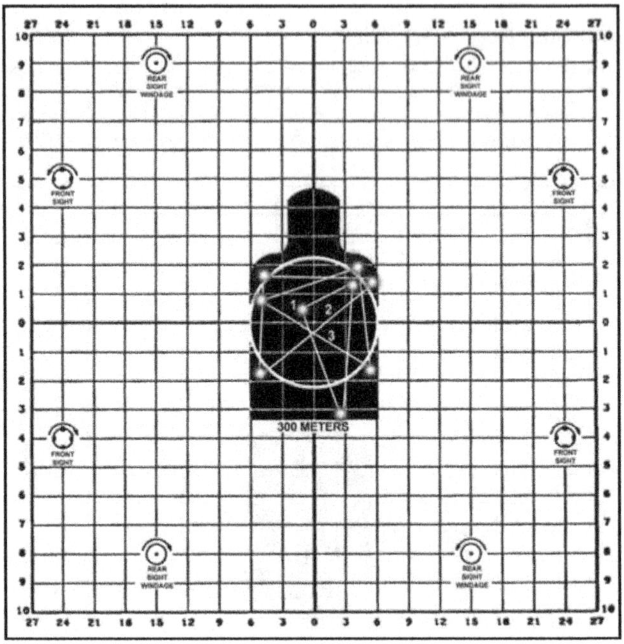

Figure 5-12. Shot groups with consistent aiming and major shooting error.

Shot Groups with Inconsistent Aiming and Major Shooting Error

5-32. The groups shown in Figure 5-13 indicate inconsistent aiming and major shooting error.

5-33. The firer should be assigned a coach to troubleshoot his application of the four fundamentals in an attempt to isolate his firing errors.

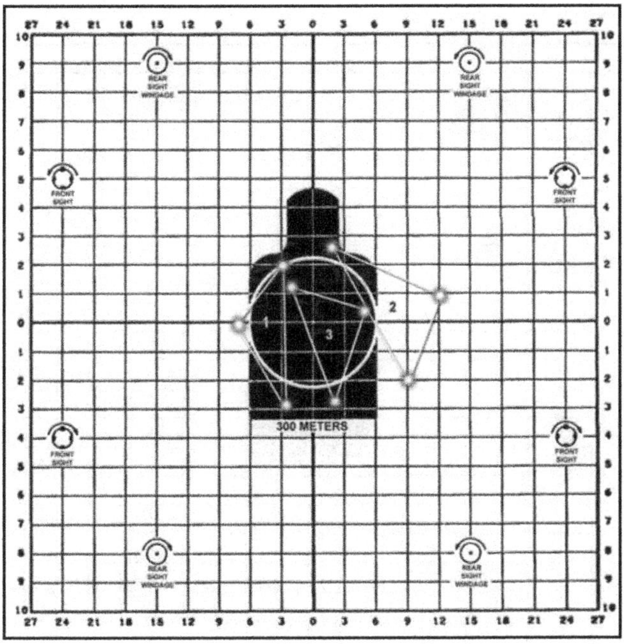

Figure 5-13. Shot groups with inconsistent aiming and major shooting error.

Shot Groups with Improper Vertical Placement

5-34. When viewed as nine shots, the shot groups shown in Figure 5-14 reflect proper horizontal placement of shots, but unsatisfactory vertical dispersion. This indicates a failure to vertically aim at the target's center of mass for each shot.

5-35. The instructor/trainer should check the Soldier's aiming procedure and adherence to marksmanship fundamentals.

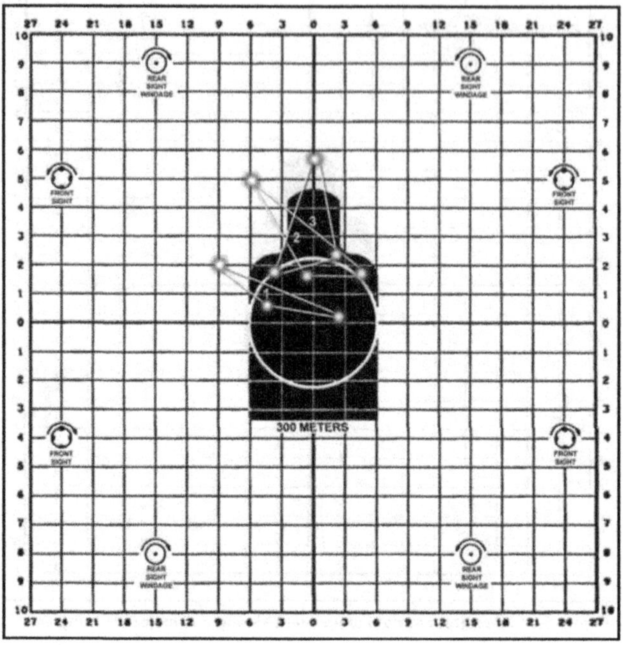

Figure 5-14. Shot groups with improper vertical placement.

IMPROPER SHOT GROUPS ON THE EDGE OF THE TARGET

5-36. The shot groups shown in Figure 5-15 are improper groups.

5-37. The Soldier should make a bold sight change to bring the groups closer to the target's center and ensure that the shot groups remain on the 25-meter zeroing target.

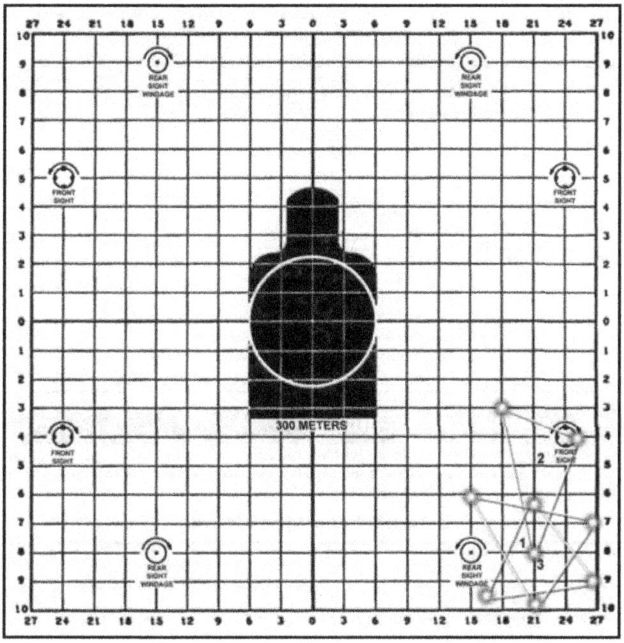

Figure 5-15. Improper shot groups on the edge of the target.

TROUBLESHOOTING THE FUNDAMENTALS

5-38. When troubleshooting the fundamentals, the coach's imagination is the only limiting factor. Table 5-2 outlines the techniques that can be used to identify errors in Soldiers' application of the fundamentals.

Table 5-2. Techniques used to identify errors in Soldiers' application of the fundamentals.

PROBLEM AREA	TECHNIQUE USED TO IDENTIFY THE PROBLEM AREA
Aiming	Attach the M16 sighting device, and observe the Soldier while he fires.
Breathing	Watch the rise and fall of the firer's chest for consistency.
Trigger squeeze	Place your finger over the firer's finger while he fires, feeling for jerking and smooth follow-through. Watch for jerking of the trigger and smooth follow-through.
Position	Observe the following areas for consistency: • Placement of the tip of the nose. • Placement of the trigger finger. • Placement of the nonfiring hand. • Placement of the legs. • Cheek-to-stock weld. • Positioning of equipment.
Other potential problem areas	Ensure that the— • Nonfiring-side eye is not shuttering. • Equipment is fitted properly. • Soldier is not flinching when the trigger is pulled. • Soldier is firing with the dominant eye. • Soldier is wearing glasses (if applicable). • Soldier is maximizing use of the supported position.

SECTION II. ZEROING PROCEDURES

This section provides guidelines for the instructor/trainer to zero M16-/M4-series weapons at 25 meters and at actual range (Table 5-3). It includes concept, organization, mechanical zero, zero recording, 25-meter sight settings, field fire sight settings, and troubleshooting of the fundamentals.

Table 5-3. Zeroing procedures.

ZEROING PROCEDURES
Instructional Intent
• Reinforce PMI while adjusting the confirmed shot groups (center of mass of the 4-centimeter circle with 5 out of 6 consecutive rounds at 25 meters).
Special Instructions
Ensure that— • The rear sight is on the proper setting (zero; M16A2/3=8/3+1; M16A4=6/3+2; M4=6/3). • The rear sight aperture is set on 300+1, not 800+1. • The M16A1's rear sight is set on the aperture marked L. • The small aperture is being used. • Proper and accurate shot group marking is enforced. • The firer's name is clearly marked on the target. • M16A2/A3/A4s zero on M16A2 zeroing targets. • M4s zero on M4 zeroing targets. • M16A1s zero on M16A1 zeroing targets.
Observables
• Coaches analyze the firer's application of the fundamentals. • The majority of the round must be inside of the circle to be counted. • Two consecutive 3-round shot groups are shot with 5 of 6 rounds in the 4- centimeter circle.

PURPOSE

5-39. The purpose of battlesight zeroing is to align the sights with the weapon's barrel given standard issue ammunition. When this is accomplished correctly, the point of aim and point of impact are the same at a given range (250 meters for the M16A1, 300 meters for the M16A2/A3/A4 and M4-series weapons). This sight setting provides the highest hit probability for most combat targets with minimum adjustment to the point of aim.

5-40. When standard zeroing procedures are followed, a properly zeroed weapon for one Soldier is close to the zero for another Soldier. When a straight line is drawn from the target's center to the tip of the front sightpost and through the center of the rear aperture, it makes little difference whose eye is looking along this line. There are many subtle factors that result in differences among individual zeros. Instructors/trainers should emphasize the similarity of individual zeros instead of the differences.

5-41. Most firers can fire with the same zeroed weapon if they properly apply marksmanship fundamentals. This information can be useful in three ways:

(1) If a Soldier has difficulty zeroing and the problem cannot be diagnosed, a good firer could zero the weapon to find the problem and eliminate the weapon as part of the problem.

(2) When a Soldier must fire another Soldier's weapon without opportunity to verify the zero by firing for example, picking up another man's weapon on the battlefield), the weapon will be closer to actual zero if the sights are left unchanged. This information is useful in deciding initial sight settings and recording zeros.

(3) All weapons in the arms room, even those not assigned, should have been previously zeroed by the last Soldier they were assigned to. Zeroing this newly assigned weapon should start with the sights left where they are.

SIGHT VARIANCE

5-42. There is no relationship between the specific sight settings a Soldier uses on his weapon to the sight settings he would use to zero another weapon, which makes it essential that each Soldier zeros the weapon that he is assigned. For example, a Soldier could zero his assigned weapon 10 clicks left of center; when zeroing another weapon, his adjustments could be 10 clicks right of center. This is due to the manufacturing difference between the weapons. Therefore, all newly assigned personnel should be required to zero their weapon as soon as possible after assignment to the unit. The same rule applies anytime a Soldier is assigned a weapon that is returned from field level or sustainment level maintenance, or when the zero is in question.

ORGANIZATION OF A 25-METER ZERO RANGE

NOTES: 1. All Soldiers should successfully group prior to zeroing.

2. If the Soldier is proficient at grouping, two shot groups should be fired to confirm proficiency prior to making any sight adjustments during zeroing procedures.

5-43. To properly conduct a 25-meter zero range, perform the following actions:

- Divide the unit into firing orders. The first order fires, while the second order coaches.
- Reserve firing points to conduct corrective instruction.
- Provide sandbags at each firing point to accommodate supported firing positions.

CONDUCT OF A 25-METER ZERO FIRING

5-44. On the 25-meter zero range, the Soldier applies the fundamentals while consistently aiming at the target's center of mass (A, Figure 5-16). The Soldier fires two separate three-round shot groups (B, Figure 5-16) and groups them. Based on the location of these groups, the Soldier makes the appropriate sight adjustments. Then, the Soldier fires two additional three-round shot groups to confirm that the adjustments have aligned the sights with the center of the target and that the bullets are in the 4-centimeter circle (Figure 5-17).

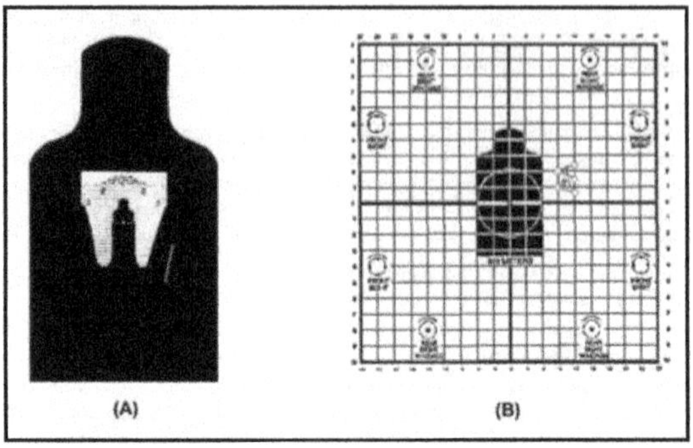

Figure 5-16. Correct aiming (A), initial shot group results (B).

Figure 5-17. Final shot group results.

5-45. To conduct a 25-meter zero range—

NOTES: 1. Each Soldier ensures that his sights are set for 25-meter zeroing.

2. Soldiers fire each shot from a supported firing position using the same point of aim (target's center of mass).

3. Ensure that the correct 25-meter zero target is being used. For M16A1s, use NSN 6920-01-167-1392 (Figure 5-18); for M16A2s, M16A3s, M16A4s, M4s, and M4As, use NSN 6920-01-395-2949 (Figure 5-19; M16A2/A3 is printed on one side, and M16A4/M4/M4A is printed on the other).

(1) The Soldier fires a three-round shot group at the 25-meter zeroing target.

(2) The firing line is cleared, and he moves downrange to examine the shot group for fundamental errors, triangulates the shot group, and puts the number 1 in the center of the shot group.

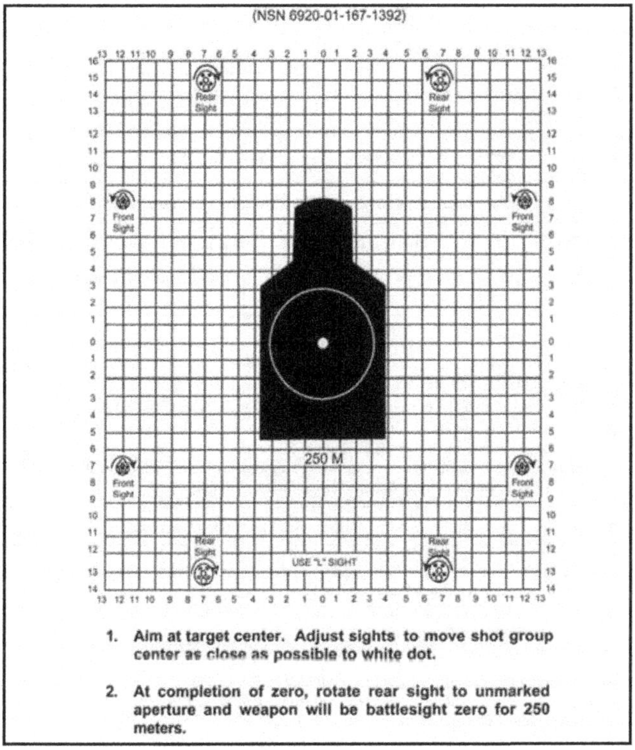

Figure 5-18. M16A1 25-meter zero target.

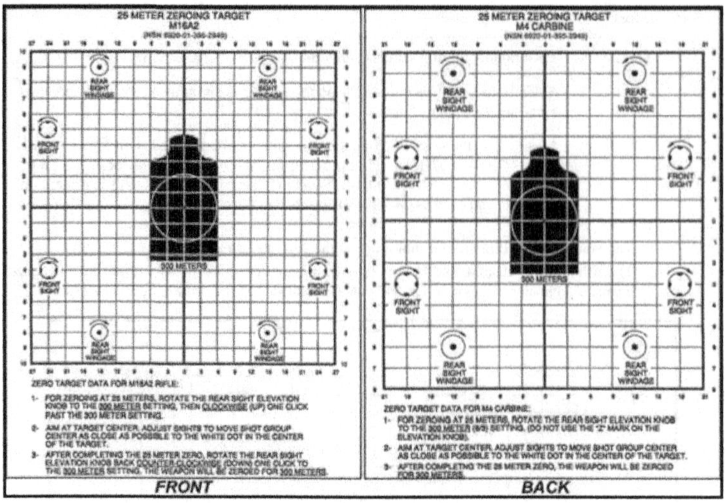

Figure 5-19. M16A2 and M4 25-meter zero target.

NOTES: 1. The Soldier fires two individual shot groups before a sight change is considered.

2. If the initial shot group is not on the target paper, the weapon should be mechanically zeroed before the Soldier fires this weapon again.

---

(3) The Soldier returns to the firing line and fires a second three-round shot group.

(4) The firing line is cleared, and the Soldier moves downrange to examine the second shot group, triangulate, and mark the center of the shot group with the number 2.

(5) The Soldier groups the two shot groups and marks the center of the two shot groups with an X.

(6) If the two shot groups fall within a 4-centimeter circle, the firer determines the sight adjustments he needs to make, identifies the horizontal and vertical lines closest to the X, and reads the 25-meter zeroing target to determine the proper sight adjustments. If the two shot groups do not fall within a 4-centimeter circle, the Soldier continues grouping.

NOTE: The majority of the round must be inside of the circle to be counted.

---

(7) The Soldier annotates any sight adjustments that need to be made to the weapon on the 25-meter zeroing target and ensures that his name is on the target.

(8) If five out of six rounds fell within the 4-centimeter circle, the Soldier is zeroed and can be removed from the firing line. If not, the Soldier returns to the firing line and makes sight adjustments.

5-46. Steps 1 through 8 are repeated until the Soldier places five out of six consecutive rounds inside the 4-centimeter circle. If the Soldier is not zeroed in 18 rounds, he should be removed from the firing line and given remedial training before attempting to zero again.

5-47. Once firing proficiency has been demonstrated from the supported firing position, zeroing exercises can be conducted from the unsupported firing position. For example, 18 rounds are allocated for the zeroing exercise. If the Soldier zeroes in 9 rounds, the Soldier can fire the remaining 9 rounds from the unsupported firing position.

Conduct of a 25-Meter Zero Firing Using the Location of Misses and Hits System

5-48. When using the LOMAH system on a KD range, zero confirmation is part of the program and will be shot as the first scenario. To conduct a 25-meter zero range using the LOMAH system, the Soldier shoots six rounds at the 175-meter/200-yard target while aiming at the target's center of mass. The outcome is evaluated using the following guidelines:

- If the shot group falls within the 11-inch circle on the LOMAH monitor, the Soldier continues the programmed scenario, which is identical to the downrange feedback scenario without LOMAH.
- If the Soldier shoots a shot group that is 11 inches or smaller but is clearly not zeroed, the instructor/trainer assists the Soldier in making sight adjustments based upon the data provided on the LOMAH monitor.
- If the shot group is not tight (greater than 11 inches), the Soldier should be removed from the firing line and given remedial training on the four fundamentals of marksmanship.

SECTION III. KNOWN DISTANCE RANGE

This section provides guidelines for the instructor/trainer to conduct a KD range and apply the effects of wind and gravity. This section also addresses three types of KD ranges: the standard KD range, the KD record fire range, and the modified field fire range.

NOTE: See Table 5-4 for the current training program.

Table 5-4. Downrange feedback.

DOWNRANGE FEEDBACK
Instructional Intent
• Reinforce PMI while shooting from the prone supported and unsupported firing positions. • Build the Soldier's confidence in his ability to hit where he aims while applying the effects of wind and gravity at range.
Special Instructions
Ensure that— • The effects of wind and gravity are thoroughly explained. • The rear sight is on the proper setting (M16A2/3=8/3; M16A4 and M4=6/3 flush; M16A1=the unmarked aperture, short-range). • The rear sight aperture is set on 300, not 800.
Observables
• Spotters provide correct feedback to firers. • Soldiers hit 8 of 10 targets at 100 meters. • Soldiers hit 14 of 20 targets at 200 meters. • Soldiers hit 5 of 10 targets at 300 meters.

CONCEPT

5-49. A KD range has three primary objectives:

(1) Fire tight shot groups at a known distance.

(2) Make sight adjustments at range (not an objective for IET Soldiers) while experiencing the effects of wind and gravity.

(3) Participate in marksmanship testing.

5-50. KD firing brings the Soldier one step closer to being able to fire during combat. The Soldier is provided information concerning the precise hit-or-miss location of every bullet fired. KD firing is conducted with a single, clearly visible target at a known distance, and the Soldier can establish a position that provides a natural point of aim on that single target. Consider the following:

- On the standard KD range, Soldiers fire at 100-, 200-, and 300-meter targets without any time constraints.
- On the KD record fire range, Soldiers fire at 100-, 200-, and 300-meter targets with time constraints.
- On the modified field fire range, Soldiers fire at 100-, 200-, and 300-meter targets on a standard 50- to 300-meter field fire qualification range.

NOTES: 1. If a qualification range is not available, this exercise may be shot on a standard 75- to 300-meter field fire range. Targets and target frames must be set up to accommodate this training.

2. On ranges that are built in yards instead of meters, the same KD targets will be used. The difference is so small that it does not need to be considered.

- The KD range does not require Soldiers to detect targets, estimate ranges to targets, scan sectors of fire, respond to surprise targets, respond to short-exposure targets, or engage multiple targets.
- An advantage of a KD range is the ability to see precisely where each bullet hits. To benefit from this training, Soldiers must clearly see the results of each firing, whether a group, single shot, or 10-round exercise.

KNOWN DISTANCE TARGET DESCRIPTION

5-51. Downrange feedback training should include detailed explanations of the targets. Consider the following:

- KD targets are large enough to capture all bullets fired. Standard E-type and F-type silhouettes can be used if standard KD targets are not available.
- The 16-centimeter circle on 100-meter targets, the 32-centimeter circle on 200-meter targets, and the 48-centimeter circle on 300-meter targets equate to the 4-centimeter zeroing target at 25 meters. If the Soldier's shot group falls within the 4-centimeter circle at 25 meters, it will fall within the circle on the target being shot. If the round falls outside of the circle, the round will clearly miss the 300-meter target (Figure 5-20).
- An X is located in the bottom portion of the circle to show the firer the target's center of mass, where he must aim when his weapon is zeroed.
- The grid system on the targets in Figure 5-20 equates to that of the 25-meter zeroing target. For example, one click on the front sightpost equals one square on the 25-meter zeroing target and also equals one square on the target being shot. Information similar to that on the zeroing target has been overprinted to help Soldiers apply sight adjustments.

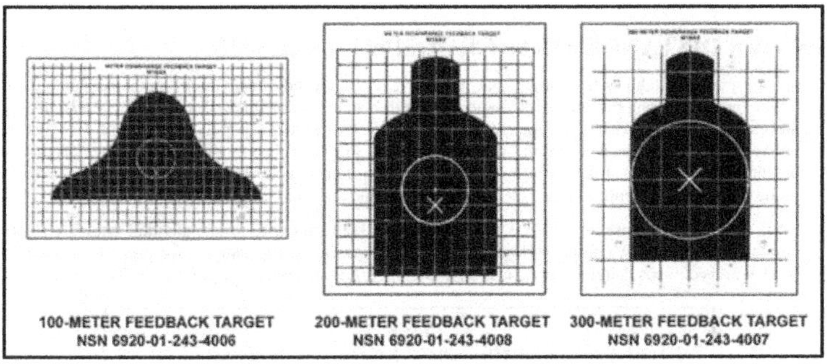

Figure 5-20. Downrange feedback targets.

MARKING KNOWN DISTANCE RANGE TARGETS

5-52. When the initial shot group is fired, target spotters/markers (Figure 5-21) should be placed in each bullet hole, placing the white side on the silhouette and the black side off of the silhouette. This procedure ensures that the firer can see where the rounds impacted and has two benefits:

- Instructors/trainers can observe the firer's performance and focus their attention on the Soldiers having the greatest problems.
- Soldiers are motivated to fire better since their peers can observe their performance.

5-53. On the second and subsequent shot groups, the target spotters/markers should be moved and placed in the holes of the new shot group. The old holes must be pasted using black pastors on black and white pastors on white. Failure to paste all bullet holes makes it difficult to determine one shot group from another.

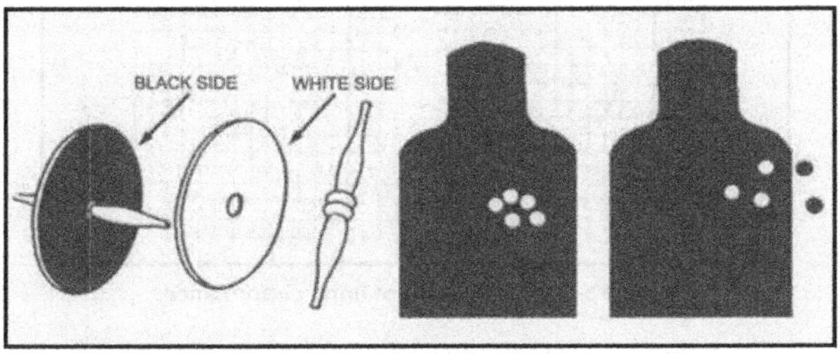

Figure 5-21. Target marking with spotters (markers).

KNOWN DISTANCE SHOT GROUPING ANALYSIS

5-54. Figure 5-22 shows two targets that were both shot with three individual rounds (A). On a pop-up target, these two firing performances would provide the same information to the firing line; each target was hit once and missed twice. Once the targets are properly marked with spotters, it becomes clear why only one round hit either target: The firer on the left is failing to properly apply the four fundamentals; the firer on the right needs to make an adjustment to his iron sights (assuming that wind was not a factor), triangulate the shot group, and read the appropriate adjustments from the target.

5-55. Figure 5-22 shows another two targets that were both shot with three individual rounds (B). On a pop-up target, these two firing performances would appear to be the same. Once properly marked with spotters, it is obvious that the firer on the left needs more training on the four fundamentals.

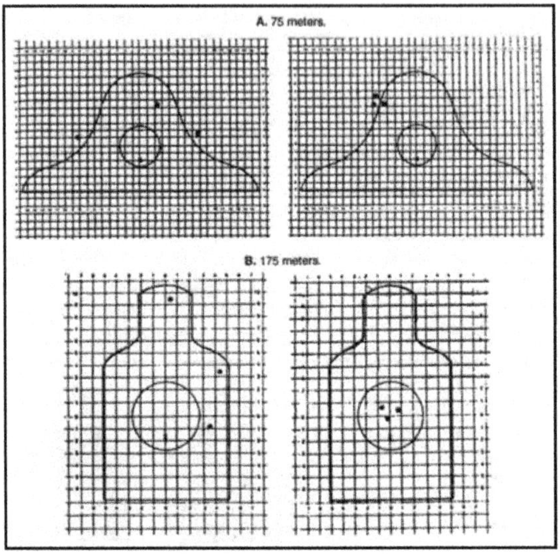

Figure 5-22. Comparison of firing performance.

KNOWN DISTANCE ZEROING

5-56. The 300-meter target can be used to confirm weapon zero or to refine the zero obtained on the 25-meter range. When Soldiers properly compensate for the wind, the zero on this target is more valid than the zero obtained on the 25-meter range. Soldiers should fire two five-round shot groups to confirm zero or three-round shot groups to refine their zero. The pit crews should spot targets after each shot group is fired. If the crosswind exceeds five miles per hour, KD zeroing should not be attempted.

NOTES: 1. IET Soldiers only zero on the 25-meter range.

2. For M16A2/3/4, M4, and M4A1 weapons only: Soldiers should use the unmarked aperture for zeroing and target engagement at all distances on the KD range. When engaging targets beyond 300 meters, the windage knob should be adjusted to the range of the target. 400-meter targets will be set on 4 flush, and 450-meter targets will be set on 4 plus two clicks.

3. For M16A1 rifles only: Soldiers should use the unmarked aperture (short-range) for refining zero at 300 meters. For target engagements beyond the 300-meter line, Soldiers should use the long-range aperture (L).

ELEVATION AND WINDAGE ADJUSTMENTS AT A DISTANCE

5-57. The elevation and windage rule states that one click of elevation or windage moves the strike of the bullet a specific distance at a specific range. At a range of 25 meters, one click of windage moves the strike of the bullet .33 centimeters, and one click of elevation on the front sight moves the strike of the bullet .83 centimeters. To compute the distance (D) one click of elevation (front sight) or windage moves the strike of a bullet at a given range (R)—

(1) Divide the range (expressed in meters) by 25.

(2) Multiply by either .33 centimeters for windage or .83 centimeters for elevation.

Windage: D = R ÷ 25 m x .33 (D = distance in cm)

Elevation: D = R ÷ 25 m x .83 (R = distance in cm)

Windage

5-58. To compute the distance one click of windage moves the strike of the bullet at a range of 300 meters—

(1) Divide 300 meters by 25 meters.

(2) Multiply by .33 centimeters.

D = 300 m ÷ 25 m = 12

12 x .33 = 3.96 cm

Round 3.96 cm up to 4.0 cm.

5-59. One click of windage moves the strike of the bullet 4 centimeters at 300 meters.

NOTE: Tables 5-5 and 5-6 show the amount of change in windage of the strike of the bullet at various ranges.

Elevation—Front Sight

5-60. To compute the distance one click of elevation (front sight) moves the strike of the bullet at a range of 300 meters—

(1) Divide 300 meters by 25 meters.

(2) Multiply by .83 centimeters.

D = 300 m ÷ 25 m = 12

12 x .83 = 9.96 cm

Round 9.96 cm up to 10 cm.

5-61. One click of elevation on the front sight moves the strike of the bullet 10 centimeters at a range of 300 meters.

NOTE: Tables 5-5 and 5-6 show the amount of change in elevation of the strike of the bullet at various ranges.

Elevation—Rear Sight

5-62. The elevation knob adjusts elevation 1.1 inch for each click at 100 meters, with each dot and each number representing one click of elevation.

Table 5-5. M16A2/3 and front sightpost of an M16A4.

DISTANCE (m)	DISTANCE 1 CLICK WILL ADJUST THE POINT OF IMPACT		
	Front Sightpost	Windage Knob	Elevation Wheel
25	.83 cm (3/8 in)	.33 cm (1/8 in)	.5 cm (1/4 in)
50	1.5 cm (5/8 in)	.5 cm (1/4 in)	1.5 cm (1/2 in)
75	2.5 cm (1 in)	1.0 cm (3/8 in)	2.0 cm (3/4 in)
100	3.5 cm (1 3/8 in)	1.5 cm (1/2 in)	2.75 cm (1 in)
150	5.0 cm (2 in)	2.0 cm (3/4 in)	4.0 cm (1 1/2 in)
175	6.0 cm (2 3/8 in)	2.25 cm (7/8 in)	5.0 cm (2 in)
200	6.5 cm (2 5/8 in)	2.5 cm (1 in)	5.5 cm (2 1/4 in)
250	8.5 cm (3 3/8 in)	3.5 cm (1 1/4 in)	7.0 cm (2 3/4 in)
300	10.0 cm (4 in)	4.0 cm (1 1/2 in)	8.5 cm (3 1/4 in)
400	13.5 cm (5 3/8 in)	5.5 cm (2 1/4 in)	11.0 cm (4 1/2 in)
500	17.0 cm (6.69 in)	6.5 cm (2 1/2 in)	14.0 cm (5 1/2 in)
600	20.5 cm (8.05 in)	8.0 cm (3 1/8 in)	16.75 cm (6 1/2 in)
700	24.0 cm (9.45 in)	9.0 cm (3 5/8 in)	19.5 cm (7 1/2 in)
800	27.5 cm (10.83 in)	10.5 cm (4 1/8 in)	22.5 cm (8 3/4 in)
NOTE: All values are rounded off.			

Table 5-6. M4/M4A1 and windage of an M16A4.

DISTANCE (m)	DISTANCE 1 CLICK WILL ADJUST THE POINT OF IMPACT		
	Front Sightpost	Windage Knob	Elevation Wheel
25	1.2 cm (1/2 in)	.5 cm (1/4 in)	.5 cm (1/4 in)
50	2.4 cm (1 in)	1.5 cm (1/2 in)	1.5 cm (1/2 in)
75	3.6 cm (1 1/2 in)	2.0 cm (3/4 in)	2.0 cm (3/4 in)
100	4.8 cm (1 7/8 in)	2.75 cm (1 in)	2.75 cm (1 in)
150	7.2 cm (2 7/8 in)	4.0 cm (1 1/2 in)	4.0 cm (1 1/2 in)
175	8.4 cm (3 3/8 in)	5.0 cm (2 in)	4.0 cm (2 in)
200	9.6 cm (3 3/4 in)	5.5 cm (2 1/4 in)	5.5 cm (2 1/4 in)
250	12.0 cm (4 3/4 in)	7.0 cm (2 3/4 in)	7.0 cm (2 3/4 in)
300	14.4 cm (5 3/4 in)	8.5 cm (3 1/4 in)	8.5 cm (3 1/4 in)
400	19.2 cm (7 1/2 in)	11.0 cm (4 1/2 in)	11.0 cm (4 1/2 in)
500	24.0 cm (9 1/2 in)	14.0 cm (5 1/2 in)	14.0 cm (5 1/2 in)
600	28.8 cm (11 1/4 in)	16.75 cm (6 1/2 in)	16.75 cm (6 1/2 in)
NOTE: All values are rounded off.			

CONDUCT OF A STANDARD KNOWN DISTANCE RANGE

NOTE: If the range is equipped with the LOMAH system, a firing order will be used to operate the LOMAH throughout the period of instruction and will be fired last.

5-63. Standard KD ranges are conducted using the following considerations:

- The standard KD range is conducted with paper targets at 100, 200, and 300 meters to obtain downrange feedback at 75, 175, and 300 meters (Figure 5-23). Soldiers fire shot groups progressively at the targets.
- Half of the bullets are fired from the supported firing position, and the other half are fired from the unsupported firing position.
- The wind speed and direction must be determined before firing, and the firer must know the distance to the target.
- Soldiers mark the targets after firing each shot group. Based on this feedback, Soldiers receive a critique from their instructor/trainer or coach and apply the adjusted point of aim (as necessary).
- The downrange feedback exercise must be conducted within the constraints of time, ammunition, and available ranges.
- If 30 rounds of ammunition are available for training, firing three-round shot groups 10 times is preferable to firing five-round shot groups 6 times.
- Once the Soldier understands the concept for adjusting the point of aim to compensate for the effects of wind and gravity, he is ready to apply his knowledge on the field fire range.

75-METER TARGETS

5-64. Instructors/trainers can provide feedback after each round, each three-round shot group, or each five-round shot group on the 100-meter feedback targets. No time limit is placed on the firer. Soldiers fire from the supported firing position and from the unsupported firing position. Then, the targets are marked and evaluated. Feedback consists of a critique of performance, adjustments to the point of aim, effects of wind and gravity, and shot placement. Target spotters mark the bullet holes so hits can be viewed from the firing line.

NOTE: IET Soldiers fire one five-round shot group from the supported firing position and one five-round shot group from the unsupported firing position. They must hit 8 out of 10 targets.

175-METER TARGETS

5-65. Firers engage the 200-meter target using the same downrange procedures as when engaging the 100-meter target.

NOTE: IET Soldiers fire 10 rounds from the supported firing position and 10 rounds from the unsupported firing position. They must hit 14 out of 20 targets.

300-METER TARGETS

5-66. Firers engage the 300-meter target using the same downrange procedures as when engaging the 100-meter target.

NOTE: IET Soldiers fire one five-round shot group from the supported firing position and one five-round shot group from the unsupported firing position. They must hit 5 out of 10 targets.

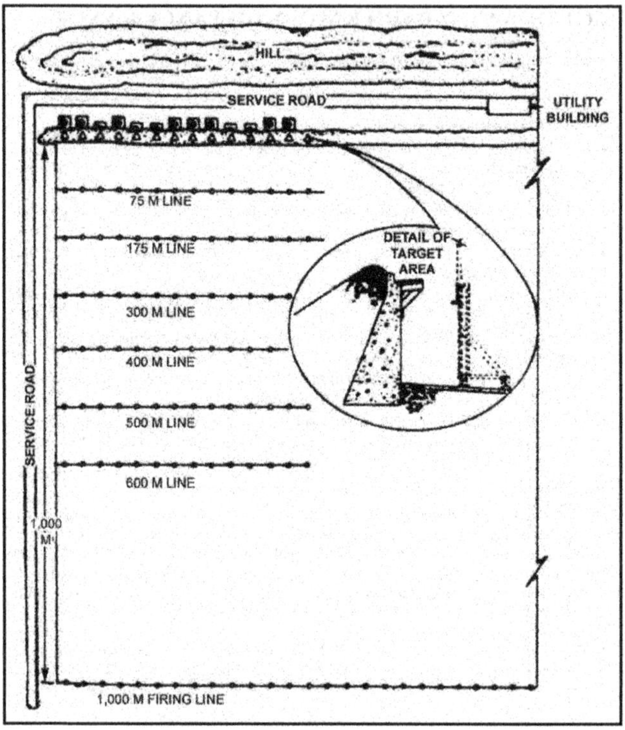

Figure 5-23. Known distance range.

KNOWN DISTANCE RECORD FIRE RANGE

NOTE: See paragraphs 6-79 through 6-82 of Chapter 6 for information about the alternate course KD record fire range.

MODIFIED FIELD FIRE RANGE

5-67. A modified field fire range can be used for downrange feedback. To conduct downrange feedback, minor changes must be made to a standard field fire range. Target frames, like those used on the 25-meter range, are placed on a standard qualification range at 100, 200, and 300 meters. The standard KD range or the KD record fire range can be conducted on the modified field fire range.

NOTE: The firing line must be cleared, moved to the targets for marking, and returned each time a firing order fires.

RECORD OF PERFORMANCE

5-68. During the conduct of downrange feedback, a record of performance should be kept on DA Form 5239-R (100-, 200-, and 300-Meter Downrange Feedback Scorecard) for the AAR.

NOTE: See Appendix B for a sample completed form and the end of this publication for a blank, reproducible copy.

5-69. As Soldiers complete each phase and achieve the performance standard for that range, they should receive a critique. Instructors/trainers must ensure that Soldiers do not progress to a greater range until they become proficient at closer ranges.

25-METER ZERO STANDARD

5-70. A standard E-type silhouette is 48.26 centimeters wide; a circle (angle) that is 48.26 centimeters at 300 meters is 4 centimeters at 25 meters. A Soldier who can fire all bullets in a 4-centimeter circle at 25 meters and adjust the sights for zero will hit the target at ranges as far away as 300 meters (Figure 5-24).

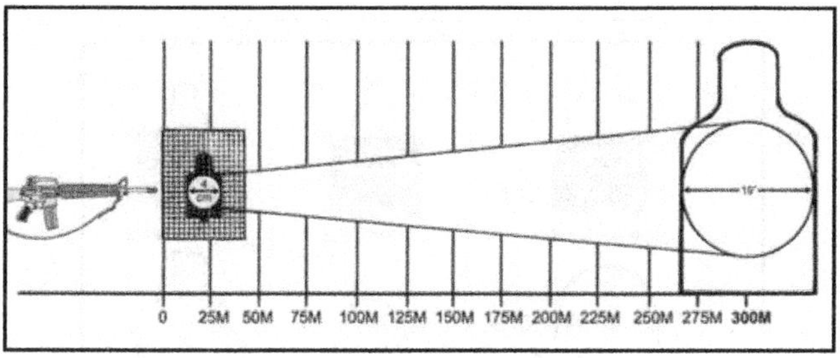

Figure 5-24. 25-meter zero standard.

SECTION IV. EFFECTS OF WIND AND GRAVITY

Marksmanship instructors/trainers should know how the effects of wind and gravity influence the flight of the bullet, and Soldiers should know how to compensate for such bullet displacement. This instruction is appropriate for all marksmanship training and concurrent training.

EFFECTS OF GRAVITY

5-71. Gases created by gunpowder push each round out of the end of the barrel. The barrel must be elevated slightly to allow the round to travel farther, creating an arc. The round will travel straight until it slows down and is gradually pulled to the ground by gravity. Each round fired will be pushed approximately the same distance and will roughly follow the same path.

NOTE: The farther the round travels, the faster it begins to fall.

5-72. When the firer zeroes his weapon, he aligns his line of sight to cross the path of the round at the distance at which he wants to zero his weapon. For example, a 300-meter zero means that the line of sight crosses the path of the round at 300 meters. If the firer engages a target at a distance other than 300 meters (excluding 25 meters), the path of the round hits the target either before or after it crosses the line of sight. If the firer wants his rounds to impact the center of mass, he must adjust his point of aim up or down to account for gravity.

ADJUSTED POINT OF AIM BASED ON GRAVITY

5-73. An adjusted point of aim (Figure 5-25) is intended to increase hit probability when properly presented. However, Soldiers can become confused, which could result in degraded performance. All Soldiers should be taught to aim at the target's center of mass unless they are confident that they know the range to the target. If adjusting the point of aim confuses the Soldier, he should aim at the target's center of mass. These points of aim place the center of each shot group in the target's center of mass (assuming a perfect zero and no firer error).

NOTES:
1. These adjustments are small and should only be applied by competent firers who wish to improve their firing performance.
2. Because the difference between M16- and M4-series weapons is so small and to avoid confusion, the same adjusted points of aim should be used regardless of the weapon being fired.

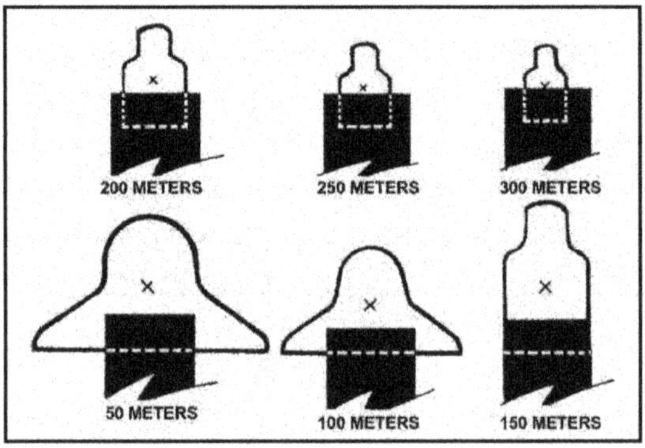

Figure 5-25. M16-/M4-series weapon aiming points.

EFFECTS OF WIND

5-74. Wind affects the bullet similar to the way gravity does: the farther the round travels, the farther the wind will push the round in the direction the wind is blowing. The faster the wind is blowing, the farther the wind will push the bullet.

WIND DIRECTION

5-75. The effects of wind vary depending on changes in wind speed and direction. Wind is classified by the direction it is blowing in relationship to the firer/target line. The clock system is used to indicate wind direction and value (Figure 5-26). This system works as follows:

- Winds that blow from the left (9 o'clock) or right (3 o'clock) are called full-value winds because they have the most effect on the bullet.
- Winds that blow at an angle from the front or rear are called half-value winds because they have about half the effect on the bullet as full-value winds.
- Winds that blow straight into the firer's face or winds that blow straight into the target are termed no-value winds because they have minimal effect on the bullet.

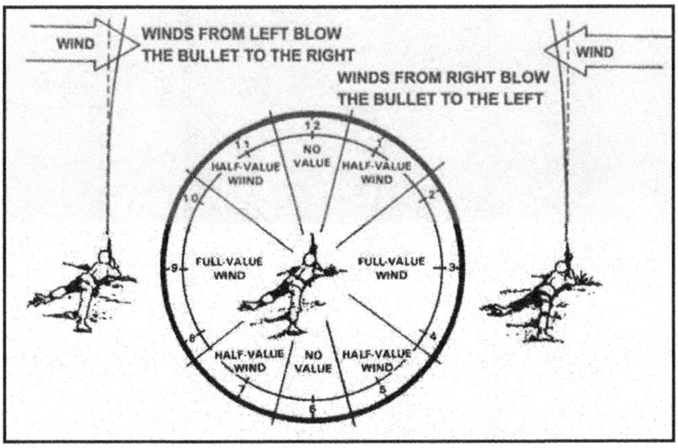

Figure 5-26. Determine wind value using the clock method.

WIND SPEED

5-76. Wind is variable and sometimes quite different at the firing position than at the target's position. Consider the following:

- When wind is blowing hard at the firing line, trees, brush, or terrain could protect the bullet's path.
- Wind can vary by several miles per hour between the time a measurement is taken and when the bullet is fired.

5-77. Therefore, training time should not be wasted trying to teach Soldiers an exact way to measure wind speed. Soldiers should understand that wind can blow a bullet off course, but they should not overcompensate and miss targets by applying too much hold-off.

5-78. A wind gauge can be used for precise measurement of wind velocity. When a gauge is not available, velocity is estimated using one of the following methods:

- Flag method.
- Pointing method.
- Observation method.

Flag Method

5-79. To perform the flag method (Figure 5-27)—

(1) Observe a flag or any cloth-like material hanging from a pole.
(2) Estimate the angle formed at the juncture of the flag and pole.
(3) Divide this angle by the number 4.

NOTE: The answer is the wind velocity expressed in miles per hour.

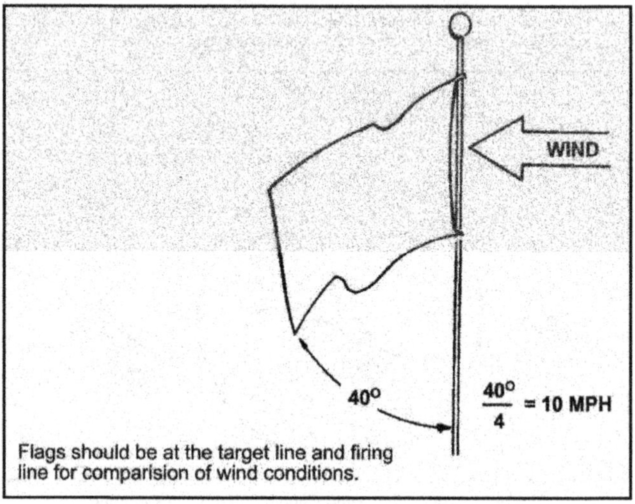

Figure 5-27. Determine wind speed using the flag method.

Pointing Method

5-80. If a flag is not visible, the firer can use the pointing method. To perform the pointing method (Figure 5-28)—

(1) Drop a piece of paper, leaf, or other light material from the shoulder.

(2) Point directly at the place where it lands.

(3) Estimate the angle created by the pointing arm.

(4) Divide this angle by the number 4.

NOTE: The answer is the approximate wind speed at the firing position expressed in miles per hour.

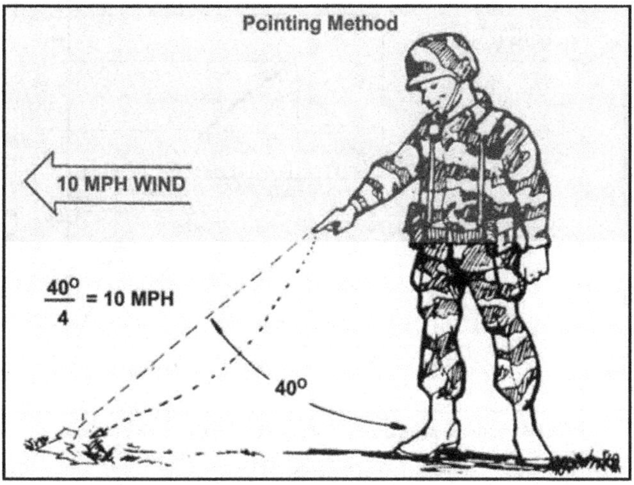

Figure 5-28. Determine wind speed using the pointing method.

Observation Method

5-81. If the flag or pointing methods cannot be used, the following information can assist in determining wind velocities:

- Winds less than 3 miles per hour can barely be felt by the firer, but the presence of slight wind can be determined by drifting smoke.
- Winds of 3 to 5 miles per hour can be felt lightly over the firer's face.
- Winds of 5 to 8 miles per hour constantly move the leaves of trees.
- Winds of 8 to 12 miles per hour raise dust and loose paper.
- Winds of 12 to 15 miles per hour cause small trees to sway.

ADJUSTED POINT OF AIM BASED ON WIND SPEED

5-82. Figure 5-29 illustrates how the effects of wind on the bullet are similar to the effects of gravity—as range increases, the effect of wind increases. For example, a 10 mile-per-hour full-value wind moves an M16A1 (M193) bullet from about ½ of an inch at 25 meters to about 15 inches at 300 meters.

5-83. Table 5-7 displays the wind effects for all conditions—a wind of greater speed increases bullet movement by a uniform amount. For example, a 15 mile-per-hour wind moves the bullet ¾ of an inch at 25 meters and about 22 ½ inches at 300 meters. A half-value wind moves the strike of the round in a 10 mile-per-hour wind ¼ of an inch at 25 meters and 7 ½ inches at 300 meters.

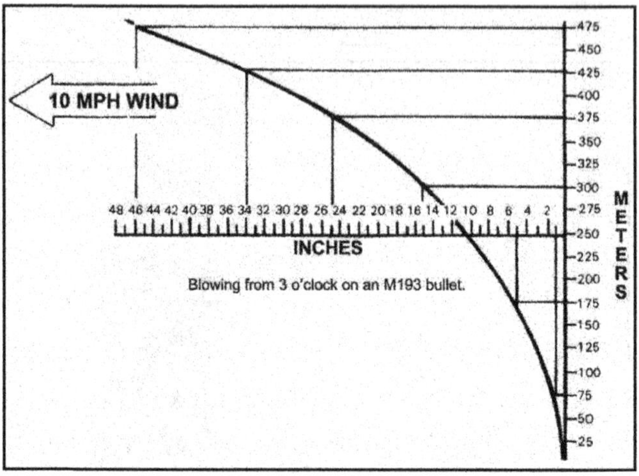

Figure 5-29. Calculate the adjusted point of aim based on wind speed.

NOTE: Table 5-7 can be used to calculate the M193 adjusted point of aim based on wind speed.

Table 5-7. M193 calculated adjusted point of aim based on wind speed (full value).

WIND SPEED (mph)	RANGE (m)								
	25	50	75	100	150	175	200	250	300
	DISTANCE MOVED (in)								
5	1/4	3/8	1/2	1	2	2.5	3.5	5	7.5
10	1/2	3/4	1	2	4	5	7	10	15
15	3/4	1-1/8	1.5	3	6	7.5	10.5	15	22.5

Drift for a 10 Mile-per-Hour Wind Using 5.56-Millimeter M855 Ammunition

5-84. Table 5-8 illustrates the drift using M855 5.56-millimeter ball ammunition fired in an M16A2 rifle with a 300-meter battlesight zero.

Table 5-8. Drift for 10 mile-per-hour wind using M855 ammunition.

DRIFT FOR 10 mph WIND USING M855 5.56-mm BALL AMMUNITION FIRED IN M16A2 RIFLE WITH A 300-METER BATTLESIGHT ZERO				
Range (m)	**Velocity (fps)**	**Trajectory (in)**	**Drop (in)**	**Drift (in)**
0	3,100	-2.5	0.0	0.0
100	2,751	+4.4	-2.3	1.1
200	2,420	+5.8	-10.2	4.9
300	2,115	0.0	-25.3	11.8
400	1,833	-15.0	-49.5	22.4
500	1,569	-42.9	-86.7	38.0
600	1,323	-88.2	-141.3	59.5
700	1,106	-156.1	-220.9	88.4
800	1,010	-267.7	-339.2	124.9

Adjusted Point of Aim Based on Gravity and Wind Speed

5-85. Wind has a minor effect on the M16 bullet (relative to the size of the target) at ranges out to 100 meters. When engaging targets in excess of 150 meters in heavy winds, Soldiers adjust the point of aim for the wind to increase the probability of a hit. Wind effects are uniform in relation to speed—that is, a 5 mile-per-hour wind has half the effect of a 10 mile-per-hour wind, and a 20 mile-per-hour wind has twice the effect of a 10 mile-per-hour wind.

5-86. Firers must adjust their points of aim into the wind to compensate for its effects. If they miss a distant target and wind is blowing from the right, they should aim to the right for the next shot. A guide for the initial adjustment is to split the front sightpost on the edge of the target facing the wind (Figure 5-30).

5-87. Newly assigned Soldiers should aim at the target's center of mass for the first shot, and then adjust for wind when they are confident that wind caused the miss. Experienced firers should apply the appropriate hold-off for the first shot, but should follow the basic rule—when in doubt, aim at the center of mass.

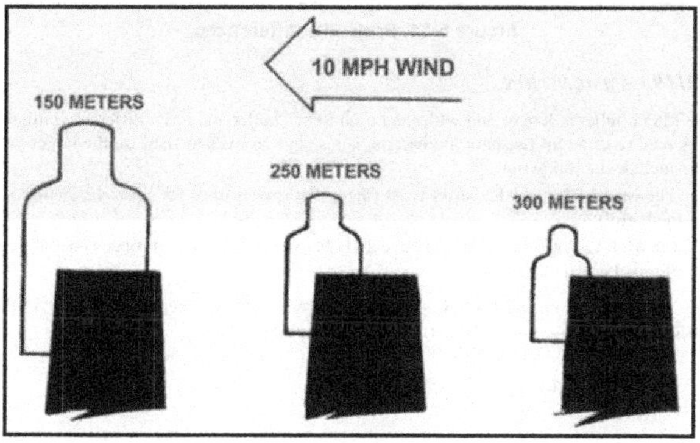

Figure 5-30. M16-/M4-series weapons adjusted point of aim based on wind speed.

SECTION V. BALLISTICS

Commanders and marksmanship trainers must understand some aspects of ballistics to teach the principles of zeroing and engagement of long-range targets. Ballistics is a science dealing with the motion and flight characteristics of projectiles. The study of ballistics in rifles and carbines is divided into three categories:

- Internal ballistics.
- External ballistics.
- Terminal ballistics.

INTERNAL BALLISTICS

5-88. Internal ballistics deals with what happens to the bullet before it leaves the weapon's muzzle.

5-89. The overall dimensions of the combat service 5.56-millimeter cartridges are the same, which allows cartridges to be fired safely in M16-series rifles and M4 carbines, but there are internal differences that affect firing accuracy (Figure 5-31).

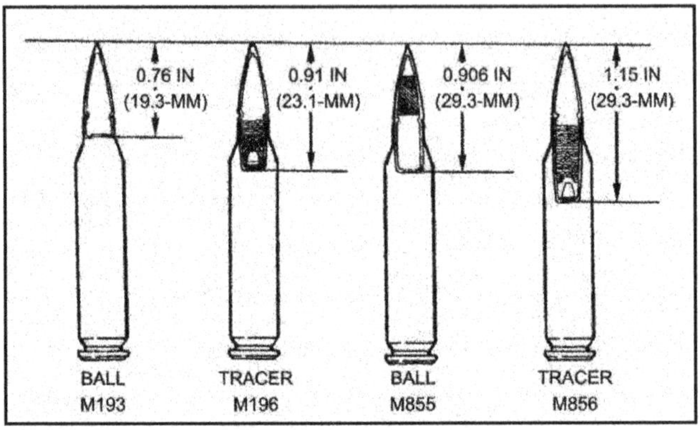

Figure 5-31. Projectile differences.

M855 AND M193 AMMUNITION

5-90. The M855 bullet is longer and wider than the M193 bullet and has a different configuration. These differences require different twists in the barrels, lands, and grooves to stabilize the bullet in flight. These differences include the following:

- The M16A1 has a 1:12 barrel twist (the bullet rotates once for every 12 inches of travel down the barrel).
- The M16A2/A3/A4 and the M4 have a 1:7 barrel twist (the bullet rotates once for every 7 inches of travel down the barrel).

5-91. The M16A1 does not put enough spin on the heavier M855 bullet to stabilize it in flight, causing erratic performance and inaccuracy (Figure 5-32). The shot groups are—

- 30.48 to 35.56 centimeters (12 to 14 inches) at 91.4 meters (100 yards).
- 182.88 centimeters (72 inches) at 274.2 meters (300 yards).

NOTE: Although firing the M855 cartridge in the M16A1 rifle is safe, it should only be used in a combat emergency, and then only for close ranges of 91.4 meters (100 yards) or less.

5-92. The M16A2/A3/A4 rifle and M4/M4A1 carbine fire both M193 and M855 ball ammunition with little difference in accuracy to a range of 500 meters. The M16A2/A3/A4 and M4/M4A1 and their ammunition are more effective than the M16A1 at ranges out to and beyond 500 meters due to better stabilization of the round.

5-93. The three 10-round shot groups in Figure 5-32 (A) were fired by a skilled marksman at a distance of 274.2 meters (300 yards) and 91.4 meters (100 yards) using the same M16A1 rifle.

- At 300 yards, the 25.4-centimeter shot group (shown on the left) was fired (and zeroed) with M193 ammunition.
- The 6-foot shot group (shown on the right) was fired with M855 ammunition.
- At 100 yards, the 35.56-centimeter (14-inch) shot group (shown in the center) was fired with M855 ammunition.

5-94. Figure 5-32 (B) shows two 25.4-centimeter (12-inch) shot groups fired by the same skilled marksman at a distance of 274.2 meters (300 yards) using an M16A2 rifle.

- The shot group on the left was fired and zeroed with M855 ammunition.
- The shot group on the right was fired using M193 ammunition.

NOTE: Both M193 and M855 ball ammunition can be used in training and accurately function in M16A2/3/4 rifles and M4/M4A1 carbines. Due to the different characteristics of each round, zero with the type of ammunition used for training. Do not switch between the types during firing. Do not zero with one type, and then fire the other for any type of training.

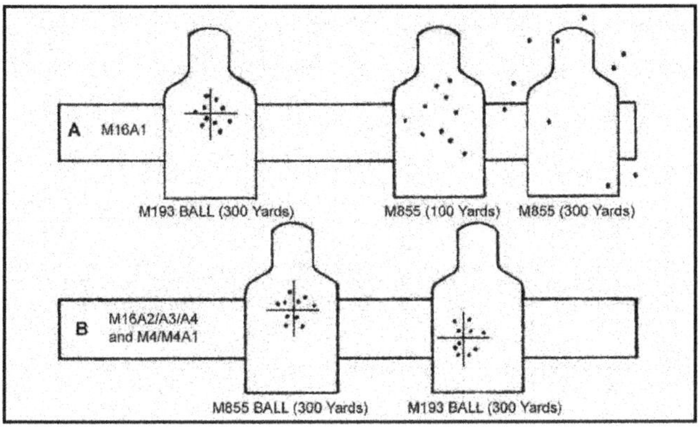

Figure 5-32. Ammunition impact comparison.

5-95. Table 5-9 and Figure 5-33 show the difference between a M16A2/A3/A4 and a M4/M4A1 weapon zeroed with M855 ammunition and then re-zeroed with M193 ammunition at 300 meters. There is practically no difference between the trajectory of the rounds and the impact of the rounds on the target.

Table 5-9. M855 zeroed at 300 meters; M193 re-zeroed at 300 meters using an M16A2/A3/A4 rifle and an M4 carbine.

RANGE (m)	FRONT SIGHT HEIGHT	THETA		BULLET DROP		BULLET LOCATION	
		M193	M855	M193	M855	M193	M855
0	2.568	0.0021987	0.00235575	0	0	-2.57	-2.57
25	2.568	0.0021987	0.00235575	-0.125	-0.14	-0.53	-0.39
50	2.568	0.0021987	0.00235575	-0.52	-0.58	1.24	1.49
75	2.568	0.0021987	0.00235575	-1.21	-1.33	2.71	3.06
100	2.568	0.0021987	0.00235575	-2.1	-2.4	3.99	4.31
125	2.568	0.0021987	0.00235575	-3.18	-3.82	5.07	5.21
150	2.568	0.0021987	0.00235575	-4.91	-5.61	5.51	5.73
175	2.568	0.0021987	0.00235575	-6.89	-7.78	5.69	5.88
200	2.568	0.0021987	0.00235575	-9.16	-10.4	5.58	5.58
225	2.568	0.0021987	0.00235575	-11.8	-13.4	5.11	4.90
250	2.568	0.0021987	0.00235575	-15.4	-16.9	3.67	3.72
275	2.568	0.0021987	0.00235575	-19.2	-20.8	2.04	2.14
300	2.568	0.0021987	0.00235575	-23.44	-25.3	-0.04	-0.04
325	2.568	0.0021987	0.00235575	-28.3	-30.4	-2.74	-2.83
350	2.568	0.0021987	0.00235575	-33.8	-36.0	-6.07	-6.11
375	2.568	0.0021987	0.00235575	-40.13	-42.3	-10.24	-10.09
400	2.568	0.0021987	0.00235575	-46.78	-49.3	-14.72	-14.77

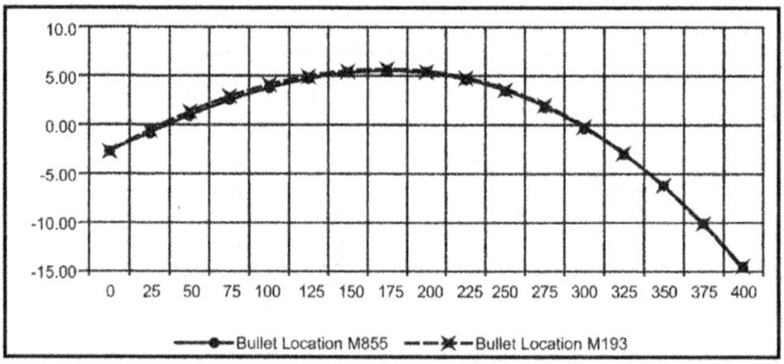

Figure 5-33. M855 zeroed at 300 meters; M193 re-zeroed at 300 meters.

5-96. When zeroing using M855 and M193 ammunition at 25 meters, the differences in the ammunition becomes apparent (Table 5-10 and Figure 5-34). If firing M193 ammunition from an M16A2/A3/A4 rifle or an M4/M4A1 carbine, the weapon should be zeroed with M193 ammunition.

Table 5-10. M855 zeroed at 25 meters; M193 re-zeroed at 25 meters.

RANGE (m)	FRONT SIGHT HEIGHT	THETA		BULLET DROP		BULLET LOCATION	
		M193	M855	M193	M855	M193	M855
0	2.568	0.0023383	0.00235575	0	0	-2.57	-2.57
25	2.568	0.0023383	0.00235575	-0.125	-0.14	-0.39	-0.39
50	2.568	0.0023383	0.00235575	-0.52	-0.58	1.51	1.49
75	2.568	0.0023383	0.00235575	-1.21	-1.33	3.13	3.06
100	2.568	0.0023383	0.00235575	-2.1	-2.4	4.54	4.31
125	2.568	0.0023383	0.00235575	-3.18	-3.82	5.76	5.21
150	2.568	0.0023383	0.00235575	-4.91	-5.61	6.33	5.73
175	2.568	0.0023383	0.00235575	-6.89	-7.78	6.65	5.88
200	2.568	0.0023383	0.00235575	-9.16	-10.4	6.68	5.58
225	2.568	0.0023383	0.00235575	-11.8	-13.4	6.35	4.90
250	2.568	0.0023383	0.00235575	-15.4	-16.9	5.05	3.72
275	2.568	0.0023383	0.00235575	-19.2	-20.8	3.55	2.14
300	2.568	0.0023383	0.00235575	-23.44	-25.3	1.61	-0.04
325	2.568	0.0023383	0.00235575	-28.3	-30.4	-0.95	-2.83
350	2.568	0.0023383	0.00235575	-33.8	-36.0	-4.15	-6.11
375	2.568	0.0023383	0.00235575	-40.13	-42.3	-8.18	-10.09
400	2.568	0.0023383	0.00235575	-46.78	-49.3	-12.52	-14.77

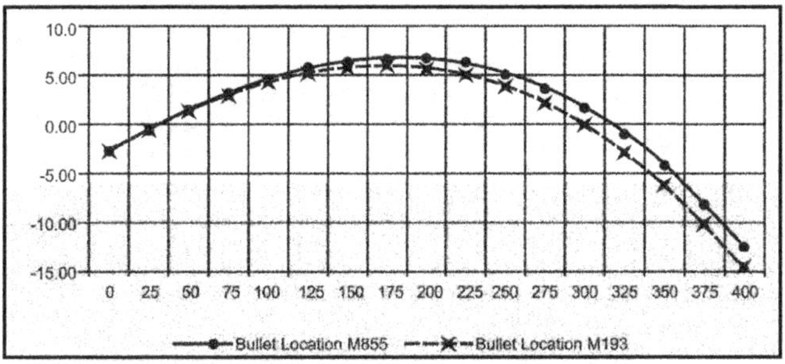

Figure 5-34. M855 zeroed at 25 meters; M193 re-zeroed at 25 meters.

5-97. If an M16A2/A3/A4 rifle or an M4/M4A1 carbine has been zeroed at 25 meters using M855 ammunition and M193 ammunition is fired without re-zeroing, the difference in the impact of the rounds on the target varies between ammunition (Table 5-11 and Figure 5-35).

Table 5-11. M855 zeroed at 25 meters; M193 fired using M855 zero.

RANGE (m)	FRONT SIGHT HEIGHT	THETA		BULLET DROP		BULLET LOCATION	
		M193	M855	M193	M855	M193	M855
0	2.568	0.00235575	0.00235575	0	0	-2.57	-2.57
25	2.568	0.00235575	0.00235575	-0.125	-0.14	-0.37	-0.39
50	2.568	0.00235575	0.00235575	-0.52	-0.58	1.55	1.49
75	2.568	0.00235575	0.00235575	-1.21	-1.33	3.18	3.06
100	2.568	0.00235575	0.00235575	-2.1	-2.4	4.16	4.31
125	2.568	0.00235575	0.00235575	-3.18	-3.82	5.85	5.21
150	2.568	0.00235575	0.00235575	-4.91	-5.61	6.43	5.73
175	2.568	0.00235575	0.00235575	-6.89	-7.78	6.77	5.88
200	2.568	0.00235575	0.00235575	-9.16	-10.4	6.82	5.58
225	2.568	0.00235575	0.00235575	-11.8	-13.4	6.50	4.90
250	2.568	0.00235575	0.00235575	-15.4	-16.9	5.22	3.72
275	2.568	0.00235575	0.00235575	-19.2	-20.8	3.74	2.14
300	2.568	0.00235575	0.00235575	-23.44	-25.3	1.82	-0.04
325	2.568	0.00235575	0.00235575	-28.3	-30.4	-0.73	-2.83
350	2.568	0.00235575	0.00235575	-33.8	-36.0	-3.91	-6.11
375	2.568	0.00235575	0.00235575	-40.13	-42.3	-7.92	-10.09
400	2.568	0.00235575	0.00235575	-46.78	-49.3	-12.25	-14.77

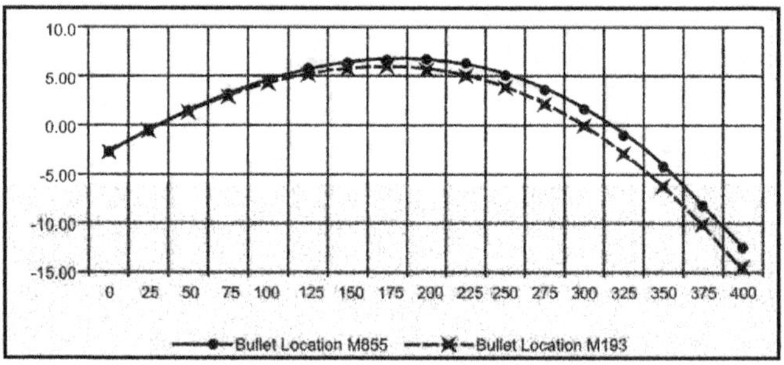

Figure 5-35. M855 zeroed at 25 meters; M193 fired using M855 zero.

5-98. A simple rule of thumb that will preclude any problem is to use only the ammunition specifically designed for each weapon (M193 ball ammunition for M16A1 rifles; M855 ball ammunition for M16A2/3/4 rifles and M4 carbines).

EXTERNAL BALLISTICS

5-99. External ballistics deals with factors affecting the flight path of the bullet between the weapon's muzzle and the target.

5-100. Soldiers must understand the basics of external ballistics so they can make necessary scope adjustments or hold compensations to allow them to hit the target. The external ballistic factors that affect bullet trajectory are:

- Gravity.
- Muzzle velocity.
- Air resistance (drag).
- Altitude or air density.
- Temperature.
- Trajectory.
- Wind.
- Angles.

GRAVITY

5-101. The force of gravity on a bullet is constant regardless of its weight, shape, or velocity. The longer a bullet is in the air or the greater its angle from the vertical, the more effect gravity will have on its trajectory.

NOTE: See paragraphs 5-71 through 5-72 for more information about the effects of gravity.

MUZZLE VELOCITY

5-102. Muzzle velocity is the speed of a bullet as it leaves the barrel, measured in feet per second. Muzzle velocity diminishes as the bullet gets farther away. The bullet reaches its maximum velocity 76 feet from the end of the weapon and slows until it reaches the target.

AIR RESISTANCE (DRAG)

5-103. Air resistance, or drag, immediately produces a slowing effect on a bullet.

ALTITUDE OR AIR DENSITY

5-104. The greater the altitude, the thinner the air and the longer the bullet will travel (with a correspondingly flatter trajectory). Each 5,000-foot elevation will raise the strike of the bullet ½ to 1 minute of angle (MOA).

TEMPERATURE

5-105. Deviation from standard daytime temperature (59 degrees Fahrenheit/15 degrees Celsius) affects bullet trajectory.

Cold Temperatures

5-106. Cold air is denser than warm air; the bullet must travel through more tightly packed air particles. This causes the bullet to lose velocity, causing the bullet to impact lower than intended. Cooler air also causes lower chamber pressure, which reduces the initial velocity.

Hot Temperatures

5-107. Warm or hot temperatures cause the strike of the round to move up.

TRAJECTORY

5-108. When a projectile exits the weapon's muzzle, it drops from the line of departure, otherwise known as the center/bore line. As the projectile travels downrange, air drag decreases the velocity, giving way to the inevitable force of gravity. This effect creates trajectory.

Line of Sight

5-109. The line of sight is an imaginary straight line extending from the firer's eye through the telescopic sight, or rear and front sight, to the target.

Line of Departure

5-110. The line of departure is an imaginary straight line extending from the center of the barrel to infinity.

Zero Range

5-111. Zero range is where the projectile intersects the line of sight. It occurs twice—once on the way up and once on the way down.

Apex

5-112. Otherwise known as midrange trajectory, the apex is the point where the projectile is at its highest in relation to the line of sight.

Bullet Path

5-113. The bullet path is the relationship of a projectile and the line of sight at any given range (normally expressed in inches).

WIND

5-114. External factors influence the trajectory relative to the point of aim, such as wind, altitude, temperature, humidity, and barometric pressure. Wind is by far the most significant. Consider the following effects of wind:

- Because the bullet is moving through the air, the air moves the bullet. Wind deflection is always in the same direction the wind is moving. A wind blowing from the left will move the bullet to the right.
- Deflection decreases as the angle of the wind to the line of flight decreases.

5-115. Effectively reading and correcting for wind effects takes practice, especially at longer ranges where accuracy in correcting is more critical. To shoot accurately in the wind, a firer must know the wind velocity, the wind direction, and the value of deflection at the range at which he is shooting.

NOTE: See paragraph 5-74 for more information on the effects of wind.

ANGLES

5-116. Firing uphill or downhill normally causes the bullet to hit high relative to a horizontal trajectory. If the firer is firing on an angle up or down at a slanted range of 100 meters, the point of impact will be higher than it would be for a level shot of 100 meters. The height depends on the angle.

5-117. Gravity acts on a bullet only during the horizontal component of its flight (the distance from the firer to the target measured as if they were both at the same level). Since the horizontal component will always be less than the slanted range, gravity will not pull the bullet down as far as it would if the range were level.

5-118. Firing uphill or downhill causes the wind to affect the shot over the entire slant range. The firer should aim at the target as if it were 25 meters away and correct for wind as if it were 400 meters away. The correct method for shooting uphill or downhill is to adjust elevation based on the horizontal range and correct for wind deflection based on the slanted range.

TERMINAL BALLISTICS

5-119. Terminal ballistics deals with what happens to the bullet when it comes in contact with the target.

5-120. Bullet penetration depends on the range, velocity, bullet characteristics, and target material. Greater penetration does not always occur at close range with certain materials, since the high velocity of the 5.56-millimeter bullet causes it to disintegrate soon after impact.

BULLET DISPERSION AT RANGE

5-121. Instructors/trainers must have a working knowledge of the effects of bullet dispersion and accuracy at various ranges.

MINUTE OF ANGLE

5-122. An MOA is the standard unit of measurement used in adjusting a weapon's sights and other ballistic-related measurements. It is also used to indicate the accuracy of a weapon.

5-123. A circle is divided into 360 degrees. Each degree is further divided into 60 minutes; therefore, a circle contains 21,600 minutes. An MOA is an angle beginning at the muzzle that covers 2.54 centimeters at a distance of 91.4 meters (Figure 5-36).

5-124. When the range is increased to 182.8 meters, the angle covers twice the distance, or 5.08 centimeters. The rule applies as range increases—7.62 centimeters at 274.2 meters, 10.16 centimeters at 365.6 meters, and so on.

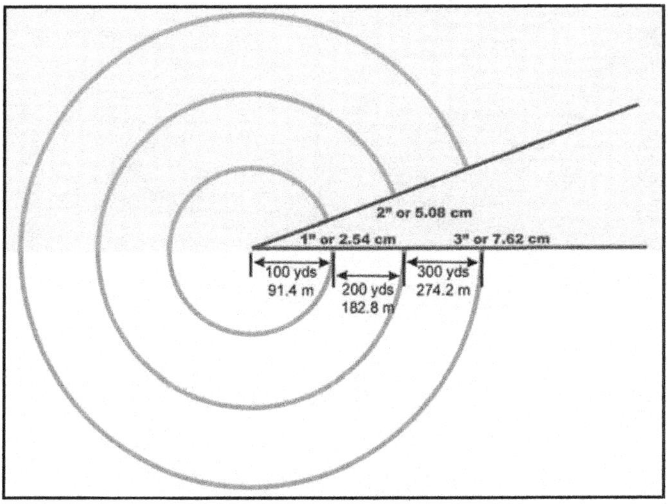

Figure 5-36. Minute of angle.

INCREASE OF SHOT GROUP SIZE

5-125. Just as the distance covered by an MOA increases each time the range increases, a shot group can be expected to do the same. If there are 2.54 centimeters between bullets on a 25-meter target, there will be an additional 2.54 centimeters of dispersion for each additional 25 meters of range. A 2.54-centimeter shot group at 25 meters (about 3.5 MOA) is equal to a 25.4-centimeter shot group at 250 meters (Figure 5-37).

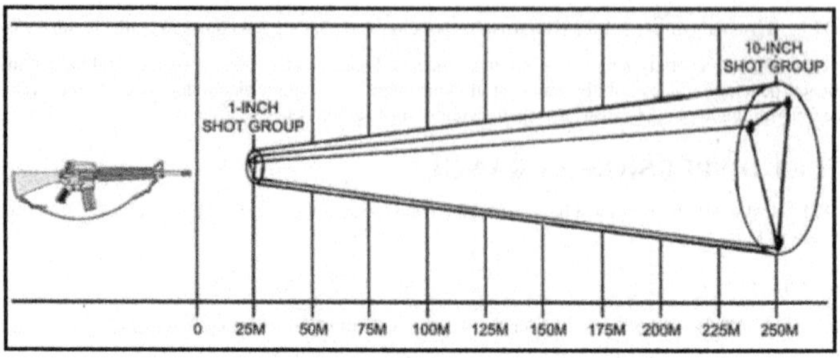

Figure 5-37. Increase in shot group size as range increases.

Chapter 6

Field Fire

Field firing is part of the continued progression in the development of combat shooting skills. This begins the Soldier's critical transition from unstressed firing at single, known distance targets to targets at various ranges for short exposures. It also requires the Soldier to practice and refine previously taught skills.

This chapter introduces the techniques for scanning the range for targets, estimating range, and firing quickly and accurately.

NOTE: See Table 6-1 for the current training program.

Table 6-1. Field Fire I and II training program.

FIELD FIRE I AND II
Instructional Intent
• Reinforce PMI and downrange feedback by detecting and engaging single and multiple timed targets with the M16-/M4-series weapon.
Special Instructions
Ensure that—
• The rear sight is on the proper setting (M16A2/3=8/3; M16A4 and M4=6/3 flush; M16A1=the unmarked aperture, short-range).
• The rear sight aperture is set on 300, not 800.
• The small aperture is being used.
• The range consists of targets at 75, 175, and 300 meters.
Observables
• Coaches are used to analyze the firer's adherence to the fundamentals—not as scorers.
• Soldiers detect and achieve 22 target hits out of 36 timed target exposures (Field Fire I).
• Soldiers detect and achieve 27 target hits out of 44 timed target exposures (Field Fire II).
• Soldiers that don't achieve the standard receive remedial training prior to refiring.

SECTION I. TARGET DETECTION

For most Soldiers, finding the target is a greater problem than hitting it. Target detection is the process of locating, marking, prioritizing, and determining the range to combat targets. Target detection must be conducted as part of individual training and tactical exercises, and must be integrated into day and night LFXs.

NOTE: Refer to TC 25-8 for construction of a target detection range.

LOCATING TARGETS

6-1. The ability to locate a combat target depends on the observer's—

- Position.
- Skill in scanning.
- Ability to observe the area and recognize the type of indicators made by the target.

Selection of a Position

6-2. A good position is one that offers maximum visibility of the area, while affording cover and concealment. In this case, the word "position" refers to both the observer's location on the ground and the position of his body at that location.

NOTE: Instructors must continuously refer to and emphasize the importance of the observer's position when conducting practical exercises.

6-3. Depending on the situation, the individual Soldier may or may not select his own position.

- In most defensive situations, the Soldier is told where to prepare his position.
- Some situations (for example, the attack and reorganization on the objective) require the Soldier to select his own position.

Scanning

6-4. To scan the area, Soldiers use three methods of search:

- Self-preservation method.
- 50-meter overlapping strip method.
- Maintaining observation of the area.

Self-Preservation Method of Search

6-5. When moving into a new area, Soldiers use the self-preservation method of search. To perform this method, use the following techniques:

(1) For approximately 30 seconds, quickly scan the area for enemy activity that may be of immediate danger.

(2) Make quick glances at specific points throughout the area, rather than just sweeping the eyes across the terrain in one continuous panoramic view.

NOTE: The eyes are sensitive to slight movements that occur within the area the eyes are focused on; panoramic views do not allow the eyes to detect the slight movements of a concealed target.

50-Meter Overlapping Strip Method of Search

6-6. If the Soldier fails to locate the enemy during the initial search, he must begin a systematic examination known as the 50-meter overlapping strip method of search. To perform this method—

(1) Begin the search with the area offering the greatest potential danger, the terrain nearest to your position.

(2) Beginning at either flank, systematically search the terrain to your front in a 180-degree arc, 50 meters in depth.

NOTE: Become familiar with the terrain as you search it. Take advantage of peripheral vision by focusing the eyes on specific points as you search from one flank to the other. Make mental notes of prominent terrain features and areas that may offer cover and concealment to the enemy.

(3) After reaching the opposite flank, search a second 50-meter strip farther out, but overlapping the first strip by approximately 10 meters.

(4) Continue in this manner until the entire area has been searched.

NOTE: This method should also be used as part of maintaining observation of the area and when the observer has been distracted from his area of responsibility.

Maintaining Observation of the Area

6-7. After completing his detailed search, the Soldier may be required to maintain observation of the area. To perform this method—

- Glance quickly at various points throughout the entire area, focusing the eyes on specific features.
- Always search the area in the same manner to ensure complete coverage of all terrain.

NOTE: Since this quick search may fail to detect the initial movement of an enemy, the observer should periodically repeat the procedures outlined in the 50-meter overlapping strip method of search.

TARGET INDICATORS

6-8. A target indicator is anything that a Soldier (friendly or enemy) does or fails to do that reveals his position. Since these indicators apply equally to both sides of the battlefield, Soldiers must learn to use target indicators to locate the enemy and to prevent the enemy from using the same indicators to locate them. For instructional purposes, these indicators can be grouped into three general areas:

- Sound.
- Movement.
- Improper camouflage.

Sound

6-9. Sounds, such as footsteps, coughing, or equipment noises, provide only a direction and general location, making it difficult to pinpoint a target by sound alone. However, detection of a sound alerts the observer, greatly increasing the possibility that he will eventually locate the target through other target indicators.

Movement

6-10. The degree of difficulty in locating moving targets depends primarily on the speed of movement. Slow, deliberate movements are much more difficult to notice than those that are quick and jerky.

Improper Camouflage

6-11. The lack or improper use of camouflage or concealment reveals the majority of targets detected on the battlefield; alert observers easily notice indicators such as light reflecting from shiny surfaces or a contrast with the background. Three general indicators may reveal a camouflaged or concealed target:

- Shine.
- Regularity of outline.
- Contrast with the background.

Shine

6-12. Metal objects, such as belt buckles, reflect light and act as a beacon to the wearer's position. This is as true at night as it is during the day.

Regularity of Outline

6-13. Humans and most types of military equipment cast outlines that are familiar to all Soldiers. The outlines of rifles, helmets, and vehicles are all easily identified. The reliability of this indicator depends on the visibility and the experience of the observer. On a clear day, most Soldiers can easily identify enemy riflemen or equipment if a distinctive outline is presented. At night or during other periods of poor visibility, seeing outlines is not only more difficult, but inexperienced troops will frequently mistake stumps and rocks for enemy Soldiers. This is an additional reason for Soldiers to become completely familiar with the terrain during periods of good visibility.

Contrast with the Background

6-14. If a Soldier wearing a dark uniform moves into a position in front of a snow bank, the contrast between the white snow and the dark uniform makes him clearly visible. However, if he was wearing a white or light-colored uniform, he would be more difficult to see.

6-15. Contrast with the background is the most difficult target indicator for a Soldier to avoid. During operations in which the Soldier is moving, he is usually exposed to numerous background colors. Since no single type of personal camouflage blends in with all areas, a moving Soldier must be continually aware of the surrounding terrain and vegetation.

MARKING TARGETS

6-16. A Soldier observes two enemy riflemen moving into completely concealed positions, one behind a bush and the other into a depression.

- By selecting a point of aim on the bush, the Soldier should hit the enemy rifleman even though he can't see him. If the target cannot be engaged, the point of aim also allows for quick and accurate engagement once the target is re-exposed.
- The enemy rifleman who moved into the depression provides no distinguishable point of aim, so the Soldier must select a nearby feature as a reference point and determine its distance and general direction from the depression. A reference point provides a general point of aim on a concealed target.

6-17. Of the two, a point of aim is usually the more effective means of delivering accurate fire. The difficulty in using reference points to mark targets moving from one location to another depends on the following factors:

- Number of targets.
- Exposure time to target.
- Spacing of targets.
- Good and poor points of aim.

Number of Targets

6-18. If several targets appear and disappear at approximately the same time, it is very difficult to note each target's point of disappearance.

Exposure Time of Target

6-19. Usually, moving targets are exposed for only a short period of time, so the observer must be alert to note the point of disappearance for all of the targets. In such situations, the Soldier should mark the location of as many targets as possible before engaging any of them. By doing so, he will know the location of several targets and can engage each of them in rapid succession.

Spacing of Targets

6-20. The greater the interval between targets, the more difficult it is to note each target's movements. When there is considerable distance between targets, the observer should accurately locate and mark the one nearest to his position and note the general area of the others.

Good and Poor Points of Aim

6-21. Good points of aim are easily distinguishable in the surrounding terrain. Targets disappearing behind good points of aim, such as manmade objects and large terrain features, can be easily marked for future reference.

6-22. Poor points of aim are not easily distinguishable within the surrounding terrain. Targets disappearing behind poor points of aim are difficult to mark accurately and are easily lost.

6-23. If two targets offer the same degree of danger to the Soldier, but one disappears behind a good point of aim and the other behind a poor point of aim, the Soldier should mark the location of the target behind the good point of aim and engage the other target first.

RANGE DETERMINATION

6-24. Range determination is the process of finding the distance between two points. In most situations, one of these points will be the Soldier's own position. The other may be a target or prominent feature.

6-25. To accomplish the mission, combat riflemen must be able to accurately determine range. Not only does this affect his combat marksmanship proficiency, it is also required to report information and adjust artillery and mortar fire.

6-26. The methods of range estimation used during this period are—

- 100-meter unit of measure method.
- Appearance of objects method.
- Front sightpost method.

100-Meter Unit of Measure Method

6-27. To perform this method, the Soldier must be able to visualize a distance of 100 meters on the ground.

- For ranges up to 500 meters, he determines the number of 100-meter increments between the two points.
- Beyond 500 meters, the Soldier selects a point halfway to the target, determines the number of 100-meter increments to the halfway point, and then doubles it to find the range to the target.

6-28. During training exercises, the Soldier must become familiar with the effect that sloping ground has on the appearance of a 100-meter increment.

- Ground that slopes upward gives the illusion of greater distance, and observers tend to underestimate a 100-meter increment.
- Ground that slopes downward gives the illusion of shorter distance, and observers tend to overestimate a 100-meter increment.

6-29. Proficiency in the 100-meter unit of measure method requires constant practice. While training this technique, comparisons should be made continually between the range as determined by the Soldier and the actual range as determined by pacing or other more accurate means of measurement.

NOTE: The best training technique is to require the Soldier to pace the range after he has visually determined it. In this way, he discovers the actual range for himself, which makes a much greater impression than if he is simply told the correct range.

6-30. The greatest limitation of the 100-meter unit of measure method is that its accuracy is directly related to the amount of terrain visible to the observer. This is particularly true at longer ranges. If a target appears at a range of 500 meters or more and the observer can see only a portion of the ground between himself and the target, it becomes very difficult to use the 100-meter unit of measure method with any degree of accuracy.

Appearance of Objects Method

6-31. The appearance of objects method is a means of determining range by the size and other details of the object observed. This is a common method of determining distances and is used by most people in their everyday living. For example, a motorist attempting to pass another car must judge the distance of oncoming vehicles based on his knowledge of how vehicles appear at various distances. Suppose the motorist knows that, at a distance of one mile, an oncoming vehicle appears to be 1 inch wide and 2 inches high. Then, any time he sees other oncoming vehicles that fit these dimensions, he knows they are about one mile away.

6-32. The rifleman can use this technique to determine ranges on the battlefield. If he knows the characteristics, size, and detail of personnel and equipment at known ranges, he can compare these characteristics to similar objects at unknown ranges. When the characteristics match, so do the ranges.

6-33. To use the appearance of objects method with any degree of accuracy, the Soldier must be thoroughly familiar with the details of objects as they appear at various ranges. For example, the Soldier should study the appearance of a man standing at a range of 100 meters. He fixes the man's appearance firmly in his mind, carefully noting details of size and the characteristics of uniform and equipment. Next, he studies the same man in a kneeling position and in a prone position. By comparing the appearance of Soldiers in these positions at known ranges from 100 to 500 meters, the Soldier can establish a series of mental images that will help him determine range on unfamiliar terrain.

NOTE: Training should also be conducted in the appearance of other familiar objects, such as weapons or vehicles.

6-34. Because the successful use of this method depends upon visibility, anything that limits visibility (for example, weather, smoke, or darkness) will limit the effectiveness of this method.

Front Sightpost Method

6-35. Using the front sightpost as a scale is another method of estimating range. This method can be used for a quick on-the-spot estimation and engagement.

- Generally, if a man-sized target is ½ of the width of the front sightpost, the target is approximately 300 meters away.
- If the target is ¼ of the width of the front sightpost, the target is approximately 600 meters away.

SECTION II. FIELD FIRE TRAINING

Field fire training provides the transition from unstressed, slow firing at known distances or feedback targets to engaging pop-up silhouettes 50 to 300 meters away. Two basic types of field firing exercises are single and multiple target timed engagements, which use 75-, 175-, and 300-meter targets. Pop-up targets are used to add stress and simulate the short exposure times of combat targets. During field fire training, the firer learns to quickly detect targets, apply SPORTS, and apply the four fundamentals simultaneously.

CONDUCT OF A FIELD FIRE RANGE

6-36. On field fire ranges, Soldiers fire (from the supported and prone unsupported firing positions) at F-type silhouettes 75 meters away and E-type silhouettes 175 and 300 meters away. Initial training begins with single exposed targets and increased time for target exposures (Field Fire I). As Soldiers become proficient, multiple target engagements with shorter exposure times are introduced (Field Fire II).

NOTE: There are two types of electronic pop-up targets used on a field fire range: those that rise to the upright position from the back and those that rise to the upright position from the side. When using targets that rise from the side, Soldiers should be instructed to wait until the target is fully raised before engaging it. If the target is engaged as it is rising, the computer will not register the hit even though the target may fall.

REMEDIAL TRAINING

6-37. Soldiers who miss most targets should be removed from the firing line for remedial training if their problem cannot be corrected. A Soldier who fires at a 300-meter target 10 times and misses it 10 times is not learning, instead he is losing confidence in his ability. The typical Soldier should hit the 300-meter target at least 7 out of 10 times.

PEER COACHING

6-38. Peer coaches should assist Soldiers in observing the strike of rounds and identifying firing problems. If the target is missed and the coach cannot observe the bullet strike, the coach should instruct the Soldier to aim lower for the next shot, expecting to see the strike of the bullet in the ground. With this information, the coach can instruct the Soldier where to aim to hit the target.

ORGANIZATION

6-39. Live-fire training can be organized in several ways. A unit is divided into two or more firing orders based on the number of personnel to be trained:

- The first order is the firer.
- The second order is the coach.
- The third order is the scorer (if required).

6-40. At the conclusion of each exercise, positions rotate until all orders have fired. Standard field fire scenarios have been developed to provide several target exposures. Although they are recommended for IET, local commanders can develop any variety of more challenging target sequences. Ammunition is allocated based on one round for each target.

RECORDING

6-41. During live-fire, the Soldier's hit-and-miss performance is recorded to facilitate the instructor/trainer's critiques or to indicate where more training, closer supervision, or remedial training is needed. Two methods are used to record firing performance:

- Manually marked scorecards.
- Automated computer printouts.

Manual Recording

6-42. When manual recording is used, the unit provides Soldiers for recording information on either DA Form 3601-R (Single Target Field Fire I Scorecard) or DA Form 5241-R (Single and Multiple Targets Field Fire II Scorecard).

NOTE: See the end of this publication for blank reproducible copies of these forms.

Automated Recording

6-43. When firing exercises are conducted using automated field fire (AFF) ranges, a computer printout is provided for each firing order. At the conclusion of each firing order, the range noncommissioned officer in charge (NCOIC) completes the printout and ensures that the Soldier identification is matched with each firing point. He adds the Soldier's numbered code to the top of each lane/firing point data column. Based on a one-round allocation for each target exposure, data should be collected on hits, misses, no-fires, and repeated shots to assist the instructor/trainer in assessing firing proficiency.

FIELD FIRE I (SINGLE TIMED TARGET)

6-44. Field Fire I is broken down into four firing tables (Table 6-3).

CONCEPT

6-45. Firing Table 1 helps the firer practice shooting skills and develop the sense of timing and rhythm required to make the transition from KD to field fire. This builds confidence prior to firing the exercises in Firing Tables 2, 3, and 4, and identifies Soldiers who are having difficulty and need reinforcement.

6-46. When firing the exercises in Firing Tables 2, 3, and 4, each Soldier demonstrates his ability to apply the fundamentals of marksmanship during the integrated act of firing by successfully detecting and engaging single timed targets.

CONDUCT

NOTE: When firing record fire, each Soldier must wear the proper uniform: the helmet, LBE, and IBA with all SAPI plates (if available). No other armor is required.

6-47. Each firer receives 54 rounds of 5.56-millimeter ball ammunition, with two 18-round magazines (one magazine each for Firing Tables 1 and 2) and two 9-round magazines (one magazine each for Firing Tables 3 and 4). Table 6-2 depicts the number of rounds that must be fired from each position. Each Soldier must achieve 22 hits out of 36 timed target exposures in Firing Tables 2, 3, and 4. Table 6-3 shows the number of target exposures, target ranges, and exposure times for each firing table.

Table 6-2. Number of rounds that must be fired from each position during Field Fire I.

POSITION	NUMBER OF ROUNDS FIRED
Supported firing position	18
Supported firing position	18
Unsupported firing position	9
Kneeling	9

Table 6-3. Field Fire I firing tables.

FIRING TABLE 1			FIRING TABLE 2			FIRING TABLE 3			FIRING TABLE 4		
ROUND	RANGE (m)	TIME (sec)	ROUND	RANGE (m)	TIME (sec)	ROUND	RANGE (m)	TIME (sec)	ROUND	RANGE (m)	TIME (sec)
1	75	6	1	75	6	1	75	6	1	75	6
2	175	8	2	175	8	2	175	8	2	175	8
3	300	10	3	300	10	3	300	10	3	75	6
4	175	8	4	175	8	4	175	8	4	175	8
5	75	6	5	75	6	5	75	6	5	75	6
6	300	10	6	300	10	6	300	10	6	175	8
7	300	10	7	300	10	7	300	10	7	75	6
8	75	6	8	75	6	8	75	6	8	75	6
9	175	8	9	175	8	9	175	8	9	175	8
10	175	8	10	175	8						
11	300	10	11	300	10						
12	175	8	12	175	8						
13	75	6	13	75	6						
14	300	10	14	300	10						
15	175	8	15	175	8						
16	75	6	16	75	6						
17	300	10	17	300	10						
18	75	6	18	75	6						

FIELD FIRE II (MULTIPLE OR SINGLE TIMED TARGETS)

6-48. Field Fire II consists of three firing tables.

CONCEPT

6-49. Firing Table 1 helps the firer practice shooting skills and develop the sense of timing and rhythm required to make the transition from single timed targets to multiple or single timed fleeting combat targets. This builds confidence prior to firing the exercises in Firing Tables 2 and 3, and identifies Soldiers who are having difficulty and need reinforcement.

6-50. When firing the exercises in Firing Tables 2 and 3, each Soldier demonstrates his ability to apply the fundamentals of marksmanship during the integrated act of firing by successfully detecting and engaging multiple and single timed targets.

CONDUCT

NOTE: When firing record fire, each Soldier must wear the proper uniform: the helmet, LBE, and IBA with all SAPI plates (if available). No other armor is required.

6-51. Every firer receives 54 rounds of 5.56-millimeter ball ammunition, with 10 rounds loaded into one magazine (for Firing Table 1) and 44 rounds loaded into two separate magazines, with 22 rounds each (for Firing Tables 2 and 3). Table 6-4 depicts the number of rounds that must be fired from each position. Each Soldier must achieve 27 hits out of the 44 timed target exposures in Firing Tables 2 and 3. Table 6-5 shows the number of target exposures, target distance, and exposure times for each firing table.

Table 6-4. Number of rounds that must be fired from each position during Field Fire II.

POSITION	NUMBER OF ROUNDS FIRED
Supported firing position	10
Supported firing position	22
Unsupported firing position	22

Table 6-5. Field Fire II firing tables.

FIRING TABLE 1			FIRING TABLE 2			FIRING TABLE 3		
ROUND	RANGE (m)	TIME (sec)	ROUND	RANGE (m)	TIME (sec)	ROUND	RANGE (m)	TIME (sec)
1	75	5	1	175	7	1	75	6
2	175	7	2	75	10	2	175	8
3	75	11	3	300		3	75	13
4	300		4	75	9	4	300	
5	75	9	5	175		5	75	11
6	175		6	300	9	6	175	
7	75	10	7	75	9	7	75	12
8	300		8	175		8	300	
9	175	11	9	175	11	9	175	13
10	300		10	300		10	300	
			11	75	9	11	75	11
			12	175		12	175	
			13	175	11	13	175	8
			14	300		14	75	6
			15	75	5	15	75	11
			16	175	11	16	175	
			17	300		17	75	12
			18	75	9	18	300	
			19	175		19	75	11
			20	75	10	20	175	
			21	300		21	175	13
			22	175	7	22	300	

SECTION III. RECORD QUALIFICATION

The objective of record fire is to access and confirm the individual proficiency of firers and the effectiveness of the training program. Important statistical data, such as qualification ratings and first-time GO rates, are obtained from record fire. These data provide goals for the Soldier and aid the commander in identifying the quality of his training.

PRACTICE RECORD FIRE I AND II

6-52. Although the Soldier receives a practice rating based on the number of target hits, practice record fire should also be considered a valuable training exercise. When practice record fire is correctly conducted, all Soldiers gain valuable experience and become more confident in engaging combat targets.

NOTE: See Table 6-6 for the current training program.

Table 6-6. Practice Record Fire I and II training program.

PRACTICE RECORD FIRE I AND II
Instructional Intent
• Reinforce PMI and KD firing and apply the techniques of target detection by engaging a more difficult course of fire with single and multiple pop-up targets and increased time constraints.
Special Instructions
Ensure that—
• The rear sight is on the proper setting (M16A2/3=8/3; M16A4 and M4=6/3 flush; M16A1=the unmarked aperture, short-range).
• The rear sight aperture is set on 300, not 800.
• The small aperture is being used.
• Peer coaching is stressed (Practice Record Fire I only).
Observables
• Soldiers apply all aspects of BRM.
• Soldiers hit 23 out of 40 target exposures.
• Soldiers that do not meet the standard receive remedial training before refiring.
• Practice record fire should be conducted on a different range than record fire.

CONCEPT

6-53. Practice Record Fire I and II should closely resemble all aspects of actual qualification.

CONDUCT

NOTES: 1. When firing record fire, each Soldier must wear the proper uniform: the helmet, LBE, and IBA with all SAPI plates (if available). No other armor is required.

2. Before engaging targets, target detection is accomplished with a dry-fire scenario.

3. If possible, Soldiers should fire Practice Record Fire I and II on different ranges. Soldiers firing Practice Record Fire II on the same range as Practice Record Fire I must fire on a different lane.

6-54. Each firer receives 40 single or multiple target exposures at ranges from 50 to 300 meters and 40 rounds of 5.56-millimeter ball ammunition. Table 6-7 depicts the number of rounds that must be fired from each position. Each Soldier must hit a minimum of 23 out of 40 target exposures.

Table 6-7. Number of rounds that must be fired from each position during Practice Record Fire I and II.

POSITION	NUMBER OF ROUNDS FIRED
Prone supported firing position or (at the unit commander's discretion) the foxhole supported firing position	20
Prone unsupported firing position	10
Kneeling unsupported firing position	10

6-55. Soldiers adhere to the following guidelines:

- Based on the total number of hits achieved in each table, Soldiers are critiqued on the practice record fire score. If a firer consistently misses targets or experiences problems with target detection and range estimation, coaches should point out the shooting error to help correct it.
- Exposure times are three to seven seconds at ranges of 50 to 300 meters. Since it requires one to two seconds for the manually activated target mechanism to raise the target, timing begins when

the target is fully exposed, rather than when the tower operator activates the target switch. When practice record fire is conducted on automated record fire (ARF) ranges, these factors are included in the computer program.

NOTE: Practice Record Fire I allows peer coaching and the use of dummy ammunition. Practice Record Fire II does not allow peer coaching, and dummy ammunition will not be used.

ALIBI FIRING

6-56. Alibi firing should be conducted at the end of each firing table IAW tower operator commands. Alibis are provided during practice record fire for three reasons:

- Malfunction of the weapon.
- Malfunction of the target mechanism.
- Faulty ammunition.

RANGE TRAINING AREAS

6-57. The three range training areas are as follows:

- Orientation area.
- Ready area.
- Retired area.

Orientation Area

6-58. The orientation area is located so that firers cannot see the firing area. Practice record fire orientation includes instructions on the conduct of fire, safety, and range operations, including the procedures used in ready and retired areas.

Ready Area

6-59. The ready area is located near the firing range, but is positioned so that firers cannot see targets on the range. While in this area, the firer blackens the weapon sights, lubricates the weapon, and checks for defects that might cause malfunctions.

Retired Area

6-60. The retired area is about 100 meters behind the ready area. Soldiers completing practice record fire move to the retired area to clean their weapons and be critiqued on their firing performance.

PRACTICE RECORD FIRE STANDARDS

6-61. A firer who fails to qualify on his first try should refire the practice record fire range after his problem has been diagnosed and remedial training has been provided. Practice qualification ratings are shown in Table 6-8.

Table 6-8. Qualification ratings for Practice Record Fire I and II.

QUALIFICATION RATINGS	NUMBER OF TARGETS HIT
Expert	36 to 40
Sharpshooter	30 to 35
Marksman	23 to 29

RECORD OF PERFORMANCE

6-62. Accurate performance data are critical. The firer's score is manually recorded using DA Form 3595-R (Record Fire Scorecard) or automatically documented using a computer printout provided on the automated range. Based on the data recorded, an AAR can be performed by range and firing position to discuss firing performance.

NOTE: See Appendix B for a sample of a completed DA Form 3595-R and the end of this publication for a blank, reproducible copy.

RECORD FIRE

6-63. The intent of record fire is to facilitate the commander's evaluation of several individual tasks and integrated marksmanship skill performances, and to provide unit readiness indicators. The qualification standards are specifically related to a prescribed procedure for conducting record fire. Individual performance must be evaluated IAW two components:

- What record fire test was used (standard, KD, or alternate scaled)?
- How were individual performances distributed (first-time GO rates, percent in each qualification rating)?

NOTE: See Table 6-9 for the current training program.

Table 6-9. Record Fire training program.

RECORD FIRE
Instructional Intent
• Reinforce all phases of BRM. • Allow Soldiers to practice and refine critical marksmanship skills. • Measure the Soldier's complete understanding of BRM.
Special Instructions
Ensure that— • The rear sight is on the proper setting (M16A2/3=8/3; M16A4 and M4=6/3 flush; M16A1=the unmarked aperture, short-range). • The rear sight aperture is set on 300, not 800. • The small aperture is being used. • All targets are operational. • Each Soldier has one 20-round magazine and two 10-round magazines.
Observables
• Soldiers apply all aspects of BRM. • Soldiers hit 23 out of 40 target exposures. • Soldiers that do not meet the standard receive remedial training before refiring.

CONCEPT

6-64. Since all Soldiers must fire the record fire course at least once a year for qualification, the course can provide excellent firing performance evaluations. It also provides excellent diagnostic information for instructor/trainers who are concerned with scheduling training to overcome the most serious firing weaknesses. The standard record fire course should be used for all Soldiers, but there are times when qualification exercises must be conducted on alternate courses.

DEVELOPMENT OF STANDARDS

6-65. Testing and development indicates that the Soldier should hit at least 39 of 40 targets if he applies the marksmanship fundamentals correctly (assuming that target mechanisms have been checked and are functioning). This probability of hit (PH) is provided as a guide to consider the capability of the typical weapon, ammunition, and Soldier firing a standard course (Table 6-10).

Table 6-10. Probability of hits.

RANGE (m)	PH	NUMBER OF TARGETS
50	1.0	6
100	1.0	8
150	1.0	11
200	.99	7
250	.95	5
300	.90	3

6-66. When the IET BRM POI or an adequate unit training program is conducted, the following PH can be expected (Table 6-11).

Table 6-11. Results from an adequate unit training program.

RANGE (m)	TARGETS	LOW PH	AVERAGE PH	HIGH PH
50	6	.80	.95	.98
100	8	.70	.90	.95
150	11	.65	.90	.95
200	7	.45	.70	.90
250	5	.35	.60	.85
300	3	.25	.50	.80
		23 hits	32 hits	37 hits

TARGET FUNCTION

6-67. The first task on a standard record fire course is to ensure that all targets function properly. When in doubt, a lane should be fired to ensure that a bullet strike will activate each target. Sometimes slapping a target with a cleaning rod can cause it to activate, but a bullet impact will not. Hot plastic targets may allow a 5.56-millimeter bullet to pass through without causing sufficient vibration to activate the mechanism, resulting in a requirement to change targets more often, use double targets, or use different silhouettes.

CONDUCT

NOTE: When firing record fire, each Soldier must wear the proper uniform: the helmet, LBE, and IBA with all SAPI plates (if available). No other armor is required.

6-68. Each firer receives 40 single or multiple target exposures at ranges from 50 to 300 meters and 40 rounds of 5.56-millimeter ball ammunition. Table 6-12 depicts the number of rounds that must be fired from each position. Each Soldier must hit a minimum of 23 out of 40 target exposures.

Table 6-12. Number of rounds that must be fired from each position during Record Fire.

POSITION	NUMBER OF ROUNDS FIRED
Prone supported firing position or (at the unit commander's discretion) the foxhole supported firing position	20
Prone unsupported firing position	10
Kneeling unsupported firing position	10

6-69. Soldiers adhere to the following guidelines:

- Credit for targets hit should not be given when rounds are "saved" from difficult targets to be used on easier targets (for example, not firing at the 300-meter target so an additional round can be fired at the 150-meter target).
- When double targets are exposed, the Soldier should fire two rounds. If the first target is missed, he may fire at that same target with the second round.
- Soldiers engage the target that poses the greatest threat first (normally assumed to be the closer target). No scoring distinction is made between near targets and far targets or the sequence in which they are engaged.
- Credit is not given if unused ammunition from one 20-round table is added to a magazine provided for the next table.

Alibi Firing

6-70. Alibi firing is reserved for Soldiers who encounter a malfunctioning target, ammunition, or weapon. A Soldier will not be issued more than 20 rounds for Table 1, 10 rounds for Table 2, or 10 rounds for Table 3. Soldiers who fire 20 rounds, despite a target malfunction, will not be issued additional alibi rounds. There are no alibis for Soldier-induced weapon malfunctions or for targets missed during the application of immediate action procedures.

NOTE: The ammunition allocation and alibi procedures for practice record fire and record fire are conducted the same. The only exception is that coaching is authorized for practice record fire.

6-71. If a weapon or target malfunction occurs—

- The Soldier must apply immediate action procedures and continue to fire the exercise.
- After firing, the Soldier notifies the NCOIC to determine if the ammunition was faulty or if the target malfunctioned.
- The NCOIC verifies the malfunction.
- The Soldier is permitted to fire at that target(s) with the exact number of rounds equal to the target malfunctions. For example, a Soldier had two confirmed target malfunctions at 250 meters. Although he may have had five rounds left from the overall exercise, the Soldier receives only two rounds to engage the two 250-meter target exposures, if repaired, or the nearest target. He is not allowed to fire all five remaining rounds at the two 250-meter target exposures.
- On a computerized range, the tower operator confirms that the target malfunctioned and indicates the number of malfunctions that occurred.

6-72. Inoperable weapons are uncorrectable malfunctions such as the following:

- A broken firing pin.
- Jam caused by a double feed, not by the Soldier.
- Failure to extract due to a broken extractor.
- Round in the bore.

6-73. The Soldier must apply correct immediate action procedures to eliminate stoppages. If a stoppage is determined to be correctable (for example, the Soldier did not apply correct immediate action procedures and, as a result, the Soldier did not engage the required number of targets), the Soldier is at fault.

Troubleshooting Performance

6-74. Onsite observation, detailed analysis and evaluation of individual results, and unit performance identify weaknesses such as the following:

- Unserviceable weapons could cause poor zeroes or failures to fire and, therefore, failures to qualify.
- Some Soldiers may not qualify because of a lack of understanding of immediate action procedures or weapon and magazine maintenance procedures.

- Soldiers who miss targets are not applying the four fundamentals or are not accurately zeroing the weapon.
- Soldiers who do not fire at exposed targets during qualification might be experiencing:
  - Failure to scan the designated area.
  - Lack of ability to detect targets.
  - Lack of ability to shift from one target to another.
  - Failure to manage ammunition.
  - A stoppage.

6-75. Training can then focus on combat tasks, skills, or other factors that address these weaknesses.

Refire

6-76. Qualified weapons personnel or the NCOIC must verify weapon malfunctions before the Soldier can refire the course. Soldiers who erroneously claim a malfunction on the firing line are considered unqualified and refire as a second-time firer. Soldiers who fail to qualify on the first attempt should be given appropriate remedial training and allowed to refire in a few days. When a Soldier refires the course—

- He remains unqualified if he hits 22 targets or less.
- A rating of marksman is awarded for a score of 23 to 40 target hits.
- If automated scoring procedures (if available) allow the Soldier's performance to be stored and retrieved before a weapon malfunction, his performance is added to the score of his first attempt after weapon repair and refire.
- If a Soldier's weapon becomes inoperable and his performance before a malfunction precludes qualification, he is considered unqualified and must refire.

QUALIFICATION RATINGS

6-77. One point is awarded for each round within or touching some part of the silhouette facing. Qualification ratings are shown in Table 6-13.

Table 6-13. Qualification ratings for Record Fire.

QUALIFICATION RATINGS	NUMBER OF TARGETS HIT
Expert	36 to 40
Sharpshooter	30 to 35
Marksman	23 to 29

RECORD OF PERFORMANCE

6-78. The record fire range is fired and recorded IAW DA Form 3595-R.

NOTE: See Appendix B for a sample completed scorecard and the end of this publication for a blank, reproducible copy.

SECTION IV. ALTERNATE QUALIFICATION COURSES

Units should conduct weapon qualification on a standard record fire range. Convenience and comfort should not be the prime consideration when choosing a range. Authorized alternate record fire courses are—

- KD record fire range.
- 25-meter scaled target alternate course.
- 15-meter scaled target alternate course.

NOTES: 1. The official records of personnel who use an alternate qualification course are noted to distinguish alternate qualification ratings from standard record fire course ratings. For example, official personnel records are annotated as follows:

JONES, John Q. 000-00-0000 Expert 36 (record fire [RF])

JONES, John Q. 000-00-0000 Expert 38 (known distance alternate course [KDAC])

JONES, John Q. 000-00-0000 Expert 38 (alternate course [AC])

2. The uniform for all alternate qualification courses is a helmet, LBE, and IBA with all SAPI plates (if available). No other armor is required.

3. Firers should engage targets from left to right, from nearest to the farthest (50-meter, 100-meter left, 100-meter center, 100-meter right, 150-meter left, 150-meter right, 200-meter left, 200-meter right, 250-meter, and 300-meter). This ensures that firers do not forget which targets they engaged during qualification. It also alleviates the possibility of shooting each target more than the prescribed number of times.

KNOWN DISTANCE RECORD FIRE RANGE

NOTE: The KD record fire range is used by all components of the U.S. Army, U.S. Army Reserve, and Army National Guard when a record fire range is not available.

6-79. The KD record fire range allows Soldiers to engage targets at range while experiencing time constraints, feedback, and the effects of wind and gravity.

CONDUCT

6-80. To complete this course, Soldiers fire three tables. Table 6-14 depicts these three tables and provides related information, such as time constraints, number of rounds that must be fired, type of target that must be used, and the distance from the firer that the target must be placed.

NOTE: Before firing the course, all Soldiers confirm the zero of their assigned weapons by assuming the prone position and firing six rounds from the 300-meter line. Zero rounds do not count for score.

Table 6-14. Known distance record fire range firing tables and related information.

TABLE	POSITION	TIME CONSTRAINTS	NUMBER OF ROUNDS	TYPE OF TARGET	DISTANCE (m)
Table 1	Prone supported firing position	2 min	20	E-type silhouette target	300
Table 2	Prone unsupported firing position	60 sec	10	E-type silhouette target	200
Table 3	Prone unsupported firing position	60 sec	10	F-type silhouette target	100

QUALIFICATION RATINGS

6-81. Scoring is conducted in the pits, with the results provided after each firing table. One point is awarded for each round within or touching some part of the silhouette facing. Qualification ratings for the KD record fire range are shown in Table 6-15.

Table 6-15. Qualification ratings for the known distance record fire range.

QUALIFICATION RATINGS	NUMBER OF TARGETS HIT
Expert	38 to 40
Sharpshooter	33 to 37
Marksman	26 to 32
Unqualified	25 and below

RECORD OF PERFORMANCE

6-82. The KD record fire range is fired and recorded IAW DA Form 5789-R (Record Fire Scorecard—Known Distance Course).

> **NOTE:** See Appendix B for a sample completed form and the end of this publication for a blank, reproducible copy.

25-METER SCALED TARGET ALTERNATE COURSE

> **NOTE:** The 25-meter scaled target alternate course is used when a standard record fire or KD range is unavailable for weapon qualification.

6-83. The 25-meter scaled target alternate course enables units to test a Soldier's weapon marksmanship proficiency, and firing at scaled silhouettes allows Soldiers to engage targets with time constraints and feedback.

CONDUCT

6-84. To complete this course, Soldiers fire three tables. Table 6-16 depicts these three tables and provides related information, such as time constraints, number of rounds that must be provided, number of silhouettes that must be engaged, and the distance from the firer that the target must be placed.

> **NOTES:** 1. Soldiers should not receive training on target detection or the effects of wind and gravity by engaging targets at 25 meters. These skills are trained by firing at longer distances.
>
> 2. If zeroing/grouping exercises are not performed on the day of record fire, all Soldiers confirm the zero of their assigned weapons by firing six rounds of training/sustainment ammunition from the 25-meter line before firing the course. Zero rounds do not count for score.

Table 6-16. 25-meter scaled target alternate course firing tables and related information.

TABLE	POSITION	TIME CONSTRAINTS	NUMBER OF ROUNDS	NUMBER OF SILHOUETTES	ADDITIONAL INFORMATION
Table 1	Prone supported firing position or foxhole supported firing position	120 sec	20-round magazine, two rounds for each silhouette if firing from the prone supported firing position or foxhole supported firing position	10 silhouettes on the same target sheet	No more than two hits for each silhouette will be scored for this table.
Table 2	Prone unsupported firing position	60 sec	10-round magazine, one round for each silhouette	10 silhouettes on the same target sheet	No more than one hit for each target will be scored for this table.
Table 3	Kneeling unsupported firing position	60 sec	10-round magazine, one round for each silhouette	5 silhouettes on the same target sheet (50 to 150 m)	No more than two hits for each target will be scored for this table.

Time Between Firing Positions

6-85. The time between each firing position is not specified, but enough time should be allotted to allow the firer to clear his weapon, quickly change firing positions, and reload before beginning the next firing table.

DUTIES OF RANGE PERSONNEL

6-86. The following personnel perform range duties:

- Officer in charge (OIC).
- Range safety officer (RSO).
- Firing line safety crew.

Officer In Charge

6-87. The OIC briefs all Soldiers on the proper scoring procedures.

Range Safety Officer

6-88. To facilitate the timely flow of the record fire qualification table, the RSO ensures that enough time is given between firing positions.

Firing Line Safety Crew

6-89. Firing line safety crew personnel—

- Perform as scorers.
- Inform the chief range officer of crossfires.
- Inform the chief range officer of allowable alibis.
- Accurately count hits and misses.
- Count only four hits for each silhouette for score.
- Complete the scorecard.
- Assist the Soldier with target repair.
- Total, sign, and return the completed scorecard to the chief range officer.

SCORING

6-90. One hit is awarded for each round that strikes within or touches some part of the silhouette. If a bullet hole does not touch some part of the scaled silhouette, it is counted as a miss. Ricochets are counted as hits or misses.

6-91. A maximum of four hits for each silhouette on the same target sheet are scored. The same target sheet is used for every 40-round qualification table that a firer completes.

6-92. DA Form 5790-R (Record Fire Scorecard—Scaled Target Alternate Course) is used to score alternate course record fire qualifications.

NOTE: See Appendix B for a sample completed form and the end of this publication for a blank, reproducible copy.

6-93. The NSNs for scaled silhouette targets are—

- 25-meter (NSN 6920-01-1671398).
- 15-meter (NSN 6920-01-1396).

RATINGS

6-94. Qualification ratings for the 25-meter scaled target alternate course are shown in Table 6-17.

Table 6-17. Qualification ratings for the 25-meter scaled target alternate course.

QUALIFICATION RATINGS	NUMBER OF TARGETS HIT
Expert	38 to 40
Sharpshooter	33 to 37
Marksman	26 to 32
Unqualified	25 and below

15-METER SCALED TARGET ALTERNATE COURSE

NOTE: Units are permitted to use the 15-meter scaled alternate course only when standard record fire and KD ranges, and 25-meter scaled target alternate courses are unavailable.

6-95. The 15-meter scaled target alternate course is conducted on a 50-foot indoor range using a .22-caliber rimfire adapter (RFA). Qualification is conducted using the 15-meter alternate course C target (NSN 6920-01-167-1396).

NOTES:
1. See Appendix A for more information about the RFA.
2. Prior to qualification, all Soldiers battlesight zero their weapons using the 15-meter battlesight zeroing target (NSN 6920-01-167-1393).
3. The conduct of fire, scoring, scorecard, and qualification ratings are the same as those used for the 25-meter scaled target alternate course.

Chapter 7

Advanced Rifle Marksmanship

The procedures and techniques for implementing the ARM training program are based on all Soldiers understanding common firing principles, being proficient marksmen, and being confident in applying their firing skills in combat. This chapter concentrates on the advanced techniques and procedures that Soldiers need to participate in collective training during unit live-fire training exercises. This chapter describes advanced firing positions, combat firing techniques, CBRN firing, unassisted night fire, moving target engagements, SRM training, and SDM training.

NOTE: Unit METL and STRAC allocation determine which ARM tasks will be trained.

SECTION I. ADVANCED FIRING POSITIONS

After mastering the four marksmanship fundamentals in the basic firing positions, Soldiers master the four fundamentals while firing from a variety of advanced firing positions. The firer's position may change, but the three remaining fundamentals never change. Ultimately, any firing position that aids the firer in applying the fundamentals is acceptable, as long as the firer applies it consistently to avoid changing his sight picture.

With minor modifications, the dry-fire exercises taught during PMI can effectively train and evaluate a Soldier's ability to apply the fundamentals while in advanced firing positions. Repetitive training (muscle memory) will teach the Soldier the corrections needed to keep the same point of aim while in different firing positions.

NOTE: The act of assuming different firing positions while keeping the same point of aim increases first-time target hits and Soldier survivability.

KNEELING SUPPORTED FIRING POSITION

7-1. This position allows the Soldier to obtain the height necessary to observe many target areas while taking advantage of available cover. This position is the same as the kneeling unsupported firing position, except the Soldier uses some form of support to stabilize his body. Solid cover that can support any part of the body or weapon assists in firing accuracy.

NOTE: See Chapter 4 for more information about the kneeling unsupported firing position.

7-2. To assume the kneeling supported firing position (Figure 7-1)—

(1) Drop to the knee toward the inside of the covered position.

(2) Place the foot toward the outside of cover, with the toes pointing toward the engagement area.

(3) Place the nonfiring hand on the edge of the cover to support the weapon.

(4) Place the firing hand on the pistol grip, with the weapon's buttstock between the SAPI plate and the bicep to stabilize the weapon and absorb recoil.

NOTE: When firing from the strong side, the firer can place the elbow of the firing side on the outside knee to provide stability. When firing from the weak side, the firer should cant the weapon approximately 45 degrees and thrust the hips forward to minimize exposure to the enemy. When placing the weapon against the cover, the firer should take care to prevent the ejection port from becoming obstructed.

Figure 7-1. Kneeling supported firing position.

STANDING UNSUPPORTED FIRING POSITION

7-3. While the standing position provides the least stability, it can be assumed quickly while moving and is a good position for target area observation. Support for any portion of the body or weapon improves stability.

7-4. To assume the standing unsupported firing position (Figure 7-2)—

(1) Face the target.

(2) Step toward the target with the foot closest to the target.

(3) Spread the feet a comfortable distance apart.

(4) Place the firing hand on the pistol grip and the nonfiring hand on the upper handguard.

(5) Place the weapon's buttstock between the SAPI plate of the IBA and the bicep.

NOTE: This action stabilizes the weapon and absorbs recoil.

(6) Shift the feet until aiming naturally at the target.

(7) Evenly distribute body weight.

NOTE: More stability can be obtained by adjusting the ammunition pouch to support the nonfiring-side elbow. This allows the weapon's magazine to rest in the nonfiring hand.

Figure 7-2. Standing unsupported firing position.

STANDING SUPPORTED FIRING POSITION AROUND OBSTACLES

7-5. To assume the standing supported firing position when firing around obstacles (Figure 7-3)—

(1) Face the target.

(2) Execute a facing movement to the firing side.

(3) Spread the feet a comfortable distance apart.

(4) Place the firing hand on the pistol grip and the nonfiring hand on the upper handguard.

(5) Place the weapon's buttstock between the SAPI plate and the bicep.

NOTE: This action helps stabilize the weapon and absorbs recoil.

(6) Lean into the wall or obstacle, with the nonfiring-side forearm, shoulder, and thigh touching the obstacle for support.

(7) Shift the feet until aiming naturally at the target.

(8) Evenly distribute body weight.

NOTE: When firing from the left side of any obstacle, the firer should take care to prevent the ejection port from becoming obstructed.

Figure 7-3. Standing supported firing position around obstacles.

MODIFIED FIRING POSITIONS

7-6. Soldiers should be encouraged to modify positions by—

- Taking advantage of available cover.
- Using anything that helps to steady the weapon.
- Making changes that allow them to hit more combat targets.

7-7. To provide maximum stability, Soldiers should use prone and supported positions when firing M16- or M4-series weapons in the automatic or burst fire mode. If the weapon is equipped with the ARS, the Soldier should use the vertical pistol grip to further increase control of the weapon. Optional modifications include the following:

- Maximize use of artificial support.
- Grip the weapon firmly, and pull it into the shoulder securely. This helps offset the progressive displacement of weapon/target alignment caused by recoil.
- Use sandbags to support the weapon.
- Position the nonfiring hand on the weapon wherever it provides the most stability and flexibility. The goal is to maintain weapon stability and minimize recoil.
- Form a 5-inch loop with the sling at the upper sling swivel. Grasp this loop with the nonfiring hand, and pull down and to the rear while firing.
- Grasp the small of the stock with the nonfiring hand, and apply pressure down and to the rear while firing.
- Assume the modified supported prone firing position (Figure 7-4). This position uses sandbags to support the handguard and frees the nonfiring hand to hold the magazine steady.

Figure 7-4. Modified supported prone firing position.

URBAN OPERATIONS FIRING POSITIONS

7-8. Although the same principles of rifle marksmanship apply, the selection of firing positions during urban operations (UO) requires some special considerations. During UO, Soldiers may be required to fire—

- Over rooftops.
- Around obstacles.
- From windows.

FIRING OVER ROOFTOPS

7-9. Long-range observation may require Soldiers to occupy positions that are high above the ground. Figure 7-5 shows a Soldier firing over a rooftop, exposing only the parts of his body necessary to engage a target.

Figure 7-5. Firing over a rooftop.

Firing Around Obstacles

7-10. In the urban environment, Soldiers may encounter man-made and natural obstacles. Figure 7-6 shows a Soldier firing around an urban obstacle.

NOTE: Firing around corners could require the Soldier to fire from the opposite shoulder to avoid exposing himself to enemy fire.

Figure 7-6. Firing around an obstacle.

FIRING FROM WINDOWS

7-11. When firing from windows, Soldiers should stay in the shadows and make sure that the weapon's muzzle does not protrude out of the opening (Figure 7-7).

Figure 7-7. Firing from a window.

SECTION II. COMBAT FIRE TECHNIQUES

Combat is the ultimate test of a Soldier's ability to apply the fundamentals of marksmanship and firing skills. Soldiers must apply the marksmanship skills mastered during training, practice, and record fire exercises to many combat situations (for example, attack, assault, ambush, or UO). Although these situations present problems, basic techniques and fundamentals require only two modifications: changes to the rate of fire and alterations in weapon/target alignment.

NOTE: The necessary changes are significant and must be thoroughly taught and practiced before performing LFXs.

RAPID SEMIAUTOMATIC FIRE

7-12. The most important firing technique during fast-moving, modern combat is rapid semiautomatic fire. It is the most accurate technique of placing a large volume of fire on poorly defined targets or target areas, such as short exposure, multiple, or moving targets. To apply rapid semiautomatic fire, the Soldier intentionally fires a quick series of shots into the target area to ensure a high probability of a hit.

NOTE: Increased speed and volume should be sought only after the Soldier has demonstrated expertise and accuracy during slow semiautomatic fire.

Effectiveness and Control of Rapid Semiautomatic Fire

7-13. With proper training, Soldiers can select the appropriate mode of fire: semiautomatic fire, rapid semiautomatic fire, or automatic/burst fire.

NOTE: Leaders must ensure that Soldiers apply proper fire discipline at all times. Even in training, unaimed fire must never be tolerated, especially unaimed automatic fire.

7-14. While Soldiers sacrifice some degree of accuracy to deliver a greater volume of fire, it is surprising how devastatingly accurate rapid semiautomatic fire can be. At ranges beyond 25 meters, rapid semiautomatic fire is superior to automatic fire in all measures: shots per target, trigger pulls per hit, and time to hit. Proper training and repeated practice increases the degree of accuracy.

7-15. Rapid application of the four fundamentals will result in a well-aimed shot every one or two seconds. This technique of fire allows a unit to place the most effective volume of fire in a target area while conserving ammunition. It is the most accurate means of delivering suppressive fire.

Modifications for Rapid Semiautomatic Fire

7-16. Trainers must consider the impact of the increased rate of fire on the Soldier's ability to properly apply the fundamentals of marksmanship and other combat firing skills, such as immediate action procedures.

Marksmanship Fundamentals

7-17. The following paragraphs describe the modifications necessary for Soldiers to apply the four fundamentals when firing in the rapid semiautomatic fire mode.

Steady Position

7-18. Consider the following modifications to achieve a steady position:

- Make sure that the weapon is well-supported to improve accuracy and reduce recovery time between shots.
- Grip the handgrip tightly to reduce recovery time and rapidly shift or distribute fire to subsequent targets.
- When possible, pivot the weapon where the nonfiring hand meets the support.
- Avoid changing the position of the nonfiring hand on the support; it is awkward and time-consuming when rapidly firing a series of shots.

Aiming

7-19. Consider the following recommendations to properly aim the weapon:

- Do not change sighting and stock weld during rapid semiautomatic fire. Keep the cheek on the stock for every shot, align the firing eye with the rear aperture, and focus on the front sightpost.
- When using slow semiautomatic fire, seek a stable sight picture.
- In the fast-moving situations that require rapid semiautomatic fire, accept target movement and unsteady sight picture, and keep firing into the target area until the target is down or there is no chance of a hit.
- Aim every shot.

Breath Control

7-20. Breath control must be modified because the Soldier does not have time to take a complete breath between shots. Consider the following modifications to achieve proper breath control:

- Hold your breath at some point in the firing process.
- Take shallow breaths between shots.

Trigger Squeeze

7-21. To maintain the desired rate of fire, the Soldier has a brief period of time to squeeze the trigger. The firer must cause the weapon to fire in about half of a second or less and still not anticipate the precise moment of firing. Consider the following modifications to achieve proper trigger squeeze:

- Apply initial trigger pressure as soon as a target is identified and while the front sightpost is being brought to the desired point of aim.
- When the front sightpost reaches the point of aim, apply final pressure to cause the weapon to fire almost at once. Apply this additional pressure, also known as final trigger squeeze, without disturbing the lay of the weapon.
- Increase the firing rate by firing, releasing enough trigger pressure to reset the sear, and then immediately firing the next shot. This technique is called rapid trigger squeeze. It eliminates the time used in fully releasing pressure on the trigger and allows the firer to rapidly deliver subsequent rounds.

NOTE: Training and practice sessions are required for Soldiers to become proficient in the technique of rapid trigger squeeze.

7-22. Repeated dry-fire training using simulators, such as the EST 2000 and LMTS, and live-fire practice ensure that the Soldier can squeeze the trigger and maintain a rapid rate of fire consistently and accurately.

Immediate Action Procedures

7-23. To maintain an increased rate of suppressive fire, Soldiers must apply immediate action quickly. Repeated dry-fire practice using blanks or dummy rounds, followed by live-fire training and evaluation, ensures that Soldiers can rapidly apply immediate action procedures while other Soldiers initiate fire.

Rapid Semiautomatic Fire Training

NOTE: Soldiers should be well-trained in all aspects of slow semiautomatic firing before attempting any rapid semiautomatic fire training. Those who display a lack of knowledge of fundamental marksmanship skills should not advance to rapid semiautomatic fire training until these skills are learned and mastered.

7-24. Initial training should focus on the modifications to the fundamentals and other basic combat skills necessary during rapid semiautomatic firing.

NOTE: See Table 7-1 for the current training program.

Table 7-1. Rapid semiautomatic fire training program.

RAPID SEMIAUTOMATIC FIRE TRAINING PROGRAM
Instructional Intent
• Soldiers learn to engage targets using rapid semiautomatic fire and practice rapid magazine changes.
Special Instructions
Ensure that—
• The M16A2/A3/A4 rifle's or M4 carbine's rear sight is set on the 0-2 aperture.
• The M16A1's rear sight is set on the unmarked aperture.
• Soldiers use a 25-meter alternate course C qualification target.
• Each Soldier is given four 5-round magazines of 5.56-millimeter ball ammunition.
• Soldiers use rapid semiautomatic fire to engage targets.
• Each Soldier fires one round at each of the 10 silhouettes on the alternate course C qualification target.
• Each Soldier does a rapid magazine change after each magazine is fired.
• The first iteration of 10 rounds is fired within a time limit of 40 seconds.
• The second iteration of 10 rounds is fired within a time limit of 30 seconds.
• Each target is inspected, and the results are posted after each iteration.
Observables
• Coaches continuously analyze the firer's application of the fundamentals.
• Each Soldier obtains 14 hits out of 20 silhouette target exposures.

Conduct

7-25. Each Soldier receives four 5-round magazines of 5.56-millimeter ball ammunition. Using rapid semiautomatic fire, the Soldier fires one round at each of the 10 silhouettes on the alternate course C qualification target. Soldiers fire two iterations, performing a rapid magazine change after each magazine is fired. The targets are inspected, and the results are posted after each iteration. Each Soldier must obtain 14 hits out of 20 silhouette target exposures.

7-26. Table 7-2 depicts the two iterations and provides related information, such as time constraints, number of rounds that must be fired, type of target that must be used, and the distance away from the firer that the target must be placed.

Table 7-2. Rapid semiautomatic fire training and related information.

ITERATION	TIME CONSTRAINTS	NUMBER OF ROUNDS	TYPE OF TARGET	DISTANCE (m)
1	40 sec	10	25-m alternate course C qualification target	25
2	30 sec	10	25-m alternate course C qualification target	25

Dry-Fire Exercises

7-27. Repeated dry-fire exercises are the most efficient means to ensure that Soldiers can apply modifications to the fundamentals. Multiple dry-fire exercises are needed, emphasizing a rapid shift in position and point of aim, followed by breath control and fast trigger squeeze.

NOTES: 1. Blanks or dummy rounds may be used to train rapid magazine changes and immediate action procedures.

2. The Soldier should display knowledge and skill during dry-fire exercises before attempting LFXs.

Live-Fire Exercises

7-28. There are two types of LFXs:

- Individual.
- Collective.

Individual

7-29. To conduct an individual LFX—

- Ensure that the emphasis is on each Soldier maintaining a heavy volume of accurate fire.
- Keep weapon downtime (during immediate action and rapid magazine changes) to a minimum.
- Begin by firing at shorter ranges, progressing to longer ranges as Soldiers display increased proficiency.
- Shorten exposure or engagement times and increase the number of rounds to simulate the need for a heavy volume of fire.
- Provide downrange feedback to determine the accuracy of fire.

Collective

7-30. Rapid semiautomatic fire should be the primary means of delivering fire during a collective LFX. To conduct a collective LFX, ensure that the emphasis is on performing staggered rapid magazine changes, maintaining a continuous volume of fire, and conserving ammunition.

AUTOMATIC OR BURST FIRE

NOTE: Automatic or burst fire should be trained only after the Soldier has demonstrated expertise during slow and rapid semiautomatic fire.

7-31. When applying automatic or burst fire, Soldiers deliver the maximum number of rounds (one to three rounds per second) into a designated target area while rapidly applying the four fundamentals. This specialized technique of delivering suppressive fire may not apply to most combat engagements.

NOTE: The M16A1/A3 rifle and M4A1 carbine have fully automatic settings. The M16A2/A4 rifle and M4 carbine use a three-round burst capability.

EFFECTIVENESS AND CONTROL OF AUTOMATIC OR BURST FIRE

7-32. Automatic or burst fire is inherently less accurate than semiautomatic fire. The first fully automatic shot fired may be on target, but recoil and a high cyclic rate of fire often combine to place subsequent rounds far from the desired point of impact. Even controlled (three-round burst) automatic or burst fire may place only one round on the target. Because of these inaccuracies, it is difficult to evaluate the effectiveness of automatic or burst fire, and even more difficult to establish absolute guidelines for its use.

FACTORS FOR USE OF SEMIAUTOMATIC VERSUS AUTOMATIC OR BURST FIRE

7-33. Trainers must ensure that Soldiers understand the capabilities and limitations of automatic or burst fire. They must know when it should and should not be used.

Semiautomatic Fire

7-34. M16 rifles and M4 carbines should normally be employed in the semiautomatic fire mode.

7-35. Depending on the tactical situation, Soldiers should employ the semiautomatic fire mode in the following conditions:

- Ammunition is in short supply, or resupply may be difficult.
- Single targets are being engaged.
- Widely spaced multiple targets are being engaged.
- The target is located more than 50 meters away.
- The effect of bullets on the target cannot be observed.
- Artificial support is not available.
- Targets may be effectively engaged using semiautomatic fire.

Automatic or Burst Fire

7-36. In some combat situations, the use of automatic or burst fire can improve survivability and enhance mission accomplishment. Clearing buildings, final assaults, FPF, and ambushes may require limited use of automatic or burst fire.

7-37. Depending on the tactical situation, Soldiers should employ automatic or burst fire in the following conditions:

- Ammunition is readily available, and there are no problems with resupply.
- Closely spaced multiple targets are located 50 meters away or less.
- Maximum fire is immediately required at an area target.
- Tracers or some other means can be used to observe the effect of bullets on the target.
- Leaders can maintain adequate control over weapons firing in the automatic fire mode.
- Good artificial support is available.
- The initial sound of gunfire disperses closely spaced enemy targets.

MODIFICATIONS FOR AUTOMATIC OR BURST FIRE

7-38. Automatic or burst fire is inherently less accurate than semiautomatic fire. Trainers must consider the impact of recoil and the high cyclic rate of fire on the Soldier's ability to properly apply the fundamentals of marksmanship and other combat firing skills, such as immediate action procedures and rapid magazine changes.

Marksmanship Fundamentals

7-39. The following paragraphs describe the modifications necessary for Soldiers to apply the four fundamentals when firing in the automatic fire mode.

Steady Position

7-40. Consider the following modifications to achieve a steady position:

- Make sure that the weapon is well-supported.
- Grip the weapon a little more firmly and pull it into the shoulder a little tighter than when in the semiautomatic fire mode.

NOTE: This support and increased grip help offset the progressive displacement of weapon/target alignment caused by recoil.

- To provide maximum stability, assume the modified supported prone firing position (Figure 7-4).

NOTE: If the weapon is equipped with the ARS, use the vertical pistol grip to further increase control of the weapon.

Aiming

7-41. Consider the following recommendations to properly aim the weapon:

- Do not change sighting and stock weld during automatic or burst fire. Keep the cheek on the stock for every shot, align the firing eye with the rear aperture, and focus on the front sightpost.
- Although recoil may disrupt this process, try to apply the aiming techniques throughout recoil.

Breath Control

7-42. Breath control must be modified because the Soldier does not have time to take a complete breath between shots. Consider the following modifications to achieve proper breath control:

- Hold your breath at some point in the firing process.
- Take shallow breaths between shots.

Trigger Squeeze

7-43. Training and repeated dry-fire practice aid the Soldier in applying proper trigger squeeze during automatic firing. LFXs enable him to improve this skill.

M16A2/3/4 Rifles and M4 Carbines

7-44. Until the weapon fires, trigger squeeze is applied in the normal manner. To use the burst fire mode—

(1) Hold the trigger to the rear until three rounds are fired.

(2) Release pressure on the trigger until it resets.

(3) Reapply pressure for the next three-round burst.

NOTES: 1. Do not slap or jerk the trigger. Squeeze it, and then quickly release pressure.

2. Depending on the position of the burst can when the selector is moved to the burst fire mode, the weapon may fire one, two, or three rounds when the trigger is held to the rear for the first time. If the weapon fires only one or two rounds, quickly release pressure on the trigger and squeeze again, holding it to the rear until a three-round burst is completed.

M16A1 Rifles

7-45. Until the weapon fires, trigger squeeze is applied in the normal manner. Because three-round bursts are the most effective rate of fire, pressure on the trigger should be released as quickly as possible. To use the burst fire mode, keep the index finger on the trigger, but quickly release pressure to prevent an excessive number of rounds from being fired in one burst. With much dry-fire practice, the Soldier can become proficient at delivering three-round bursts with the squeeze/release technique.

Immediate Action

7-46. To maintain an increased rate of suppressive fire, Soldiers must apply immediate action quickly. Repeated dry-fire practice using blanks or dummy rounds, followed by live-fire training and evaluation, ensures that Soldiers can rapidly apply immediate action procedures.

Rapid Magazine Changes

7-47. Rapid magazine changes are vital in maintaining automatic or burst fire. Rapid magazine changes must be correctly taught and practiced during dry-fire and live-fire exercises until the Soldier becomes proficient.

Automatic or Burst Fire Training

NOTE: Soldiers should be well-trained in all aspects of slow semiautomatic firing before attempting any automatic training. Those who display a lack of knowledge of fundamental skills should not advance to automatic or burst fire training until these skills are learned.

7-48. Initial training should focus on the modifications to the fundamentals and other basic combat skills necessary during automatic firing.

7-49. Unit training is vital to properly applying this technique. Soldiers must be taught the advantages and disadvantages of automatic and burst firing so they know when it should be used. Without this knowledge, Soldiers tend to switch to the automatic or burst fire mode in life-threatening situations.

NOTE: See Table 7-3 for the current training program.

Table 7-3. Automatic or burst fire training program.

AUTOMATIC OR BURST FIRE TRAINING PROGRAM
Instructional Intent
• Soldiers learn the advantages and disadvantages of automatic or burst fire.
Special Instructions
Ensure that—
• The M16A2/A3/A4 rifle's or M4 carbine's rear sight is set on the 0-2 aperture.
• The M16A1's rear sight is set on the unmarked aperture.
• Soldiers use a 25-meter alternate course C qualification target.
• Each Soldier is in a proper modified automatic/burst firing position.
• Each Soldier is given two 15-round magazines of 5.56-millimeter ball ammunition.
• Each Soldier fires one 3-round burst at each of the 10 silhouettes on the alternate course C qualification target.
• Each Soldier does a rapid magazine change after each magazine is emptied.
Observables
• Each Soldier obtains five target hits.
• Soldiers demonstrate control of the weapon in the automatic/burst fire mode.

Conduct

7-50. Each Soldier receives two 15-round magazines of 5.56-millimeter ball ammunition. Each Soldier fires one 3-round burst at each of the 10 silhouettes on the alternate course C qualification target, performing a rapid magazine change after each magazine is emptied. Each Soldier must obtain five target hits.

7-51. Table 7-4 depicts automatic or burst fire training and provides related information, such as number of rounds that must be fired, type of target that must be used, and the distance away from the firer that the target must be placed.

Table 7-4. Automatic or burst fire training and related information.

FIRING POSITION	NUMBER OF ROUNDS	TYPE OF TARGET	DISTANCE (m)
Modified automatic/burst firing position	30, one 3-round burst at each of the 10 silhouettes	Alternate course C qualification target	25

Dry-Fire and Live-Fire Exercises

7-52. Repeated dry-fire exercises are the most efficient means to ensure that Soldiers can apply modifications to the fundamentals. Multiple dry-fire exercises are needed, emphasizing a stable position and point of aim, followed by breath control and appropriate trigger squeeze.

NOTES: 1. Blanks or dummy rounds may be used to train trigger squeeze, rapid magazine changes, and immediate action procedures.

2. The Soldier should display knowledge and skill during dry-fire exercises before attempting LFXs.

SUPPRESIVE FIRE

7-53. Suppressive fire is precisely aimed at a definite point or area target. Some situations may require a Soldier to place suppressive fire into a wide area (for example, wood line, hedgerow, or small building) while, at other times, the target may be a smaller area (for example, a bunker or window). Suppressive fire is used to control the enemy and the area he occupies. It is employed to kill the enemy or to prevent him from observing the battlefield, effectively using his weapons, or moving.

EFFECTIVENESS AND CONTROL OF SUPPRESSIVE FIRE

7-54. Many Soldiers have difficulty delivering effective suppressive fire when they cannot see a definite target, only likely locations or general areas where the enemy is known to exist. Even though definite targets cannot be seen, most suppressive fire should be well-aimed.

7-55. When controlling suppressive fires, two factors must be considered:

- Point of aim.
- Rate of fire.

Point of Aim

7-56. Suppressive fire should be well-aimed, sustained, semiautomatic fire. Although lacking a definite target, the Soldier must be taught to control and accurately deliver fire within the limits of the suppressed area. As when engaging a point target, the weapon sights are used, with the front sightpost placed so each shot impacts within the desired area.

Rate of Fire

7-57. During most phases of live-fire training (for example, grouping, zeroing, qualifying), shots are delivered using slow semiautomatic fire (one round every 3 to 10 seconds). During training, this allows a slow and precise application of the fundamentals. Successful suppressive fire requires a faster, but sustained, rate of fire. Soldiers may need to fire full automatic or bursts (13 rounds per second) for a few seconds to gain initial fire superiority. Rapid semiautomatic fire (one round every one or two seconds) allows the firer to sustain a large volume of accurate fire while conserving ammunition.

MODIFICATIONS FOR SUPPRESSIVE FIRE

7-58. The tactical situation dictates the most useful rate of fire, but the following must be considered:

- Marksmanship fundamentals.
- Rapid magazine changes.
- Ammunition conservation.

Marksmanship Fundamentals

7-59. As the stress of combat increases, some Soldiers may fail to apply the fundamentals of marksmanship. This factor contributes to reduced accuracy and effectiveness. While some modifications are appropriate, the basic fundamentals should be applied and emphasized—regardless of the rate of fire or combat stress. Strategies to enhance marksmanship skills under combat stress include shooting in the prone position, as opposed to standing.

7-60. Factors that contribute to combat stress are:

- Environmental.
- Operational.

Environmental

7-61. Environmental stressors have been shown to degrade marksmanship accuracy up to 20 percent. Such stressors include—

- Heat.
- Altitude.

Operational

7-62. Operational stressors have been shown to degrade marksmanship accuracy from 17 percent to 136 percent. Such stressors include—

- MOPP gear.
- Tasks that require carrying rucksacks, litter patients, and other equipment on the body.
- Sleep deprivation.

Rapid Magazine Changes

7-63. One of the keys to sustained suppressive fire is reloading the weapon rapidly. Rapid magazine changes must be correctly taught and practiced during dry-fire and live-fire exercises until the Soldier becomes proficient. Small-unit training exercises must be conducted so Soldiers who provide suppressive fire practice staggered magazine changes.

Ammunition Conservation

7-64. Automatic or burst fire should be used sparingly and only to gain initial fire superiority. Depending on the tactical situation, the rate of fire should be adjusted so that a minimum number of rounds are expended. Accurate fire conserves ammunition, while preventing the enemy from placing effective fire on friendly positions.

SUPPRESSIVE FIRE TRAINING

NOTE: See Table 7-5 for the current training program.

Table 7-5. Suppressive fire training program.

SUPPRESSIVE FIRE TRAINING PROGRAM
Instructional Intent
• Soldier learns to suppress targets using suppressive fire.
Special Instructions
Ensure that— • The M16A2/A3/A4 rifle's or M4 carbine's rear sight is set on the 0-2 aperture. • The M16A1's rear sight is set on the unmarked aperture. • Soldiers use a 25-meter scaled landscape target. • Each Soldier is given two 9-round magazines and one 12-round magazine of 5.56-millimeter ball ammunition. • Each Soldier is in a proper supported firing position. • Each Soldier fires 9 rounds at the open window area of the target using rapid semiautomatic fire with the first 9-round magazine. • Each Soldier fires 12 rounds at the fence or hedgerow area of the target using rapid semiautomatic fire with the 12-round magazine. • Each Soldier fires three 3-round bursts at the tank turret area of the target using the automatic/burst fire mode with the second 9-round magazine.
Observables
• Each Soldier achieves 5 hits inside the open window area within 18 seconds. • Each Soldier achieves 10 hits inside the dotted lines surrounding the fence or hedgerow area within 24 seconds. • Each Soldier achieves 3 hits inside the tank turret area within 24 seconds.

7-65. Figure 7-8 shows a landscape target suitable for suppressive fire training. When this type of target is used, trainers must develop a firing program to include areas of engagement and designated target areas. At 25 meters, this target provides the firer with an area to suppress without definite targets to engage.

Figure 7-8. Landscape target.

Conduct

7-66. Each Soldier receives two 9-round magazines and one 12-round magazine of 5.56-millimeter ball ammunition. The Soldier engages three areas of a 25-meter scaled landscaped target: the open window area, the fence or hedgerow area, and the tank turret area. Each Soldier achieves 5 hits inside of the open window area, 10 hits inside of the dotted lines surrounding the fence or hedgerow area, and 3 hits inside of the tank turret area.

7-67. Table 7-6 depicts suppressive fire training and provides related information, such as number of rounds that must be fired, type of target that must be used, and the distance away from the firer that the target must be placed.

Table 7-6. Suppressive fire training and related information.

FIRING POSITION	TYPE OF TARGET	AREA OF TARGET ENGAGED	NUMBER OF ROUNDS	TYPE OF FIRE	TIME CONSTRAINTS
Supported firing position	25-m scaled landscape target	Open window	9	Rapid semiautomatic	18 sec
		Fence or hedgerow	12	Rapid semiautomatic	24 sec
		Tank turret	9, in three 3-round bursts	Automatic/burst	24 sec

QUICK FIRE

7-68. The two main techniques of directing fire with a rifle or carbine are—

- Aim using the sights.
- Use weapon alignment, instinct, bullet strike, or tracers to direct the fire.

7-69. The preferred technique is to use the sights, but sometimes quick reflex action is required. Quick fire, also known as instinctive firing or quick kill, is a technique used to deliver fast, effective fire on surprise personnel targets 25 meters away or less.

Effectiveness and Control of Quick Fire

7-70. Quick fire techniques are appropriate when Soldiers are presented with close, suddenly appearing, surprise enemy targets; or when close engagement is imminent.

NOTE: Fire may be delivered in the SEMIAUTO or AUTOMATIC/BURST fire mode. For example, a point man in a patrol may carry the weapon on AUTOMATIC/BURST. This may also be required when clearing a room or bunker. Initial training should be in the SAFE mode.

7-71. Two techniques of delivering quick fire are:

- Aimed.
- Pointed.

7-72. The difference in the speed of delivery of these two techniques is small. Pointed quick fire can be used to fire a shot about one-tenth of a second faster than aimed quick fire. The difference in accuracy, however, is more pronounced:

- A Soldier well-trained in pointed quick fire can hit an E-type silhouette target at 15 meters, although the shot may strike anywhere on the target.
- A Soldier well-trained in aimed quick fire can hit an E-type silhouette target at 25 meters, with the shot or burst striking 5 inches from the center of mass.

7-73. This variance of target hit for this type of engagement reinforces the need for well-aimed shots.

7-74. Pointed and aimed quick fire should be used only when a target cannot be engaged fast enough using the sights in a normal manner. These techniques should be limited to targets appearing at 25 meters or less. Modern short-range combat (SRC) techniques emphasize carrying the weapon with the buttstock high so that the weapon sights can be brought into display as quickly as firing a hasty unaimed shot. In extremely dangerous moments, special reaction teams (SRTs) commonly advance with weapons shouldered, aiming as they advance.

Aimed

7-75. When using this technique, a Soldier can accurately engage a target at 25 meters or less in one second or less.

7-76. To use aimed quick fire (Figure 7-9)—

(1) Bring the weapon to the shoulder.

(2) With the firing eye, look through or just over the rear sight aperture.

(3) Use the front sightpost to aim at the target.

(4) Quickly fire a single shot.

Figure 7-9. Aimed quick fire.

Pointed

7-77. When using this technique, a Soldier can engage a target at 15 meters or less in less than one second.

7-78. To use pointed quick fire (Figure 7-10)—

(1) Keep the weapon at your side.

(2) Keeps both eyes open, and use instinct and peripheral vision to line up the weapon with the target.

(3) Quickly fire a single shot or burst.

Figure 7-10. Pointed quick fire.

Factors for Use of Controlled Pairs Versus Burst Fire

7-79. Tactical considerations dictate whether controlled pairs or burst fire is most effective in a given situation.

Modifications for Quick Fire

7-80. Trainers must consider the impact of the increased rate of fire on the Soldier's ability to properly apply the fundamentals of marksmanship and other combat firing skills.

Marksmanship Fundamentals

7-81. Quick fire techniques require major modifications to the four fundamentals of marksmanship. Initial training in these differences, followed by repeated dry-fire exercises, will be necessary to prepare the Soldier for live-fire.

Steady Position

7-82. The quickness of shot delivery prevents the Soldier from assuming a stable firing position. Consider the following modifications:

- Fire from the present position when the target appears.
- If moving, stop.
- Do not make adjustments for stability and support before the round is fired.

Aimed

7-83. Consider the following modifications:

(1) Pull the weapon's buttstock into the pocket of the shoulder as the cheek comes in contact with the stock.

(2) Firmly grip the weapon with both hands, applying rearward pressure.

(3) Place the firing eye so that it looks through or just over the rear sight aperture.

(4) Place the sight on the target.

Pointed

7-84. Consider the following modifications:

- Pull the weapon into the side.
- Firmly grip the weapon with both hands, applying rearward pressure.

Aiming

7-85. This fundamental must be highly modified because the Soldier may not have time to look through the rear sight, find the front sight, and align it with the target.

NOTE: When using either aiming technique, bullets may tend to impact above the desired location. Repeated live-fire practice is necessary to determine the best point of aim or the best focus. Such practice should begin with the Soldier using a center of mass aim.

Aimed

7-86. Consider the following modified procedure:

(1) Initially focus on the target.

(2) Place the firing eye so that it looks at the target through or just over the rear sight aperture.

(3) Using peripheral vision, locate the front sightpost and bring it to the center of the target.

NOTE: Focus remains on the front sightpost throughout the aiming process.

---

(4) When the front sightpost is in focus, fire a controlled pair.

Pointed

7-87. Consider the following modifications:

- Place the focus on or slightly below the center of the target as you align the weapon with it, and fire the weapon.
- Use your instinctive pointing ability and peripheral vision to aid in proper alignment.

Breath Control

7-88. This fundamental has little application to the first shot of quick fire. The round must be fired before a conscious decision can be made about breathing. If subsequent shots are necessary, breathing must not interfere with the necessity of firing quickly. When possible, use short, shallow breaths.

Trigger Squeeze

7-89. Consider the following modifications:

(1) Apply initial pressure as weapon alignment is moved toward the target.

(2) Exert trigger squeeze so when weapon/target alignment is achieved, the rounds are fired at once.

7-90. Perfecting rapid trigger squeeze requires much training and practice.

QUICK FIRE TRAINING

NOTE: Only Soldiers in basic training will conduct quick fire training. SRM will be conducted at the unit level. See Section VI of this chapter for more information about SRM training.

7-91. The key to the successful employment of both quick fire techniques is practice. Both pointed and aimed quick fire must be repeatedly practiced during dry-fire training. LFXs provide further skill enhancement and illustrate the difference in accuracy between the two techniques.

NOTE: See Table 7-7 for the current training program.

Table 7-7. Quick fire training program.

QUICK FIRE TRAINING PROGRAM
Instructional Intent
• Soldiers learn how to engage targets using the quick fire techniques.
Special Instructions
Ensure that— • The M16A2/A3/A4 rifle's or M4 carbine's rear sight is set on the 0-2 aperture. • The M16A1's rear sight is set on the unmarked aperture. • Each Soldier is given two 10-round magazines. • Each Soldier engages 10 target exposures of 2 seconds each at 15 meters using the first 10-round magazine. • Each Soldier moves to the 25-meter line and engages 10 target exposures of 2 seconds each at 25 meters using the second 10-round magazine.
Observables
• Each Soldier achieves 7 hits out of 10 target exposures at 15 meters. • Each Soldier achieves 5 hits out of 10 target exposures at 25 meters.

Conduct

7-92. Each Soldier receives two 10-round magazines. Each Soldier must achieve 7 target hits out of 10 target exposures at 15 meters and 5 target hits out of 10 target exposures at 25 meters.

7-93. Table 7-8 depicts quick fire training and provides related information, such as the number of target exposures, distance from the firer, number of rounds that must be fired, and time constraints.

Table 7-8. Quick fire training and related information.

NUMBER OF TARGET EXPOSURES	DISTANCE	NUMBER OF ROUNDS	TIME CONSTRAINTS
10	15	10	2 sec per target exposure
10	15	10	2 sec per target exposure

SECTION III. CHEMICAL, BIOLOGICAL, RADIOLOGICAL, AND NUCLEAR FIRING

All Soldiers must effectively fire their weapons to accomplish combat missions in a CBRN environment. With proper training and practice, Soldiers gain confidence in their ability to effectively hit targets in full MOPP equipment. MOPP firing proficiency must be part of every unit's training program.

MISSION-ORIENTED PROTECTIVE POSTURE EQUIPMENT FIRE

7-94. Firing weapons is only part of overall CBRN training. Soldiers must be familiar with CBRN equipment, its use, and proper wear before they progress to learning the techniques of MOPP firing.

MODIFICATIONS FOR MISSION-ORIENTED PROTECTIVE POSTURE FIRE TRAINING

7-95. Trainers must consider the impact of MOPP equipment (for example, hood or mask, gloves, overgarments) on the Soldier's ability to properly apply the fundamentals of marksmanship and combat firing skills.

Operation and Function

7-96. Many actions are affected by MOPP equipment: handling the weapon, performing operation and function checks, loading and unloading, and cleaning. Consider the following modifications:

- Movements are slowed; tasks take longer to complete and often require more effort.
- Vision is impaired.
- Care is needed to avoid damaging MOPP equipment and possible exposure to lethal agents.

7-97. Because of the great differences between MOPP Level 1 and MOPP Level 4, Soldiers must be trained in all aspects of weapon operation and maintenance while practicing at the highest MOPP level. Only through repeated training and practice can the Soldier be expected to perform tasks efficiently.

Immediate Action

7-98. Consider the following modifications:

- Under normal conditions, a Soldier should be able to clear a stoppage in 3 to 5 seconds. Under increased MOPP levels, however, this may take as long as 10 seconds to successfully complete.
- Mask (with or without hood) and gloves must be worn. Care must be taken not to snag or damage the gloves or dislodge the hood or mask during movements.
- Vision is limited to what can be seen through the mask lenses or faceplate. Peripheral vision is severely restricted. The lenses or faceplate may be scratched or partly fogged, further restricting vision.

NOTE: Soldiers requiring corrective lenses must be issued insert lenses before training.

- Scanning movements may be restricted by the hood or mask.

7-99. These factors could adversely affect the Soldier's ability to quickly and accurately detect targets. Dry-fire practice under these conditions is necessary to reduce time and streamline actions. Until Soldiers can instinctively apply immediate action to stoppages, they should practice using dummy or blank ammunition.

Marksmanship Fundamentals

7-100. Although the four marksmanship fundamentals remain valid during MOPP firing, some modifications may be needed to accommodate the equipment.

Steady Position

7-101. Consider these modifications:

- Due to the added bulk of the overgarment, adjust firing positions for stability and comfort.
- Stand, crouch, or squat during dry- and live-fire to reduce bodily contact with contaminated ground or foliage.
- Understand that a consistent spot or stock weld is difficult to maintain due to the shape of the protective mask. This requires the firer to hold his head in an awkward position to place the eye behind the sight.

Aiming

7-102. Wearing a protective mask may force firers to rotate (cant) the weapon to see through the rear aperture. This ideal aiming procedure (Figure 7-11) should be the initial procedure taught and practiced. If this cannot be achieved, a canted sight picture may be practiced.

Ideal Aiming Procedure

7-103. To perform the ideal aiming procedure—

(1) Rotate the weapon as little possible to see through and line up the sights.

(2) Place the center tip of the front sightpost on the ideal point of aim.

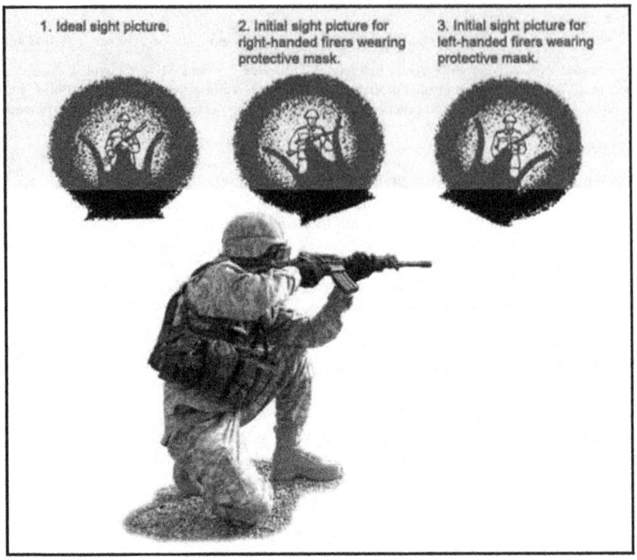

Figure 7-11. Sight picture when canting the rifle while wearing a protective mask (75-meter target).

Canted Sight Picture

NOTE: The normal amount of cant has a limited influence on rounds fired at ranges of 75 meters or less. Rifle ballistics cause the strike of the bullet to impact low in the direction of the cant at longer ranges.

7-104. Due to a shift in bullet strike and the many individual differences in sight alignment when wearing a protective mask, it is important to conduct downrange feedback training at ranges beyond 75 meters on KD ranges. This allows Soldiers to determine the aiming adjustments needed to hit the center of the target.

7-105. Figure 7-12 shows what might be expected for a right-handed firer engaging a target at 175 meters with a certain amount of cant; it depicts the adjustment in point of aim needed to move the bullet strike to the center of the target. Figure 7-13 shows what might be expected for a right-handed firer engaging a 300-meter target.

NOTE: The adjustments in point of aim for left-handed firers are the opposite of those shown in Figures 7-12 and 7-13.

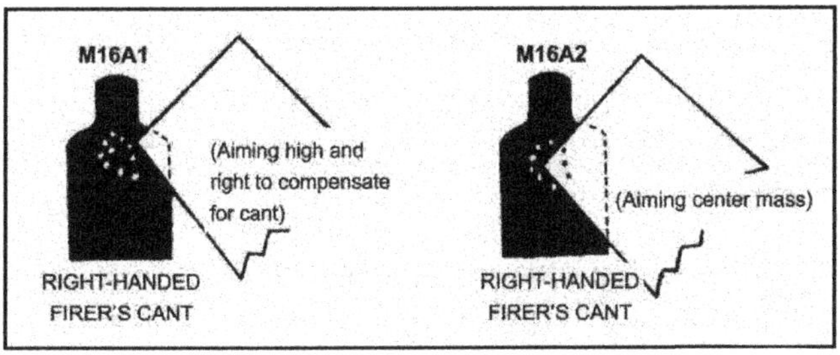

Figure 7-12. Engagement of 175-meter target.

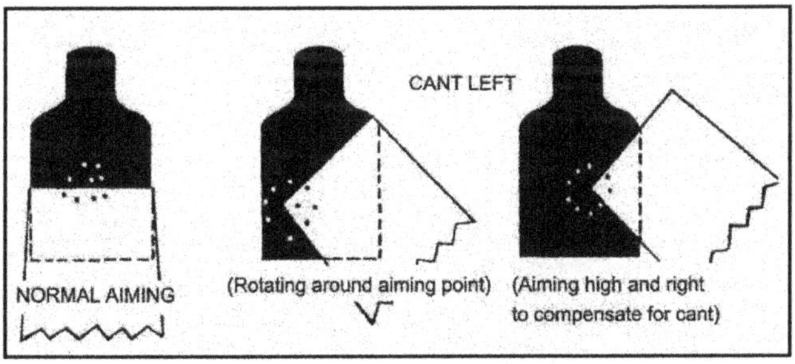

Figure 7-13. Engagement of 300-meter target.

7-106. Although bullet strike is displaced when using a cant, individual differences vary; center-of-mass aiming should be used until the individual knows what aiming adjustment is needed. When distant targets are missed, a right-handed firer should usually adjust his point of aim to the right and high; a left-handed firer should adjust to the left and high. Actual displacement of the point of aim must be determined by using downrange feedback targets positioned more than 75 meters from the firer.

Breath Control

7-107. Moving when encumbered by MOPP equipment requires more physical effort, and breathing is restricted while wearing the protective mask. Physical exertion can produce labored breathing and make settling into a normal breathing rhythm much more difficult, which, in turn, can increase the breath rate. All of these factors make holding and controlling the breath to produce a well-aimed shot more energy- and time-consuming. Emphasis must be placed on rapid target engagement during the limited amount of time a firer can control his breath.

Trigger Squeeze

7-108. Consider the following modifications:

- MOPP gloves complicate the act of grasping the pistol grip and squeezing the trigger with the index finger. The action of the trigger finger is restricted, and the fit of the glove may require the release of the swing-down trigger guard.
- Because the trigger feels different, control differs from that used in barehanded firing. This difference cannot be accurately predicted.

7-109. Dry-fire training using dime/washer exercises or simulators such as the EST 2000 or LMTS may be necessary to ensure that the firer knows the changes he will encounter during live-fire.

Mission-Oriented Protective Posture Equipment Training

7-110. Repeated dry-fire training and LFXs are the most efficient means to prepare a Soldier wearing MOPP Level 4 equipment for successful target engagements at any range. The Soldier must follow these procedures and applications to be combat-effective in a CBRN environment.

Conduct

NOTE: See Table 7-9 for the current training program.

Table 7-9. Chemical, Biological, Radiological, and Nuclear fire training program.

CBRN FIRE TRAINING PROGRAM
Instructional Intent
• Develop the Soldier's confidence and ability to engage targets while wearing any level of MOPP equipment.
Special Instructions
Ensure that— • The M16A2/A3/A4 rifle's or M4 carbine's rear sight is set on the 0-2 aperture. • The M16A1's rear sight is set on the unmarked aperture. • Soldiers have inserted lenses before firing, if required. • Soldiers have a proper seal on the mask to prevent fogging and loss of visibility. • Each Soldier is issued 20 rounds of ammunition to be loaded in two magazines (10/10). • Each Soldier engages 20 targets at 50 meters. • Target exposures consist of 10 from the right and 10 from the left from the foxhole supported firing position using Table 1 of the Record Fire Qualification firing table (DA Form 3595-R).
Observables
• Each Soldier obtains 11 hits out of 20 target exposures.

Dry-Fire Exercises

7-111. As with all marksmanship training, Soldiers must start at the basics in order to become proficient at CBRN fire. Modified fundamentals can be taught anywhere, and Soldiers must learn them before participating in an LFX. The dry-fire exercises used during CBRN training are the same as those used during initial rifle marksmanship (dime/washer exercise, target box, SPORTS, EST 2000, and LMTS). Soldiers can also practice by using MILES equipment during force-on-force training.

NOTE: Soldiers should wear MOPP Level 4 when participating in dry-fire exercises; training at the highest degraded level allows Soldiers to adjust their shooting technique to increase their marksmanship ability in a CBRN environment.

Downrange Feedback

7-112. CBRN downrange feedback gives Soldiers the confidence, knowledge, and skills required to consistently deliver accurate, well-aimed fire against combat targets as far away as 300 meters. On a KD range, the Soldier will perform the following scenario:

- The Soldier will be issued six magazines.
  - The first and second magazine will have 5 rounds each.
  - The third and fourth magazine will have 10 rounds each.
  - The fifth and sixth magazine will have 5 rounds each.

- The Soldier engages the appropriate targets.
  - The Soldier engages the 75-meter target with the first and second 5-round magazine from the foxhole, standing, or kneeling supported position.
  - The Soldier engages the 175-meter target with both 10-round magazines from the foxhole, standing, or kneeling supported firing position.
  - The Soldier engages the 300-meter target with the remaining two 5-round magazine from the foxhole, standing, or kneeling supported firing position.
- In order to receive a GO, the Soldier must obtain a minimum of—
  - 8 hits out of 10 shots on the 75-meter target.
  - 14 hits out of 20 shots on the 175-meter target.
  - 5 hits out of 10 shots on the 300-meter target.

NOTES: 1. DA Form 5789-R is used to record the firer's results.

2. Soldiers can use ammunition allocated by DA PAM 350-38 for advanced skill training for the CBRN downrange feedback scenario.

50-Meter Live-Fire Exercise

7-113. The basic CBRN LFX allows all Soldiers to gain confidence in their CBRN firing abilities. Practice and proficiency firing can be conducted on any range; however, when a Remote Electronic Target System (RETS) range is used for this exercise, the two 50-meter mechanisms are used. For CBRN LFXs, Soldiers perform the following scenario.

- Each Soldier is issued 20 rounds of ammunition to be loaded in two 10-round magazines.
- On the command "GAS–GAS–GAS," each Soldier engages twenty 50-meter targets. Target exposures consist of 10 from the right and 10 from the left from the foxhole supported or the kneeling firing position.
- Each Soldier must achieve 11 hits out of 20 target exposures.
- This is a GO/NO GO exercise.

Alternate Fire Exercise

7-114. The CBRN alternate fire course uses the 25-meter scaled silhouette timed-fire target. This silhouette can be used on any 25-meter range, and the target provides feedback to the firer on where the strike of the round impacts the target. This exercise is conducted in the same manner as the 25-meter alternate course record fire.

- Each Soldier will be in MOPP Level 4.
- Each Soldier will be issued 20 rounds of ammunition to be loaded two 10-round magazines.
- Each Soldier will engage each silhouette with two rounds from the foxhole supported or kneeling position using Table I of DA Form 5790-R.
- Each Soldier must achieve 11 hits out of 20 shots.
- This is a GO/NO GO exercise.

NOTE: At the commander's discretion, Table II of DA Form 3595-R may be used as an alternate course of fire. This table allows Soldiers to engage targets from 50 to 300 meters.

SECTION IV. NIGHT FIRE TRAINING

Soldiers should be able to fire their weapons effectively in total darkness, in bright sunlight, and under all conditions between these two extremes. Scheduled marksmanship training should provide a variety of night and limited visibility conditions. The battlefield may be illuminated by ground flares, handheld flares, M203 flares, mortar and artillery illumination, aerial flares, searchlights, exploding rounds, and burning vehicles, or obscured by smoke, fog, and various environmental conditions. The well-trained Soldier should have experienced a number of these conditions and be confident that he can effectively employ his weapon when required. This section provides guidance on training Soldiers to be effective in total darkness without using iron sights and in limited visibility when using iron sights.

NOTE: See Appendix C for more detailed information about night fighting.

UNASSISTED NIGHT FIRE TRAINING

7-115. All Soldiers must be able to effectively employ their weapons during limited visibility. Soldiers must experience the various conditions of night combat—from total darkness to the many types of artificial illumination.

NOTE: See Table 7-10 for the current training program.

Table 7-10. Unassisted night fire training program.

UNASSISTED NIGHT FIRE TRAINING PROGRAM
Instructional Intent
• Develop the Soldier's confidence in his ability to hit targets when he cannot see through his weapon sights and does not have night vision capability.
Special Instructions
Ensure that— • The M16A2/A3/A4 rifle's or M4 carbine's rear sight is set on the 0-2 aperture. • The M16A1's rear sight is set on the unmarked aperture. • Each Soldier is given two 15-round magazines with 10 rounds of ball ammunition and 5 rounds of tracer ammunition in each magazine. • Each Soldier engages the 50-meter F-type silhouette target from the prone supported or foxhole supported fighting position with one magazine. • Each Soldier engages the 50-meter F-type silhouette target from the prone unsupported fighting position or kneeling position with the second magazine.
Observables
• Each Soldier achieves 7 hits out of 30 target exposures.

MODIFICATIONS FOR NIGHT FIRE

7-116. During limited visibility, a firer cannot effectively use his iron sights in most situations, and without artificial illumination, the sights block his field of vision. Trainers must consider the impact of limited visibility on the Soldier's ability to properly apply combat firing skills and the fundamentals of marksmanship.

Operation and Maintenance of the Weapon

7-117. Many actions are affected by nighttime conditions: handling the weapon, performing operation and function checks, loading and unloading, and maintenance. Consider the following modifications:

- Because combat conditions and enforcement of noise and light discipline restrict the use of illumination, Soldiers must be trained to operate, service, and clean their weapons in total darkness.
- Movements are slower, tasks take longer to complete, vision is impaired, and equipment is more easily misplaced or lost.

NOTE: Although initial practice of these tasks should occur during daylight to facilitate control and error correction, repeated practice during actual nighttime conditions should be integrated with other training. Only through repeated practice and training can the Soldier be expected to perform all tasks efficiently.

Immediate Action

7-118. Under normal conditions, a Soldier should clear a stoppage in three to five seconds. After dark, this task usually takes longer. Identifying the cause of the stoppage may be difficult and frustrating for the Soldier. A dry-fire practice (applying SPORTS) under these conditions using dummy or blank rounds is necessary to reduce time and build confidence. Once the Soldier is confident in applying immediate action in darkness, he can perform such actions rapidly during actual firing.

NOTES:

1. Training should be practiced first during daylight for better control and error correction by the trainer.
2. To learn the hands-only technique of identifying a stoppage, the firer must practice applying immediate action with his eyes closed. This is a technique that Soldiers should be able to master with practice.

Firing Positions

7-119. The firing position recommended for use during limited visibility is the supported firing position.

NOTE: This unassisted night fire supported position differs slightly from the supported position taught in earlier periods of instruction because the firer cannot use his sights during limited visibility.

7-120. To effectively engage targets during limited visibility—

(1) Assume a supported firing position.

(2) Establish a raised stock weld (look 2 to 3 inches above the sights, level with the barrel).

(3) Point the weapon at the target.

(4) Fire in the semiautomatic fire mode.

NOTE: To obtain optimum results, keep the eyes open, and move the head, arms, and weapon as one unit.

Marksmanship Fundamentals

7-121. The four marksmanship fundamentals apply to night firing, but some modifications are needed depending on the conditions.

Steady Position

7-122. When applying unassisted night fire, the Soldier must change his head position or stock weld, especially when using weapon/target alignment techniques. Consider the following modifications:

- Position the head high so that the weapon is aligned on the target.
- Look just over the iron sights.
- Keep the cheek in contact with the stock.

7-123. Repeated dry-fire practice, followed by live-fire training, is necessary to learn and refine these modifications and still achieve the steadiest position.

Aiming

7-124. Modifications to the aiming process vary. Consider these modifications:

- When firing unassisted, use off-center vision instead of pinpoint focus.
- Open both eyes to gather the maximum available light.
- Focus both eyes downrange.

Breathing

7-125. This fundamental is not affected by unassisted night fire conditions.

Trigger Squeeze

7-126. This fundamental is not affected by unassisted night fire conditions. The objective is to not disrupt weapon's alignment with the target when squeezing the trigger.

Target Detection

7-127. Trying to detect a target during the day is difficult enough, but at night, it becomes even more so. The firer must detect and engage targets without artificial illumination or NVDs. Consider these modifications:

- Scan potential target areas and when a target is detected, engage it using a modified quick fire position.
- Take a few seconds to improve weapon/target alignment by pointing slightly low to compensate for the usual tendency to fire high (Figure 7-14).
- Apply the three principles of night vision:
  - Dark adaptation.
  - Off-center vision.
  - Scanning.

7-128. Tracer ammunition may provide feedback on the line of trajectory and facilitate any adjustments in the weapon/target alignment.

Figure 7-14. Lower weapon—target alignment.

Dark Adaptation

7-129. This process conditions the eyes to see under low levels of illumination. The eyes of the average person take about 30 minutes to acquire 98 percent night vision in a completely darkened area. Moving from illuminated to darker areas will decrease night vision until the eyes have adjusted to the surrounding area again.

Off-Center Vision

7-130. When an individual looks at an object during the daytime, he looks directly at it. However, if he did this at night he would only see the object for a few seconds. To see this object for any length of time, he must look 6 to 10 degrees from this object (Figures 7-15 and 7-16) while concentrating his attention on the object. This allows the light sensitive area of the eye to detect faint light sources or reflection.

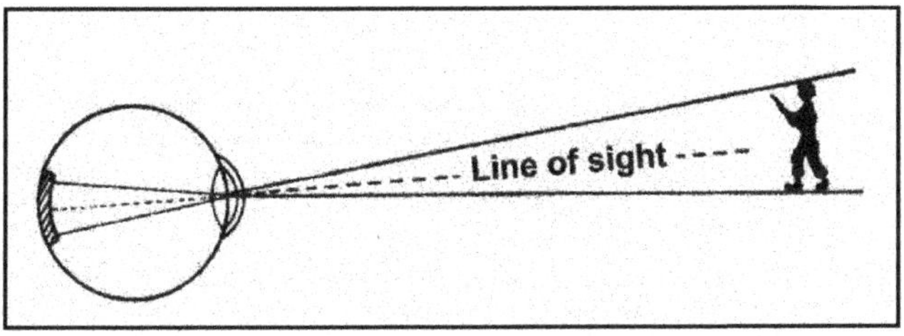

Figure 7-15. Daytime field of view using pinpoint focus.

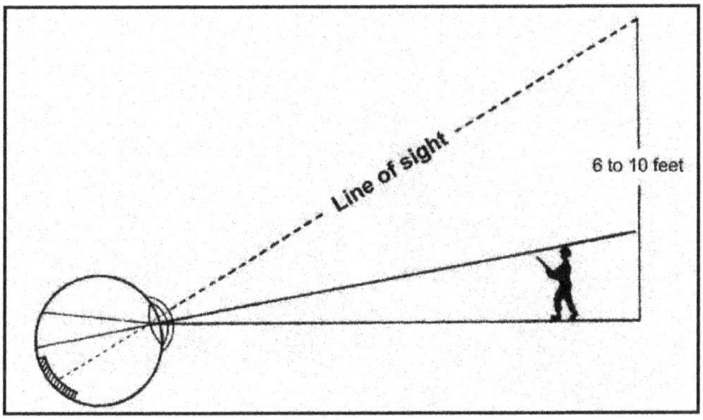

Figure 7-16. Nighttime field of view using off-center vision.

Scanning

7-131. The act of scanning relates to the short, abrupt, irregular movement of the firer's eyes every 4 to 10 seconds around an object or area. Be aware that scanning ranges vary according to levels of darkness.

NIGHT FIRE TRAINING

7-132. All units must include basic unassisted night fire training annually in their unit marksmanship programs.

7-133. Combat units should conduct tactical night fire training at least quarterly. This tactical training should include MILES, during force-on-force training, as well as live-fire training.

Live-Fire Exercises

7-134. The basic unassisted LFX allows all Soldiers to apply night fire principles and to gain confidence in their ability to effectively engage targets out to 50 meters. Practice and proficiency firing can be conducted on any range equipped with mechanical lifters and muzzle flash simulators (Figure 7-17). The muzzle flash simulator provides the firer with a momentary indication that a target is presenting itself for engagement. When a RETS range is used for this exercise, the two 50-meter mechanisms are used. For the unassisted night LFX, the Soldier will perform the following scenario:

- Each Soldier will be issued two 15-round magazines with a tracer and ball combination.
  - The Soldier engages the F-type silhouette target at 50 meters while in the prone supported or foxhole supported firing position. The Soldier uses one magazine of 15 rounds (10 rounds ball; 5 rounds tracer). The Soldier will detect and engage 15 target exposures at 50 meters.
  - The Soldier engages the F-type silhouette target at 50 meters while in the prone unsupported position or kneeling position. The Soldier uses a second magazine of 15 rounds (10 rounds ball; 5 rounds tracer). The Soldier will detect and engage 15 target exposures at 50 meters.
- Each Soldier must achieve 7 hits out of 30 target exposures.
- When the automated range is used, the Soldier's performance is recorded in the tower. If automatic scoring is not available, a coach can observe and score the number of target hits the firer achieves using NVDs.

Figure 7-17. Night-fire target.

ARTIFICIAL ILLUMINATION TRAINING

7-135. Artificial illumination allows for better target detection and long-range accuracy than unassisted night vision. When the artificial light is gone, time must be spent regaining night vision and adapting. Only when the level drops enough so that the target cannot be seen through the iron sights should the firer resume short-range scanning, looking just over the sights.

NOTE: See Table 7-11 for the current training program.

Table 7-11. Artificial illumination training program.

ARTIFICIAL ILLUMINATION TRAINING PROGRAM
Instructional Intent
• Develop the Soldier's confidence in his ability to locate, mark, prioritize and engage targets at night using artificial illumination.
Special Instructions
Ensure that—
• M16A2/A3/A4 or M4-series weapon's rear sight is set on the 0-2 aperture.
• M16A1 rear sight is set on the unmarked aperture.
• Each Soldier is given two 15-round magazines with 10 rounds of ball and 5 rounds of tracer ammunition.
• Each Soldier engages 15 target exposures from the foxhole supported fighting position with the first 15-round magazine.
• Each Soldier engages 15 target exposures from the prone unsupported fighting position with the second 15-round magazine.
Observables
• Each Soldier achieves 15 hits out of 30 target exposures.

EFFECTIVENESS AND CONTROL OF ARTIFICIAL ILLUMINATION

7-136. When artificial illumination is used, the eyes lose most of their night adaptation, and off-center vision is no longer useful. Artificial illumination allows the firer to use the iron sights with the 0-2 rear sight aperture, as he does during the day.

7-137. To preserve night vision while artificial illumination is being used, the Soldier closes his firing eye and scans his sector for enemy targets with his nonfiring eye. This allows the Soldier to have night vision in at least one eye to keep scanning his sector for enemy targets after the artificial illumination has burned out. However, keeping one eye closed to preserve its night vision results in a drastically altered sense of perception when both eyes are opened.

NOTE: Repeated dry-fire training and target detection practice are the keys to successful engagement of targets out to 250 meters or more during live-fire under artificial illumination.

LIVE-FIRE EXERCISE

7-138. The unassisted LFX with artificial illumination allows all Soldiers to apply night fire principles and to gain confidence in their abilities to effectively detect and engage targets 150 meters away and beyond when using artificial illumination. Soldiers use the night record fire scenario:

- Each Soldier is issued two 15-round magazines with the appropriate tracer and ball combination.
  - During the night, each Soldier engages 15 E-type silhouette target exposures from 50 to 250 meters with one magazine of 15 rounds (10 rounds ball, 5 rounds tracer) while in the foxhole supported or prone supported firing position.
  - During the night, each Soldier engages 15 E-type silhouette target exposures from 50 to 250 meters with the second magazine of 15 rounds (10 rounds ball, 5 rounds tracer) while in the prone unsupported or kneeling firing position.
- Each Soldier must achieve 15 hits out of 30 shots.

SECTION V. MOVING TARGET ENGAGEMENTS

In combat situations, enemy Soldiers do not stand still; they rush from one covered or concealed position to another. While making the rush, enemy Soldiers present rapidly moving targets. Moving targets are open to aimed fire at two points in the rush: as the target begins to gain speed (at the beginning) and as it slows down to a new position.

NOTE: See Table 7-12 for the current training program.

Table 7-12. Moving target engagement training program.

MOVING TARGET ENGAGEMENT TRAINING PROGRAM
Instructional Intent
• Soldiers learn to detect and engage moving and stationary targets with the M16 rifle or M4 carbine.
Special Instructions
Ensure that—
• The M16A2/A3/A4 rifle's or M4 carbine's rear sight is set on the 0-2 aperture.
• The M16A1's rear sight is set on the unmarked aperture.
• Soldiers get into a proper semisupported firing position.
• Soldiers understand and apply lead guidance rules.
• Each Soldier receives two magazines with 25 rounds each of 5.56-millimeter ball ammunition.
• Each Soldier engages 34 moving target exposures at ranges from 35 to 185 meters, and 15 stationary target exposures at ranges from 50 to 300 meters.
Observables
• Each Soldier achieves 18 hits out of 50 target exposures.

MODIFICATIONS FOR MOVING TARGET ENGAGEMENTS

7-139. Soldiers in combat do not know if their next target will be stationary or moving; they must fire immediately at whatever target presents itself. Trainers should consider the following when conducting moving target engagement instruction:

- More dispersion and erratic shots are expected when Soldiers are trained to hit moving targets.
- Considering the environment and the variables of the weapon and ammunition, well-trained Soldiers should be able to hit 300-meter stationary silhouette targets.
- When the target is moving laterally, well-trained Soldiers may only hit 150-meter targets 7 out of 10 times. This is considered an acceptable performance.

7-140. Further, trainers should consider modifications to the fundamentals for engaging stationary targets:

- Steady position.
- Aiming.
- Breath control.
- Trigger squeeze.

STEADY POSITION

7-141. When firing at moving targets, firers should assume the standard supported firing position, but be flexible so that they can track any target in the sector. When a target is moving directly at the firer, directly away from the firer, or at a slight angle, the Soldier engages the target without changing his firing position. Consider the following aspects of firing at moving targets:

- When targets are moving laterally, only minor changes are needed to allow effective target engagement.
- Most moving targets are missed in the horizontal plane (firing in front of or behind the target) and not in the vertical plane (firing too low or too high).

7-142. The Soldier must make other adjustments, as highlighted in Table 7-13.

Table 7-13. Modifications for a steady position when firing at moving targets.

PART OF BODY	MODIFICATION
Nonfiring hand	Grip the weapon more tightly with the nonfiring hand, and apply more pressure to the rear. This helps to maintain positive control of the weapon and steady it for rapid trigger action.
Nonfiring-side elbow	Lift the nonfiring-side elbow from the support position only to maintain a smooth track.
Firing hand	Apply more rearward pressure to the pistol grip to steady the weapon during trigger squeeze.
Firing-side elbow	Lift the firing-side elbow from support only to help maintain a smooth track. NOTE: The weapon pocket in the shoulder and the stock weld are the same as for stationary targets.

AIMING

7-143. When aiming at moving targets, Soldiers must apply precise lead rules, and in turn, Soldiers must accurately estimate speed, angle, and range to the target to apply precise lead rules. Then, he must apply the single-lead rule in order to place effective fire on combat targets.

7-144. The procedures used to engage moving targets vary as the angle and speed of the target vary. For example, when a target is moving directly at the firer, stationary target procedures apply. However, for a close, fast-moving target at a 90-degree angle, the weapon and firer's entire upper body must be free from support so the target can be tracked.

Lead Requirements

7-145. Aiming directly at a 300-meter target moving 8 miles per hour at a 90-degree angle would result in missing it; this type of target covers 4 ½ feet while the bullet is traveling toward him. To hit the target, the Soldier must apply target lead (Figure 7-18) and understand how target lead and bullet speed relate to the range, angle, and speed of the target. To hit the target depicted, the Soldier must aim and fire at position D when the target is at position A.

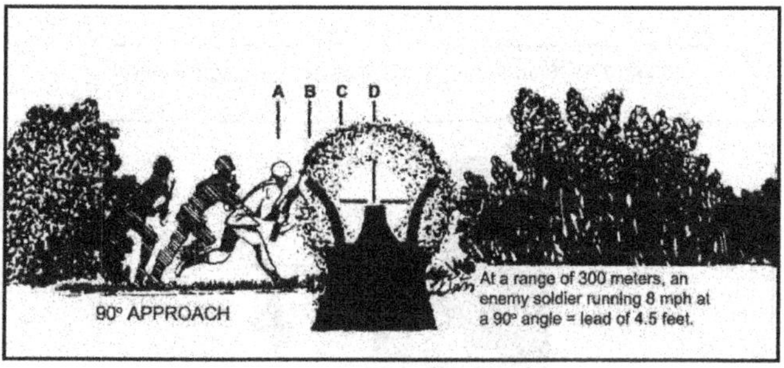

Figure 7-18. Lead requirement based on distance and approach angle.

Single-Lead Rule

7-146. The single-lead rule says: To hit a target moving laterally, place the trailing edge of the front sightpost at the target's center (Figure 7-19). This rule also provides that the lead increases as the range to the target increases (Figure 7-20).

NOTE: At 100 meters, the rule begins to break down for targets moving at slight and large angles.

EXAMPLE

As Figure 7-20 depicts, the front sightpost covers about 1.6 inches at 15 meters and about 16 inches at 150 meters. Since the center of the front sightpost is the actual point of aim, placing the trailing edge of the front sightpost at the target's center provides a .8-inch lead on a 15-meter target and an 8-inch lead on a target at 150 meters.

This provides a dead-center hit on a 15-meter target moving at 7 miles per hour at a 25-degree angle because the target moves .8 inches between the time that the weapon is fired and the time that the bullet arrives at the target. A 150-meter target moving at 7 miles per hour at a 25-degree angle moves 8 inches between the time the weapon is fired and the bullet arrives.

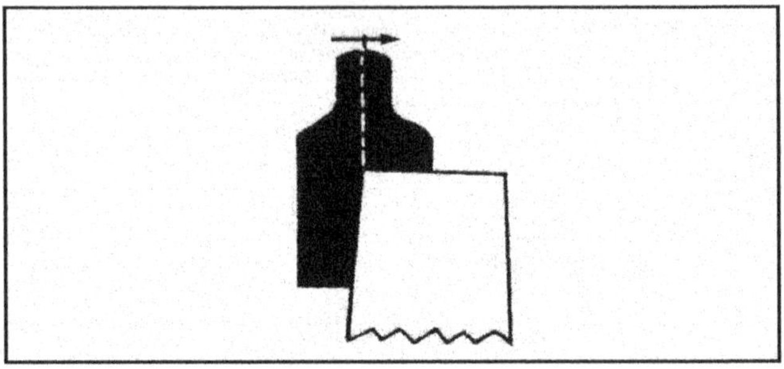

Figure 7-19. Sight-target relationship for the single-lead rule.

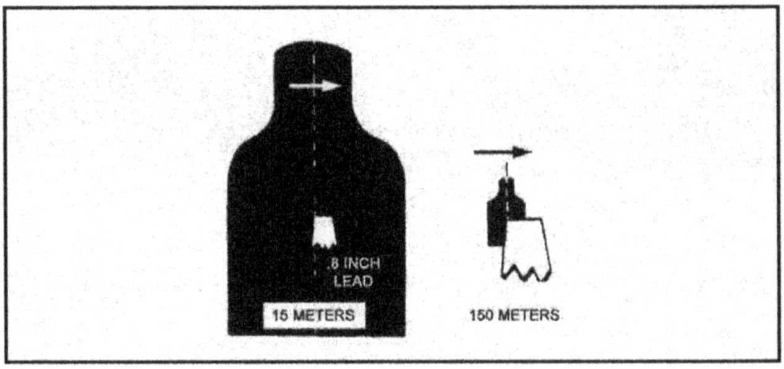

Figure 7-20. Lead increasing at greater ranges.

Target Speed

7-147. Figure 7-21 reflects the differences in lateral speeds for various angles of target movement for a target traveling at 8 miles per hour at a distance of 150 meters from the firer. The angle of target movement is the angle between the target/firer line and the target's direction of movement. An 8-mile-per-hour target moves 24 inches during the bullet's flight time. If the target is moving on a 15-degree angle, it moves 6 inches (the equivalent of 2 miles per hour).

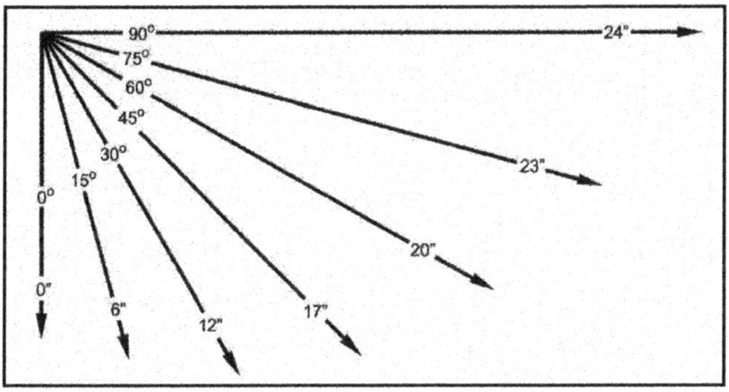

Figure 7-21. Target movement (distance) at various angles.

7-148. Since the target lead is half the perceived width of the front sightpost, at 100 meters, the standard sight provides 5.4 inches of lead for M16A1/2/3/4 rifles and M4 carbines (Table 7-14).

Table 7-14. Angle of target movement.

ANGLE OF TARGET MOVEMENT (Degrees)	RANGE: 100 m (STANDARD SIGHT) TARGET SPEED		
	4 mph	6 mph	8 mph
5	+4.9"	+4.5"	+4.3"
10	+4.1"	+3.5"	+2.7"
15	+3.5"	+2.5"	+1.5"
20	+2.8"	+1.5"	+.2"
25	+2.2"	+.7"	-1.0"
30	+1.7"	-.2"	-2.0"
35	+1.1"	-1.1"	-3.2"
40	+.6"	-1.9"	-4.3"
45	0"	-2.7"	-5.4"
50	-.4"	-3.3"	-6.2"
55	-.8"	-4.0"	-7.0"
60	-1.2"	-4.5"	-7.7"
65	-1.5"	-4.9"	-8.4"
70	-1.7"	-5.3"	-8.8"
75	-1.9"	-5.6"	-9.2"
80	-2.0"	-5.9"	-9.6"
85	-2.1"	-5.9"	-9.7"
90	-2.1"	-6.0"	-9.8"
NOTE: Plus (+) indicates bullet strike in the direction of movement; minus (-) indicates bullet strike behind the target's center.			

Target Distance

7-149. The front sightpost covers only a small part of close-in targets, providing hits on close targets moving at any angle and any speed. However, if the lead rule is applied on more distant targets moving at a slight angle—for example, 5 degrees at 100 meters—the bullet strikes about 4 inches forward of the target's center. Soldiers must be taught to fire at targets as though they are stationary until lateral movement is observed (15 degrees).

7-150. The rule provides for many speed/angle combinations that place the bullet within 2 inches of the target's center (Table 7-14). Since the Soldier is expected to fire a 12-inch group on moving targets at 100 meters, the rule provides for hits on the majority of targets. Even the worst case (a 90-degree target moving at 8 miles per hour) would result in the shot group's center being located 9.8 inches behind the target's center. If bullets were evenly distributed in a 12-inch group, this would result in hitting the target 40 percent of the time.

7-151. Soldiers should be taught to increase their lead if they miss the target, which increases their probability of hitting all targets. For example, if target exhibits much lateral movement and the Soldier feels that he has missed the target by applying the lead rule and firing fundamentals, he should increase his lead.

7-152. The training program must be simple and provide Soldiers with information relevant to improving their performance in combat. It should address the following topics:

- Soldiers should understand and apply the single-lead rule in the absence of more information.
- Soldiers should engage moving targets coming toward them or on a slight angle (0 to 15 degrees) as stationary targets.
- Information should be presented and practice allowed on applying additional lead to targets for Soldiers who demonstrate this aptitude.

Target Angle

7-153. The single-lead rule does not apply to targets moving at small and large angles (Table 7-15). A walking enemy Soldier at 250 meters is hit dead center when he is moving at 40 degrees. Hits can be obtained if he is moving on any angle between 15 and 75 degrees. When he is running, a center hit is obtained when the target is on an angle of 18 degrees; misses occur when he exceeds an angle of 30 to 35 degrees.

7-154. The information provided in Figure 7-21 and Table 7-15 is designed to enhance instructor understanding so proper concepts are presented during instruction. For example, a target at 100 meters moving at 6 miles per hour receives a center hit when moving at 29 degrees. When moving at an angle less than 29 degrees, the bullet strikes somewhat in front of the target's center. When moving at an angle of more than 29 degrees, the bullet strikes somewhat behind the target's center.

Table 7-15. Target angle when dead center; hits occur using the single-lead rule.

STANDARD SIGHT			
RANGE (m)	4 mph	6 mph	8 mph
25	48°	30°	22°
50	47°	30°	22°
100	45°	29°	21°
150	44°	28°	20°
200	41°	27°	19°
250	40°	26°	18°
300	33°	21°	16°
350	38°	24°	18°
400	35°	22°	17°
450	33°	21°	16°

Tracking

7-155. Tracking is a more accurate technique of engaging targets used by experienced firers. It involves establishing and maintaining a point of aim in relationship to the target and moving with the target to maintain that sight picture, while squeezing the trigger. As the target moves, this technique puts the firer in position for a second shot if the first one misses.

Trapping

7-156. Trapping involves setting up a point of aim forward of the target and along the target path. The trigger is squeezed as the target comes into the sights. This technique works on targets with slow lateral movement. It does not require tracking skills, but the firer must know precisely when the weapon is going to fire.

NOTE: Soldiers who can squeeze the trigger without reacting to the weapon firing may fire better using this technique.

Breath Control

7-157. This fundamental is unchanged.

Trigger Squeeze

7-158. To use proper trigger squeeze—

- Apply rearward pressure on the handguard and pistol grip to hold the weapon steady while applying pressure to the trigger.
- Squeeze the trigger quickly (almost a controlled jerk).
- Apply heavy pressure on the trigger (at least half of the pressure it takes to make the weapon fire) before squeezing.

MOVING TARGET LIVE-FIRE EXERCISE

7-159. Soldiers engage in a firing scenario once for practice and then for qualification. Soldiers who fail to qualify on the initial day of qualification receive only one refire on the same day.

SECTION VI. SHORT-RANGE MARKSMANSHIP TRAINING

SRM training allows Soldiers to quickly and effectively engage targets at ranges less than 50 meters. Although normally associated with UO, SRM techniques are also used during operations in restrictive terrain, such as clearing a trench line, the final assault across an objective during an attack or raid, or when fighting in dense vegetation or during periods of limited visibility. SRM instruction consists of four components:

- Phase I—Reflexive Fire Training (blank-fire day and night).
- Phase II—Target Discrimination (blank-fire day and night).
- Phase III—Short-Range Marksmanship Qualification (day and night live-fire) and Barricade Transition Qualification.
- Phase IV—Shotgun and Automatic or Burst Firing Familiarization.

CONDUCT OF SHORT-RANGE MARKSMANSHIP TRAINING

7-160. SRM requires individual Soldiers to be trained to standard on reflexive firing, target discrimination, and all necessary BRM fundamentals prior to semiannual qualification.

NOTE: An explanation of the base level proficiency requirements is provided with each course of fire.

7-161. At a minimum, Soldiers should be qualified on their individual weapon within the previous six months. Shotgun and automatic firing is required for annual familiarization only. Reflexive MILES dry-fire drills are an essential part of the training process and should be conducted by the team leader or squad leader during troop-leading procedures and before any SRC or SRM training.

NOTE: See Table 7-16 for the current training program.

Table 7-16. Short-range marksmanship training program.

SHORT-RANGE MARKSMANSHIP TRAINING PROGRAM
Instructional Intent
• Soldiers gain confidence and knowledge in SRM fundamentals.
Special Instructions
Ensure that—
• The M16A2/A3/A4 rifle's or M4 carbine's rear sight is set on the 0-2 aperture.
• The M16A1's rear sight is set on the unmarked aperture.
• Each Soldier is given one 20-round magazine of 5.56-millimeter ball ammunition for marksmanship qualification.
• Each Soldier is given one 28-round magazine and three 4-round magazines of 5.56-millimeter ball ammunition for barricade transition firing.
NOTE: The round must impact within the lethal zone to be scored a hit.
Observables
• Each Soldier achieves 20 target hits during the day marksmanship qualification.
• Each Soldier achieves 16 target hits during the night marksmanship qualification.
• Each Soldier achieves 14 target hits during the day marksmanship qualification while wearing a protective mask.
• Each Soldier achieves 12 target hits during the night marksmanship qualification while wearing a protective mask.

NOTE: All SRC and SRM training should begin with a review of the principles of safe weapon handling: Assume the weapon is always loaded, and never point the weapon at anything that you do not intend to shoot.

FUNDAMENTALS OF SHORT-RANGE MARKSMANSHIP

7-162. During SRC, there is little margin for error. Too slow a shot at the enemy, too fast a shot at a noncombatant, or inaccurate shots can all be disastrous for the Soldier. Further, the risk of fratricide or noncombatant casualties is greatest during SRC. To survive and accomplish missions in close quarters, Soldiers must master four fundamentals of SRM:

- Firing stance, weapon ready positions, and movement techniques.
- Aiming technique.
- Point of aim.
- Trigger manipulation.

FIRING STANCE, WEAPON READY POSITIONS, AND MOVEMENT TECHNIQUES

7-163. Regardless of the weapon ready position used, Soldiers must always assume the correct firing stance to ensure stability and accuracy when engaging targets.

Firing Stance

7-164. To assume the correct fighting stance—

- Keep the feet approximately shoulder-width apart.
- Point the toes straight to the front (direction of movement).
- Stagger the firing-side foot slightly to the rear of the nonfiring-side foot.
- Bend the knees slightly.

- Lean the upper body slightly forward.
- Square the shoulders and pull them back; don't roll them over or slouch.
- Keep the head up and both eyes open.

7-165. When engaging targets—

- Hold the weapon with the buttstock firmly against the shoulder.
- Hold the firing-side elbow close against the body.

7-166. Although short-range engagements generally take place from the standing position, a Soldier may be required to engage targets from the kneeling position. The kneeling position is generally used when correcting a weapon malfunction.

Weapon Ready Positions

7-167. There are two weapon ready positions: high ready and low ready.

High Ready Position

7-168. The high ready position (Figure 7-22) is best suited for the lineup outside of a building, room, or bunker entrance.

7-169. To hold the weapon in the high ready position—

(1) Hold the weapon's buttstock under the armpit, with the barrel pointed slightly up so that the top of the front sightpost is just below the line of sight, but still within peripheral vision.

(2) Grasp the handguards toward the front sling swivel with the nonfiring hand. Keep the trigger finger outside of the trigger well and the thumb of the firing hand on the selector lever.

7-170. To engage a target from the high ready position—

(1) Push the weapon forward, as if to bayonet the target.

(2) Bring the buttstock firmly against the shoulder as it slides up the body.

Figure 7-22. High ready position.

Low Ready Position

7-171. The low ready position (Figure 7-23) is best suited for movement inside of buildings.

7-172. To hold the weapon in the low ready position—

(1) Place the weapon's buttstock firmly in the pocket of the shoulder, with the barrel pointed down at a 45-degree angle.

(2) Grasp the handguards toward the front sling swivel. Keep the trigger finger outside of the trigger well and the thumb of the firing hand on the selector lever.

7-173. To engage a target from the low ready position, bring the weapon up until the proper sight picture is achieved.

Figure 7-23. Low ready position.

Movement Techniques

7-174. Soldiers must practice moving with their weapons up until they no longer look at the ground, but concentrate on their sectors of responsibility and move without stumbling over their own feet.

NOTE: The low ready position is the best position to use when moving or turning.

7-175. To engage a target to the left—

(1) Step toward the target with the left foot.

(2) Move the right foot even with the left, assuming a proper standing firing position.

7-176. To engage a target to the right—

(1) Step toward the target with the right foot.

(2) Move the left foot even with the right, assuming a proper standing firing position.

7-177. To turn to the rear—

(1) Position the firing-side foot forward.

(2) Place the body weight on the firing-side foot.

(3) Pivot the body, similar to the drill movement "rear march."

AIMING TECHNIQUES

7-178. Four aiming techniques are used during SRC:

- Slow aimed fire.
- Rapid aimed fire.
- Aimed quick kill.
- Instinctive fire.

7-179. Each has advantages and disadvantages, and the Soldier must understand when, how, and where to use each technique.

Slow Aimed Fire

7-180. Slow aimed fire is the slowest, but most accurate, technique. When using this technique, Soldiers take a steady position, properly align the sight picture, and squeeze off rounds. This technique should only be used to engage targets more than 25 meters away, when good cover and concealment is available, or when the need for accuracy overrides the need for speed.

Rapid Aimed Fire

7-181. The rapid aimed fire technique utilizes an imperfect sight picture. When using this technique, the Soldier focuses on the target and raises his weapon until the target is obscured by the front sightpost. Elevation is less critical than windage when using this technique. This aiming technique is extremely effective on targets 0 to 15 meters away.

Aimed Quick Kill

7-182. The aimed quick kill technique is the quickest and most accurate method of engaging targets up to 12 meters away. As Soldiers become more experienced at using this technique, they may use it at greater ranges. When using this technique, the Soldier aims over the rear sight, down the length of the carrying handle, and places the top ½ to ¾ of an inch of the front sightpost on the target.

Instinctive Fire

7-183. Instinctive fire is the least accurate technique and should only be used in emergencies. It relies on instinct, experience, and muscle memory. To use this technique, the firer concentrates on the target and points the weapon in the general direction of the target. While gripping the handguards with the nonfiring hand, he extends the index finger to the front, automatically aiming the weapon on a line toward the target.

POINT OF AIM

7-184. Most short-range engagements will be decided by who hits his target with the first round. During this type of engagement, it is more important to put the target down as quickly as possible than it is to kill him immediately.

7-185. Soldiers must aim at the lethal zone (center of mass) of the body. Although shots to the center of the body may prove to be eventually fatal, they may not immediately incapacitate the target. A shot that does not immediately incapacitate the target may be no better than a clean miss. Because of this, and the possible presence of military equipment or protective vests, Soldiers must also be able to engage targets with incapacitating shots.

Lethal Shot Placement

7-186. The target's lethal zone (Figure 7-24) is its center of mass, between the waist and the chest. Shots in this area maximize the hydrostatic shock of the shot pellets. Due to the nature of SRC, Soldiers must continue to engage targets until they go down.

Figure 7-24. Lethal zone.

Incapacitating Shot Placement

7-187. Only one shot placement guarantees immediate and total incapacitation: roughly centered in the face, below the middle of the forehead and the upper lip, and from the eyes in. Shots to the side of the head should be centered between the crown of the skull and the middle of the ear opening, from the center of the cheekbones to the middle of the back of the head (Figure 7-25).

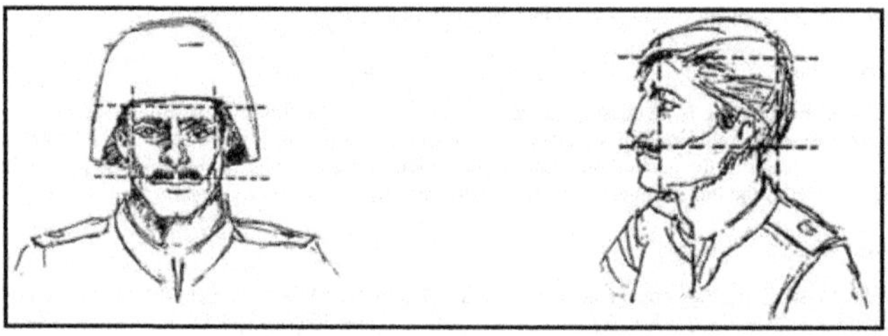

Figure 7-25. Incapacitation zone.

Trigger Manipulation

7-188. SRC engagements are usually quick, violent, and deadly. Due to the reduced reaction time, imperfect sight picture, and requirement to effectively place rounds into targets, Soldiers must fire multiple rounds during each engagement in order to survive. Multiple shots may be fired either through the use of a controlled pair or automatic weapon fire.

Controlled Pair

7-189. A controlled pair is two rounds fired in rapid succession. Controlled pairs should be fired at single targets until they go down. When multiple targets are present the Soldier must fire a controlled pair at each target, and then reengage any targets left standing. To fire a controlled pair—

(1) Fire the first round, and allow the weapon to move in its natural arc without fighting the recoil.

(2) Rapidly bring the weapon back on target, and fire a second round.

7-190. Soldiers must practice firing the controlled pair until it becomes instinctive.

Automatic Fire

7-191. While rapid, aimed, semiautomatic fire is the most accurate method of engaging targets during SRC and controlled three-round bursts are better than automatic fire, automatic weapon fire may be necessary to maximize violence of action or gain fire superiority when gaining a foothold in a room, building, or trench. When properly trained, Soldiers should be able to fire six rounds (two three-round bursts) in the same time it takes to fire a controlled pair. With practice, the accuracy of engaging targets can be equal to that of semiautomatic fire at 10 meters.

NOTE: The key to burst or automatic firing is to squeeze the trigger, not jerk it.

7-192. For the majority of Soldiers, fully automatic fire is rarely effective and can lead to unnecessary noncombatant casualties or fratricide. Not only is fully automatic fire inaccurate and difficult to control, but it also rapidly empties ammunition magazines. A Soldier who finds himself out of ammunition with an armed, uninjured enemy Soldier during SRC will become a casualty unless a fellow Soldier intervenes.

Failure Drill

7-193. To make sure that a target is completely neutralized, Soldiers should be trained to execute the failure drill. The firer will fire a controlled pair into the lethal zone, followed by a third round placed into the incapacitation zone. This type of target engagement is particularly useful when engaging targets wearing body armor.

PRELIMINARY SHORT-RANGE MARKSMANSHIP INSTRUCTION

7-194. As with all other forms of marksmanship training, preliminary SRM instruction must be conducted to establish a firm foundation. Soldiers must be taught, and must understand, the fundamentals of SRM. Blank-fire drills are conducted to ensure a thorough understanding of the fundamentals, as well as to provide the trainers with valuable feedback about each Soldier's level of proficiency.

NOTE: To maximize safety during training and in combat situations, it is important to emphasize muzzle awareness and selector switch manipulation during preliminary SRM instruction The risk of fratricide or noncombatant casualties is greatest during SRC.

7-195. Table 7-17 outlines the tasks that preliminary SRM instruction should include (at a minimum).

Table 7-17. Preliminary SRM tasks and explanation.

TASK	EXPLANATION
Weapon ready positions and firing stance	Ensure that each Soldier understands and can properly carry his weapon.
Magazine changing drills	Have Soldiers perform magazine changes using the two methods: • Parallel method. • L-shaped method. NOTE: See the paragraph below for more information about magazine changing drills.
Moving with a weapon	Ensure that Soldiers can walk, run, and turn left, right, and to the rear, as well as move from the standing to kneeling firing position and the kneeling back to the standing firing position.
Weapon malfunction drills	Ensure that Soldiers instinctively drop to the kneeling firing position, clear a malfunction using SPORTS, and continue to engage targets. This drill can be performed by issuing each Soldier a magazine loaded with six to eight rounds of blank ammunition with one expended blank round.
Target engagement drills	Teach Soldiers to move from the weapon ready position to the firing stance, emphasizing speed and precision. Soldiers must be observed to ensure that the finger is outside of the trigger well and that the selector switch remains on SAFE until the weapon is raised to the firing position. NOTE: This is a force protection issue and must be drilled until all Soldiers can perform to standard.

MAGAZINE CHANGING DRILLS

NOTE: Before performing magazine changing drills, have Soldiers configure their LBE or magazine racks to allow for the easiest access for the magazine change. The time limit to perform any of the magazine changing drills is 8 seconds.

7-196. There are two methods of changing the magazine:

- Parallel method.
- L-shaped method.

Parallel Method

7-197. When shooting from the right side, perform the right-side parallel magazine changing method (Figure 7-26):

(1) Engage the target until the magazine is empty.
(2) Take a knee.
(3) Place the weapon at a 45-degree angle, with the magazine well facing in, the muzzle pointed in a safe direction, and the buttstock in the crease of the right elbow.
(4) Slide the nonfiring hand down the handguard to the receiver, and press the magazine release catch.
(5) Secure a full magazine with two fingers and the thumb of the nonfiring hand.
(6) Bring the full magazine (top of the magazine toward the magazine well) next to and parallel with the empty magazine.
(7) Slide the empty magazine out with the nonfiring hand and insert the full magazine.
(8) Hit the bolt release.
(9) Stow the empty magazine in the ammunition pouch.

Figure 7-26. Right-side parallel magazine changing method.

7-198. When shooting from the left side, perform the left-side parallel magazine changing method (Figure 7-27):

(1) Engage the target until the magazine is empty.
(2) Take a knee.
(3) Place the weapon at a 45-degree angle, with the magazine well facing in, the muzzle pointed in a safe direction, and the buttstock in the crease of the left elbow.
(4) Secure a full magazine with two fingers and the thumb of the nonfiring hand.
(5) Bring the full magazine (top of the magazine toward the magazine well) next to and parallel with the empty magazine.
(6) Press the magazine release catch with the trigger finger.
(7) Slide the empty magazine out with the nonfiring hand, and insert the full magazine.
(8) Hit the bolt release.
(9) Stow the empty magazine in the ammunition pouch.

Figure 7-27. Left-side parallel magazine changing method.

L-Shaped Method

7-199. When shooting from the right side, perform the right-side L-shaped magazine changing method (Figure 7-28):

(1) Engage the target until the magazine is empty.
(2) Take a knee.
(3) Place the weapon at a 45-degree angle, with the magazine well facing in, the muzzle pointed in a safe direction, and the buttstock in the crease of the right elbow.
(4) Slide the nonfiring hand down the handguard to the receiver, and press the magazine release catch.
(5) Secure a full magazine with two fingers and the thumb of the nonfiring hand.
(6) Bring the full magazine (top of the magazine facing to the rear) next to and perpendicular to the empty magazine, forming an L-shape.
(7) Slide the empty magazine out with the nonfiring hand, rotate the full magazine so the top of the magazine faces the magazine well, and insert the full magazine.
(8) Hit the bolt release.
(9) Stow the empty magazine in the ammunition pouch.

Figure 7-28. Right-side L-shaped magazine changing method.

7-200. When shooting from the left side, perform the left-side L-shaped magazine changing method (Figure 7-29):

(1) Engage the target until the magazine is empty.
(2) Take a knee.
(3) Place the weapon at a 45-degree angle, with the magazine well facing in, the muzzle pointed in a safe direction, and the buttstock in the crease of the left elbow.
(4) Press the magazine release catch with the trigger finger.
(5) Secure a full magazine with two fingers and the thumb of the nonfiring hand.
(6) Bring the full magazine (top of the magazine facing to the rear) next to and perpendicular to the empty magazine, forming an L-shape.
(7) Slide the empty magazine out with the nonfiring hand, rotate the full magazine so the top of the magazine faces the magazine well, and insert the full magazine.
(8) Hit the bolt release.
(9) Stow the empty magazine in the ammunition pouch.

Figure 7-29. Left-side L-shaped magazine changing method.

PHASE I—REFLEXIVE FIRE TRAINING

7-201. Reflexive fire training involves the practical application of all four of the fundamentals of SRM. All Soldiers must receive a GO on the task, Conduct Reflexive Firing, before proceeding with training. This is a perishable skill that must be constantly reinforced, and reflexive fire training should be conducted frequently as refresher training to ensure that the Soldiers' skills are always at the highest level possible.

Reflexive Firing Targets

7-202. Targets can be purchased locally (FBI style) or manufactured by the unit (bowling pin targets). E-type silhouettes may be painted as shown in Figure 7-30.

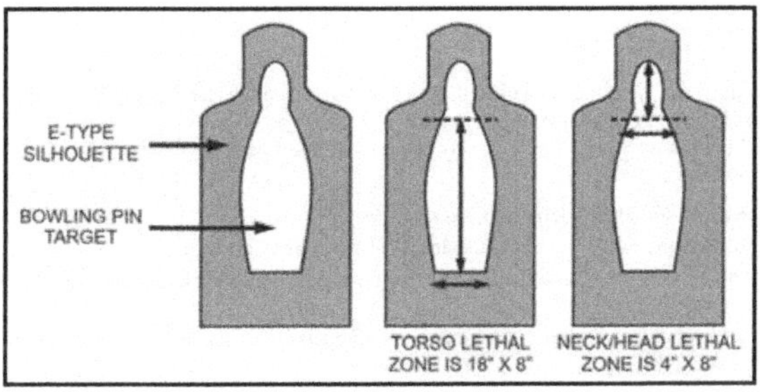

Figure 7-30. Dimensions and placement of bowling pin targets.

RANGE SETUP

7-203. The range must be at least 25 meters in length, with identification marks at the 5-, 10-, 15-, and 25-meter distances. Each lane should be marked in a way that prevents crossfiring between lanes. A lane safety coach is assigned to each lane to observe and evaluate the Soldiers' performance, as well as ensure the safe conduct of firing. All firing cues are given by the tower or line safety.

CONDUCT OF TRAINING

NOTE: Unit commanders should conduct training continually to first establish and then sustain levels of proficiency in reflexive firing.

7-204. Each Soldier conducts a dry-fire exercise and a blank-fire exercise prior to conducting the LFX. The dry-fire and blank-fire exercises give the Soldier the repetition needed to successfully engage targets quickly and accurately.

7-205. During the dry-fire exercise, Soldiers identify and engage the proper targets at ranges from 5 to 25 meters from the stationary position, while turning and walking. The tower or line safety gives all firing cues. All tables are fired at night, with and without protective masks, using automatic fire for familiarization, and while using NVDs. The dry-fire exercise proceeds as follows:

(1) The Soldier starts at the 25-meter line at the low ready position, facing the targets.

(2) The Soldier is then told the engagement position (for example, facing left, turns right) and, once in position, is given the cue to fire.

(3) On cue, the Soldier assumes the proper firing position and stance, places the selector lever on SEMI, uses the correct aiming technique for the target's distance, and engages the target.

(4) After engaging the target, the Soldier continues to cover the target to reinforce firing until the threat is eliminated.

NOTE: If Soldiers will be engaging targets using lasers, optics, or the protective mask, they should complete all steps using the same equipment. Do not have Soldiers familiarize with iron sights and then fire the LFX with optics.

Evaluation

7-206. Soldiers are evaluated on a GO/NO GO basis, based on the standards in the training and evaluation outline (T&EO) and scoring table. Soldiers must score a GO on the familiarization firing tables (Table 7-18 and Table 7-19) before attempting to qualify. The standard for protective mask firing is 60 percent day and 50 percent night.

7-207. Use the following guidelines for scoring:

- Rounds fired after the time standard will be scored as a miss.
- The number of rounds fired after the time standard will be subtracted from the total number of hits the Soldier has scored.
- All rounds must impact on the E-type silhouette. Hits are defined as being in the lethal zone (bowling pin).

Table 7-18. Familiarization (stationary).

POSITION	ROUNDS FIRED	DISTANCE (m)	METHOD	TIME STANDARD	LETHAL ZONE HIT STANDARD
Straight ahead	4	25	Single shot	None	3
Straight ahead	4	10	Single shot	None	3
Straight ahead	4	25	Controlled pair	None	3
Straight ahead	4	10	Controlled pair	None	3

Table 7-19. Familiarization (moving).

POSITION	ROUNDS FIRED	DISTANCE (m)	METHOD	TIME STANDARD	LETHAL ZONE HIT STANDARD
Facing left; turning right	4	25	Controlled pair	None	3
Facing right; turning left	4	25	Controlled pair	None	3
Straight ahead walking	4	10, start at 15	Controlled pair	None	3
Straight ahead	4	5	Controlled pair	None	3

PHASE II—TARGET DISCRIMINATION TRAINING

7-208. During SRC, there is little or no margin for error. Too slow a shot at the enemy, too fast a shot at a noncombatant or friendly Soldier, or inaccurate shots can all be disastrous for the Soldier. Target discrimination is the act of distinguishing between threat and nonthreat targets during SRC. Target discrimination is an inescapable responsibility and must be stressed in all situations regardless of mission.

NOTE: It is essential that target discrimination training be aimed at instilling fire control and discipline in Soldiers. The first priority is always the safety of a Soldier.

TARGET DISCRIMINATION TARGETS

7-209. Targets used to conduct target discrimination training include—

(1) Two or more E-type silhouettes with bowling pins painted on each side of the silhouette (for example, brown side and green side).

(2) Multiple E-type silhouettes with different painted shapes (for example, squares, triangles, and circles).

(3) A series of 3-inch circles on E-type silhouettes.

(4) Pop-up targets (for example, electrical or pull targets).

(5) Target's hands and center of the uniform.

Two or More E-Type Silhouettes with Bowling Pins Painted on Each Side

7-210. Target discrimination is best taught using two or more E-type silhouettes with bowling pins painted on each side of the silhouette. To conduct training using this type of target—

(1) The instructor calls out a color for the firer to identify.

(2) On the command "READY, UP" or at the whistle blast, the firer quickly scans all targets for the color and engages the correct target using a controlled pair.

7-211. This exercise trains Soldiers to accomplish missions under the expected ROE. During force-on-force training, the OPFOR wears distinctive uniforms; this exercise prepares Soldiers to eliminate threats based on enemy uniforms and reduces the chances of a Soldier hesitating and becoming a casualty. Using realistic targets displaying threat and nonthreat personnel is another variation.

Multiple E-Type Silhouettes with Different Painted Shapes

7-212. To conduct training using this type of target—

(1) The instructor calls out a shape for the firers to identify.

(2) On the command "READY, UP" or at a whistle blast, the firers quickly scan all targets, searching for the shape, and engage the correct target using the controlled pair technique.

7-213. This is repeated until one shape is mastered. Subsequently, a sequence of shapes is announced, and the firers engage accordingly.

Series of 3-Inch Circles on E-Type Silhouettes

7-214. To conduct training using this type of target—

(1) Instructors call out which circle to engage (for example, top left).

(2) Firers react accordingly.

7-215. Marksmanship is emphasized using this technique.

Target's Hands and Center of the Uniform

7-216. If a target is a threat, the first and most obvious indicator is a weapon in the target's hands. A good technique for teaching Soldiers target discrimination is to have them focus on the target's hands and the center of the uniform. To conduct this exercise, the Soldier must visualize the entire target because an armed target could be a fellow Soldier or other friendly. This level of target discrimination should not be trained until Soldiers are thoroughly proficient in basic SRC and SRM tasks.

Range Setup

7-217. The range must be at least 25 meters in length, and each lane should be at least 5 meters wide. Each lane should have target holders and should be marked in a way that prevents crossfiring between lanes. A coach/safety is assigned to each lane to observe and control the Soldier's performance. The tower, lane safety, or senior instructor gives all firing commands.

Conduct

7-218. Regardless of the type of target used, the exercise proceeds as follows:

NOTE: Each Soldier must complete a dry-fire exercise and a blank-fire exercise before moving to the live-fire portion.

(1) The Soldier faces away from the target.

NOTE: This action requires the Soldier to identify and discriminate targets, and reinforces skills used during reflexive firing training.

(2) The Soldier is given a target description.

(3) On the command "READY," the Soldier begins to scan for the target.

(4) On the command "UP," voice command, or whistle blast, the Soldier turns toward and engages the target.

NOTES: 1. Instructors should vary commands and targets so that the Soldier does not fall into a pattern. Intermixing "NO FIRE" commands will add to realism.

2. The first priority is always the safety of the Soldier.

7-219. All Soldiers must receive a GO on this task before SRM qualification. Targets must be scored and marked after each firing distance. A Soldier will be scored as a NO GO if he fails to engage a target or engages a target other than the one called for by the instructor. Soldiers will also receive a NO GO if, at any time, they point their weapon at another Soldier or they fail to keep their weapon on SAFE before acquiring and engaging the targets.

NOTE: Initial training and sustainment training may be conducted by changing the uniform in the standards statement.

EVALUATION

7-220. Table 7-18 will be used to score this exercise.

PHASE III—SHORT-RANGE MARKSMANSHIP QUALIFICATION

7-221. Soldiers should conduct SRM qualification semiannually. In addition to qualification, commanders should conduct familiarization using the same qualification standards while altering the conditions. Firing the qualification tables in protective masks and during periods of limited visibility should be included; Soldiers should train as they fight—with all assigned equipment.

NOTE: Although the qualification is intended to be fired with open sights only, iterations using laser aiming devices, CCOs, and NVDs are highly encouraged.

RECORD AND PRACTICE FIRE

Conduct

NOTES:
1. Soldiers must complete a blank-fire iteration of the qualification tables before conducting live-fire qualification.
2. Each Soldier will have a coach to ensure that he acquires the target, keeps the weapon on SAFE until time to engage the target and is then places it back on SAFE, and maintains muzzle awareness throughout the exercise.
3. If Soldiers will be engaging targets with lasers, optics, or the protective mask, they should complete all steps using the same equipment. Do not have Soldiers familiarize with iron sights and fire the exercise while wearing the protective mask.

7-222. Each Soldier engages the target IAW the firing table (Table 7-20).

Table 7-20. Record and practice fire.

POSITION	ROUNDS FIRED	DISTANCE (m)	METHOD	TIME STANDARD
Straight ahead	2	4	Controlled pair	3 sec from command "UP"
Left turn	2	7	Controlled pair	3 sec from command "UP"
Right turn	2	10	Controlled pair	3 sec from command "UP"
Walk straight	2	5, start at 10	Controlled pair	3 sec from command "UP"
Walk straight	2	10, start at 15	Controlled pair	3 sec from command "UP"
Run, stop, and shoot	2	10, start at 20	Controlled pair	3 sec from command "UP"
Straight ahead	2	25	Controlled pair	3 sec from command "UP"
Run, stop, kneel, and shoot	2	15, start at 25	Controlled pair	3 sec from command "UP"
Walk laterally right	2	7	Controlled pair	3 sec from command "UP"
Walk laterally left	2	7	Controlled pair	3 sec from command "UP"

Evaluation

NOTE: For scoring purposes, a hit is a round that impacts within the lethal zone.

7-223. Each Soldier scores 16 hits day and night. The standard when wearing a protective mask is 14 day and 12 night. In addition to achieving a qualifying score, all 20 rounds must hit the E-type silhouette in order to qualify.

BARRICADE TRANSITION QUALIFICATION FIRING

Range Setup

7-224. The range must be at least 50 meters in length, with tall barricades at the 35- and 50-meter distances and short barricades at the 25- and 40-meter distances. Each lane should be marked in a way that prevents crossfiring between lanes. A lane safety coach is assigned to each lane to observe and evaluate the Soldier's performance, as well as ensure that firing is conducted safely. The tower or line safety gives all firing cues.

Conduct

NOTE: Soldiers must complete a dry-fire and a blank-fire iteration of the qualification tables before conducting live-fire qualification. The dry-fire and blank-fire exercises gives the Soldier the repetition needed to successfully engage targets quickly and accurately.

7-225. The exercise proceeds as follows:

(1) The Soldier receives one 28-round and three 4-round magazines.
(2) The Soldier loads one 28-round magazine.
(3) The Soldier assumes the low ready position at the 50-meter line, facing the targets.
(4) The Soldier engages the targets as shown in Table 7-21.
(5) On cue, the Soldier assumes the proper firing position and stance, places the selector lever on SEMI, uses the correct aiming technique for the target's distance, and engages the target.

NOTE: If a Soldier is having difficulty during the blank-fire exercise, he must be retrained before continuing with the qualification.

Evaluation

7-226. Use the following guidelines for scoring:

- All rounds must impact on the E-type silhouette.
- Hits are defined as being in the lethal zone (bowling pin).

Table 7-21. Barricade transition fire.

BARRICADE/DISTANCE	POSITION	ROUNDS FIRED	METHOD
Start with 28-round magazine in the open.			
Open/50 m	Standing	2	Controlled pair
	Kneeling	2	Controlled pair
	Prone	2	Controlled pair
	Sitting	2	Controlled pair
Tall/50 m	Left	2	Controlled pair
	Right	2	Controlled pair
Short/40 m	Left	2	Controlled pair
	Top	2	Controlled pair
	Right	2	Controlled pair
Tall/35 m	Left	2	Controlled pair
	Right	2	Controlled pair
Short/25 m	Left	2	Controlled pair
	Top	2	Controlled pair
	Right	2	Controlled pair
Transition back to the tall 35-m barricade. Perform parallel or L-shaped magazine change. Insert 4-round magazine.			
Tall/35 m	Left	2	Controlled pair
	Right	2	Controlled pair
Transition back to short 40-m barricade. Perform parallel or L-shaped magazine change. Insert 4-round magazine.			
Short/45 m	Left	2	Controlled pair
	Right	2	Controlled pair
Transition back to tall 50-m barricade. Perform parallel or L-shaped magazine change. Insert 4-round magazine.			
Tall/50 m	Left	2	Controlled pair
	Right	2	Controlled pair

PHASE IV—SHOTGUN AND AUTOMATIC OR BURST FIRING FAMILIARIZATION

7-227. Shotgun and automatic or burst firing familiarization is no different for SRM than for BRM.

NOTE: Publications for shotgun firing familiarization are currently being developed.

SECTION VII. SQUAD DESIGNATED MARKSMAN TRAINING

Advances in technology have led to the development of weapon systems that are increasingly more accurate and able to engage targets at much longer ranges. Conversely, today's rifleman is trained to engage targets only out to 300 meters. This 300-meter limit is well short of the weapon/ammunition combination's capability. Snipers engage targets at 600 meters and beyond. The SDM engages targets with direct small arms fire in the gap between the engagement range of the average combat Soldier and the sniper. Possessing the ability to estimate range, detect targets, and place effective, well-aimed fire on intermediate range targets, the SDM plays a vital role on the modern battlefield.

MISSION OF THE SQUAD DESIGNATED MARKSMAN

7-228. The SDM program provides the squad with a designated marksman that has been trained to engage targets from 300 to 500 meters. He will operate and maneuver as a rifleman, but will have the added responsibility of engaging targets out to 500 meters with effective, well-aimed fires. The SDM is a vital member of his individual squad, not a squad sniper. The SDM has neither the equipment nor training to engage targets at extended ranges with precision fires while operating individually or in a small team. He can also be used to help direct the fires of other squad members into enemy positions. Due to the increased skill level required for his position, the SDM must maintain a high level of proficiency through continued training of the required skills. The SDM must possess a thorough understanding and mastery of—

- The fundamentals of rifle marksmanship.
- Ballistics.
- Elevation and windage.
- Hold-off (adjusted points of aim).
- Sight manipulation.
- Range estimation.

SELECTION

7-229. The platoon sergeant and squad leaders must take special consideration in selecting the SDM. The SDM must have—

- A solid marksmanship performance.
- A clear understanding of the fundamentals.
- The ability to apply these fundamentals consistently during dry-fire and live-fire training.

SQUAD DESIGNATED MARKSMAN SKILLS PROGRESSION

7-230. The skills progression program assesses the Soldier's ability to apply the fundamentals of marksmanship and trains and assesses the Soldier's proficiency in several key areas. Firing events will serve to both reinforce and assess these areas.

CONDUCT

7-231. While conducting the skills progression program, instructors-trainers will adhere to the following guidelines:

- The skills progression program for the SDM is based on the M16-/M4-series weapon system and a 98-rounds-per-man ammunition requirement.
- Soldiers will use their assigned weapon during the training.
- The firing events will be conducted with the iron sights or BUISs only.
- The firing events will be conducted on a KD range that enables firing out to 600 meters at a minimum.

QUALIFICATION

7-232. SDM qualification requires the completion of five phases:

(1) Position evaluation.
(2) Dry-fire training.
(3) Range estimation and sight manipulation.
(4) Hold-off.
(5) Field fire.

7-233. Each phase stresses marksmanship fundamentals and specific skill areas required to perform as an SDM. To continue training, Soldiers must receive a GO in each phase. Soldiers who fail in any area should be removed from training.

NOTE: If an optic is issued for use, the phases dealing with hold-off and field record fire will be removed and relevant optics training and testing will be substituted.

PHASE I—POSITION EVALUATION

7-234. Phase I of the training consists of demonstrating the ability to consistently assume proper firing positions. The foxhole supported and prone unsupported firing positions will provide the Soldier with the smallest target exposure to the enemy and will be used during this training cycle. The prone supported position can be substituted for the foxhole supported position dependent on range configurations. Prior to this phase of training, trainers ensure that the—

- Weapon is cleared and that no ammunition is loaded prior to training.
- Weapon is zeroed prior to training.
- Soldier is able to assume a steady firing position.

Foxhole Supported

7-235. The Soldier must be able to successfully assume a proper supported position while firing from a foxhole. The trainer must ensure that the Soldier has a good steady position. After the Soldier has assumed a good supported position in the foxhole—

(1) The trainer uses DA Form 7650-R (Squad Designated Marksman—Position Evaluation) to evaluate his position and take notes on all of the following characteristics:

NOTE: See the end of this publication for a blank, reproducible copy of DA Form 7650-R.

- Eye relief.
- Trigger finger.
- Elbows.
- Nonfiring hand.
- Legs.

NOTE: The main areas that will differ between the foxhole supported and the prone supported positions are in the placement of the elbows, legs, and nonfiring hand. These body positions will be similar to those of the prone unsupported position.

---

(2) After all characteristics have been noted, the trainer has the Soldier lay his weapon down, relax, and then assume another supported position in the foxhole.

(3) The trainer evaluates this position by comparing his notes from the original supported position. The Soldier should maintain the same characteristics in the second evaluation as he did in the first.

(4) Once the trainer is satisfied that the Soldier has demonstrated the proper position and is able to show it in two consecutive attempts, the Soldier moves to the unsupported prone position.

Eye Relief

7-236. To evaluate the Soldier's eye relief, ensure that the Soldier

- Demonstrates a consistent eye relief by checking the placement of the Soldier's cheek on the weapon's buttstock.
- Places his eye is the same distance from the rear sight each time he is evaluated.

Trigger Finger

7-237. To evaluate the placement of the Soldier's trigger finger, ensure that the Soldier—

- Uses his own style; not all Soldiers place their finger on the trigger in the same place.
- Places his finger on the trigger the same way each time he is evaluated.

Elbows

7-238. To evaluate the placement of the Soldier's elbows, ensure that the Soldier—

- Places his elbows firmly a comfortable distance apart on the outside edge of the foxhole.
- Uses a sandbag, and not his arms, to support the weapon's weight.
- Assumes a stable position each time he is evaluated by slightly nudging him.

Nonfiring Hand

7-239. To evaluate the placement of the Soldier's nonfiring hand, ensure that the Soldier—

- Places the nonfiring hand in a position that is comfortable and provides the best weapon stability and support. Show the Soldier different ways this can be done.
- Is supporting the weapon properly by nudging him after the weapon has been stabilized.

Legs

7-240. To evaluate the placement of the Soldier's legs, ensure that the Soldier—

- Places the legs inside the foxhole while firing.
- Plants the legs firmly for a stable position while firing. Slightly nudge the Soldier to make sure that his legs are firmly planted in the foxhole.

Prone Unsupported

7-241. The Soldier must be able to successfully assume a proper unsupported firing position. The trainer must ensure that the Soldier has a good steady position. After the Soldier has assumed a good unsupported firing position —

(1) Taking special care to observe the positioning of the elbows, the nonfiring hand, and the legs, the trainer uses DA Form 7650-R (Squad Designated Marksman—Position Evaluation) to evaluate the same characteristics as with the supported firing position with the exception of the—

- Elbows.
- Nonfiring hand.
- Legs.

NOTE: See the end of this publication for a blank, reproducible copy of DA Form 7650-R.

(2) Once the trainer has noted the Soldier's position, he has the Soldier lay his weapon down, stand up, relax, and then get back down into another unsupported prone position.

(3) The trainer evaluates this position by comparing his notes from the original position. The Soldier should maintain the same characteristics in the second evaluation as he did in the first evaluation.

(4) The trainer lets the Soldier hold this firing position for approximately 15 seconds to check for shaking. If the Soldier starts to shake, have him relax and reposition himself.

(5) Once the trainer is satisfied that the Soldier has demonstrated the proper position and is able to accomplish it in two consecutive attempts, the Soldier moves on to the next phase of training.

Elbows

7-242. To evaluate the placement of the Soldier's elbows, ensure that the Soldier—

- Places the elbows a comfortable distance apart on the ground.
- Uses the bone, not the muscles, to support the weapon's weight. This will prevent any unnecessary muscle fatigue and will allow for a steadier firing position.
- Assumes a stable position. Slightly nudge the Soldier to ensure that his position is stable.

Nonfiring Hand

7-243. To evaluate the placement of the Soldier's nonfiring hand, ensure that the Soldier—

- Places his nonfiring hand in a comfortable position on the handguards.
- Does not support his nonfiring hand on the ground, sandbag or anything that would create a supported position.

Legs

7-244. To evaluate the placement of the Soldier's legs, ensure that the Soldier—

- Positions his legs in such a way that he has a stable position; not all Soldiers position their legs the same way while shooting from the prone position.
- Spreads his legs a comfortable distance apart, with the heels on the ground or as close as possible without causing strain.

Zero Confirmation

7-245. After completing Phases I and II, the Soldier conducts a firing event (Table 7-22) to zero or confirm the zero on his weapon and reinforce the fundamentals of marksmanship. This firing event will be conducted on a 25-meter range. If the Soldier cannot zero within 18 rounds, the trainer recommends re-training, retesting, or possible removal from the course. After the weapon is zeroed, any additional rounds will be fired and the coach will observe the Soldier for deficiencies in his marksmanship fundamentals.

Table 7-22. Zero/zero confirmation firing event.

FIRING EVENT	ROUNDS	TARGET RANGE (m)
Zero/Zero Confirmation	18	25

PHASE II—DRY-FIRE TRAINING

7-246. SDMs must have a solid grasp on the fundamentals to successfully engage targets at longer ranges. During this phase of training, the Soldier must demonstrate that he can apply the fundamentals of marksmanship correctly. If the Soldier does not receive a GO in this phase of training, he will be dropped from the course. The components of this phase of training are—

- Follow-through.
- Borelight exercise.
- Target box exercise.
- Dime/washer drill.
- Zero confirmation.

7-247. Prior to this phase of training trainers ensure that the—

- Weapon is cleared and no ammunition is loaded prior to training.
- Weapon is zeroed prior to training.
- Soldier is able to consistently apply the fundamentals of marksmanship.

Follow-Through

7-248. Follow-through involves applying all of the marksmanship fundamentals while and after the weapon fires. A good follow-through ensures that the weapon is allowed to fire and recoil naturally. The Soldier/weapon combination reacts as a single unit to such actions. It consists of the following:

- Keeping the cheek in firm contact with the stock (stock weld).
- Keeping the finger on the trigger all the way to the rear.
- Continuing to look through the rear aperture.
- Keeping muscles relaxed.
- Avoiding reaction to recoil or noise.
- Releasing the trigger only after the recoil has stopped.

Borelight Exercise

7-249. The borelight dry-fire exercise provides evaluation of the Soldier throughout the integrated act of firing.

NOTE: If a borelight is not available, the target box exercise will be used.

7-250. To perform a borelight exercise—

(1) The trainer attaches a 25-meter zero target to a flat surface.

(2) The trainer positions the Soldier 10 meters away, facing the target.

(3) The Soldier assumes a good prone supported firing position with the borelight inserted in the weapon's barrel and with the weapon placed in the dry-fire mode.

(4) The trainer uses DA Form 7650-R (Squad Designated Marksman—Position Evaluation) to evaluate the Soldier's position:

NOTE: See the end of this publication for a blank, reproducible copy of DA Form 7650-R.

---

(5) The Soldier aims at the silhouette's center of mass on the 25-meter zero target and squeezes the trigger.

(6) The borelight is activated as the trigger is fired. The laser is seen on the 25-meter zero target.

(7) The trainer marks the 25-meter zero target exactly were the borelight laser hit the target.

(8) The Soldier gets out of position and then back into a prone supported firing position.

7-251. This process will be done until a three-round shot group has been achieved. The Soldier will do the same from the prone unsupported firing position. To receive a GO, the Soldier must place a three-round shot group in a 3-centimeter circle from both prone positions.

Target Box Exercise

7-252. The target box exercise checks the consistency of aiming and placement of three-round shot groups in a dry-fire environment. To conduct the exercise—

(1) The target man places the silhouette on a plain sheet of paper 25 meters away from the firer and moves the silhouette target as directed by the Soldier.

(2) When the Soldier establishes proper aiming, he squeezes the trigger.

(3) The target man marks through the silhouette with a pen or pencil at the target's center of mass.

(4) The target man moves the silhouette to another spot on the paper and tells the firer to repeat the process twice more to obtain a shot group.

NOTE: A simulated shot group covered within a 1/2-centimeter circle indicates consistent aiming.

7-253. Since no weapon or ammunition variability is involved and shot groups do not have to be placed in a certain location, a 1/2-centimeter standard may be compared to obtaining a 4-centimeter shot group on the 25-meter live-fire zero range.

Dime/Washer Drill

7-254. The dime/washer drill is an effective way of measuring the Soldier's trigger squeeze. To conduct the exercise—

(1) The Soldier takes aim and squeezes the trigger.

(2) If the dime or washer remains in place, he has successfully squeezed the trigger.

7-255. The Soldier must successfully obtain five out of five consecutive shot groups within 1 centimeter and without allowing the dime or washer to drop. The trainer evaluates the Soldier's performance and gives the Soldier a GO or NO GO. If the Soldier receives a NO GO, the trainer recommends retraining, retesting, or possible removal from the course.

Zero Confirmation

7-256. Zero confirmation is conducted as shown in Table 7-6.

PHASE III—RANGE ESTIMATION AND SIGHT MANIPULATION 100 TO 500 METERS

7-257. SDMs must use range estimation methods to determine the distance between their position and the target. Trainers ensure that the—

- Weapon is cleared and no ammunition is loaded prior to training.
- Weapon is zeroed prior to training.
- Soldier knows how to adjust for wind and gravity.
- Soldier can manipulate the rear sight for different ranges.

7-258. The trainer sets up a range estimation course using E-type silhouettes at ranges from 100 meters to 700 meters. Soldiers practice on this course until they find the method that works best for them.

7-259. Once the Soldiers have had time to practice, trainers test their ability to estimate range. The Soldier is given six targets. He must estimate the range within 50 meters of the actual range to receive a GO. The Soldier must estimate range correctly six out of six targets to move on to the next portion of this phase.

Methods of Range Determination

7-260. SDMs can use five different methods of range determination:

- 100-meter unit-of-measure method.
- Range card method.
- Front sightpost method.
- Appearance of objects method.
- Combination method.

100-Meter Unit-of-Measure Method

7-261. To use this method, the SDM must be able to visualize a distance of 100 meters on the ground.

- For ranges up to 500 meters, the SDM determines the number of 100-meter increments between the two objects he wishes to measure.
- Beyond 500 meters, he must select a point halfway to the object, determine the number of 100-meter increments to the halfway point, and then double the number.

NOTES: 1. See Chapter 6 for more information about this method of range determination.

2. For example, terrain with much dead space limits the accuracy of the 100-meter method.

Range Card Method

7-262. SDMs use a range card to quickly determine ranges throughout the target area. Once a target is detected, the SDM determines its location on the card and then reads the proper range to the target.

Front Sightpost Method

7-263. Using the front sightpost as a scale is another method of estimating range. This method can be used for a quick on-the-spot estimation and engagement.

- If a man-sized target is approximately ½ of the width of the front sightpost, he is approximately 300 meters away.
- If the target is 1/2 the width of the front sightpost, then the target is approximately 600 meters away.

NOTE: See Chapter 6 for more information about this method of range determination.

Appearance of Objects Method

7-264. This method of range determination is based on the size and visible characteristics of an object. To use this method with any degree of accuracy, the SDM must be familiar with the appearance and visible detail of an object at various ranges. Some common guidelines can be used to determine the range of a human target:

- At 200 meters, a human target is clear and details can be seen.
- At 300 meters, the target is still clear, but no details can be seen.
- At 400 meters, the target's outline is clear; however, the target itself is blurry.
- At 500 meters, the body tapers and the head disappears.
- At 600 meters, the body resembles a wedge shape.

NOTE: See Chapter 6 for more information about this method of range determination.

Combination Method

7-265. In a combat environment, perfect conditions rarely exist. Therefore, only one method of range estimation may not be enough for the SDM's specific mission. By using a combination of two or more methods to determine an unknown range, an experienced SDM should arrive at an estimated range close to the true range.

Factors Affecting Range Estimation

7-266. Three factors affect range estimation:

- Nature of the target
- Nature of the terrain.
- Light conditions.

Nature of the Target

7-267. The nature of the target affects its perceived range:

- An object with a regular outline, such as a house, appears closer than one with an irregular outline, such as a clump of trees.
- A target that contrasts with its background appears to be closer than it actually is.
- A partly exposed target appears more distant than it actually is.

Nature of the Terrain

7-268. The contour of the terrain affects the observer's ability to estimate range:

- As the observer's eye follows the contour of the terrain, he tends to overestimate distant targets.
- Observing over smooth terrain, such as sand, water, or snow, causes the observer to underestimate distant targets.
- Looking downhill, the target appears farther away.
- Looking uphill, the target appears closer.

Light Conditions

7-269. Light conditions affect range estimation:

- The more clearly a target can be seen, the closer it appears.
- When the sun is behind the observer, the target appears to be closer.
- When the sun is behind the target, the target is more difficult to see and appears to be farther away.

Elevation Knob Training

7-270. Elevation knob training involves nothing more than being able to adjust the rear elevation knob (front sightpost on the M16A4) for the various ranges that the SDM must engage. With this knowledge, he can better determine his range settings for the different distances between the 100-meter adjustments.

7-271. The rear elevation knob adjusts the point of aim—

- From 300 to 800 meters on the M16A2.
- From 300 to 600 meters on the M16A4 and M4.

Conduct

7-272. During elevation knob training, the Soldier determines the number of adjustments (clicks) between the different range settings on his rear elevation adjustment knob (Tables 7-23 and 7-24). Once the Soldier understands how to set the proper point of aim for his target using his rear elevation knob, the instructor/trainer has him conduct another range estimation course. This time, the instructor/trainer has him estimate the range and set the rear elevation for the range that he has estimated. The Soldier must estimate range and set his rear elevation knob properly six out of six times to receive a GO. If the Soldier receives a NO GO, the trainer recommends retraining, retesting, or possible removal from the course.

7-273. Once the Soldier has an understanding of range estimation and sight manipulation, he is able to begin the live-fire training exercise (Table 7-25). The Soldier will be given 20 rounds in which to engage 20 targets at ranges from 100 to 500 meters using mechanical sight adjustments.

Table 7-23. Elevation knob, M16A2/3 and front sightpost, M16A4.

DISTANCE (m)	DISTANCE ONE CLICK WILL ADJUST THE POINT OF IMPACT		
	FRONT SIGHTPOST	WINDAGE KNOB	ELEVATION WHEEL
25	.83 cm (3/8 in)	.33 cm (1/8 in)	.5 cm (1/4 in)
50	1.50 cm (5/8 in)	.5 cm (1/4 in)	1.5 cm (1/2 in)
75	2.50 cm (1 in)	1.0 cm (3/8 in)	2.0 cm (3/4 in)
100	3.50 cm (1 3/8 in)	1.5 cm (1/2 in)	2.75 cm (1 in)
150	5.00 cm (2 in)	2.0 cm (3/4 in)	4.0 cm (1 1/2 in)
175	6.00 cm (2 3/8 in)	2.25 cm (7/8 in)	5.0 cm (2.0 in)
200	6.50 cm (2 5/8 in)	2.5 cm (1 in)	5.5 cm (2 1/4 in)
250	8.50 cm (3 3/8 in)	3.5 cm (1 1/4 in)	7.0 cm (2 3/4 in)
300	10.0 cm (4 in)	4.0 cm (1 1/2 in)	8.5 cm (3 1/4 in)
400	13.5 cm (5 3/8 in)	5.5 cm (2 1/4 in)	11.0 cm (4 1/2 in)
500	17.0 cm	6.5 cm (2 1/2 in)	14.0 cm (5 1/2 in)
600	20.5 cm	8.0 cm (3 1/8 in)	16. 75 cm (6 1/2 in)
700	24.0 cm	9.0 cm (3 5/8 in)	19.5 cm (7 1/2 in)
800	27.5 cm	10.5 cm (4 1/8 in)	22.5 cm (8 3/4 in)
NOTE: All values were rounded off.			

Table 7-24. Elevation knob, M4/M4A1 and windage, M16A4.

DISTANCE (m)	DISTANCE ONE CLICK WILL ADJUST THE POINT OF IMPACT		
	FRONT SIGHTPOST	WINDAGE KNOB	ELEVATION WHEEL
25	1.2 cm (1/2 in)	.5 cm (1/4 in)	.5 cm (1/4 in)
50	2.4 cm (1 in)	1.5 cm (1/2 in)	1.5 cm (1/2 in)
75	3.6 cm (1 1/2 in)	2.0 cm (3/4 in)	2.0 cm (3/4 in)
100	4.8 cm (1 7/8 in)	2.75 cm (1 in)	2.75 cm (1 in)
150	7.2 cm (2 7/8 in)	4.0 cm (1 1/2 in)	4.0 cm (1 1/2 in)
175	8.4 cm (3 3/8 in)	5.0 cm (2.0 in)	5.0 cm (2.0 in)
200	9.6 cm (3 3/4 in)	5.5 cm (2 1/4 in)	5.5 cm (2 1/4 in)
250	12.0 cm (4 3/4 in)	7.0 cm (2 3/4 in)	7.0 cm (2 3/4 in)
300	14.4 cm (5 3/4 in)	8.5 cm (3 1/4 in)	8.5 cm (3 1/4 in)
400	19.2 cm (7 1/2 in)	11.0 cm (4 1/2 in)	11.0 cm (4 1/2 in)
500	24.0 cm (9 1/2 in)	14.0 cm (5 1/2 in)	14.0 cm (5 1/2 in)
600	28.8 cm (11 1/4 in)	16. 75 cm (6 1/2 in)	16.75 cm (6 1/2 in)
NOTE: All values were rounded off.			

Table 7-25. Known distance (mech. adj.) firing event.

FIRING EVENT	ROUNDS	TARGET RANGE (m)
Known Distance (Mech. Adj.)	20	100 to 500

PHASE IV—HOLD-OFF 100 TO 500 METERS

7-274. Hold-off is shifting the point of aim to achieve a desired point of impact. Hold-off is used only when the SDM does not have time to change his sight setting. Certain situations, such as multiple targets at varying ranges and rapidly changing winds, do not allow proper windage and elevation adjustments. Familiarization and practice of elevation and windage hold-off techniques prepares the SDM to meet these situations. The SDM rarely achieves pinpoint accuracy when holding off; a minor error in range determination or a lack of a precise point of aim might cause the bullet to miss the desired point.

7-275. Prior to this phase of training, trainers ensure that the—

- Weapon is cleared and that no ammunition is loaded prior to training.
- Weapon is zeroed prior to training.
- Soldier knows how to adjust for wind and gravity.
- Soldier can manipulate the rear sight for different ranges.

Elevation

7-276. When a Soldier aims directly at a target at ranges greater than the set range, his bullet will hit below the point of aim. At lesser ranges, his bullet will hit higher than the point of aim. If the SDM understands this and knows about trajectory and bullet drop, he will be able to hit the target at ranges other than that for which the weapon was adjusted.

7-277. For example, the SDM adjusts the weapon for a target located 500 meters downrange, and another target appears at a range of 600 meters. The hold-off would be 25 inches; that is, the SDM should hold off 25 inches above the center of visible mass in order to hit the center of mass of that particular target. If another target were to appear at 400 meters, the SDM would aim 14 inches below the center of visible mass in order to hit the center of mass.

7-278. The chart in Figure 7-31 shows the projectile's trajectory when fired from the M4 carbine and the M16A2 rifle. This demonstrates the drop of the round at various ranges.

NOTE: This diagram will assist the trainer in teaching vertical hold-off during this phase.

7-279. As the chart in Figure 7-31 shows, the hold-off at 400 meters is about half the height of the standard E-type silhouette; to hold-off at 400 meters, the firer aims half the height of the target over the target to hit it. The drop at 500 meters is considerably larger, so holding off will not be practical. The firer will have to adjust his rear elevation knob to get the proper point of aim for that distance.

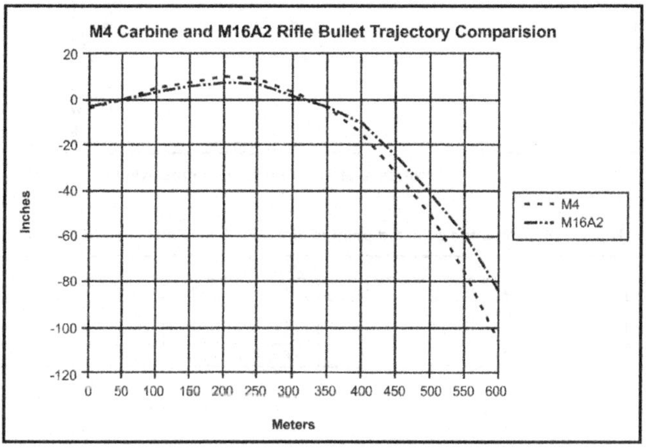

Figure 7-31. Bullet trajectory comparison.

Windage

7-280. When firing during windy conditions and there is no time to make sight adjustments, the SDM must use hold-off to adjust for windage (Figure 7-32).

- When holding off, the SDM aims into the wind. If the wind is moving from the right to left, his point of aim is to the right. If the wind is moving from left to right, his point of aim is to the left.
- If the SDM misses the target and the point of impact of the round is observed, he notes the lateral distance of his error and re-fires, holding off that distance in the opposite direction.

7-281. Table 7-26 shows calculated adjusted points of aim based on wind speed.

7-282. Table 7-27 shows the drift for a 10-mph wind using 5.56-millimeter M855 ball ammunition fired in a M16A2 rifle with a 300-meter battlesight zero.

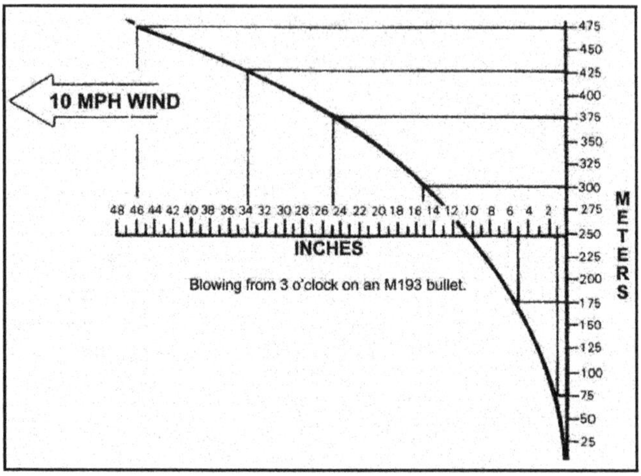

Figure 7-32. Windage effects of a 10-mph crosswind.

Table 7-26. Calculated adjusted point of aim based on wind speed (full value).

WIND SPEED	RANGE (m)								
	25	50	75	100	150	175	200	250	300
5 mph	1/4 in	3/8 in	1/2 in	1 in	2 in	2.5 in	3.5 in	5 in	7.5 in
10 mph	1/2 in	3/4 in	1 in.	2 in	4 in	5 in	7 in	10 in	15 in
15 mph	3/4 in	1 1/8 in	1.5 in	3 in	6 in	7.5 in	10.5 in	15 in	22.5 in

Table 7-27. Drift for 10-mph wind using M855 ammunition when fired from M16A2 rifle with 300-meter battlesight zero.

RANGE (m)	VELOCITY (fps)	TRAJECTORY (in)	DROP (in)	DRIFT (in)
0	3,100	-2.5	0.0	0.0
100	2,751	4.4	-2.3	1.1
200	2,420	5.8	-10.2	4.9
300	2,115	0.0	-25.3	11.8
400	1,833	-15.0	-49.5	22.4
500	1,569	-42.9	-86.7	38.0
600	1,323	-88.2	141.3	59.5
700	1,106	-156.1	-220.9	88.4
800	1,010	-267.7	-339.2	124.9

7-283. Firers use the M15 sighting device to demonstrate that they understand holding off. To do so, the firer aligns the sights on the silhouette on the proper adjusted point of aim.

NOTE: See Appendix A for information about the M15 sighting device.

7-284. Once the firer has an understanding of elevation and windage hold-off, he can begin the live-fire training exercise (Table 7-28). The firer will be given 20 rounds in which to engage 20 targets at ranges from 100 to 500 meters using elevation and windage hold-off.

Table 7-28. Firing event, known distance (hold off).

FIRING EVENT	ROUNDS	TARGET RANGE (m)
Known Distance (Hold Off)	20	100 to 500

PHASE V—FIELD FIRE 100 TO 500 METERS

7-285. The field fire events (Table 7-29) will test the individual's marksmanship, range estimation, and target detection skills. Field fire will consist of both a Record Fire I and a Record Fire II course. The Record Fire I course requires the individual to use mechanical elevation and windage adjustments. The Record Fire II course requires the individual to use elevation and windage hold-off (adjusted points of aim).

Conduct

7-286. To complete this course, Soldiers fire two firing events: Record Fire I and Record Fire II.

7-287. During each event, each Soldier will engage a total of 20 targets with 20 rounds. Soldiers must attain a total of 14 hits out of 20 targets on each record fire to pass. Table 7-29 depicts these two events and provides related information, such as number of rounds that must be fired, position that must be used, and the distance away from the firer that the target must be placed.

Table 7-29. Firing event, Record Fire I and II.

FIRING EVENT	POSITION	ROUNDS	NUMBER OF TARGETS	TARGET RANGE (m)
Record Fire I	Foxhole supported or prone supported position and the prone unsupported firing position	20	20	100 to 500
Record Fire II	Foxhole supported or prone supported position and the prone unsupported firing position	20	20	100 to 500

NOTE: If the SDM is issued an optic, the Record Fire II course will substitute use of that optic instead of using adjusted points of aim.

7-288. Prior to training, ensure that—

- The weapon is zeroed prior to training.
- Each Soldier assumes a steady firing position.
- Each Soldier consistently applies the fundamentals of marksmanship.
- Each Soldier knows how to adjust for wind and gravity.
- Each Soldier manipulates the rear sight for different ranges.

NOTE: See Appendix B for a sample completed form and the end of this publication for a blank, reproducible copy.

CERTIFICATION

7-289. Once the firer has successfully completed the SDM program, he is designated as an SDM and will be able to perform all duties and responsibilities set forth by these guidelines.

NOTE: SDM skills are highly perishable, and sustainment training should be conducted to ensure retention of the skills. At a minimum, sustainment training should be conducted semiannually.

Chapter 8

Advanced Optics, Lasers, and Iron Sights

BRM teaches Soldiers how to effectively engage targets with their weapons using the iron sights. ARM adds other marksmanship situations that a combat Soldier may encounter. This chapter discusses how to enhance marksmanship skills through proper training, using the Army's newest optics and lasers to ensure that Soldiers can fight as well at night as they can during the day. This chapter implements new night qualification standards to complement current Army training strategies.

SECTION I. BORELIGHT

The borelight is an accurate means of zeroing weapons and most aided-vision equipment without the use of ammunition. The time and effort required to ensure a precise boresight will, in turn, save time and ammunition. Table 8-1 outlines weapon/aided-vision device combinations that can be zeroed using the borelight.

NOTE: See Table 8-2 for the current training program.

Table 8-1. Weapon/aided-vision device combinations.

WEAPON / AIDED-VISION DEVICE	M16A2	M16A3/A4	M4/MWS
IRON SIGHT	X	N/A	N/A
BUIS	N/A	X	X
AN/PAQ-4B/C	X	X	X
AN/PEQ-2A/B	X	X	X
AN/PAS-13B/C/D	X	X	X
M68 CCO	X	X	X
ACOG	X	X	X

NOTES:
1. Precise boresighting of a laser will allow direct engagement of targets without a 25-meter zero. If a borelight is not available, a 25-meter zero must be conducted to zero the laser.
2. All optics must be 25-meter zeroed; a borelight only aids in zeroing.

Table 8-2. Borelight training program.

BORELIGHT
Instructional Intent
• Soldiers align the optic, laser, or iron sight to the bore of the weapon to reduce or eliminate the time and ammunition it requires to live-fire zero.
Special Instructions
Ensure that Soldiers— • Zero the borelight. • Use only approved 10-meter boresighting targets from Picatinny Arsenal. • Use the proper 10-meter boresighting target for weapon configuration. • Boresight 10 meters from the end of the barrel. • Stabilize both the weapon and the target (or the boresight will not be accurate). • Install filters for aiming lasers to reduce blooming.
Observables
Soldiers confirm that the— • Borelight spins on itself when zeroed at 10 meters. • Weapon configuration is boresighted using the official and proper targets only. • Target and weapon do not move during the boresighting procedure. • Borelight is centered on the circle on the target. • Aiming device is aiming at the center of the crosshair on the offset.

CONCEPT

8-1. Boresighting is a simple procedure that saves time and ammunition. The visible laser of the borelight is aligned with the barrel of a designated weapon. Then, using a 10-meter boresighting target, the weapon can be boresighted with any optic, laser, or iron sight that the Soldier is assigned to fire.

ZEROING THE BORELIGHT

WARNINGS

Before using the borelight, ensure that the weapon is clear and on SAFE, and that the bolt is locked in the forward position.

When rotating the borelight to zero it, ensure that the mandrel is turning counterclockwise (from the firer's point of view) to avoid loosening the borelight from the mandrel.

DANGERS

DO NOT STARE INTO THE VISIBLE LASER BEAM.

DO NOT LOOK INTO THE VISIBLE LASER BEAM THROUGH BINOCULARS OR TELESCOPES.

DO NOT POINT THE VISIBLE LASER BEAM AT MIRROR-LIKE SURFACES.

DO NOT SHINE THE VISIBLE LASER BEAM INTO OTHER INDIVIDUALS' EYES.

8-2. Before boresighting the weapon system, the borelight must first be zeroed to the weapon. To zero the borelight to the weapon—

(1) Stabilize the weapon by placing it in a rifle box rest or by laying two rucksacks side by side and placing another rucksack on top of the weapon.

NOTE: The weapon does not have to be perfectly level with the ground when boresighting.

(2) Align the visible laser with the weapon's barrel.

CAUTION

Do not over-adjust the laser. Do not point the laser at Soldiers or reflective material.

(3) Attach the 5.56-millimeter mandrel to the borelight.

(4) Insert the mandrel into the weapon's muzzle.

NOTE: The borelight is seated properly when the mandrel cannot be moved any further into the muzzle and the mandrel spins freely.

(5) Measure 10 meters with the 10-meter cord that comes with the borelight, or pace off eleven paces.

(6) Draw a zeroing mark (small dot) on a piece of paper or tree bark, or use the borelight reference point on the 10-meter boresighting target (Figure 8-1).

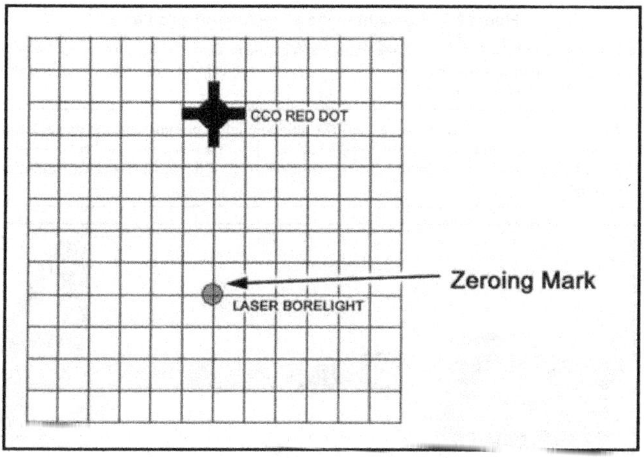

Figure 8-1. Example of a zeroing mark.

(7) Rotate the borelight until the battery compartment is facing upward and the adjusters are on the bottom (Figure 8-2).

NOTE: Once the Soldier performs step (7), the borelight's position and the place where the visible laser is pointing are identified as the start point. The command "START POINT" is given to ensure clear communication between the Soldier at the weapon and the Soldier at the boresighting target.

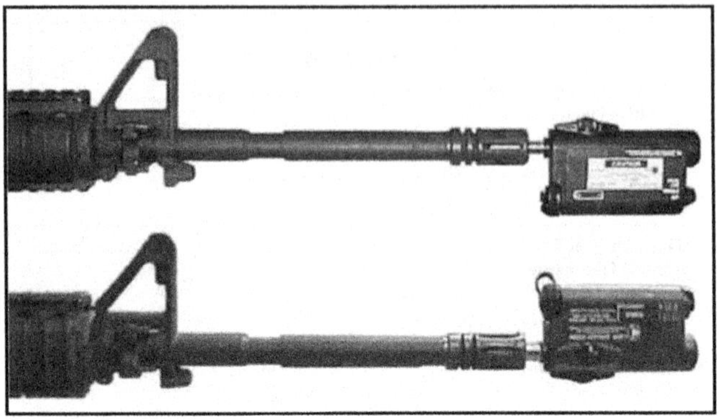

Figure 8-2. Borelight in the START POINT position.

(8) Rotate the borelight until the battery compartment is facing downward and the adjusters are on top to allow for easy access to the adjusters and help with communication and stabilization of the weapon (Figure 8-3).

NOTE: Once the Soldier performs step (8), the borelight's position and the place where the visible laser is pointing are identified as the half-turn position. The command "HALF-TURN" is given to ensure clear communication between the Soldier at the weapon and the Soldier at the boresighting target.

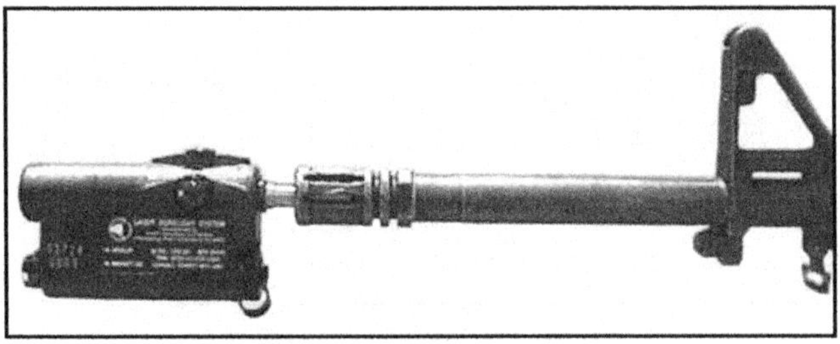

Figure 8-3. Borelight in the HALF-TURN position.

(9) Identify the point approximately halfway between the start point and the half-turn point. This is the reference point (Figure 8-4).

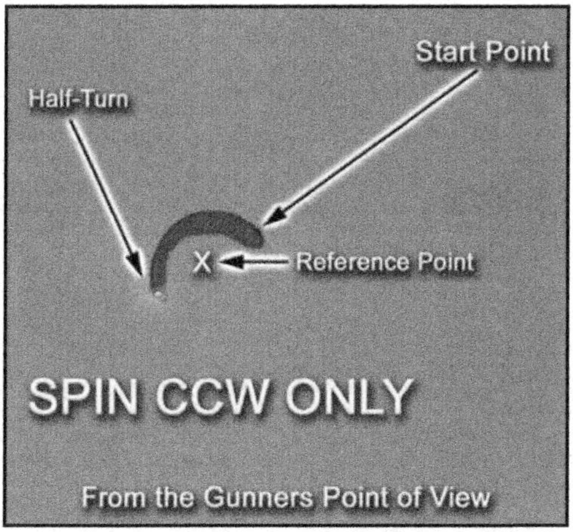

Figure 8-4. Example of a start point, half-turn, and reference point.

(10) Turn the borelight on and spin it until it is in the start point position.

(11) Place the zeroing mark approximately 10 meters from the end of the barrel so that the visible laser strikes the zeroing mark.

(12) Slowly rotate the borelight 180 degrees while watching the visible laser made by the borelight.

NOTE: If the visible laser stops on the zeroing mark, the borelight is zeroed to the weapon. If the borelight does not stop on the zeroing mark, elevation and windage adjustments must be made to the borelight.

---

(13) From the start point, realign the zeroing mark with the visible laser, rotate the borelight 180 degrees to the half-turn position, and identify the reference point.

(14) Using the adjusters on the borelight, move the visible laser to the reference point.

(15) Rotate the borelight back to the start point; move the zeroing mark to the visible laser.

NOTE: If the visible laser cannot be located when the borelight is spun to the half-turn position, start this procedure at 2 meters instead of 10 meters. When the visible laser is adjusted to the reference point at 2 meters, restart the procedure at 10 meters.

---

(16) Repeat steps (13) through (15) until the visible laser spins on itself.

NOTE: Every barrel is different, so steps (10) through (15) must be performed with every weapon to ensure that the borelight is zeroed to that barrel. If the borelight is zeroed, go directly to the boresighting procedures.

BORESIGHTING

8-3. Two Soldiers are required to properly boresight a weapon: a firer and a target holder. Their duties are as follows:

- The firer's primary duty is to zero the borelight and make all adjustments on the aided-vision device being used.
- The target holder secures the 10-meter boresighting target straight up and down 10 meters from the borelight and directs the firer in making necessary adjustments to the aiming device. The target holder must wear night vision goggles when boresighting IR aiming lasers.

NOTE: Appendix F shows the most current 10-meter boresighting target. The 10-meter boresighting target grids are 1x1-centimeter squares, unlike those on 25-meter zeroing targets. Contact the proponent of this publication (C Co, 2/29 IN, Fort Benning, GA) for information about the availability of boresighting targets.

CAUTION

Weapon stability is crucial in boresighting. The weapon should be in the bolt forward position and must not be canted left or right during boresighting procedures. If the weapon is boresighted using field-expedient methods (for example, sandbags, rucksacks) and the weapon is laid on its side for stability, ensure that the boresighting target is also oriented in the same manner.

Boresighting the Weapon Using the Backup Iron Sights

8-4. The BUIS can be boresighted to a new user to expedite 25-meter zeroing. To boresight the weapon using the BUIS—

(1) Align the BUIS with the Canadian bull on the 10-meter boresighting target.

(2) Make adjustments to the windage and elevation of the BUIS until the borelight is centered with the circle on the boresighting target.

Boresighting the Weapon Using the M68 Close Combat Optic

NOTES: 1. Before boresighting, ensure that the borelight has been zeroed to the weapon. The more accurate the boresight of the M68 CCO to the assigned weapon, the closer to a battlesight zero the weapon will be.

2. 25-meter zeroing must be conducted to ensure that the M68 CCO is properly zeroed.

3. The M68 CCO is a parallax-free sight beyond 50 meters; boresighting is conducted at 10 meters. To get a solid boresight, ensure that the red dot is centered within both the CCO and the target's center of mass.

8-5. To boresight the weapon using the M68 CCO—

(1) Select the proper 10-meter boresighting target for the weapon/M68 configuration.

(2) With the help of an assistant, place the boresighting target 10 meters in front of the weapon.

(3) Turn the M68 CCO to the desired setting by turning it clockwise to position 10, turning it counterclockwise one click at a time until you can no longer see the dot, and then turning it clockwise two clicks.

NOTE: The desired setting may need to be adjusted throughout the day depending on ambient light conditions.

(4) Get behind the weapon in a stable supported firing position, and look through the M68 CCO.
(5) Aim the red dot of the M68 CCO on the crosshair located on the 10-meter boresighting target.
(6) Make adjustments to the M68 CCO until the visible laser of the borelight is centered on the borelight circle on the 10-meter boresighting target.
(7) Turn the borelight off.
(8) Move the weapon off of the crosshair, realign the M68 CCO's red dot on the crosshair, and turn the borelight back on.
(9) If the borelight is on the circle and the M68 CCO's red dot is on the crosshair, the weapon system is boresighted.
(10) Turn the laser off, and carefully remove the borelight and the mandrel from the weapon so that the borelight device is not damaged.

NOTE: If the firer does not get the same sight picture after the second realignment, he more than likely has a fundamentals problem with his firing position and sight picture. To save time on the range, a coach should troubleshoot the Soldier before continuing to boresight the M68 CCO.

BORESIGHTING THE ADVANCED COMBAT OPTICAL GUNSIGHT

NOTES: 1. Before boresighting, ensure that the borelight has been zeroed to the weapon. The more accurate the boresight of the ACOG to the assigned weapon, the closer to a battlesight zero the weapon will be.

2. 25-meter zeroing must be conducted to ensure that the ACOG is properly zeroed.

8-6. To boresight the ACOG—

NOTE: Use the 10-meter boresighting target that is used for boresighting the M68 CCO.

(2) With the help of an assistant, place the boresighting target 10 meters in front of the weapon.
(4) Get behind the weapon in a stable supported firing position, and look through the ACOG.
(5) Aim the tip of the 300-meter post on the crosshair located on the 10-meter boresighting target.
(6) Make adjustments to the ACOG until the visible laser of the borelight is centered on the borelight circle on the boresighting target.
(7) Turn the borelight off.
(8) Move the weapon off of the crosshair, realign the tip of the 300-meter post on the crosshair, and turn the borelight back on.
(10) If the borelight is on the circle and the tip of the 300-meter post is on the crosshair, the weapon system is boresighted.
(10) Turn the laser off, and carefully remove the borelight and the mandrel from the weapon so that the borelight device is not damaged.

BORESIGHTING THE THERMAL WEAPON SIGHT

NOTES: 1. Before boresighting the TWS, ensure that the borelight has been zeroed to the weapon. The more accurate the boresight of the TWS to the assigned weapon, the closer to a battlesight zero the weapon will be.

2. 25-meter zeroing must be conducted to ensure that the TWS is properly zeroed.

3. Both the narrow field of view (NFOV) and wide field of view (WFOV) must be boresighted and zeroed.

8-7. To boresight the TWS—

(1) Select the proper 10-meter boresighting target for the weapon/TWS configuration.

(2) With the help of an assistant, place the boresighting target 10 meters in front of the weapon.

(3) Ensure that the weapon's reticle is displayed.

(4) Get behind the weapon in a stable supported firing position, and look through the TWS.

(5) Place a finger on each oval on the 10-meter boresighting target.

(6) Aim between the fingers with the 300-meter point of aim.

(7) Make adjustments to the TWS until the visible laser of the borelight is centered on the borelight circle on the 10-meter boresighting target.

(8) Move off of the aiming block, realign the TWS to the center of the heated block, and then turn the borelight back on.

(9) If the boresight is properly aligned, the weapon system is boresighted; otherwise, you will need remedial training on your sight picture.

(10) Change the field of view (FOV) on the sight by rotating the FOV ring, and repeat steps (1) through (9).

(11) Turn the laser off, and carefully remove the borelight and the mandrel from the weapon so that the borelight device is not damaged.

BORESIGHTING THE AN/PAQ-4B/C

NOTE: Before boresighting the AN/PAQ-4B/C, make sure that the borelight has been zeroed to the weapon. The more accurate the boresight of the AN/PAQ-4B/C to the assigned weapon, the closer to a battlesight zero the weapon will be.

DANGERS

DO NOT STARE INTO THE IR LASER BEAM WITH THE NAKED EYE OR THROUGH BINOCULARS OR TELESCOPES.

DO NOT POINT THE IR LASER BEAM AT MIRROR-LIKE SURFACES OR OTHER INDIVIDUALS' EYES.

ALTHOUGH THIS LASER IS EYE-SAFE, IT IS A SAFE PRACTICE TO TREAT ALL LASERS AS NOT EYE-SAFE.

3x EXTENDERS MAGNIFY THE LASER AIMING LIGHT. WHEN USING THE 3x EXTENDERS, THE AN/PAQ-4B/C IS NOT CONSIDERED EYE-SAFE AT ANY DISTANCE.

DO NOT STORE THE AN/PAQ-4B/C WITH THE BATTERIES INSTALLED.

8-8. To boresight the AN/PAQ-4B/C —

(1) Select the proper 10-meter boresighting target for the weapon/AN/PAQ-4B/C configuration.
(2) With the help of an assistant, place the boresighting target 10 meters in front of the weapon.
(3) Install the borelight filter, and turn the AN/PAQ-4B/C on.
(4) Align the 10-meter boresighting target with the visible laser of the borelight.
(5) Adjust the adjusters on the AN/PAQ-4B/C until the IR laser is centered on the crosshair located on the 10-meter boresighting target.

NOTE: For windage and elevation, one click equals 1 centimeter at 25 meters. A click is defined as the sound or feel of the positive detent movement.

CAUTION

Do not turn the adjustment screws too much. Over-turning the adjustment screws will cause them to break.

NOTES: 1. The boresighting target and zeroing mark must be kept stable during the boresighting procedure.

2. Regardless of the mounting location, the adjuster that is on top or bottom will always be the elevation adjuster, and the one on the side will always be the windage adjuster.

BORESIGHTING THE AN/PEQ-2A/B

NOTES: 1. Before boresighting the AN/PEQ-2A/B, ensure that the borelight has been zeroed to the weapon. The more accurate the boresight of the AN/PEQ-2A/B to the assigned weapon, the closer to a battlesight zero the weapon will be.

2. The boresighting target and zeroing mark must be kept stable during the boresighting procedure.

DANGERS

THE AN/PEQ-2A/B EMITS INVISIBLE LASER RADIATION. AVOID DIRECT EXPOSURE TO THE BEAM.

DO NOT STARE INTO THE IR LASER BEAM WITH THE NAKED EYE OR THROUGH BINOCULARS OR TELESCOPES.

DO NOT POINT THE IR LASER BEAM AT MIRROR-LIKE SURFACES OR OTHER INDIVIDUALS' EYES.

EYE DAMAGE CAN OCCUR IF CARELESS HANDLING OF THE LASER OCCURS. EYE-SAFE DISTANCE IS BEYOND 25 METERS (IN DUAL LO MODE) IN TRAINING MODE AND BEYOND 220 METERS IN TACTICAL MODE.

3x EXTENDERS MAGNIFY THE LASER AIMING LIGHT. WHEN USING THE 3x EXTENDERS, THE AN/PEQ-2A/B IS NOT CONSIDERED EYE-SAFE AT ANY DISTANCE.

DO NOT STORE THE AN/PEQ-2A/B WITH THE BATTERIES INSTALLED.

8-9. To boresight the AN/PEQ-2A/B—

(1) Select the proper 10-meter boresighting target for the weapon/AN/PEQ-2A/B configuration.

(2) With the help of an assistant, place the boresighting target 10 meters in front of the weapon.

(3) Install the filter on the aiming laser, and turn the AN/PEQ-2A/B on.

(4) Align the 10-meter boresighting target with the visible laser of the borelight.

(5) Adjust the adjusters on the AN/PEQ-2A/B until the IR laser is centered on the crosshair located on the 10-meter boresighting target.

NOTE: For windage and elevation, one click equals 1 centimeter at 25 meters. A click is defined as the sound or feel of the positive detent movement.

CAUTION

Do not turn the adjustment screws too much. Over-turning the adjustment screws will cause them to break.

NOTES: 1. Each click of elevation and windage is 1 centimeter and each square of the 25-meter zeroing target is .9 centimeter. For ease of use, round up to one square per click of elevation. This will have a negligible effect on the zero unless a large adjustment is needed.

2. Regardless of the mounting location, the adjuster that is on top or bottom will always be the elevation adjuster, and the one on the side will always be the windage adjuster.

(6) Adjust the illuminator in the same manner.

(7) Turn the laser off, and carefully remove the borelight and the mandrel from the weapon so that the borelight device is not damaged.

BORESIGHTING THE AN/PVS-4

NOTES: 1. Before boresighting, ensure that the borelight has been zeroed to the weapon. The more accurate the boresight of the AN/PVS-4 to the assigned weapon, the closer to a battlesight zero the weapon will be.

2. 25-meter zeroing must be conducted to ensure that the AN/PVS-4 is properly zeroed.

(1) Select the proper 10-meter boresighting target for the weapon/AN/PVS-4 configuration.

(1) With the help of an assistant, place the boresighting target 10 meters in front of the weapon.

(2) Ensure that the M16 reticle is displayed.

(3) Get behind the weapon in a stable supported firing position, and look through the AN/PVS-4.

(4) Turn the borelight laser on.

(5) Align the borelight laser with the circle on the 10-meter target offset.

(6) Keeping the laser in place, adjust the windage and elevation until the reticle of the AN/PVS-4 is aligned with the circular crosshair.

NOTE: If there is not enough ambient light to see the 10-meter target offset circular crosshair, shine a flashlight indirectly at the target to provide ambient light.

(7) Turn the borelight off.

(8) Move the reticle off of the circular crosshair, and then realign on the target.

(9) Turn the borelight laser back on.

(9) If the borelight is in the circle, then the AN/PVS-4 is boresighted.

(10) Turn the laser off, and carefully remove the borelight and mandrel from the weapon so that the borelight device is not damaged.

SECTION II. TRAINING STRATEGIES AND QUALIFICATION STANDARDS

Before beginning a night marksmanship program, Soldiers must qualify on their assigned weapons during the day, as outlined in the previous chapters of this manual. Commanders should follow these training strategies and abide by the qualification standards set forth. Although some courses of fire may seem redundant or inappropriate, numerous tests show that these training strategies work, and the qualification standards are achievable if the strategies are followed.

BACKUP IRON SIGHT

8-10. The BUIS is a semipermanent flip-up sight equipped with a rail-grabbing base. The BUIS provides a backup capability effective out to at least 600 meters and can be installed on M16A4 rifles and M4 carbines.

NOTE: See Table 8-3 for the current training program.

Table 8-3. Backup iron sights training program.

BACKUP IRON SIGHTS
Instructional Intent
• Soldiers zero and qualify with the BUIS.
Special Instructions
Ensure that Soldiers— • Apply the marksmanship fundamentals. • Place the BUIS in the full vertical position and lock them prior to firing. • Install the plastic insert in the BUIS during boresighting and zeroing.
Observables
• Soldiers attain the same day standards for zeroing and qualification as with the standard iron sights (See Chapter 5).

CONCEPT

8-11. The BUIS is adjusted for a 300-meter battlefield zero to provide backup in the event an optic or laser device fails to function. The BUIS is zeroed on the M4/M4A1 target on the backside of the M16A2 zeroing target (NSN 6920-01-395-2949).

NOTE: The 25-meter zeroing procedures are the same as for conventional rear sight assembly on M16 rifles and M4 carbines.

CONDUCT OF TRAINING

NOTES: 1. The BUIS training strategy is the same as the iron sight training strategy.

2. All procedures for the BUIS are the same as with standard iron sights.

8-12. BUIS equipment training should familiarize the Soldier with the proper operation and characteristics of the BUIS.

Boresight the Backup Iron Sights

8-13. Optional.

Zero

8-14. The zeroing standards for the BUIS are the same as with iron sights.

- To zero the BUIS for M4 carbines, set the range selector to 300 meters.
- To zero the BUIS to an M16A4, place the range selector on the white line below the 300-meter mark.

Target Detection

8-15. Target detection procedures for the BUIS are the same as with standard iron sights.

Practice Qualification

8-16. A practice qualification must always precede an actual qualification. Practice qualification allows the Soldier to practice and refine the skills needed to succeed during qualification. Practice qualification standards for the BUIS are the same as with standard iron sights. If the Soldier qualifies during the practice qualification, it may be counted as the record qualification.

Record Qualification

8-17. Qualification with the BUIS is conducted on a standard record fire range, and the standards for qualification are the same as the record fire day standards.

M68 CLOSE COMBAT OPTIC

8-18. The M68 CCO is a reflex (nontelescopic) sight. It uses a red aiming reference (collimated dot) and is designed for the two eyes open method of sighting. The dot follows the horizontal and vertical movement of the firer's eye, while remaining fixed on the target. No centering or focusing is required.

NOTE: See Table 8-4 for the current training program.

Table 8-4. M68 close combat optic training program.

M68 CLOSE COMBAT OPTIC
Instructional Intent
• Soldiers qualify with the M68 CCO.
Special Instructions
Ensure that Soldiers— • Are proficient with the M68 CCO. • Install the half-moon spacer when using the M68 CCO with the M4, M16A4, and MWS. • Do not install the half-moon spacer when using the M68 CCO with the M16A1/A2. • Use the proper offset during boresighting procedures. • Confirm 10-meter boresight with a 25-meter zero. • Retighten the rail grabber after the initial 3 rounds are fired. • Sets the M68 CCO's dot for best sight picture. • Apply the marksmanship fundamentals. • Zero and qualify with the same sight picture (one eye or two eyes open method). • Zero on the M16A2 25-meter zeroing target. Ensure that the designated impact zone is 1.4 centimeters down from the center of mass of the 300-meter silhouette on the 25-meter zeroing target.
Observables
• Soldiers zero the M68 CCO to the same standards as with iron sights. • Rounds impact in the 4x4-centimeter square designated impact zone. • Soldiers achieve the same practice qualification and qualification standards as with day record fire.

CONCEPT

8-19. Soldiers must qualify on their assigned weapons during daylight conditions, as outlined in this manual. The integrated act of firing with the M68 CCO is identical to the iron sights except for the change in the sight picture.

CONDUCT OF TRAINING

NOTES: 1. The M68 CCO training strategy is the same as the iron sight training strategy.

2. All procedures for the M68 are the same as with standard iron sights.

8-20. M68 equipment training should familiarize the Soldier with the proper operation and characteristics of the M68 CCO IAW TM 9-1240-413-13&P.

MODIFIED FUNDAMENTALS

8-21. M68 CCO operation requires modifications to the fundamentals of marksmanship:

- Steady position.
- Aiming.
- Breath control.
- Trigger squeeze.

Steady Position

8-22. When operating the M68 CCO, Soldiers no longer need a good stock weld to get a good sight picture after the M68 is zeroed at 25 meters. The M68's reflexive sight allows the Soldier to fire the weapon with his cheek in a comfortable position, but since the CCO is parallax-free beyond 50 meters only, the Soldier must zero and fire using the same cheek position.

Aiming

8-23. When using the M68 CCO, the two eyes open method is the preferred method of aiming.

NOTE: The aiming method used to zero must also be used to engage targets. When using the M68 CCO, the weapon must not be canted during aiming or firing.

Two Eyes Open Method (Preferred)

8-24. This method allows a much greater field of view and makes scanning for targets much easier; however, getting accustomed to the two eyes open method takes practice. The Soldier must keep the rifle and M68 in a vertical alignment each time he fires.

8-25. To use the two eyes open method—

(1) Position the head so that one eye can focus on the red dot and the other eye can scan downrange.

(2) Place the red dot on the target's center of mass, and engage.

One Eye Open Method

8-26. To use the one eye open method—

(1) With the nonfiring-side eye closed, look through the M68 CCO to ensure that the red dot can be seen clearly.

(2) Place the red dot on the target's center of mass, and engage.

Breath Control

8-27. This fundamental does not change.

Trigger Squeeze

8-28. This fundamental does not change.

Zeroing Without a Borelight

8-29. To zero without a boresight—

(1) Begin with a securely installed and live-fire zeroed BUIS.

(2) Mount the CCO to the front of the receiver rail or to the top ARS, as preferred.

(3) Adjust windage and elevation on the reflex sight until the center of the aiming dot is at the tip of the front sightpost when viewed through the BUIS while assuming a normal firing position.

25-Meter Zeroing Procedures

NOTE: Conduct zeroing only on the M16A2 25-meter target.

8-30. When zeroing the CCO at 25 meters, a designated point of impact zone must be identified on the 25-meter zeroing target (Figure 8-5). To zero the CCO at 25 meters—

(1) Starting from center of mass of the 300-meter silhouette on the 25-meter zeroing target, count down 1½ squares or 1.4 centimeters. This is now the point of impact when zeroing the M68 CCO.

(2) Continue to aim at the center of mass of the 300-meter silhouette, and make adjustments to the M68 CCO so that the rounds impact in the secondary 4x4-centimeter circular box, 1½ squares or 1.4 centimeters down from the point of aim.

8-31. Other procedures are the same as standard iron sight procedures:

NOTE: A click is defined as the sound or feel of the positive detent movement.

- For windage and elevation, two clicks equal 1 centimeter at 25 meters.
- For elevation, one clockwise click moves the bullet strike down.
- For windage, one clockwise click moves the bullet strike left.

NOTES: 1. At ranges of 50 meters and beyond, the effects of parallax are minimal. However, at ranges of 50 meters and closer, parallax exists and the firer must ensure that the red dot is centered while zeroing.

2. The aiming method (two eyes open or one eye open) used to zero must be used to engage targets.

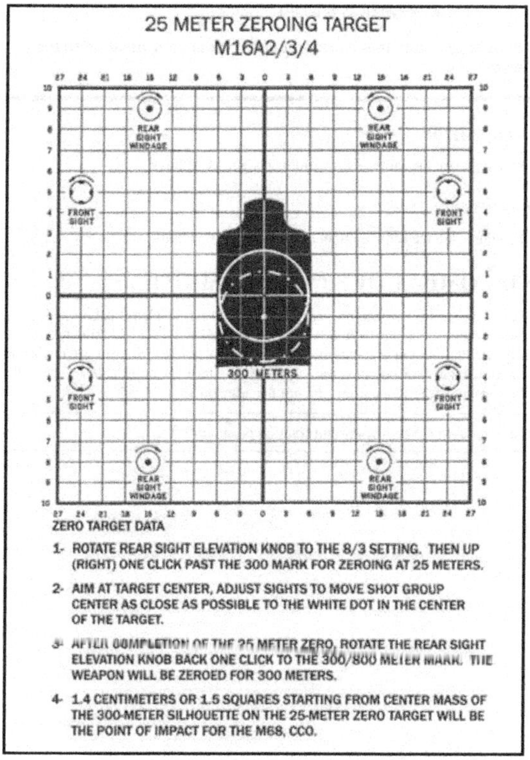

Figure 8-5. Close combat optic, 25-meter zeroing target.

Target Detection

8-32. Target detection procedures for the M68 are the same as with standard iron sights.

WARNING

In Positions 4 and above, the red dot is visible through the front of the sight. For night vision operations—

- **Close the front lens cover before turning the rotary switch clockwise to Positions 2 and 3.**
- **Check the light for proper intensity before opening the front lens cover.**
- **Close the front lens cover before turning the rotary switch counterclockwise to the OFF position.**

Failure to follow this warning could reveal your position to the enemy.

Practice Qualification

8-33. The procedures are the same as standard iron sight procedures.

Record Qualification

8-34. The procedures are the same as standard iron sight procedures.

ADVANCED COMBAT OPTICAL GUNSIGHT

8-35. The ACOG scope is designed to provide enhanced target identification and hit probability for the M16/M4 series weapons out to 800 meters. It is designed with dual illuminated technology using fiber optics for daytime illumination and tritium for night and low-light use. The ACOG scope is a lightweight rugged, fast and accurate 3.5 power optic scope. It is internally adjustable to allow the shock of rough handling to be carried by the scope body and not the adjustment mechanism.

NOTE: See Table 8-5 for the current training program.

Table 8-5. Advanced combat optical gunsight training program.

ADVANCED COMBAT OPTICAL GUNSIGHT
Instructional Intent
• Soldiers qualify with the ACOG.
Special Instructions
Ensure that Soldiers— • Are proficient with the ACOG. • Use the proper offset during boresighting procedures. • Confirm the 10-meter boresight with 100-meter zero (or 25-meter zero if the 100-meter range is not available). • Retighten the rail grabber after the initial 3 rounds are fired. • Apply the marksmanship fundamentals. • Zero and qualify with the same sight picture. Ensure that— • The point of aim and point of impact are the same at 100 meters. • The designated impact zone is 1 centimeter down from the center of mass of the 300-meter silhouette on the 25-meter zeroing target.
Observables
• Soldiers zero the ACOG to the same standards as with iron sights. • Soldiers achieve the same practice qualification and qualification standards as with day record fire.

CONCEPT

8-36. Soldiers must qualify on their assigned weapons during the day, as outlined in this manual. The integrated act of firing with the ACOG is identical to the iron sights except for the change in sight picture.

CONDUCT OF TRAINING

NOTES: 1. The ACOG training strategy is the same as the iron sight training strategy.

2. All procedures for the ACOG are the same as with standard iron sights.

8-37. ACOG equipment training should familiarize the Soldier with its proper operation and characteristics.

OPERATION

8-38. The ACOG scope is internally adjustable. Adjustment is made using the adjuster mechanisms located inside the adjuster caps on the top and right-hand side of the scope. Adjustment can be made with a small screwdriver, coin, or other hard object that fits the adjustment screws. The caps are very tight to ensure a waterproof seal with the O-rings inside. The caps should only be off the scope when adjustments are being made.

CAUTION

The ACOG contains an internal adjustment mechanism to allow zeroing. Adjustments to the extreme ends of the range can result in damage to the internal prism assembly. Do not continue to adjust the windage and elevation mechanisms if you encounter resistance.

8-39. The ACOG scope is shipped with a pre-centered setting. Normally this means that only small adjustments are necessary. Do not adjust the scope to the extremes. It is possible that over-adjustment will damage the precise alignment of the prism assembly inside the rifle scope.

8-40. The ACOG scope is adjusted at the factory to be as parallax-free as possible at 100 meters. This in no way affects the accuracy of the scope.

8-41. The reticle patterns in the scope have been designed to provide many features while retaining simplicity of operation. The user does not need to make any manual adjustments between shots at different ranges. Ranging capability is built into the reticle patterns.

8-42. The outside legs of the chevron reticle in the ACOG correspond to 19 inches (average width of a man's shoulders) at 300 meters.

8-43. The widths of the horizontal hash marks on the BDC reticles in all ACOG scopes correspond to the width of a .5 meter (19 inches) silhouette (man-size) at that range (Figure 8-6).

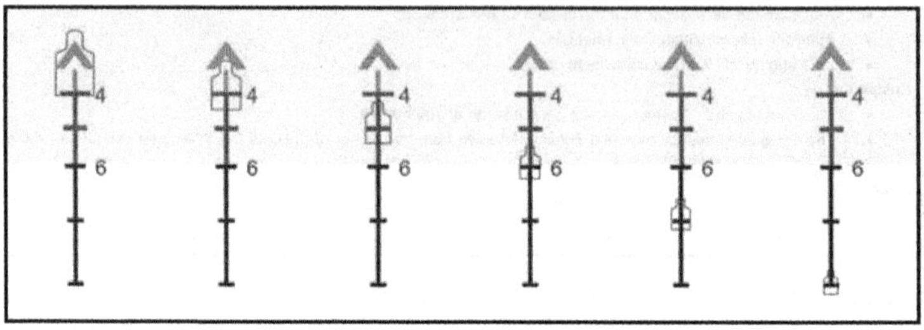

Figure 8-6. Width of horizontal hash marks.

Zeroing the Advanced Combat Optical Gunsight

8-44. The adjustment increments for the ACOG are 1/4 inch per click at 100 meters. A click is defined as the sound or feel of the positive detent movement. This means that 4 clicks are required to move the bullet approximately one inch on the target at 100 meters. At 25 meters, 16 clicks move the bullet one inch. Turning the screw in the direction of the arrow moves the bullet impact in the direction marked (up/right).

NOTE: To ensure a consistent zero, tap the scope with the palm of your hand to stabilize the adjustment mechanism after an adjustment has been made, and then fire a three-shot group on the target.

8-45. The ACOG scope with chevron reticle is designed to be zeroed at 100 meters, using the tip of the chevron reticle for POA/POI (Figure 8-7).

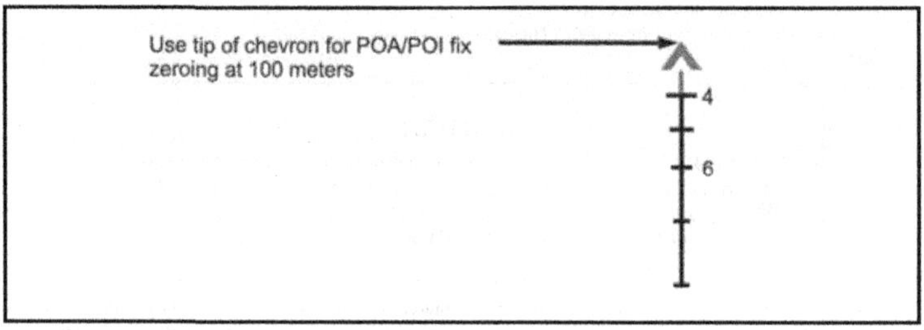

Figure 8-7. Advanced combat optical gunsight reticle point of aim at 100 meters.

8-46. For a quick combat zero, the scope may be zeroed at 25 meters, using the tip of the 300-meter post as POA/POI. A 25-meter zero is less precise than a 100-meter zero and should be verified at longer distances once time and a range is available (Figure 8-8).

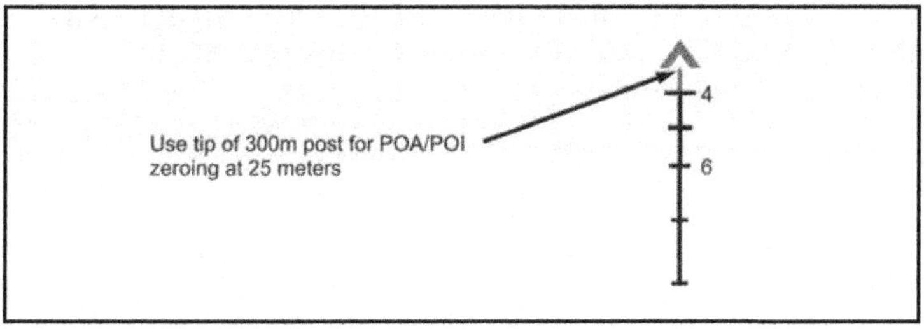

Figure 8-8. Advanced combat optical gunsight reticle point of aim at 25 meters.

Modified Fundamentals

8-47. ACOG operation requires modifications to the following fundamentals of marksmanship:

- Steady position.
- Aiming.
- Breath control.
- Trigger squeeze.

Steady Position

8-48. This fundamental does not change.

Aiming

8-49. When the TA11F ACOG is attached to the carrying handle of an M16A1/A2/A3 rifle, the BDC reticle will be off slightly at extended distances. Beginning at 400 meters, the BDC reticle will be approximately one MOA off to the extent of the BDC reticle, once properly zeroed.

Example: At 500 meters, the impact will be 5 inches high; at 600 meters, the impact will be 6 inches high.

Once the target is ranged beyond 400 meters, hold slightly low.

Breath Control

8-50. This fundamental does not change.

Trigger Squeeze

8-51. This fundamental does not change.

Target Detection

8-52. Target detection procedures for the ACOG are the same as with standard iron sights.

Practice Qualification

8-53. The procedures are the same as standard iron sight procedures.

Record Qualification

8-54. The procedures are the same as standard iron sight procedures.

AN/PAS-13B/C/D (V1) LIGHT WEAPON THERMAL SIGHT AND AN/PAS-13B/C/D (V3) HEAVY WEAPON THERMAL SIGHT

8-55. The AN/PAS-13B/C/D (V1, V3) TWS is an IR imaging sensor used for target acquisition under conditions of low visibility. IR light is received through the telescope, detected by an IR sensor, converted to digital data, processed, and displayed for the user.

NOTE: See Table 8-6 for the current training program.

Table 8-6. AN/PAS-13B/C/D thermal weapon sight training program.

AN/PAS-13B/C/D TWS
Instructional Intent
• Soldiers qualify with the ANPAS-13B/C/D TWS.
Special Instructions
Ensure that Soldiers— • Are proficient with the TWS. • Use the spacer with the M4, M16A4, and MWS. • Use the proper 10-meter boresighting target during boresighting procedures. • Boresight both FOVs. • Confirm the 10-meter boresight with a 25-meter zero. • Use the M16A2 zeroing target with a 4x4-centimeter square cut out of the center of the silhouette. • Thermalize the zero range and qualification range (inspect for targets that aren't thermalized). • Use every other lane during zero and qualification.
Observables
• Soldiers zero the TWS to the same standards as with iron sights. • Soldiers achieve the same practice qualification and qualification standards as with day record fire.

WARNING

Ensure that the weapon is not loaded and is on SAFE before installing the TWS on the weapon. A loaded weapon may accidentally discharge, causing severe injury or death.

Concept

8-56. The AN/PAS-13B/C/D TWS training strategy is much the same as that of aiming lights. The TWS does not require the use of NVDs. The course of fire for the TWS is the same scenario as the day qualification tables, with the same standards of fire as for current day standards. Qualification standards are the same for day and night.

Conduct of Training

8-57. AN/PAS-13B/C/D equipment training should familiarize the Soldier with the proper operation and characteristics of the TWS IAW the TM.

MODIFIED FUNDAMENTALS

8-58. AN/PAS-13B/C/D TWS operation requires modifications to the fundamentals of marksmanship:

- Steady position.
- Aiming.
- Breath control.
- Trigger squeeze.

Steady Position

8-59. This fundamental slightly changes due to the height of the sight. Consider the following modifications:

- Soldiers must adjust their position so they can properly look through the sight.
- In most cases, the cheek-to-stock weld no longer exists.

Aiming

8-60. To properly aim with the TWS, Soldiers must ensure that the correct reticle is selected in the sight.

NOTE: Refer to TM 11-5855-312-10 for reticle selection and point of aim for use with the TWS.

Breath Control

8-61. This fundamental is not affected by night firing conditions when using the TWS.

Trigger Squeeze

8-62. This fundamental of marksmanship does not change during night firing.

25-METER ZEROING PROCEDURES

NOTE: Refer to TM 11-5855-312-10 for target preparation.

8-63. To zero the AN/PAS-13B/C/D TWS at 100 meters, use the same procedures and standards as with iron sights, along with the following:

- At the 25-meter range, each incremental adjustment to the azimuth or elevation setting moves strike of the round as follows:
  - 1 ½ centimeters for LWTS on WFOV.
  - ½ centimeter for LWTS on NFOV.
  - ¾ centimeter for HWTS on WFOV.
  - ¼ centimeter for HWTS on NFOV.
- Retighten the rail grabber after firing the first three rounds.
- Zero both FOVs (Figure 8-9).

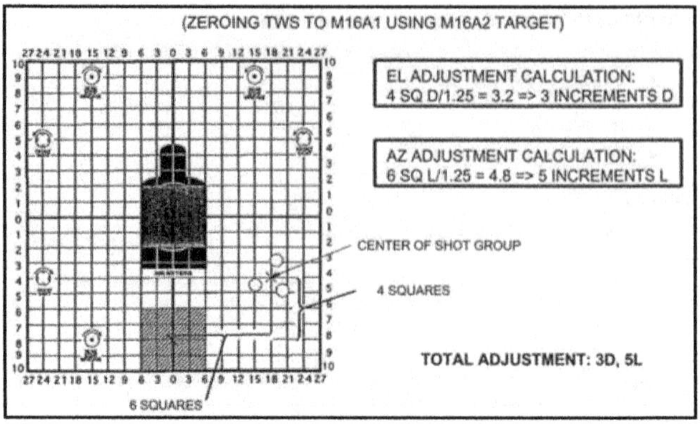

Figure 8-9. Example of thermal weapon sight zeroing adjustments.

TARGET DETECTION

8-64. With NVDs, the field of view is much smaller, scanning becomes much more deliberate, and, with the TWS, camouflage becomes less of a factor. Even though NVDs greatly enhance the Soldier's ability to acquire a target at night, increased awareness of target detection must be trained to allow the Soldier to recognize the visual cues of IR imagery.

Select Position

8-65. The TWS is a large device; selecting a position that allows for good fields of view, but does not silhouette the Soldier and his equipment, might be a challenge. Since the TWS detects thermal energy (heat) emitted from an object, a position near an object emitting a vast amount of thermal energy (for example, a vehicle with the engine running or a fire) may affect the Soldier's ability to acquire a target.

Scanning

8-66. With earlier versions of the TWS, scanning too fast causes a stuttering on the screen, which might cause the Soldier to miss or overlook a target. With these versions, Soldiers must scan slowly in order to maintain a good thermal image on the screen.

8-67. With the newer version, this stuttering is not as obvious. One advantage of the TWS is its two FOVs—wide and narrow. Each has advantages and disadvantages:

- The NFOV increases magnification but decreases the field of view.
- The WFOV decreases magnification but increases the field of view.

Target Indicators

8-68. While scanning the sector or lane with the TWS, the Soldier should recognize thermal cues that allow him to detect and identify targets. The engine compartment, exhaust, and tires of a vehicle that has been moving are all examples of thermal cues. Adjusting the brightness, contrast, and polarity helps enhance the thermal cues of a target, allowing for quicker detection and identification.

Sound

8-69. Use the same techniques outlined in day and night target detection.

Movement

8-70. Thermal cues become much more obvious on a moving object than on an object standing still. A good example is a vehicle's tires. When the vehicle is not moving, the tires are cold. On a moving vehicle, the friction between the road and the tires causes the tires to heat up and become prominent when observed through the TWS. The same is true with the human body—a moving person generates more heat than someone standing still.

Camouflage

8-71. The TWS allows Soldiers to see thermal signatures not completely masked by camouflage, such as paint, foliage, and camouflage netting, thereby increasing both day and night target detection abilities.

WARNING

If the TWS is operated with the eyecup removed, light emitting from the eyepiece may be visible to the enemy's NVDs.

PRACTICE QUALIFICATION

8-72. Practice qualification with the TWS is the same as day practice qualification with iron sights. Dry-fire exercises are performed to allow Soldiers to make adjustments to the TWS. Every other firing lane should be used so that the Soldier engages only the targets in his lane.

RECORD QUALIFICATION

8-73. Record qualification with the TWS is the same as day record qualification with iron sights.

NOTES:
1. Record qualification with the TWS can be done day and or night. Regardless of the qualification, the standard day record fire for the iron sights will be used. The standards for qualification with the TWS, either day or night, are 23 out of 40.
2. During practice qualification and qualification, the Soldier chooses his own polarity and FOV.

AN/PAQ-4B/C AND AN/PEQ-2A/B INFRARED AIMING LASERS

8-74. The newest IR aiming lasers greatly increase the night firing accuracy of all infantry weapons. IR aiming lasers complete the transition from day optics to night optics. Their effectiveness is limited by the capability of the image-intensifier (I2) sight with which they are used.

NOTE: See Table 8-7 for the current training program.

Table 8-7. AN/PAQ-4B/C or AN/PEQ-2A/B infrared aiming laser training program.

AN/PAQ-4B/C OR AN/PEQ-2A/B INFRARED AIMING LASERS
Instructional Intent
• Soldiers qualify with the AN/PAQ-4B/C or AN/PEQ-2A/B IR aiming laser.
Special Instructions
Ensure that Soldiers— • Are proficient with the AN/PAQ-4B/C or AN/PEQ-2A/B IR aiming laser. • Use the proper 10-meter boresighting target during boresighting procedures. • Use the borelight filter. • Set the AN/PEQ-2A/B IR aiming laser to AIM LO. • Boresight the illuminator on the AN/PEQ-2A/B. • Use the M16A2 25-meter zeroing target for 25-meter zero. Ensure that a 3x3-centimeter hole is cut in the center of the 25-meter zeroing target and E-type silhouette.
Observables
• Soldiers conduct either 10-meter boresight or a 25-meter zero. • Soldiers display good scanning, IR discipline, and IR walking technique. • Soldiers achieve at least 17 target hits out of 40 target exposures.

Concept

8-75. Two training strategies have been devised to adequately train Soldiers in the use of AN/PAQ-4B/C and AN/PEQ-2A/B IR aiming lasers: the night initial training strategy and the night sustainment training strategy. Units should always review the night initial training strategy prior to the night sustainment training strategy.

Night Initial Training Strategy

8-76. The night initial training strategy is used for Soldiers who have little or no previous experience with night vision goggles, or for units beginning a night training program.

Night Sustainment Training Strategy

8-77. The night sustainment training strategy is for Soldiers who are familiar with night vision goggles, and for units that have already implemented a night training program.

Conduct of Training

8-78. AN/PAQ-4B/C and AN/PEQ-2A/B IR aiming laser equipment training should familiarize Soldiers with the proper operation and characteristics of AN/PAQ-4B/C and AN/PEQ-2A/B IR aiming lasers IAW the TM.

Modified Fundamentals

8-79. NVD operation requires modifications to the fundamentals of marksmanship:

- Steady position.
- Aiming.
- Breath control.
- Trigger squeeze.

Steady Position

8-80. Consider the following modifications:

- Realize that a good cheek-to-stock weld is not possible with NVGs mounted on the head.
- Ensure that the weapon's buttstock is firmly pulled into the pocket of the shoulder to prevent the laser from wobbling.
- When ready to fire, plant the elbows firmly on the ground to prevent the laser from wobbling excessively.

Aiming

8-81. Consider the following modifications:

- Practice raising the head just enough to clear the weapon with the NVGs.
- Acquire a good sight picture by walking the laser onto the target and then aiming at the center of mass.

Breath Control

8-82. This fundamental is not modified for night firing conditions.

Trigger Squeeze

8-83. Do not disrupt the laser/target alignment by jerking the trigger.

25-METER ZEROING PROCEDURES

8-84. If the borelight is not available, a 25-meter zero must be conducted. A 25-meter zeroing target is shown in Figure 8-10.

AN/PAQ-4B/C Infrared Aiming Laser

8-85. The 25-meter zeroing procedures for the AN/PAQ-4B/C IR aiming laser are as follows:

NOTE: The zero standards are the same as with iron sights.

(1) Set the adjusters to their zero preset position.

NOTE: See TM 11-5855-301-12&P for more information about setting the adjusters to their zero preset position.

(2) Prepare a 25-meter zeroing target by cutting a 3x3-centimeter square out of the center of the silhouette.

(3) Adjust for windage and elevation.

- For windage and elevation, one click equals 1 centimeter at 25 meters.
- For elevation, one clockwise click moves the bullet strike up.
- For windage, one clockwise click moves the bullet strike left.

(4) Retighten the rail grabber after the first three rounds are fired.

NOTE: When cutting the 3x3-centimeter square out of the target, some of the strike zone may be cut out. Take care when annotating the impact of the rounds. When the weapon is close to being zeroed, some of the shots may be lost through the hole in the target.

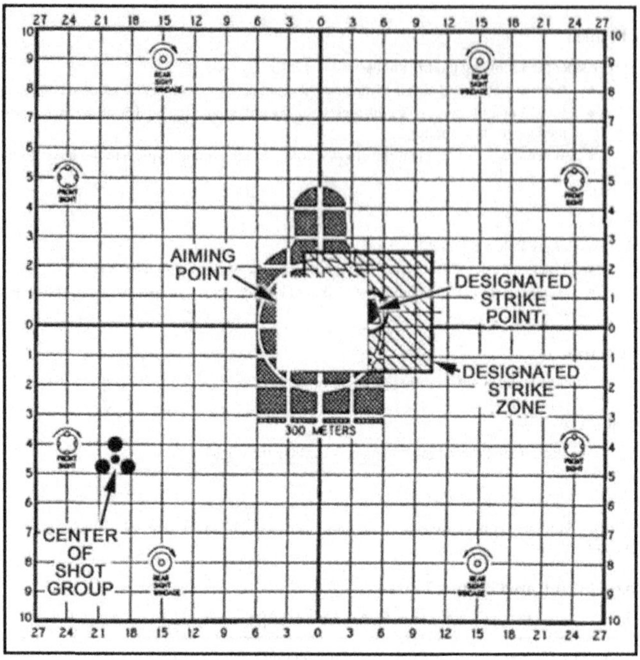

Figure 8-10. Example of shot group adjustment with strike zone.

AN/PEQ-2A/B Infrared Aiming Laser

8-86. The 25-meter zeroing procedures for the AN/PEQ-2A/B IR aiming laser are as follows:

NOTE: The zero standards are the same as with iron sights.

---

(1) Set the adjusters to their zero preset position.

NOTE: See TM 11-5855-308-12&P for more information about setting the adjusters to their zero preset position.

---

(2) Prepare the 25-meter zeroing target by cutting out a 3x3-centimeter square in the center of the target and E-type silhouette.
(3) Turn the aiming beam on in the low power setting (AIM LO).
(4) Install the aim point filter to eliminate excessive blooming.
(5) Adjust the AN/PEQ-2A/B (top-mounted) aiming point as follows:
- For windage and elevation, one click equals 1 centimeter or one square at 25 meters.
- For elevation, one clockwise click moves the bullet strike up.
- For windage, one clockwise click moves the bullet strike right.

(6) Adjust the AN/PEQ-2A/B (top-mounted) target illuminator as follows:
- For windage and elevation, one click equals 1 centimeter or one square at 25 meters.
- For elevation, one clockwise click moves the bullet strike down.
- For windage, one clockwise click moves the bullet strike right.

(7) Retighten the rail grabber and the AN/PEQ-2A/B.

(8) Once the aiming beam is zeroed, rotate the selector knob to the DUAL LO, DUAL LO/HI, or DUAL HI/HI mode to observe both aiming and illumination beams.

(9) Rotate the illumination beam adjusters to align the illumination beam with the aiming beam.

NOTES: 1. Failure to fully tighten the mounting brackets and AN/PEQ-2A/B thumbscrew may cause zero retention problems. Confirm that equipment is tight prior to zeroing.

2. To retain zero, remove the TPIAL and rail grabber as a whole assembly and place back onto the same notch as removed.

TARGET DETECTION

8-87. Soldiers should receive in-depth instruction on the proper use and fit of night vision goggles, to include characteristics and capabilities, maintenance, and mounting procedures. Extensive testing has proven that the average Soldier does not properly use NVDs. Unit leaders must be proficient in the train-the-trainer strategy. At night, Soldiers should conduct a terrain walk to become more familiar and build confidence using the night vision goggles.

Scanning for Targets

8-88. NVDs have a 40-degree field of view, which causes the average firer to miss easy targets of opportunity. Soldiers must be trained to aggressively scan their sectors of fire for targets.

8-89. The art of target detection at night is only as good as the Soldier practices. Regular blinking during scanning relieves some of the eyestrain that Soldiers experience when trying to spot distant targets. Regular blinking must be reinforced during training. After Soldiers have mastered the art of scanning, they will find that targets are more easily detected by acknowledging the flicker or movement of a target.

Infrared Discipline

8-90. Soldiers must be taught that what they can see downrange or on the battlefield through NVGs, the enemy can also see. Soldiers must train to activate the laser at the base of the target and engage the target as soon as the target is detected. After the target has been engaged, the laser is deactivated.

8-91. When a Soldier uses proper IR discipline while scanning for targets, he must keep his weapon oriented within his sector of fire. When the target is detected, the Soldier orients his weapon around the base of the target, activates his laser, and walks the laser to the target's center of mass for engagement.

FIELD FIRE

8-92. During dry-fire exercises, Soldiers acquire a sight picture on all exposed silhouette targets before conducting the field fire scenario. This allows Soldiers to focus on the targets at range. The procedures for field fire include the following:

- Conduct a dry-fire exercise prior to conducting a live-fire.
- Conduct Field Fire in the same manner as Field Fire II.
- Engage targets at 50, 150, and 250 meters.
- Fire 36 rounds:
  - 18 rounds from the supported firing position.
  - 18 rounds from the prone unsupported firing position.

Practice Qualification

8-93. The procedures for practice qualification include the following:

- Conduct a dry-fire exercise.
- Use coaches.
- Fire 40 rounds:
  - 20 rounds from the prone supported firing position.
  - 10 rounds from the unsupported firing position.
  - 10 rounds from the kneeling firing position.
- Engage targets from 50 to 250 meters.
- Meet the standards (17 hits out of 40 target exposures).

Record Qualification

8-94. The procedures for record qualification include the following:

- Conduct a dry-fire exercise.
- Fire 40 rounds:
  - 20 rounds from the prone supported firing position.
  - 10 rounds from the unsupported firing position.
  - 10 rounds from the kneeling firing position.
- Engage targets from 50 to 250 meters.
- Meet the standards (17 hits out of 40 target exposures).

AN/PVS-4 NIGHT VISION DEVICE

8-95. The AN/PVS-4 NVD is a portable, battery-operated electro-optical instrument used for observation and aimed fire of weapons at night. It amplifies reflected light, such as moonlight, starlight, and sky glow, so that the viewed scene becomes clearly visible to the operator. It can be mounted on the M16A2/A3/A4 rifle and M4/M4 MWS. Mounting brackets are provided for each type of weapon.

NOTE: See Table 8-8 for the current training program.

Table 8-8. AN/PVS-4 night vision device training program.

AN/PVS-4 NIGHT VISION DEVICE
Instructional Intent
• Soldiers qualify with the AN/PVS-4 NVD.
Special Instructions
Ensure that Soldiers— • Are proficient with the AN/PVS-4 NVD. • Install the spacer and Picatinny rail grabber when mounting on the MWS. • Use the proper 10-meter boresighting target during boresighting procedures. • Use the proper reticle. • Confirm 10-meter boresight with a 25-meter zero.
Observables
• Soldiers zero the AN/PVS-4 to the same standard as with the iron sight. • Soldiers achieve the same practice and qualification standards as with day record fire.

CONCEPT

8-96. Training strategy for the AN/PVS-4 is much the same as that for aiming lights, and the course of fire and qualifications standards are the same as those for aiming lasers.

CONDUCT OF TRAINING

8-97. This training should familiarize the Soldier with the proper operation and characteristics of the AN/PVS-4 IAW the TM.

MODIFIED FUNDAMENTALS

8-98. NVD operation requires modifications to the fundamentals of marksmanship:

- Steady position.
- Aiming.
- Breath control.
- Trigger squeeze.

Steady Position

8-99. This fundamental slightly changes due to the height of the sight. Consider the following modifications:

- Adjust your position so you can properly look through the sight.
- Recognize that, in most cases, the cheek-to-stock weld no longer exists.

Aiming

8-100. Consider the following modifications:

- To properly aim the AN/PVS-4, ensure that the proper reticle is inserted in the sight.
- Place the aiming point on the target's center of mass.

NOTE: Refer to TM 11-5855-213-10 for more information about inserting the proper reticle.

Breath Control

8-101. This fundamental of marksmanship does not change during night firing.

Trigger Squeeze

8-102. This fundamental of marksmanship does not change during night firing.

25-METER ZEROING PROCEDURES

8-103. To zero the AN/PVS-4 at 25 meters, use the same procedures and standards as with the iron sights, along with the following:

- At 25-meter range, each increment of azimuth or elevation setting moves the strike of the round .63 centimeters or ¼ mil.
- Two clicks of the windage or elevation will move the strike of the round approximately one square on the M16A2 zeroing target.
- Retighten the thumbscrew on the rail grabber after firing the initial three rounds.

NOTE: If there is not enough ambient light to see the boresight mark at ten meters or the silhouette on the zeroing target during boresighting or zeroing procedures, shine a flashlight indirectly at the target to provide ambient light.

TARGET DETECTION

8-104. Target detection with the AN/PVS-4 is very similar to target detection with the night vision goggles. The AN/PVS-4 has a 14.5-degree field of view, causing the average firer to miss easy targets of opportunity, more commonly the 50-meter left or right.

8-105. Soldiers must be trained to aggressively scan their sectors of fire for targets. The art of target detection at night is only as good as the Soldier practices. Regular blinking during scanning relieves some of the eyestrain that Soldiers experience when trying to spot distant targets. Regular blinking must be reinforced during training. After Soldiers have mastered the art of scanning, they will find that targets are more easily detected by acknowledging the flicker or movement of a target.

FIELD FIRE

8-106. During dry-fire exercises, Soldiers acquire a sight picture on all exposed silhouette targets before conducting the field fire scenario. This allows Soldiers to focus on the targets at range. The procedures for field fire include the following:

- Conduct a dry-fire exercise prior to conducting a live-fire.
- Conduct Field Fire in the same manner as Field Fire II.
- Engage targets at 50, 150, and 250 meters.
- Fire 36 rounds:
  - 18 rounds from the supported firing position.
  - 18 rounds from the prone unsupported firing position.

PRACTICE QUALIFICATION

8-107. The procedures for practice qualification include the following:

- Conduct a dry-fire exercise.
- Use coaches.
- Fire 40 rounds:
  - 20 rounds from the prone supported firing position.
  - 10 rounds from the unsupported firing position.
  - 10 rounds from the kneeling firing position.
- Engage targets from 50 to 250 meters.
- Meet the standards (17 hits out of 40 target exposures).

RECORD QUALIFICATION

8-108. The procedures for record qualification include the following:

- Conduct a dry-fire exercise.
- Fire 40 rounds:
  - 20 rounds from the prone supported firing position.
  - 10 rounds from the unsupported firing position.
  - 10 rounds from the kneeling firing position.
- Engage targets from 50 to 250 meters.
- Meet the standards (17 hits out of 40 target exposures).

Appendix A

Training Aids, Devices, and Exercises

Training aids and devices must be included in a marksmanship program. This appendix lists those available and provides information on how to obtain them for marksmanship training.

SECTION I. ENGAGEMENT SKILLS TRAINER 2000

The EST 2000 (Figure A-1) supports realistic and comprehensive gated rifle marksmanship instruction, identifies Soldiers' needs by requiring them to satisfy gate requirements in order to progress, and facilitates any necessary remedial training prior to qualification. This unit/institutional, indoor, multipurpose, multilane, small arms, crew-served, and individual antitank training simulator is used to—

- Train and evaluate individual marksmanship training for initial entry Soldiers (BCT/OSUT).
- Provide unit sustainment training for active and reserve components in preparation for qualification on individual and crew small arms live-fire weapons.
- Provide unit collective tactical training for static dismounted infantry, scout, engineer, military police squads, and sustainment elements.
- Simulate training events which lead to live-fire individual and crew weapon qualification.
- Simulate training events that contribute to increased weapon, crew, fire team, and squad combat effectiveness.
- Simulate squad collective, defensive, ambush, gunnery, and tactical tasks.
- Train leaders of fire teams and squads in the command, control, and distribution of fires.
- Save ammunition, travel time, transportation costs, and other range support resources.
- Support functional gunnery training strategies and standards in weapon training.

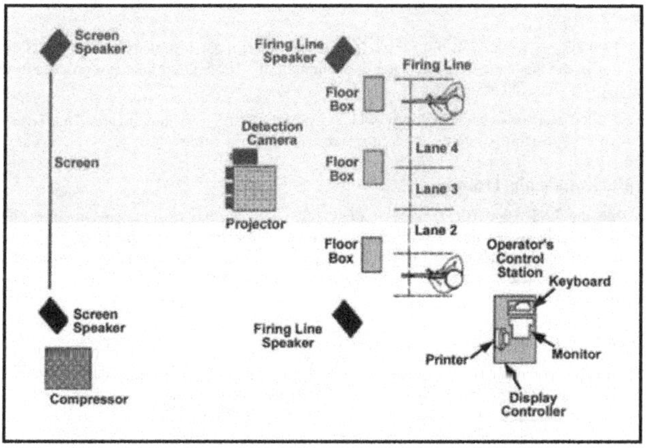

Figure A-1. Engagement skills trainer (five-lane configuration).

GENERAL CHARACTERISTICS

A-1. The EST 2000 replicates eleven weapons, including the rifle, carbine, pistol, grenade launcher, all machine guns, MK19, shotgun, and AT4. It has three modes of training:

- Marksmanship training.
- Tactical collective training.
- Shoot/don't shoot rules of engagement training.

A-2. It also includes a scenario editor for generating or tailoring new scenarios.

MARKSMANSHIP TRAINING

A-3. When used in the marksmanship training mode, the EST 2000—

- Uses Army standard courses of fire for all small arms weapons.
- Accurately simulates live-fire ranges in daylight and limited visibility conditions using precision-scaled targets, high-resolution imagery, and the essential weapon's system accuracy to compensate for errors (drift, parallax).
- Isolates, captures, and displays shots with replay that highlights firer's errors in the application of the fundamentals of marksmanship. Replay of the point of aim trace (before the shot, during the shot, and after the shot) diagnoses firer problems with aiming, breathing, steady hold, trigger control, and shot recovery for on-the-spot corrections. Cant sensors visually indicate firer-induced right or left cant, possibly resulting in missed shots.

Basic Rifle Marksmanship Training

NOTE: The marksmanship core scenarios can be found in the EST 2000 Operator's Manual, TD-07-6910-702-10.

A-4. The EST 2000 begins training the fundamentals of marksmanship right from the beginning, before the Soldier has a chance to develop bad habits. Using EST 2000 technology, Soldiers and units can reduce their rate of marksmanship failures and increase Soldiers' confidence in being able to fire their assigned weapons.

A-5. EST 2000 is particularly useful for teaching BRM, where a gated strategy requires the Soldier to pass certain gates in the simulation before firing live ammunition. The Soldier does not proceed or pass a gate scenario until he meets the standard.

A-6. EST 2000 marksmanship training provides basic range firing and qualification and is accomplished in 5-, 10-, or 15-lane configurations. Each firer is restricted to one firing lane.

Remedial Marksmanship Training

A-7. While the EST 2000 BRM gated strategy often reduces the requirements for remedial live-fire training, it is highly useful in diagnosing and correcting problems before the Soldier fires live rounds. Using the EST 2000 technology of rifle cant, trigger pressure, and before-the-shot, during-the-shot, and after-the-shot AARs, trainers can quickly identify and correct problems, thus raising confidence and first-time qualifications.

Individual and Unit Sustainment Training

A-8. Sustainment training and prequalification refresher training can be conducted with the EST 2000. During individual and unit sustainment training, grouping, zeroing, and Field Fire I and II should be conducted to standard in the EST 2000 prior to live-fire qualification.

NOTE: EST 2000 will not be used in lieu of live-fire qualification except for those outlined in DA Pam 350-38.

Tactical Collective Training

A-9. When used in the tactical collective training mode, the EST 2000—

- Presents tactical collective exercises that train squad, team, and element leaders in fire distribution and control.
- Provides fully articulated interactive targets with variable outcomes based on a squad's action or inaction.
- Uses realistic 3D-modeled battlefield terrain and variable environmental effects (for example, day/night and dawn/dusk, variable weather conditions, and illumination).
- Uses other special effects, such as a weapon's effects, explosions, and vehicle damage, to enhance the static eye point of the battlefield.
- Allows trainers and units to build scenarios as they would fight.
- Provides shot-by-shot feedback that is tied to each firer's lane of fire.
- Will soon include an entry-level indirect fire capability (product improvement).

NOTE: Tactical collective training core scenarios can be found in the EST 2000 Operator's Manual, TD-07-6910-702-10.

A-10. Tactical collective training is conducted on two networked five-lane subsystems. This configuration can support up to 11 weapons, including tandem weapons.

NOTE: The tandem weapons capability is available in collective training only. This capability allows the use of an extra weapon connected to the fifth lane (port 6) in the third floor box, allowing a firer to manage two weapons. For further instructions, refer to the EST 2000 Operator's Manual.

Shoot/Don't Shoot Rules of Engagement Training

A-11. When used in the shoot/don't shoot rules of engagement training mode, the EST 2000—

- Can be configured to enhance special operations and counterterrorism training.
- Is the premier training simulation for stability and support operations training.
- Uses video-based graphic overlays with multiple escalation or de-escalation points that require the firer to justify his actions based on his situational awareness.

NOTE: Shoot/don't shoot core scenarios can be found in the EST 2000 Operator's Manual, TD-07-6910-702-10.

A-12. Shoot/don't shoot rules of engagement training is conducted on a five-lane subsystem. This training uses video-based graphic overlays that provide important clues (for example, facial expressions and body language) for the firer to cue on. Multiple escalation or de-escalation points require the firer to justify his actions based on his situational awareness.

Scenario Editor

A-13. The scenarios currently available in the EST 2000 meet 90 percent of a unit's training requirements. As a unit's mission changes, weapons and TTPs are changed, or additional training requirements occur, the unit can use the scenario editor to generate or tailor new scenarios. The scenario editor can be used to enhance the individual Soldier's skills and, collectively, the squad's ability to engage and destroy an enemy threat.

NOTE: Detailed instructions on how to create and modify scenarios can be found in the EST 2000 training support package.

WEAPON SAFETY

A-14. The following general safety precautions should be adhered to:

- Fire simulated weapons only if they are pointed downrange.
- Post warning signs at all entry doors.
- Do not allow personnel to stand downrange from the firing line.

WARNING

No one should be allowed beyond the firing line.

- Instruct weapon handlers never to look directly into a barrel.
- Take the weapon off-line for testing and service at the first indication of malfunction and refer to the troubleshooting procedures.

DANGER

Each simulated weapon has the same appearance as a fully functioning weapon, with the exception of the trainer-peculiar umbilical cable. Under certain circumstances, especially in the subdued light of a training room, it is possible to mistake a live firearm for a simulated weapon. This creates the potential for personal injury or damage to property. To avoid confusion, no live or blank ammunition and no live weapons should be allowed in the training room.

Simulated weapons will not accept live or blank ammunition. Any attempt to chamber a live or blank round may damage the simulated weapon and create an unsafe situation.

LASER SAFETY

A-15. The lasers used in the simulated weapons meet ANSI Standard Z136.1-1993 Class I Standards for single laser pulse power. This classification is commonly referred to by the industry rating of "eye-safe." However, even eye-safe lasers may be dangerous under extraordinary circumstances. To ensure personnel safety, weapon handlers should not stare directly down a simulated weapon's barrel. Serious eye injury could result if a laser malfunctioned while a user was staring into the weapon's muzzle.

WARNING

Laser light is used in the operation of this equipment. Injury may result if personnel fail to observe safety precautions.

Never stare into the laser beam, look down the barrel of the simulated weapon, or directly view the laser beam with optical instruments.

Avoid direct eye exposure.

The instructor should ensure that all persons entering the training room are aware that laser radiation is present.

SECTION II. LASER MARKSMANSHIP TRAINING SYSTEM

The LMTS supports training with a Soldier's own weapon without the use of live ammunition. It is not designed to replace live-fire training or to eliminate the need for knowledgeable instructors. Major components include a battery-powered laser transmitter mounted to a mandrel inserted in the rifle barrel or affixed to the front sight with a mounting bracket, and a variety of laser-sensitive targets. The exercise is performed in the same manner as an LFX, except that the ammunition is a laser beam. The target senses shot locations, which are shown on a laptop screen. Unit commanders should expect and require the following:

- A training process that focuses on the four fundamentals of marksmanship (steady position, sight alignment and picture, breath control, and trigger squeeze).
- Opportunity for experienced marksmen to "test out" and serve as peer trainers or return to other duties.
- Real-time feedback.
- All-season training.
- Soldiers trained on their assigned weapon throughout the process.

Using LMTS technology, units can consistently reduce the rate of first-time marksmanship failures and increase new Soldiers' confidence in their ability to fire their basic weapon.

EQUIPMENT

A-16. Software enhancements continue to optimize the training process and minimize computer requirements by enabling an instructor to control up to 10 targets with only one computer. This feature reduces overall system costs and provides maximum throughput with a minimum number of instructors.

A-17. Minimum LMTS systems consist of a basic laser transmitter with a rod to fit the weapon and a laser target. Systems can be expanded to include a variety of components.

NOTE: Table A-1 provides a complete component list.

Table A-1. Laser marksmanship training strategy parts list.

PART	INCLUDES
110 system (110v or 220v)	TR-700 target, LT100C laser, transmitter rod, mask set, AC power adapter, and user's manual
330A system (110v or 220v)	TR-900 target, LT100C laser, transmitter rod, cable, software, AC power adapter, and user's manual
360 system (available in 3-, 4-, and 5-target array)	TR-900 target, LT100 laser, transmitter rod, cable, software, AC power adapter, control box, and user's manual
430 system (110v or 220v)	TR-900 target, MP400 laser, 556C rod, software, AC power adapter, and user's manual
Mini-range	TR-700 target, LT100C laser, transmitter rod, transceiver unit with RS-232 cable, software, E-tat, AC power adapter, and user's manual
Sound and recoil replicator system	M16A2 upper receiver M4 upper receiver CO_2-powered weapon simulator (M4 or M16)
Borelight kit	MP400 laser, LTA-556C transmitter rod, and carrying case
Targets	TR-700 electronic TR-900 electronic 25-m zero reflective (small) 25-m zero reflective (large) Other reflective targets available
Laser transmitters	LT-100 laser transmitter LT-500 in-barrel laser transmitter

Table A-1. Laser marksmanship training strategy parts list (continued).

PART	INCLUDES
System software	330A system 360 system Mini-range system
Instruction booklets	110 system 330A system 360 system
Laser transmitter rods	LTA-1200 Type B 12-gauge LTA-190 Type B Cal .177 LTA-220 Type B Cal .22 LTA-240 Type B Cal .25 LTA-310-2 Type A Cal .30, 2" barrel LTA-380 Type A Cal .38 and .357 LTA-380-2 Type A Cal .38 and .357, 2" barrel LTA-410 Type A Cal .40 and .41 LTA-440 Type A Cal .44 LTA-450-2 Type A Cal .45, 2" barrel LTA-500 Type A Cal .50 LTA-556C Cal 5.56-mm (special order) LTA-762C Cal 7.62-mm (special order)
Mask sets for TR-700 targets	Various E-type silhouettes; military mask set to simulate 300 m, 450 m, and 600 m for training conducted at 25 m
System components and accessories	Training vest for TR-700 target Sound and Recoil Replicator Consumable—5.56-mm, 7.62-mm, and 9-mm Notebook computer Safety rod CO2 4 oz., 7.5 oz., and 20 lbs. AC-600 100v adapter for TR-700 target AC-610 220v adapter for TR-700 target AC-910 110v adapter for TR-900 target AC-920 220v adapter for TR-900 target RC-260 remote cable BL-265 daisy chain cable BL-640 long cable BL-654 daisy chain cable RS-232 15-m cable extension CB-440 control boxTR-700 RF control (E-tag) Mini-range transceiver unit (tag-receiver)
Carrying cases	110 system, 330A system, and 360 system (per 3 targets) M16/M4 sound and recoil replicator Borelight kit

MARKSMANSHIP TRAINING

A-18. For initial skill development (for example, IET), Exercises 1 through 4 should be conducted sequentially. After grouping and zeroing standards, the Soldier moves to the LMTS alternate course C target where the course of fire replicates the live-fire course, except that the ammunition is a laser beam. Failure to meet the standards for this course of fire identifies the Soldier as a candidate for remedial training.

REMEDIAL TRAINING

A-19. Failure to achieve the standards set forth in this manual identifies the Soldier as a candidate for remedial training. Using the LMTS technology, trainers can quickly identify and correct problems, significantly raising qualification rates after subsequent attempts at qualification. After remedial training, the Soldier moves to the LMTS alternate course C target where the course of fire replicates that of the live-fire course, except that the ammunition is a laser beam.

SUSTAINMENT TRAINING

A-20. The training model in the exercises provides commanders and unit trainers with a sustainment training system that can be employed year-round, ideally as integrated concurrent training to cause the least disruption to other planned training.

A-21. Soldiers are administered a skill test at a regular frequency (current training guidance recommends quarterly). The results of this test allow commanders to focus training efforts on those Soldiers least able to demonstrate the minimum skills required.

A-22. For quarterly sustainment training, Soldiers should be pretested to determine the extent of training required. The pretest should begin with the grouping exercise (from Exercise 3) followed by the electronic alternate course C or mini-RETS (Exercise 4). Soldiers unable to meet pretest standards are given refresher training in the four fundamentals of rifle marksmanship, followed by completion of Exercises 1 through 4.

EXERCISES

A-23. LMTS exercises define procedures for using LMTS equipment to train and sustain basic marksmanship fundamentals. They may be conducted as independent stations or combined on a single station as appropriate for the training scenario. Trainers should employ LMTS equipment in a manner that accounts for:

- Space and time available at the training site.
- Unit size and composition.
- Remedial training requirements.
- Equipment availability.

NOTE: Check the LMTS operator's manual for specific information about equipment setup and operation.

A-24. Training in Exercises 1 through 3 should be conducted using the Soldier's own service rifle in the dry-fire mode. Exercise 4 may be conducted in the dry-fire mode, but an optional sound and recoil replicator should be employed for added realism.

NOTE: Sound and recoil replicators provide nearly 100 percent of the recoil felt with full rifle function. They require the Soldier to properly load magazines and enable the trainer to cause the rifle to misfeed or misfire to verify a Soldier's ability to perform immediate action procedures to reduce a stoppage.

A-25. If LMTS training immediately precedes a live-fire grouping and zeroing exercise and time permits, trainers may wish to use the system's prezeroing capability during Exercise 3 by using calibrated or "spun" lasers Using the calibrated lasers, Soldiers make adjustments to their own rifle sights during Exercise 3, resulting in a savings of time and ammunition on the grouping and zeroing range. All LMTS-based zeros must be confirmed by live-fire. If no live-firing is planned, calibrated lasers need not be used, and adjustments are made to the laser in Exercise 3.

NOTE: See the LMTS operator's manual for a description of the laser calibration process.

Exercise 1: Reflective Target Exercise

Table A-2. Action, conditions, and standards for a reflective target exercise.

ACTION	Demonstrate the four fundamentals of rifle marksmanship while using the LMTS reflective zeroing target.
CONDITIONS	Given an M16-/M4-series weapon, laser transmitter with mandrel, and reflective target.
STANDARDS	Demonstrate the four fundamentals of marksmanship by: • Achieving a good steady position. • Applying the proper sight alignment and sight picture. • Applying proper breath control. • Applying proper trigger squeeze.

A-26. Exercise 1 introduces Soldiers to the four fundamentals of marksmanship, outlines how to diagnose and correct firer problems, and reinforces proper application of the fundamentals.

A-27. This exercise requires a high degree of instructor involvement, but one instructor may effectively train up to 20 lanes. Decreased trainer/firer ratio will result in decreased efficiency and effectiveness. One trainer per 10 lanes is the optimum ratio.

A-28. The exercise requires little time to complete; to retain group integrity, it should be combined with Exercise 2 to allow more advanced firers to progress while problem firers receive remedial training.

A-29. A reflective zeroing target with MP400 laser/mandrel provides a simple but effective tool for remedial training during LFXs. Problem firers should be sent to a remedial station for a quick check of the application of the fundamentals and remedial training.

Exercise Performance

A-30. To perform Exercise 1, use the following procedures:

(1) The Soldier assumes a proper supported position using sandbags. The trainer inserts the MP-400/LTA-556C assembly into the rifle barrel and uses laser windage and elevation adjustments to achieve a bold sight adjustment with laser spot on front sight (Figure A-2). Then, the Soldier turns the laser to the ON position.

NOTE: Soldiers should become familiar with both supported and unsupported firing positions.

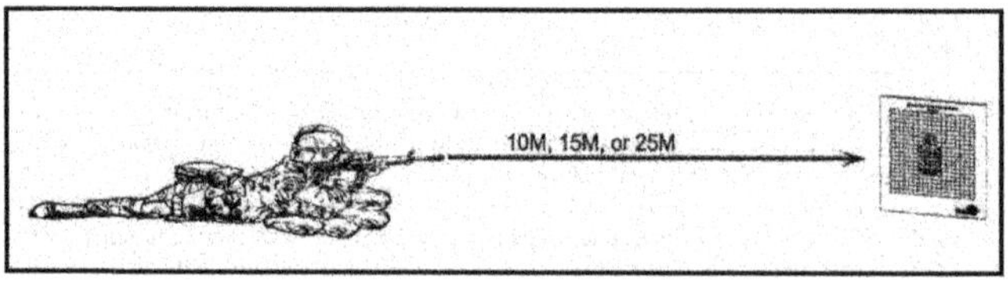

Figure A-2. Exercise 1.

(2) Under trainer supervision, the Soldier establishes a proper sight alignment and sight picture on a reflective zeroing target set at 10 meters, 15 meters, or 25 meters. With the MP-400 laser turned to ON, the trainer or coach blocks the beam with his finger.

NOTE: Use the appropriate target with the corresponding distance.

(3) When the Soldier is confident with the sight alignment and picture, the trainer removes his finger and observes the location of the red laser dot on the target.

(4) If the laser dot is in the 4-centimeter circle, proceed to Step 5. If the laser dot is outside of the 4-centimeter circle, the trainer instructs the Soldier regarding correct aiming techniques to bring the dot inside the circle, and repeats Step 2.

NOTE: If the trainer is reasonably certain that the laser and sights are aligned, the visible laser dot may be used to help the firer understand correct sight picture and alignment. The firer should be instructed to bring the laser dot to the target silhouette's center of mass, and then observe the relationship of the front and rear sights to the target.

(5) With the MP-400 in constant ON mode, the trainer or coach uses the red dot trace to confirm steady hold and proper breathing and trigger control.

NOTE: This trace can also be used to show the effects of improper steady position breath control and trigger control, and reinforce proper techniques.

(6) The trainer or coach turns the MP-400 to the training (TRN) mode and instructs the Soldier to fire six shots into the target's center of mass. The trainer or coach observes the laser hits to confirm proper application of the four fundamentals of rifle marksmanship.

NOTE: Failure to achieve this standard provides an early indication of the need for more intense instruction in the fundamentals of marksmanship or remedial training.

EXERCISE 2: INTERACTIVE DRY-FIRE EXERCISE

Table A-3. Action, conditions, and standards for an interactive dry-fire exercise.

ACTION	Demonstrate the integrated act of firing while using the LMTS 130-target system.
CONDITIONS	Given an M16-/M4-series weapon, laser transmitter with mandrel, and TR-700 targets with military masks.
STANDARDS	• Achieve 8 hits out of 10 shots two times on an open-faced target from the prone unsupported position. • Achieve 8 hits out of 10 shots two times on a 300-meter masked target from the supported position.

A-31. This exercise provides Soldiers with an opportunity for practicing the four fundamentals of rifle marksmanship in the integrated act of firing and may easily be conducted concurrently with Exercise 1 on the same station. TR-700 targets may be used both indoors and outdoors in various environments and arrangements to meet the training requirement.

Exercise Performance

A-32. To perform Exercise 2, use the following procedures:

(1) The Soldier assumes a proper firing position using sandbags for a supported position (Figure A-3).

NOTE: Sleeping mats should be used on hard floors.

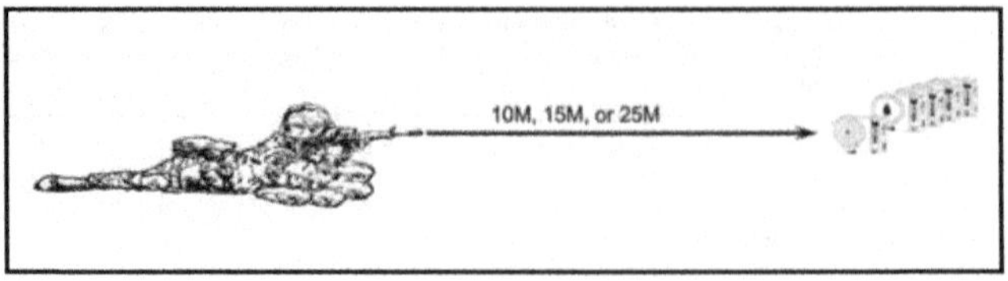

Figure A-3. Exercise 2.

(2) The Soldier applies the four fundamentals of marksmanship to engage a TR-700 open-faced target with 10 shots from the prone unsupported position. The Soldier cocks the rifle after each shot, forcing a break and reestablishing a proper stock weld to build muscle memory.

NOTE: The TR-700 open-faced target at 25 meters equals a doublewide E-silhouette target at 300 meters.

---

(3) The trainer inspects the target score for the number of hits. If the number is less than eight, the trainer should perform a visual laser/sight alignment check. If the laser/sight alignment is correct, the trainer reconfirms the Soldier's understanding of the four fundamentals of marksmanship and directs the Soldier to repeat Step 2. If the number of hits is less than eight after several tries, the Soldier reports for remedial training. If the number of hits is eight or more, the Soldier repeats Step 2 to confirm, and then proceeds to Step 4.

(4) After the Soldier completes Step 3, a 300-meter scaled E-silhouette mask is installed over the face of a TR-700 target. The Soldier repeats Step 2 from the supported position and repeats Step 4 to confirm.

(5) OPTIONAL: As time allows, the Soldier increases the number of shots to 20 and/or installs smaller masks for an additional skill challenge. Additional firing positions may also be reinforced, if needed.

NOTES:

1. The largest mask presents a 300-meter E-target size scaled for 25 meters.
2. The middle mask presents a 300-meter E-target size scaled for 15 meters or a 450-meter E-target size scaled for 25 meters.
3. The smallest mask presents a 300-meter E-target size scaled for 10 meters, a 450-meter E-target size scaled for 15 meters, or a 600-meter E-target size scaled for 25 meters.

Exercise 3: Grouping and Zeroing Exercise

Table A-4. Action, conditions, and standards for a grouping and zeroing exercise.

ACTION	Group and zero an M16-/M4-series weapon using the TR-900 target system with military mask.
CONDITIONS	Given an M16-/M4-series weapon, laser transmitter with mandrel, and TR-900 target system with military mask.
STANDARDS	From the supported firing position— • Grouping. Fire up to 27 shots (dry-fire) in three-round shot groups and achieve two consecutive shot groups within a 4-centimeter circle (25 meters), 2.4-centimeter circle (15 meters), or 1.6-centimeter circle (10 meters). • Battlesight Zero. Adjust the sights so that five out of six rounds in two consecutive shot groups strike within the zeroing circle in the silhouette on the zeroing target.

A-33. This exercise evaluates a Soldier's ability to apply the four fundamentals of rifle marksmanship in the integrated act of firing through shot grouping. The exercise is conducted in the same manner as live-fire grouping and zeroing exercises and can make those exercises more efficient and effective. All normal range commands should be used to reinforce training in proper range procedures. Up to 10 targets may be grouped together for scoring on one computer. This exercise is most efficient with one trainer to run the control and scoring console, plus one trainer for every five lanes. Training distance must correspond to the distance used in Exercise 4.

NOTE: Whenever this exercise is conducted prior to an LFX, calibrated lasers should be used to support prezeroing. Adjustments to the rear sight of an M16A2/A3/A4 or M4-series weapon, associated optics, and the front sight of an M16A1 must be made when training at 10 meters or 15 meters to compensate for parallax error. See the LMTS operator's manual for a detailed description of these adjustments.

Exercise Performance

A-34. To perform Exercise 3, use the following procedures:

(1) From the supported firing position (Figure A-4), the Soldier fires three-round shot groups at the target overlay's center of mass, continuing until two consecutive groups fall within a 4-centimeter circle anywhere on the target (maximum 27 shots). Trainers should provide feedback to the Soldier between each shot group. If the Soldier is unable to achieve the standard within 27 shots, the trainer attempts remedial actions or sends the Soldier to the remedial training station.

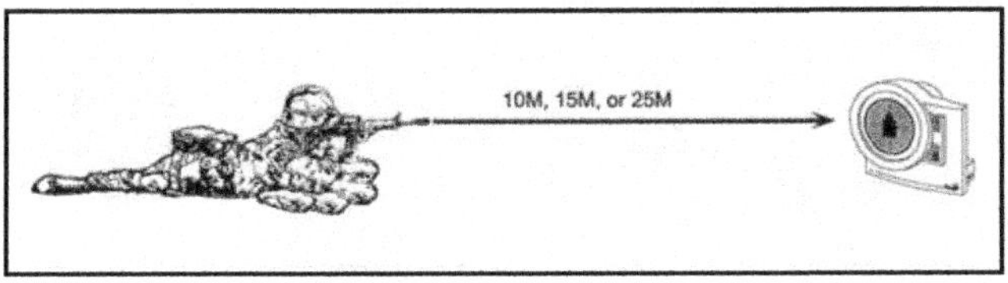

Figure A-4. Exercise 3.

(2) When the grouping standard is met, the Soldier makes appropriate sight changes as instructed by the trainer, who begins a new session for the zeroing process. The Soldier continues to fire three-round shot groups, adjusting the sights as instructed by the trainer to bring the shot groups (maximum 18 shots) within the zeroing circle on the target silhouette. When a shot group falls within the zeroing circle, the Soldier fires an additional shot group for confirmation. Five of six shots must fall within the zeroing circle.

NOTE: Failure to achieve the standard identifies the Soldier as a candidate for remedial training. The Soldier should not progress to Exercise 4 until the standard is met.

Exercise 4: Laser Marksmanship Training Strategy Prequalification Exercise

Table A-5. Action, conditions, and standards for a laser marksmanship training strategy prequalification exercise.

ACTION	Engage 10-, 15-, or 25-meter alternate course C scaled silhouettes with an M16-/M4-series weapon
CONDITIONS	Given an M16-/M4-series weapon, laser transmitter with mandrel, and electronic alternate C target system with ten 10-, 15-, or 25-meter scaled silhouettes. Engage each silhouette with two shots from the supported position and two shots from the prone unsupported position.
STANDARDS	Without assistance, the Soldier engages 10 target silhouettes using the M16-/M4-series weapon with laser transmitter and achieves a minimum of 30 hits out of 40 shots.

A-35. Exercise 4 is used as a skill test to determine the need for training or the results of training, and serves as an accurate predictor of live-fire alternate course C performance.

NOTE: Soldiers failing to meet the standards of this exercise should receive remedial training prior to live-fire qualification.

A-36. The exercise may be conducted in the dry-fire mode, with Soldiers recocking the rifle between shots. When performing the dry-fire method, Soldiers should use a magazine with the follower and spring removed. Another option for the dry-fire mode involves removing the charging handle and attaching a piece of cord (looped on both ends with the free end about 3 inches from the end of the stock) to the bolt. The coach can recock the rifle between shots by pulling the cord directly to the rear.

A-37. The sound and recoil replicator options include the M16A2 Blazer and the alternate laser-mounting bracket used with a standard blank firing attachment (BFA). Both options offer added realism by providing full rifle function, sound, and recoil. When using the special safe Blazer blanks, the Blazer option may be employed indoors without hearing protection. Standard M200 blanks may only be used with the BFA while outdoors and while wearing hearing protection. When available, the LMTS mini-RETS range should be employed for added training realism and to prepare Soldiers for firing on pop-up targets.

Exercise Performance

A-38. To perform Exercise 4, use the following procedures:

(1) The Soldier assumes the proper supported firing position using sandbags (Figure A-5).

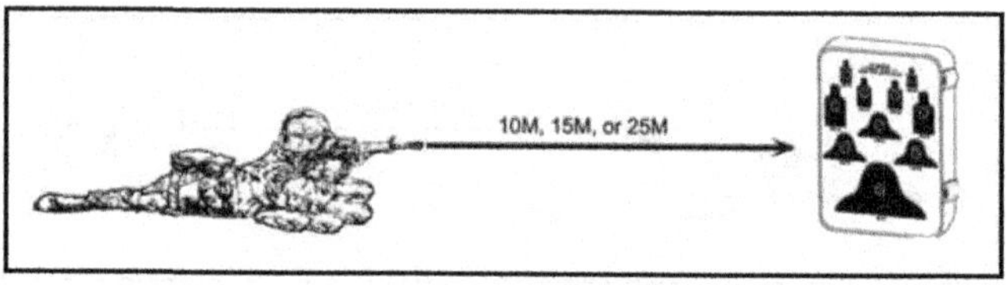

Figure A-5. Exercise 4.

(2) The trainer prepares the electronic 25-meter alternate C target and computer control station for the prequalification skill test. The course may be conducted at 25 meters, 15 meters, or 10 meters using the appropriate target overlay.

NOTE: The trainer should use appropriate range commands and enforce all range safety procedures.

(3) To fire the alternate course C, the Soldier applies the four fundamentals of rifle marksmanship using the service rifle (laser-zeroed) during Exercise 3.

(4) The computer automatically times the test and can print a score sheet.

SOUND AND RECOIL REPLICATOR

A-39. To add realism to the training, a special upper receiver sound and recoil replicator provides full live-fire functionality (without the projectile). It supplies nearly 100 percent of the recoil with 50 percent of the noise using a specially-designed nontoxic theatrical blank.

SECTION III. OTHER TRAINING DEVICES AND EXERCISES

This section provides the classification and nomenclature for training aids, devices, and targets.

TRAINING SUPPORT CENTERS

A-40. Training support centers (TSCs) are located throughout the world to provide training aids and devices. Each TSC provides training aid services to customers (for example, active Army units and schools, reserve components, and ROTC units) in their geographic area of support.

NOTE: For more information concerning TSC operations, write:

Commander
United States Army Training Support Center
ATTN: ATIC-DM
Fort Eustis, VA 23604

SELECTION OF TRAINING AIDS AND DEVICES

A-41. After training requirements have been established, appropriate training aids and devices can be selected from the TSC. Table A-6 lists many of those aids and devices available and their identification numbers.

Table A-6. Training aids and devices.

TYPE/NOMENCLATURE	IDENTIFICATION NUMBER
Rifle Marksmanship Trainer (Weaponeer)	DVC 07-57/DVC 07/57B
M15A2 Blank Firing Attachment	Supply Item (see TM)
Chamber Block (M16A1/A2)	TAD-0001
M16 Sighting Device (A1 0r A2) (Left and Right)	DVC-T 7-84
Target Box Paddle	DVC-T 7-86
Riddle Device	DVC-T 7-87
M16 Rifle Brass Deflector	DVC-T 23-30
M15A1 Aiming Card	DVC-T 07-26
M16A1 Display Mat (canvas)	TAD-0034 (locally)
Rifle Rest (for target box exercise)	TAD-12 (locally)
Front and Rear Sight, M16 Rifle	TAD-26 (locally)
Front and Rear Sight, M16A2 Rifle	TAD-0026A
Graphic Training Aids (GTA)	
M16A1 Disassembly Mat (paper)	GTA 09-06-43
Rifle, 5.56-mm, M16A1 Mechanical	GTA 7-1-26
Rifle, M16 Disassembly (M16A1)	GTA 9-6-43
M16A1 Rifle Malfunction	GTA 9-6-44
M16A1 Rifle Maintenance Card	GTA 21-1-3
Training Films	
*Rifle, M16A1 Part I, Care, Cleaning, Lubrication	TF 21-3907
*Rifle, M16A1 Part II Field Expedients	TF 21-3908
*Also available in videotape.	
Videotapes	
Engagement of Moving Personnel Targets with the M16A1 Rifle Team from the Foxhole Position	2E/010-071-1271-B
Cycle of Functioning M16A1 Rifle	2E/010-071-0444-B
Overview of BRM Training	2E/010-071-0086-B
TVT 7-13 (Feb 87)	2E/010-071-0725-B
TVT 7-1 Teaching Rifle Marksmanship: Part I	
TVT 7-2 Teaching Rifle Marksmanship: Part II	

TARGET ORDERING NUMBERS

A-42. Table A-7 lists the description and NSN to use when ordering marksmanship targets.

Table A-7. Target ordering numbers.

DESIGNATION	DESCRIPTION	NSN
D Prone	Full-length face with V through two scoring areas	6920-00-922-7450
D Prone	Repair center with V through two scoring areas	6920-00-922-7451
E-Silhouette	Full-length face, solid color paper	6920-00-600-6874
E-Silhouette	Full-length, pop-up, solid color plastic	6920-00-071-4780
E-Silhouette	Full-length face, cardboard, kneeling	6920-00-079-1806
F-Silhouette	Short-length face, solid color paper	6920-00-610-9086
F-Silhouette	Short-length, pop-up, solid color plastic	6920-00-071-4589
F-Silhouette	Short-length face, pasteboard	6920-00-795-1807
25-Meter Alternate Course Scaled Qualification Target	50- to 300-meter scaled silhouette target	6920-01-167-1398
15-Meter Battlesight Zeroing Target (.22 Caliber RFA)	250-meter scaled silhouette target (50-foot indoor range)	6920-01-167-1393
15-Meter Alternate Course C (.22 Caliber RFA)	50- to 300-meter scaled silhouette target (50-foot indoor range)	6920-01-167-1396
25-Meter M16A1 Zeroing Target	250-meter scaled silhouette target	6920-01-167-1392
25-Meter M16A2 Zeroing Target	300-meter scaled silhouette target	6920-01-253-4005
25-Meter M16A1 Slow Fire Target	75- to 300-meter scaled silhouette target	6920-01-167-1391
25-Meter M16A1 Timed Fire Target	50- to 300-meter scaled silhouette target	6920-01-167-1397
75-Meter M16A1 Feedback Target	75-meter scaled F-type silhouette	6920-01-169-6921
75-Meter M16A2 Feedback Target	75-meter scaled F-type silhouette	6920-01-253-4006
175-Meter M16A1 Feedback Target	175-meter scaled E-type silhouette	6920-01-167-1395
175-Meter M16A2 Feedback Target	175-meter scaled E-type silhouette	6920-01-167-1395
Pasters, Black		6920-00-165-6354
Pasters, Buff		6920-00-172-3572
Landscape target		6920-00-713-8253
Spindle, Target Spotter, Wood		6920-00-713-8257
Spotters, 1 1/2 inches in diameter		6920-00-789-0869
Spotters, 3 inches in diameter		6920-00-713-8255
Spotters, 5 inches in diameter		6920-00-713-8254
Thermal Blankets		6920-01-516-9912

TRAINING DEVICES

A-43. Several marksmanship training devices are available to aid in PMI and sustainment training. They are beneficial when ammunition is limited for training or during practice exercises, such as field firing on the EST 2000 or zeroing and qualifying with short-range training ammunition (SRTA). Some training devices are complex, costly, and in limited supply, while others are relatively simple, cheap, and in large supply. Individuals or squads can sustain and practice basic marksmanship skills and fundamentals using devices and aids alone or in combinations.

Aiming Card

A-44. The M15A1 aiming card (Figure A-6) determines if the Soldier understands how to aim at a target's center of mass, how to adjust the point of aim, how to allow for gravity, and how to engage a moving target.

Figure A-6. M15A1 aiming card.

A-45. To use the aiming card—

(1) The card is misaligned, the Soldier is instructed to establish the correct point of aim, and a trainer checks it. Each Soldier demonstrates six out of six of the points of aim.

NOTE: The sight/target relationship on the card is the same visual perception that the Soldier should have when he is zeroing on a standard silhouette target.

---

(2) The Soldier shows the side alignment technique three times—place the front sightpost on the left or right edge of the target and bring the front sightpost to the target's center of mass.

(3) The Soldier shows the bottom-up alignment technique—place the front sightpost at the bottom of the target and bring the front sightpost to target's center of mass.

Riddle Sighting Device

A-46. The Riddle sighting device (Figure A-7) indicates if the Soldier understands the aiming process while using the rifle. It is a small plastic plate with a magnet and a drawing of an E-type silhouette target. A two-man team is required for its use.

A-47. To use the Riddle sighting device—

(1) The Soldier assumes a supported or prone firing position. The assistant places the Riddle device on the front sight assembly and adjusts the plastic plate in the direction of the firer until he reports the proper sight picture.

(2) Without disturbing the plastic plate, the trainer or coach aims through the sights to determine if the Soldier has aligned the target and sight properly.

A-48. Many sightings are conducted, and the trainer may include variations to ensure that the Soldier understands the process. Each Soldier demonstrates six out of six points of aim, starting with the plastic plate offset to the front sightpost.

NOTE: This device has a small metal clip that slips over the front sight assembly to allow a smooth surface for attachment of the magnet. The device may also be used without the metal clip.

This device is provided with a small metal clip that slips over the front sight assembly. It all a smoother surface for attachment of the magnet; however, the device can be used without the metal clip.

Figure A-7. Riddle sighting device.

M16 SIGHTING DEVICE

A-49. The M16 sighting device (Figure A-8) is made of metal, with a tinted square of glass placed at an angle. When the device is attached to the rear of the M16A1 carrying handle, an observer can look through the sight to see what the firer sees.

NOTE: The M16 sighting device can be mounted on the M16A2 rifle, but the charging handle must be pulled to the rear first. Then, the M16 sighting device is mounted on the rear of the carrying handle, and the charging handle is returned forward.

A-50. When using the M16 sighting device, observe the following recommendations:

- The M16 sighting device can be used in a dry-fire or live-fire environment, but a brass cartridge deflector must be used during live-fire.
- The observer must practice with the sight to be effective. For example, if the observer looks at a reflected image and the Soldier is aiming to the right, it appears left to the observer.
- The device must be precisely positioned on the rifle (it may need to be bent to stay on).
- The observer's position must remain constant.
- The observer must talk with the firer to ensure a correct analysis of the aiming procedures.

A-51. The Soldier must achieve six out of six proper sight alignment drills.

NOTE: The M16 sighting device is made for left- and right-handed firers.

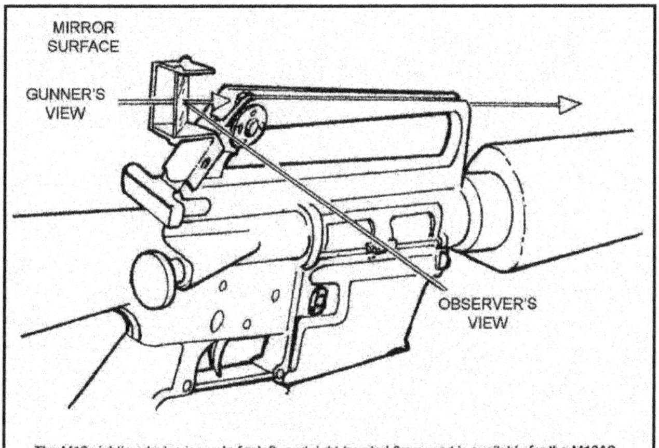

The M16 sighting device is made for left- and right-handed firers, and is available for the M16A2.

Figure A-8. M16 sighting device.

M15A2/M23 Blank Firing Attachment

A-52. The M15A2/M23 BFA (Figure A-9) attaches to the muzzle of M16-/M4-series weapons. It is designed to keep sufficient gas in the weapon's barrel to allow for semiautomatic, automatic, or burst firing with blank ammunition (M200 only).

A-53. When using the M15A2/M23 BFA, observe the following recommendations:

- After firing 50 rounds, check the attachment for a tight fit.
- Continuous blank firing results in a carbon buildup in the bore, gas tube, and carrier key. If this occurs, follow the cleaning procedures outlined in TM 9-1005-249-10.

NOTE: The M15A2 is painted red and is used on the M16-series weapons. The M23 is painted yellow and is used on the M4-series weapons. For identification, the M23 is stamped "M4 Carbine Only."

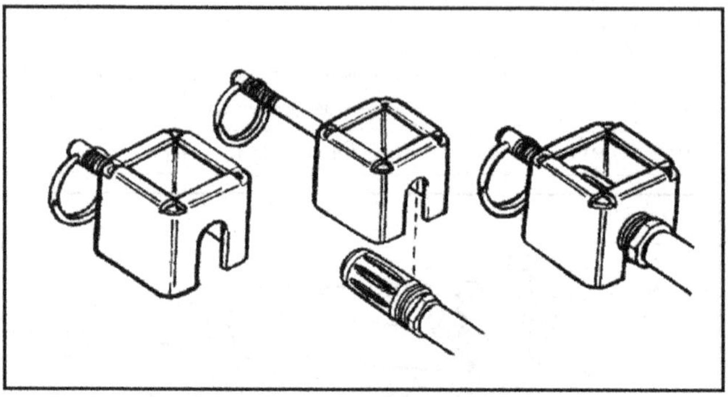

Figure A-9. Blank firing attachment.

CAUTION

Hand tighten the M15A2/M23 only.

Location of Miss-and-Hit System

A-54. The LOMAH system is a range aid used during downrange feedback exercises. The device uses acoustical triangulation to compute the exact location of a supersonic bullet as it passes through a target. The bullet impact is displayed instantly on a video monitor at the firing line. More importantly, it shows the location of a bullet miss, allowing the firer to make either a sight adjustment or a hold-off for subsequent shots.

NOTE: LOMAH, like other devices, is only an aid. When the Soldier uses LOMAH, he must understand the weapon and firing techniques and have a coach/instructor.

A-55. LOMAH ranges have been fielded in USAREUR and Korea. In locations where KD ranges are not available and restrictions prohibit walking downrange, the LOMAH system is a practical alternative to essential downrange feedback.

NOTE: Requests for LOMAH devices should be sent to:

Commander
US Army Training Support Center
ATIC-DM
Fort Eustis, VA 23604.

M261 .22-CALIBER RIMFIRE ADAPTER

A-56. The .22-caliber RFA can contribute to a unit's marksmanship program when 5.56-millimeter ammunition is not available or when ranges that allow firing 5.56-millimeter ammunition are not available. The RFA can be useful for marksmanship training such as night fire, quick fire, and assault fire, but is not recommended for primary marksmanship training.

Training Considerations

A-57. When service ammunition is in short supply, the RFA can be used to complement a unit's training program.

Rifle Performance

A-58. The RFA cannot be depended on to fire in the same place as 5.56-millimeter ammunition. The RFA and its .22-caliber ammunition cannot replicate the exact ballistics of the 5.56-millimeter ammunition, but efforts to match RFAs with specific rifles can result in reasonable replication.

NOTE: Under ideal training conditions, the RFA should be used with dedicated rifles. It is not necessary for the Soldier to use his own weapon during RFA training.

A-59. Finding the right match of RFA and rifle can eliminate some variability. A trial-and-error technique can match RFAs to rifles, which results in weapons that fire well.

Rifle Zero

A-60. RFA ammunition differs from 5.56-millimeter ammunition in the following ways:

- RFA ammunition will not usually group in the same location as 5.56-millimeter ammunition at 25 meters and cannot be used for weapon zero.
- RFA ammunition normally fires a slightly larger shot group than 5.56-millimeter ammunition.

A-61. When a Soldier uses an RFA in his rifle, he must be careful not to lose his 5.56-millimeter zero. This can be accomplished by using hold-off while firing .22-caliber ammunition or keeping a record of sight changes so the sights can be moved back.

NOTES: 1. The .22-caliber round approximates the 5.56-millimeter trajectory out to 25 meters.

2. The correct zeroing target or appropriate scaled-silhouette targets can be used for practice firing exercises at 15 meters (50 feet) or 25 meters.

Advantages and Disadvantages

A-62. If the RFA is used as a training aid, the advantages and disadvantages must be considered during training.

Advantages

A-63. The .22-caliber ammunition is cheaper and may be available in larger quantities than 5.56-millimeter ammunition. It can be fired on all approved indoor ranges and in other close-in ranges where 5.56-millimeter ammunition is prohibited. RFA training can be used to sustain marksmanship skills during periods when full-caliber 5.56-millimeter ammunition training cannot be conducted.

Disadvantages

A-64. Some negative training aspects exist because of differences in the weapon's functioning when using the RFA. These differences include—

- The forward assist does not work.
- The bolt does not lock to the rear after the last round is fired.
- More malfunctions can occur with the RFA than with 5.56-millimeter ammunition.
- Immediate action procedures are different.

SHORT-RANGE TRAINING AMMUNITION

A-65. SRTA is a plastic practice cartridge (M862) that enables a unit to conduct realistic firing training at shorter distances with reduced danger areas. The M862 has a maximum range of 250 meters. The blue plastic projectile reduces the risk of overpenetration and ricochet, which makes it ideal for UO training.

A-66. To fire the M862 SRTA from an M16-/M4-series weapon, the standard bolt and bolt carrier must be replaced by the M2 practice bolt. The M2 practice bolt consists of a bolt carrier, which is a fixed bolt. The practice bolt changes the weapon from a gas-operated action to a blow-back action that permits cyclic fire with the lower-powered M862.

NOTES: 1. Because of the design of the M2 practice bolt, standard 5.56-millimeter rounds cannot be fired from the weapon while it is installed.

2. See TM 9-6920-746-12&P for more information on the M862 SRTA and the M2 practice bolt.

WEAPONEER

A-67. The Weaponeer is an effective rifle marksmanship training device that simulates the live-firing of the M16-series rifle. The system can be used for developing and sustaining marksmanship skills, diagnosing and correcting problems, and assessing basic skills.

NOTE: The EST 2000 has replaced the Weaponeer, but some units still use the Weaponeer.

Operation

A-68. Figure A-10 shows the Weaponeer in the standing supported firing position. With the exception of smoke and cartridge ejection, the rifle operates normally and has the same weight and balance as the standard weapon. An IR aiming sensor simulates round trajectory and hit point to an accuracy of better than one MOA. The recoil rod that attaches at the rifle's muzzle end simulates recoil.

NOTE: Recoil is provided in both semiautomatic and automatic modes of fire and is adjustable from no-net force to 30 percent more than that of a live M16.

A-69. Sound is provided through headphones and is adjustable from 115 to 135 decibels. Special magazines are used. One magazine simulates a continuous load; the other is used to train rapid magazine change and can be loaded with 1 to 30 simulated rounds. Selectable misfire can be used to detect gun-shyness and drill immediate action. The front and rear sights are zeroed the same as standard rifles.

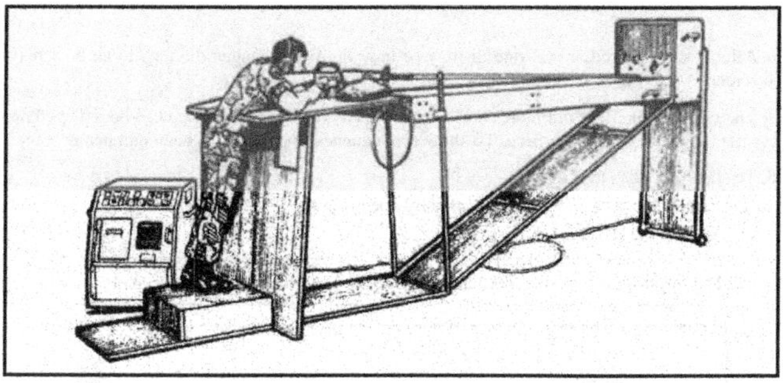

Figure A-10. Weaponeer set up in the standing supported position.

A-70. The Weaponeer range can be raised or lowered to accommodate all firing positions. The target assembly contains four targets: a scaled 25-meter zeroing target and three pop-up targets. E-type and F-type silhouettes at ranges from 75 meters can be used on the Weaponeer. Known distance and various other types of targets displayed in fixed or random sequences can be used. Target exposure times may be set to unlimited or from 1 to 30 seconds. The fall-when-hit mode can be selected with the KILL button.

A-71. The operator's console contains the system control buttons, graphics printer, and video feedback monitor. The back of the console has counters that total rounds and hours, and a storage bin for storing magazines, printer paper and ribbon, headphones, two wrenches for assembling the Weaponeer, and a small hex-head wrench for aligning the rifle sensor. A remote control, which attaches to the back of the console, enables a trainer or firer to operate select functions away from the console.

Feedback

A-72. The Weaponeer provides feedback to help trainers to teach and Soldiers to learn marksmanship skills. The tools used for feedback include—

- Fall-when-hit mode.
- Real-time aiming point display.
- Immediate shot impact display.
- Replay.
- Shot groups.
- Printer.

Fall-When-Hit Mode

A-73. Lighting the KILL button enables the fall-when-hit mode. When the button is activated, targets fall when hit. This feedback provides the same hit-or-miss information as a train-fire (RETS) range.

Real-Time Aiming Point Display

A-74. When a firer aims on or near a target, his aiming point relative to the target is continuously displayed on the video screen. The aiming point display allows the trainer to teach and verify aiming techniques, and to continuously monitor the firer's steadiness, technique, time on target, trigger squeeze, and recovery from recoil.

Immediate Shot Impact Display

A-75. When a shot is fired, its impact relative to the target is immediately displayed on the video screen as a blinking white dot (Figure A-11, left target).

Replay

A-76. After a shot is fired, a real-rate display of how the firer engaged the target can be replayed on the video screen.

A-77. The target to the right in Figure A-11 shows the type of information that can be replayed on the video screen after a series of shots are fired. To show the sequence, the dots have been numbered.

A-78. To show a replay, the firer—

(1) Selects the shot he wishes to replay by operating the EACH SHOT button.

(2) Presses the REPLAY button.

NOTE: Some Weaponeers record and store replays for only the first three shots.

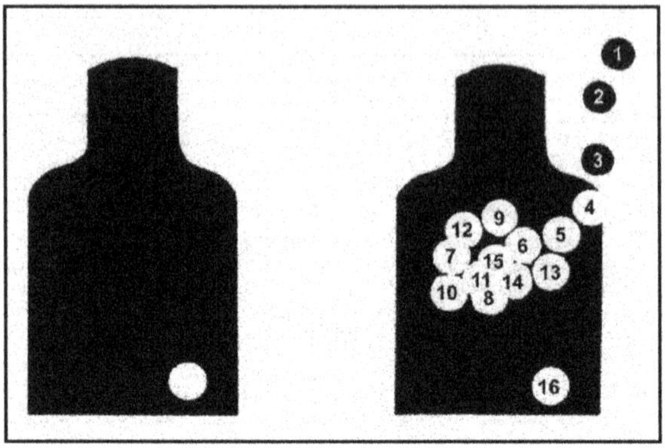

Figure A-11. Replay of shot.

Shot Groups

A-79. The impact location of up to 32 shots is automatically stored in the Weaponeer's memory and displayed on the video screen. Each impact is indicated by a white dot, and the last shot blinks for indication. All 32 shots can be fired and displayed on a single target or split among a combination of targets. The CLEAR button erases all shots from the Weaponeer's memory.

Printer

A-80. A hard copy printer is provided for postfiring analysis, for firer progress tracking, and for record keeping. Pressing the PRINT button causes the target displayed on the video to print (Figure A-12). Some Weaponeers can print the three pop-up targets at the same time by holding in the REPLAY button and pressing the PRINT button.

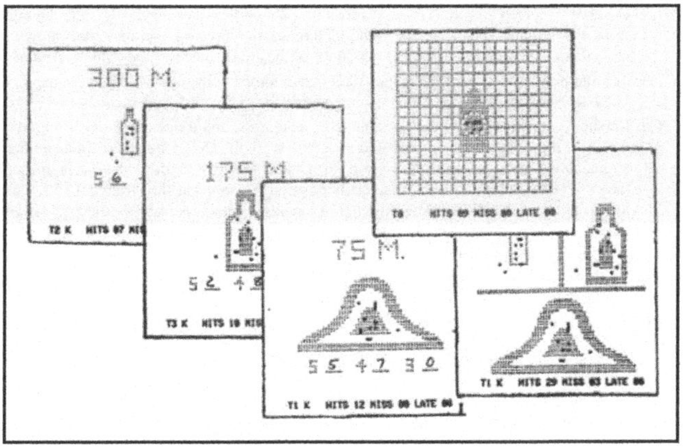

Figure A-12. Weaponeer printouts.

Use of the Weaponeer

A-81. The Weaponeer is used to evaluate the firer's ability to apply the four fundamentals. It is used throughout the program to help diagnose and remediate problems. In the unit, the Weaponeer should be used much like it is used in BRM.

A-82. When training Soldiers on the Weaponeer—

- Proceed at a relaxed pace, and emphasize accuracy before speed.
- If possible, train with small groups, allowing each Soldier several 10- to 15-minute turns on the device.
- For remedial training, try to relax the Soldier. A nervous Soldier will have trouble learning and gaining confidence in his marksmanship skills. For sustainment training, encourage competition between individuals or units.

Diagnosis of Firing Problems

A-83. The Weaponeer assists in the diagnosis of firing problems. The seven-step program is recommended as a guide. These seven steps are designed to diagnose and show the Soldier his firing errors. This could be enough to correct the error.

NOTES: 1. Diagnosis needs to be followed up with remedial exercises with the Weaponeer, target box exercise, or dime/washer exercise.

2. Depending on the extent of the firer's problems and time constraints, the number of shots may be increased.

A-84. To diagnose firing problems—

(1) Tell the Soldier to assume a good firing position, aim at a target, and hold steady (supported and prone unsupported positions).

(2) Visually check the firer's position and correct any gross errors.

(3) Observe the video screen. If there is no aiming dot on the video screen or if the aiming dot is far from the target's center, teach sight picture to the firer. If the light dot shows excessive movement, check and correct the techniques of the steady position and natural point of aim.

(4) Tell the Soldier to fire a three-round shot group aimed at the target's center of mass. Watch the video screen and Soldier as he fires. Note violations of the four fundamentals.

(5) Replay each shot to show the firer his aim, steadiness, and trigger squeeze. In Figure A-11, the target on the right shows a numbered series of 16 shots. Dots 1 through 4 indicate that the firer approached the target from the high right. Dots 5 through 15 show that he is aiming near the target's center, but does not have a steady position. The sudden shift from dot 15 to 16 (dot 16 is the hit point of the shot) indicates that gun-shyness or improper trigger squeeze caused the firer to pull his point of aim down and to the right just before firing. Replay helps the firer understand and correct his firing errors.

(6) Confirm and refine the diagnosis by allowing the Soldier to fire additional three-round shot groups. Use replay to show the firer his faults.

(7) Summarize and record the Soldier's basic firing problems.

TRAINING EXERCISES

A-85. Marksmanship training exercises are designed to aid in PMI and sustainment training. They are beneficial when ammunition is limited for training. Training exercises can be used alone or in combinations.

TARGET BOX EXERCISE

A-86. The target box exercise checks the consistency of aiming and placement of three-round shot groups in a dry-fire environment.

A-87. To conduct the exercise (Figure A-13)—

(1) The target man places the silhouette anywhere on the plain sheet of paper and moves the silhouette target as directed by the firer.

NOTE: The two positions (separated by 15 yards or 25 meters) must have already been established so the rifle is pointed at some place on the paper.

(2) When the firer establishes proper aiming, he signals the target man to "Mark."

NOTE: Only hand signals are used since voice commands would be impractical when training several pairs of Soldiers at one time.

Figure A-13. Target box exercise.

(3) The target man places the pencil through the hole in the silhouette target and makes a dot on the paper. Then, he moves the silhouette to another spot on the paper and indicates to the firer that he is ready for another shot.

(4) When the three shots are completed, the target man triangulates the three shots and labels it "Shot Group 1." The firer and instructor view the shot group.

(5) The Soldier fires several shot groups. After two or three shot groups are completed in one location, the rifle, paper holder, or paper is moved so shots fall on a clean section of the paper.

NOTE: Any movement of the rifle or paper between the first and third shots of a group voids the exercise. Two devices are available to hold the rifle. Rifle-holding devices (Figures A-14 and A-15) are positioned on level ground, or are secured by sandbags or stakes to ensure the rifle does not move during the firing of the three shots. Movement of the paper is eased by using a solid backing (Figure A-16). Any movement of either is reflected in the size of the shot group. Several varieties of wooden target boxes have been locally fabricated.

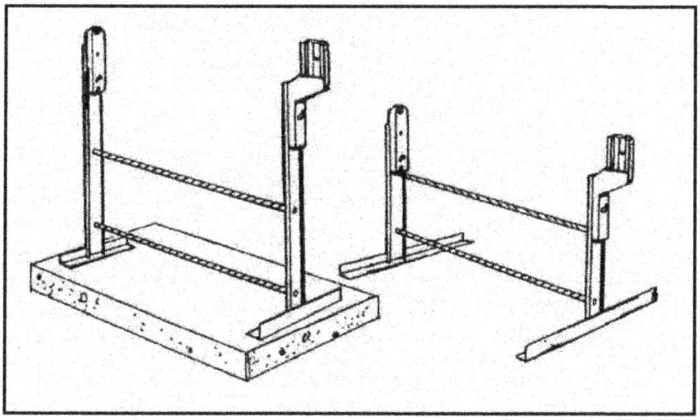

Figure A-14. Rifle-holding device (TA-G-12A).

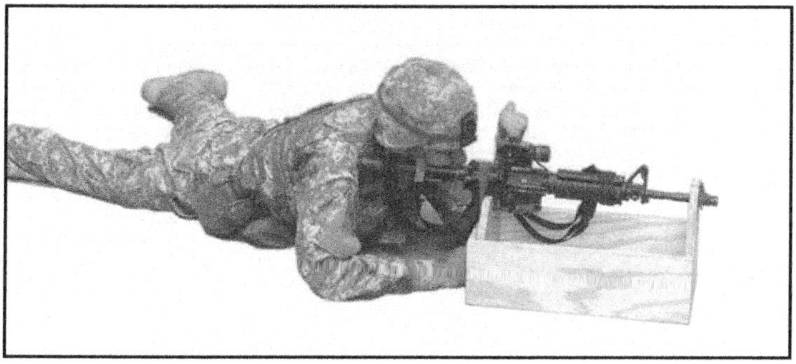

Figure A-15. Staked rifle holding box.

Figure A-16. Paper being placed on a stationary object.

A-88. Each Soldier dry-fires the exercise until he has demonstrated six out of six of the points of aim within the plastic target box paddle's 4-centimeter template. The exercise should be repeated as many times as necessary to achieve two consecutive shot groups that will fit into the same 2-centimeter circle.

A-89. A simulated shot group covered with a 1-centimeter (diameter) circle indicates consistent aiming. Since no rifle or ammunition variability is involved and since there is no requirement to place the shot group in a certain location, a 1-centimeter standard may be compared to obtaining a 4-centimeter shot group on the 25-meter live-fire zero range.

NOTE: The silhouettes on the plastic paddle (Figure A-17) are scaled to represent an E-type silhouette target at 250 meters. The visual perception during the target box exercise is similar to what a Soldier sees while zeroing on a standard zeroing target. The small E-type silhouette is the same scale at 15 yards as the larger silhouette is at the 25-meter range (some training areas are set up at 15 yards; others are set up at 25 meters). While there are some benefits to representing a 250-meter target, the main benefit of this exercise can be obtained at any distance. In place of the paddle, a standard zeroing target can be used at 25 meters by placing a small hole in the center (dot), moving the target sheet over the paper, and marking as previously outlined.

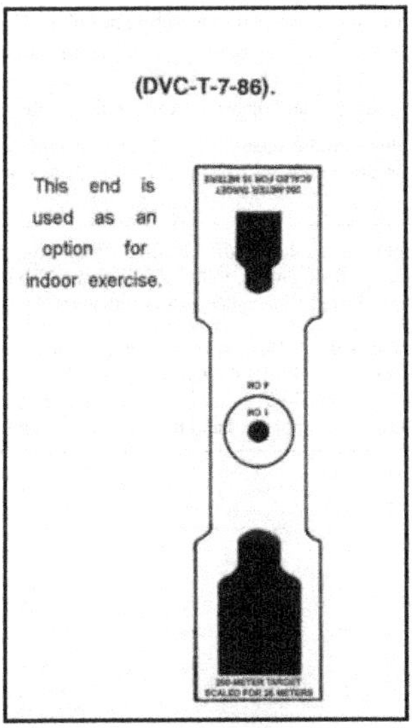

Figure A-17. Target box paddle (DVC-T-7-86).

A-90. The shot group exercise provides a chance for the trainer to critique the Soldier on his aiming procedures, aiming consistency, and placement of shot groups. Assuming the rifle and paper remain stationary and the target man properly marks the three shots, the only factor to cause separation of the dots on the paper is error in the Soldier's aiming procedure. When the Soldier can consistently direct the target into alignment with the sights on this exercise, he should be able to aim at the same point on the zero range or on targets at actual range.

Ball and Dummy Exercise

A-91. This exercise is conducted on a live-fire range and is used to detect if the firer is anticipating the shot or using improper trigger squeeze.

A-92. To perform this exercise—

(1) The coach or designated assistant inserts a dummy round into a magazine of live rounds. The firer must not know when a dummy round is in the magazine.

(2) When the hammer falls on a dummy round (which the firer thought was live), the firer and his coach may see movement. The firer anticipating the shot or using improper trigger squeeze causes this. Proper trigger squeeze results in no movement when the hammer falls.

A-93. The Soldier demonstrates the ability to properly utilize the fundamentals of marksmanship six consecutive times.

Dime/Washer Exercise

A-94. This dry-fire technique is used to teach or evaluate the skill of trigger squeeze and is effective when conducted from an unsupported position. To perform this exercise—

(1) The Soldier cocks the weapon.

(2) The Soldier assumes an unsupported prone firing position.

(3) The Soldier aims at the target. An assistant places a dime or washer on the rifle's barrel between the flash suppressor and front sightpost assembly.

(4) The Soldier tries to squeeze the trigger naturally without causing the dime or washer to fall off.

A-95. Several repetitions of this exercise must be conducted to determine if the Soldier has problems with trigger squeeze. The Soldier is a GO if he can dry-fire six of six consecutive shots without causing the dime or washer to fall.

NOTES: 1. If the dime or washer is allowed to touch the sight assembly or flash suppressor, it may fall off due to the jolt of the hammer. The strength of the hammer spring on some rifles can make this a difficult exercise to perform.

2. When using the M16A2 rifle, the dime/washer exercise is conducted the same, except that a locally fabricated device must be attached to the weapon. A piece of 3/4-inch bonding material is folded into a clothespin shape and inserted in the weapon's flash suppressor and the dime or washer is placed on top of it.

Appendix B
Scorecards

During live-fire events, a Soldier's hit-and-miss performance is recorded to facilitate the instructor/trainer's critiques or to indicate where more training is needed. The following are examples of completed scorecards.

EXAMPLES OF COMPLETED SCORECARDS

B-1. Figures B-1 through B-8 show examples of completed scorecards.

REPRODUCIBLE FORMS

B-2. Blank copies of the following can be found at the end of this publication:

- DA Form 3595-R (Record Fire Scorecard).
- DA Form 3601-R (Single Target—Field Fire I Scorecard).
- DA Form 5239-R (100-, 200-, and 300-Meter Downrange Feedback Scorecard).
- DA Form 5241-R (Single and Multiple Targets—Field Fire II Scorecard).
- DA Form 5789-R (Record Fire Scorecard—Known-Distance Course).
- DA Form 5790-R (Record Fire Scorecard—Scaled Target Alternate Course).
- DA Form 7489-R (Record Night Fire Scorecard).
- DA Form 7649-R (Squad Designated Marksman—Record Fire I and II).
- DA Form 7650-R (Squad Designated Marksman—Position Evaluation).

NOTE: These forms, scorecards, and position evaluation sheets are not available through the normal supply channels. You may reproduce them locally on 8 1/2- x 11-inch paper or download them from the Army Publishing Directorate at http://www.apd.army.mil/.

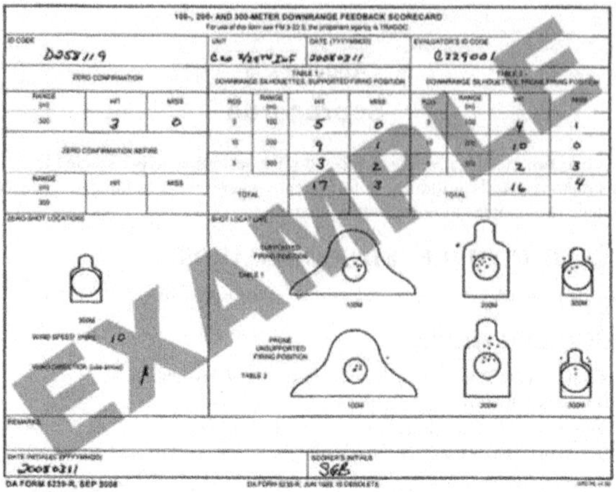

Figure B-1. Example of completed DA Form 5239-R (100-, 200-, and 300-Meter Downrange Feedback Scorecard).

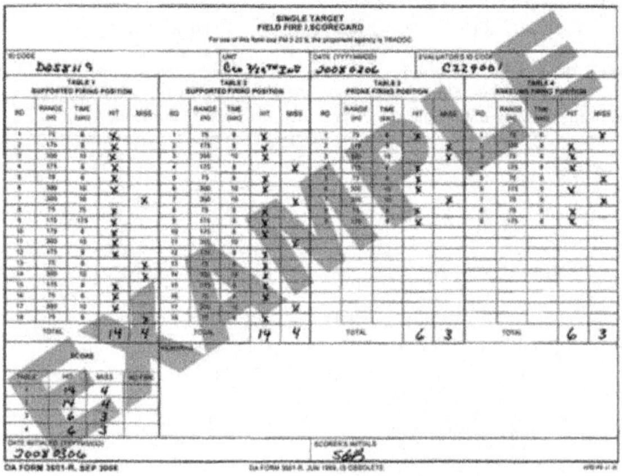

Figure B-2. Example of completed DA Form 3601-R (Single Target—Field Fire I Scorecard).

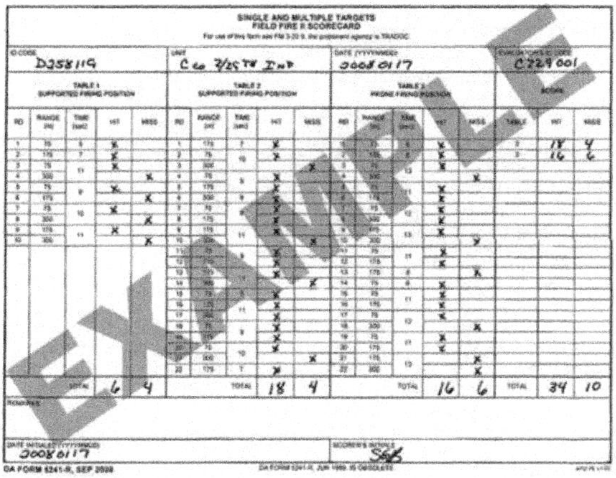

Figure B-3. Example of completed DA Form 5241-R
(Single and Multiple Targets—Field Fire II Scorecard).

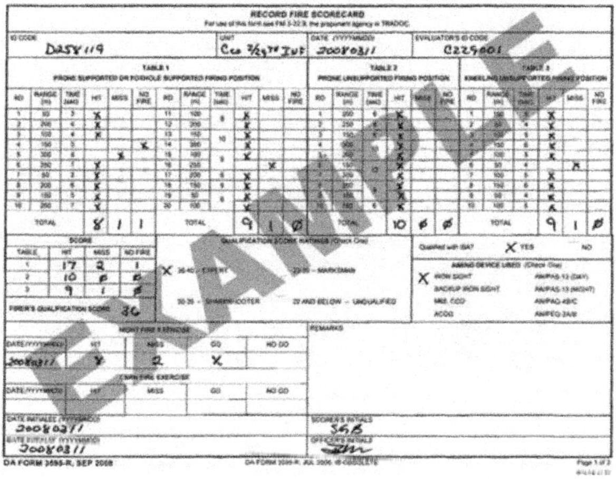

Figure B-4. Example of completed DA Form 3595-R
(Record Fire Scorecard).

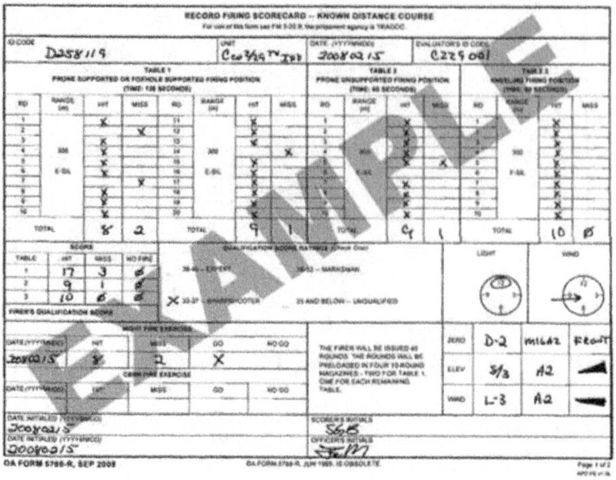

RECORD FIRING SCORECARD – KNOWN DISTANCE COURSE

ID CODE	UNIT	DATE (YYYYMMDD)	EVALUATOR'S ID CODE
D258119		20080215	C225001

TABLE	HIT	MISS	NO FIRE
1	17	3	Ø
2	9	1	Ø
3	10	Ø	Ø

DA FORM 5789-R, SEP 2008

Figure B-5. Example of DA Form 5789-R (Record Fire Scorecard—Known Distance Course).

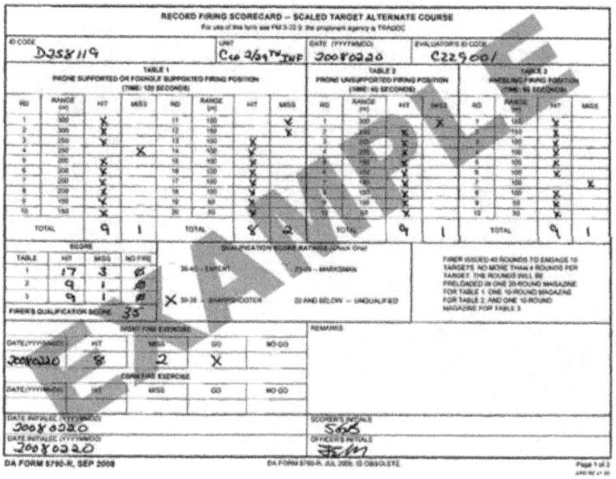

RECORD FIRING SCORECARD – SCALED TARGET ALTERNATE COURSE

ID CODE	UNIT	DATE (YYYYMMDD)	EVALUATOR'S ID CODE
D258119		20080220	C225001

TABLE	HIT	MISS	NO FIRE
1	17	3	Ø
2	9	1	Ø
3	9	1	Ø

FIRER'S QUALIFICATION SCORE 35

DA FORM 5790-R, SEP 2008

Figure B-6. Example of completed DA Form 5790-R (Record Fire Scorecard—Scaled Target Alternate Course).

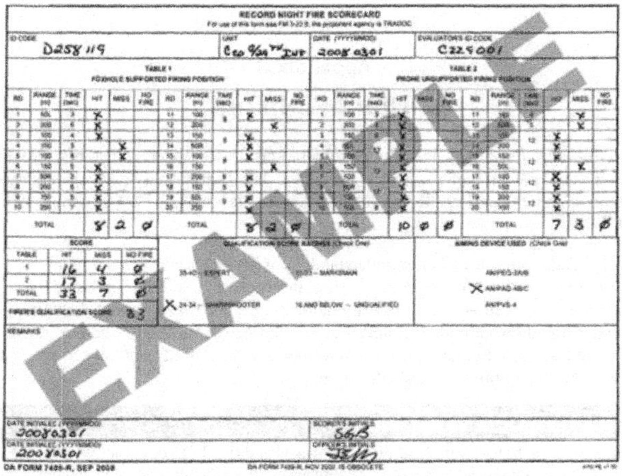

RECORD NIGHT FIRE SCORECARD

ID CODE: D258119 | UNIT: Co 4/39th Inf | DATE (YYYYMMDD): 20080301 | EVALUATOR'S ID CODE: C225001

TABLE 1 FOXHOLE SUPPORTED FIRING POSITION — TOTAL: HIT 8, MISS 2, NO FIRE Ø; TOTAL: HIT 8, MISS 2, NO FIRE Ø

TABLE 2 PRONE UNSUPPORTED FIRING POSITION — TOTAL: HIT 10, MISS Ø, NO FIRE Ø; TOTAL: HIT 7, MISS 3, NO FIRE Ø

SCORE			
TABLE	HIT	MISS	NO FIRE
1	16	4	Ø
2	17	3	Ø
TOTAL	33	7	Ø

FIRER'S QUALIFICATION SCORE: 33

X 24-34 – SHARPSHOOTER

AIMING DEVICE USED: X AN/PAQ-4B/C

DATE INITIALED (YYYYMMDD): 20080301

SCORER'S INITIALS: SGB

DA FORM 7489-R, SEP 2008

Figure B-7. Example of completed DA Form 7489-R (Record Night Fire Scorecard).

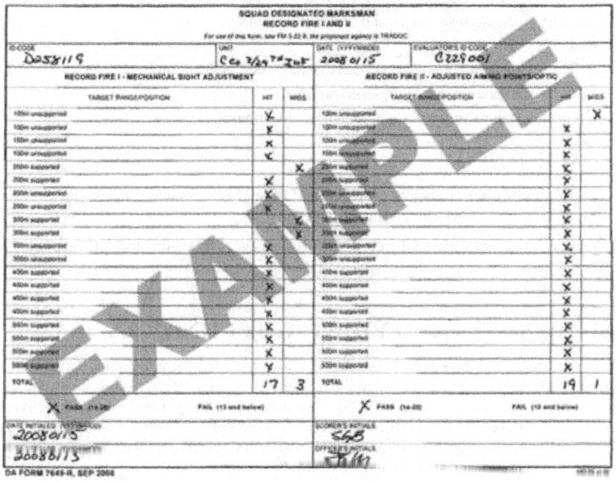

SQUAD DESIGNATED MARKSMAN RECORD FIRE I AND II

ID CODE: D258119 | UNIT: Co 2/39th Inf | DATE (YYYYMMDD): 20080115 | EVALUATOR'S ID CODE: C225001

RECORD FIRE I - MECHANICAL SIGHT ADJUSTMENT — TOTAL: HIT 17, MISS 3 — X PASS (14-20)

RECORD FIRE II - ADJUSTED AIMING POINTS/OPTIC — TOTAL: HIT 19, MISS 1 — X PASS (14-20)

DATE INITIALED (YYYYMMDD): 20080115

SCORER'S INITIALS: SGB

DA FORM 7649-R, SEP 2008

Figure B-8. Example of completed DA Form 7649-R (Squad Designated Marksman—Record Fire I and II).

Appendix C

Night Fighting

With the Army's emphasis on night operations, commanders must be sure that riflemen are lethal at night as well as during the day. That lethality depends largely on whether riflemen can fire effectively with today's technology: NVGs, aiming lights, and TWSs.

This appendix provides a better understanding of how eyes adapt to the night, as well as information about night devices. The information has been provided through continuous observation during operational testing, the Army Research Institute, and lessons learned by individual Soldiers across the Army. Be it with the naked eye or with night devices, Soldiers must learn to increase their ability not only to see better at night, but their ability to "own the night."

SECTION I. UNAIDED NIGHT VISION

Although operating at night has definite advantages, it is also difficult. Eyes do not work as well as they do during the day, yet they are crucial to effective performance; 80 percent of sensory input comes through them. Soldiers need to be aware of the constraints that their eyes place upon them at night. These constraints include—

- Reduced ability to see crisp and clear images (visual acuity).
- Inability to distinguish colors under certain conditions.
- Reduced depth perception.
- Difficulty in seeing objects at certain distances due to the night blind spot.
- Lost dark adaptation due to exposure to light.
- Confusion (your eyes may seem to play tricks on you).

NORMAL BLIND SPOTS

C-1. The normal blind spot is always present, day and night. It is caused by the lack of light receptors where the optic nerve inserts into the back of the eye.

C-2. The normal blind spot occurs when you use just one eye. When you close the other eye, objects 12 to 15 degrees away from where you are looking will disappear. When you uncover your eye, the objects will reappear.

NIGHT BLIND SPOTS AND VIEWING TECHNIQUES

C-3. When you stare at an object under starlight or lower levels of illumination, it can disappear or fade away. This is a result of the night blind spot. The night blind spot affects both eyes at the same time and occurs when using the central vision of both eyes. Consequently, when looking directly at an object, Soldiers miss larger objects as the distances increase. A hand grenade 2 meters away might not be seen; an enemy Soldier at 50 meters may be missed. An M1 tank at 300 meters can even be missed.

Avoiding Night Blind Spots

C-4. To avoid night blind spots—

- Look to all sides of objects you are trying to find or follow.
- **DO NOT STARE.**

C-5. Diamond viewing, a technique similar to the off-center vision technique taught in rifle marksmanship, is a good technique for peripheral viewing. To use the diamond viewing technique—

- Move your eyes just slightly, a few degrees, in a diamond pattern around the object you wish to see.
- Do not move your head; use your peripheral vision.

DARK ADAPTATION

C-6. In order for your visual system to work efficiently at night, you need to dark-adapt, or get your eyes used to seeing things under low light conditions. Dark adaptation is similar to walking into a movie theater when it's very dark; you can't see things at first, but as your eyes gradually adapt, you can see better.

C-7. It takes about 30 to 45 minutes to fully dark-adapt when going from a brightly lighted area into the dark. However, people dark-adapt at varying rates. People who are older, people who smoke, or people who are not in great physical shape will take longer to dark-adapt.

C-8. There are three stages of dark adaptation:

- Daylight vision.
- Twilight vision.
- Night vision.

Daylight Vision

C-9. Daylight vision occurs under maximum lighting conditions, such as when the sun is shining or in a well-lit room. Under these conditions, Soldiers have—

- Optimal visual acuity. In daylight conditions, Soldiers use both their central and peripheral vision, which provides optimal visual acuity—20/10, 20/15, and 20/20 vision.
- Optimal color vision. Colors look most vivid under daylight conditions.
- Quickest reaction time.

Twilight Vision

C-10. Twilight vision occurs during many military night operations and when driving around in a car at night. It occurs at dawn and dusk, down to full moonlight; when there is artificial illumination; and when snow is on the ground at night. It can occur in the daytime with several layers of jungle canopy. Under these conditions, Soldiers have—

- Poorer visual acuity. Visual acuity can be as poor as 20/100. Under twilight conditions, optimal visual acuity is between 20/50 and 20/100.
- Poorer color vision. Colors will not be as vivid.
- Slower reaction times.

Night Vision

C-11. Night vision occurs under starlight, as well as on moonless and cloudy nights when there are no stars or cultural lighting. Remember, there is a night blind spot, as discussed earlier. Under these conditions, Soldiers have—

- The worst visual acuity—from 20/200 to 20/400 and possibly much worse. Soldiers can recognize silhouettes, but not details of the objects. This is why knowing the silhouettes of vehicles and critical natural and man-made objects is important.
- Poor color vision. Soldiers cannot see colors, only various shades of gray. The longer wavelengths of light, such as the reds and oranges, are hard to see and appear dark. Unless a dark color is bordered by two lighter colors, it becomes totally invisible. On the other hand, greens and blues appear brighter, but Soldiers may not be able to determine their color.

NOTE: Reds are almost invisible at night. Red crosses are on white backgrounds on tents or vehicles so they can be seen more easily at night.

PROTECTING (BEFORE OPERATION)

C-12. Soldiers must protect their eyes before night operations so they can dark-adapt in an efficient manner. To efficiently dark-adapt, use the following suggestions:

- Don't smoke before night operations. Not smoking four to six hours before night operations will aid in dark adaptation.
- Wear sunglasses if you plan to spend time in the sun. Without sunglasses, it will take longer to dark-adapt.
- Watch what you eat. Maintain adequate levels of Vitamin A.
- Use dim white lighting or red lighting before night operations.

PROTECTING (DURING OPERATION)

C-13. Once a Soldier has dark-adapted, it is important to maintain that dark adaptation. To maintain dark adaptation, use the following suggestions:

- Minimize your use of unnecessary lighting.
- Close one eye before being flashed by flares and other bright lights.

ILLUSION (APPARENT MOVEMENT OF LIGHT)

C-14. The illusion of movement, which a static light exhibits when stared at in the dark, is related to the loss of surrounding visual references that normally serve to stabilize visual perceptions. Consequently, very small eye movements are perceived by the brain as movement of the light. To avoid illusions of movement—

- Begin a scan pattern, and control the eye movement.
- Use large movements, and scan.
- Try to find another light, and shift your gaze back and forth between the lights.

SECTION II. AIDED NIGHT VISION

There are three devices available to dismounted Soldiers that will help increase their lethality at night: night observation devices (NODs), aiming lasers, and TWSs. Each provides the dismounted Soldier with different views of the IR spectrum. Before Soldiers can fully operate these devices, they must receive training on how the systems work within the IR range and the electromagnetic (light) spectrum. Soldiers must also know what constraints and advantages each piece of equipment provides so that they can determine when to employ each device.

ELECTROMAGNETIC (LIGHT) SPECTRUM

C-15. To understand why some night devices cannot be used in conjunction with others, Soldiers must understand the electromagnetic spectrum. The electromagnetic spectrum is a spectrum of energy (light) containing x-rays, gamma rays, radio waves, cosmic rays, and ultraviolet rays. Also within this spectrum is visible light, light visible with the naked eye. Just beyond red visible light is IR light. IR light is broken down into three different ranges: near IR, middle IR, and far IR.

C-16. Two types of night devices will increase the Soldier's vision into the IR range:

- Image-intensification devices.
- Thermal sight.

IMAGE-INTENSIFICATION DEVICES

C-17. Image-intensification devices rely on ambient light and energy within the near IR range, such as light emitted from natural and artificial sources (for example, moonlight or starlight). Image-intensification devices include the AN/PVS-4, PVS-5, PVS-7A/B/C/D, and PVS-14s. The Army also has aiming devices that emit near IR energy in a colliminated beam, such as the AN/PAQ-4B/C and the AN/PEQ-2A/B. Since image-intensification devices and aiming lasers work within the same range of near IR energy, they can be used in conjunction with each other.

THERMAL WEAPON SIGHT

C-18. In the past, thermal technology has been solely reserved for tanks, fighting vehicles, and antiarmor specialists (for example, TOW and Dragon gunners). These devices were bulky, heavy, and impractical for the dismounted Soldier. Now, the Army has a thermal device that can be mounted on a dismounted Soldier's weapon or handheld. The TWS operates within the middle/far IR range. It can detect IR light emitted from friction, combustion, or objects radiating natural thermal energy. Since the TWS and other thermal devices operate within the middle/far IR range, they cannot be used in conjunction with image intensifiers or other image-intensification devices.

IMAGE-INTENSIFICATION DEVICES

C-19. As the name implies, image-intensification devices are designed to amplify light. To be effective, some degree of light must be available. When light enters the image-intensification tube, the light releases electrons, which the tube accelerates until the light is much brighter. Under optimal conditions, second-generation devices, such as the PVS-5-series, intensify ambient light up to about 1,500 times. Third-generation devices, such as the PVS-7/14-series NODs, double that level of intensification.

ADJUSTMENTS

C-20. To acquire and engage a target at night, image-intensification devices must be adjusted properly.

NOTE: Soldiers will not be able to obtain the acuity level that they have during the day. Under optimal night conditions, a Soldier with 20/20 daytime vision can expect no better than 20/50 vision with second-generation NODs and 20/40 vision with third-generation NODs. To approach these levels of acuity, you must be able to adjust your NOD for optimal clarity.

C-21. To adjust the NOD—

(1) Mount the NOD. Mount the head mount or helmet mount IAW the appropriate TM. If using the helmet mount, ensure that the tilt is adjusted until you have a comfortable viewing angle. Use the nape strap to maintain proper acuity with the NODs.

NOTE: If the mounting bracket is permanently attached to the helmet, ensure that the nape strap rear bracket is also permanently attached. (See TM 11-5855-306-10 for more information about attaching the nape strap rear bracket.) Use of the nape strap will prevent the NOD's weight from pulling the helmet downward, causing the NOD to rest on the bridge of your nose. The nape strap allows for proper acuity of the sight and allows you to engage targets with more ease and accuracy.

(2) Set the eye-relief. Move the goggles so that the eyecups cover the eye, but not so close that the eyepiece touches your eyelashes or glasses.

(3) Turn the goggles on.

(4) Set the inner-pupillary distance (AN/PVS-7 series). Move each eyepiece until they are centered over each eye. Close one eye, and make adjustments until the eye that is open is viewing a complete circle and not an oval. Continue to make adjustments to the other eye.

(5) Adjust the diopter focus ring. Close one eye, and with the eye that is open, turn the diopter focus ring in one direction until the diopter is totally out of focus. Then, turn the diopter focus ring the opposite direction until the display is focused to your eye. Follow the same procedures for the other eye if using the AN/PVS-7 series. No further adjustments should have to be made to the diopter focus ring.

NOTE: Adjust the diopter focus ring before adjusting the objective focus ring. The diopter focus ring focuses the display lens to your eye, while the objective focus ring focuses the target. You cannot focus the sight to the target without your eye being focused to the display first.

(6) Adjust the objective focus ring. While looking at an object, turn the objective focus ring until the objective lens is out of focus and then slowly turn the objective focus ring in the opposite direction until the object becomes as clear as possible.

NOTE: Adjustments will have to be made for targets at different ranges using the objective focus ring.

(7) Adjust the variable gain control (AN/PVS-14 only). Turn the variable gain control to the point that both eyes are almost receiving the same amount of light.

NOTE: The AN/PVS-14 has a variable gain control that controls the illumination input to the eye. Keeping the variable gain turned up will cause your brain to form two separate images, one darker and one very bright. With the variable gain turned down to the point that both eyes are almost receiving the same amount of light, the brain will produce one image, making it seem like both eyes are looking through the same sight.

AIMING LASERS

C-22. AN/PAQ-4-series and the AN/PEQ-2A/B aiming lasers operate within the electromagnetic spectrum, specifically near the IR range, and are seen through image-intensification devices. Aiming lasers emit a highly colliminated beam of IR energy that allows for quick "point and shoot" capability at night. Even though aiming lasers provide a quick and easy means of engaging the enemy at night, special attention must be given to the following:

- Proper adjustments to the image intensifiers.
- 10-meter boresighting procedures or 25-meter zeroing procedures.
- Scanning.
- Walking.
- IR discipline.

Scanning

C-23. NVDs have a 40-degree field of view, leaving the average firer to miss easy targets of opportunity, more commonly targets 50 meters left or right.

C-24. Soldiers must train to aggressively scan their sectors of fire for targets. Training must reinforce regular blinking during scanning, which relieves some of the eyestrain from spotting far targets. After Soldiers master the art of scanning, they find that targets are easier to detect by acknowledging the flicker or movement of a target.

Walking

C-25. Once a Soldier has located a target, he must be aware of the placement of the aiming laser. If he activates his laser while it is pointing over the target and into the sky, he wastes valuable time trying to locate exactly where the laser is pointing and increases his chances of being detected and fired upon by the enemy. Walking the laser to the target is a quick and operationally secure means of engaging the enemy with the aiming laser. To walk the laser to a target—

(1) Aim the laser at the ground just in front of the target.
(2) Walk the aiming laser along the ground and up to the target's center of mass.
(3) Engage the target.

Infrared Discipline

C-26. To exercise proper IR discipline—

- While on the range, actively scan for targets with the laser off.
- Once a target is located, walk the laser to the target, and engage.
- Once a target has been located and engaged with the aiming laser, deactivate the laser.

THERMAL WEAPON SIGHT

C-27. To understand how the TWS is able to convert energy into an image suitable for viewing, Soldiers must understand the electromagnetic spectrum and the range of IR light in which the TWS operates. The TWS absorbs all available light into the lens, and then filters out all light except for middle/far IR (thermal) light. Then, the TWS converts the thermal light into an image and creates a video that is displayed on the raster.

C-28. The TWS can convert thermal energy that is reflected, radiated, or generated from an object. All objects, such as trees, metal, plastic, and living creatures, display a quality that allows them to be seen with thermal technology. How well the objects display these qualities will determine how well they are seen.

ABSORPTION

C-29. During the day, all inanimate objects absorb thermal energy from the sun to varying degrees. Metal objects have a much higher rate of absorption than wood, leaves, or grass; therefore, a metal object sitting in the sun will stand out more than the grass surrounding it when viewed through the TWS.

EXPOSURE

C-30. The amount of time an object is exposed to thermal energy determines how well that object will be seen. Naturally, an object with a long exposure time will have absorbed more thermal energy than an object exposed to the same thermal energy for a shorter period of time.

EMISSIVITY

C-31. Emissivity is the rate at which an object emits the thermal energy it has absorbed or generates. Usually, objects with a high absorption rate will have a high emissivity factor. Although the human body does not have a high absorption rate, it has a high emissivity factor due to the fact that it generates a high amount of thermal energy. An object that has a high emissivity factor will be much hotter, and, therefore, when seen through the TWS, much easier to see and recognize.

REFLECTION

C-32. Items such as glass and water have virtually no absorption rate. Instead, they reflect thermal energy, which makes it very difficult to see objects through glass and water.

C-33. Snow and ice have the same effect, especially during the day with no clouds present. Snow and ice reflect most of the thermal energy from the sun, so it is difficult to acquire a good thermal image on objects that are close to the ground.

DIURNAL CYCLE

C-34. There are two times during the day when motionless objects that do not generate their own thermal energy, such as trees, rocks, and man-made objects, become the same temperature as the surrounding air: once in the morning and once in the evening. This is known as the diurnal cycle. The specific times that this cycle will take effect are based on the time of year, but it usually occurs shortly after sunrise and shortly after sunset.

C-35. The diurnal cycle contains crossover points. During the day, a motionless object will absorb thermal energy from the sun; the crossover point is the time when that object stops absorbing thermal energy (day) and starts radiating thermal energy (night). As the night goes on, that same object will come to a point where it stops radiating thermal energy and will once again start absorbing thermal energy (day).

ADJUSTMENTS

C-36. Rain, snow, fog, smoke, and the diurnal cycle are just a few environmental or combat situations that may affect your thermal image. These conditions may cause objects to be difficult to see, so adjustments must be made to the TWS in order to refine the thermal image.

C-37. To allow Soldiers to maximize the capability of the sight, the TWS is equipped with—

- A diopter focus ring.
- Two FOVs.
- An objective focus ring.
- A brightness knob.
- An auto and manual contrast switch.
- A polarity switch.

Diopter Focus Ring

C-38. When making adjustments to the sight, Soldiers begin with the diopter focus ring. The diopter focus ring focuses the display screen (raster) to the eye. To adjust the diopter focus ring, adjust the diopter focus ring until everything on the display screen is clear and easily read.

NOTE: This is best done with the objective lens cover closed.

C-39. Once you have adjusted the diopter focus to your eye, no other adjustments to the diopter focus ring should be necessary.

Fields of View

C-40. The TWS has two operating FOVs—wide and narrow. The Soldier should be allowed to select the FOV that suits him best. Through use, Soldiers learn to use the appropriate FOV under different combat situations.

NOTE: When selecting an FOV, make sure that the FOV ring is turned completely to the left or to the right. If the FOV ring is turned only halfway, you will not be able to see through the sight.

Wide Field of View

C-41. The WFOV has the least magnification, but a greater FOV and is great for scanning.

Narrow Field of View

C-42. The NFOV has greater magnification, but less degrees of FOV.

Objective Focus Ring

C-43. The objective focus ring focuses the sight to the target. Adjustments to the objective focus ring are based on the range of the object being viewed.

NOTES:
1. Over-adjustment to the objective focus ring will lock the FOV ring to the point that the FOV cannot be changed.
2. Make adjustments to the objective focus ring only after focusing the diopter focus ring.

Brightness Knob

C-44. The dual-function brightness knob is used to turn on the TWS, adjust the brightness of the raster, and refine the thermal image. Used in conjunction with the contrast knob, it helps to combat the effects of the diurnal cycle and other conditions that might require fine-tuned adjustment to the thermal image.

Contrast Switch

C-45. The dual-function contrast switch has automatic and manual contrast modes. Used in conjunction with the brightness knob, the contrast switch allows Soldiers to obtain the best possible thermal image.

Automatic Contrast Mode

C-46. The automatic contrast mode is used under normal operating conditions.

Manual Contrast Mode

C-47. The manual contrast is used under conditions other than normal (for example, during 10-meter boresighting or 25-meter zeroing; during rain, fog, smoke, or snow; during the diurnal cycle; or when trying to obtain as much detail of a target as possible).

Polarity Switch

C-48. The polarity switch has two modes: white hot or black hot. Use of the polarity switch is a user's preference. Through continued use, Soldiers decide which polarity setting works best under different combat or environmental conditions.

White Hot Mode

C-49. When in the white hot mode, hotter objects appear white, while cooler objects have shades of gray to black.

Black Hot Mode

C-50. When using the black hot mode, hotter objects appear black, while the cooler objects are shades of gray to white.

Appendix D

Range Safety and Risk Management

All personnel training on a rifle range should be briefed on the safety and local requirements for that range. This briefing fulfills the minimum requirements for a rifle range safety briefing. Information may be added to conform to local requirements and safety regulations. ARs 385-10 and 385-63 and DA Pam 385-63 should be reviewed by all range personnel before operating any range.

RECOMMENDED BRIEFING

D-1. The first priority on any range is training, but safety must be at the forefront of the training program. The safety program prescribes the safety precautions necessary to minimize accidents when firing and using ammunition during training and range operations. The safety program should identify—

- Surface danger zones (SDZs), as described in AR 385-63.
- The location of medical personnel.
- Left and right limits of the range. Firers never fire outside of these limits.
- The designated smoking area (if applicable).

D-2. Further, the safety program should include the following information:

- Inspect for objects located near the weapon's muzzle before firing, especially during unassisted night fire.
- When not on the firing line, ensure that the weapon's selector lever is on SAFE and the bolt is locked to the rear.
- Ensure that firers enter and exit the firing line at the entry or exit point.
- Before occupying a firing position, inspect it for wildlife or obstructions.
- Always keep the weapon's muzzle pointed downrange when on the firing line, and keep the finger outside of the trigger housing area.
- Never touch a weapon while personnel are downrange or in front of the firing line.
- Load the weapon only on command from the tower or control point.
- Never fire without using hearing protection.
- Ensure that left-handed firers attach left-handed brass deflectors to their weapon before firing.
- When entering or exiting the firing line, clear the weapon with a cleaning rod.
- Consider the rifle loaded at all times, even in break areas. Never point the weapon at anyone.
- Anyone observing an unsafe act will immediately call "CEASE FIRE," place the weapon on SAFE, place it in the V-notched stake or lay it on the sandbags, and give the verbal and visual command of cease fire.
- Once cleared off of the firing line, firers report to the ammunition point and turn in all brass and ammunition.
- No one will leave the range until they have been inspected for live ammunition and brass.
- In case of an electrical storm, personnel will be directed to lock and clear all weapons, ground their equipment (except wet weather gear), and disperse into a predetermined area.
- Eating and drinking are not permitted on the firing line unless the tower operator permits drinking from the canteen.

NOTE: Drink water often to prevent heat injuries.

PERSONNEL AND DUTIES

D-3. To provide a safe and efficient range operation and effective instruction, certain duties may be required of personnel. The personnel may include—

- OIC.
- RSO.
- NCOIC.
- Ammunition detail.
- Unit armorer.
- Assistant instructor.
- Medical personnel.
- Control tower operators.
- Maintenance detail.

OFFICER IN CHARGE

D-4. The OIC is responsible for the overall operation of the range before, during, and after live-firing.

RANGE SAFETY OFFICER

D-5. The RSO—

- Is responsible for the safe operation of the range.
- Conducts a safety orientation before each scheduled LFX.
- Ensures that a brass and ammunition check is made before the unit leaves the range.
- Ensures that all personnel comply with the safety regulations and procedures prescribed for the conduct of an LFX.
- Ensures that all left-handed firers use left-handed firing devices.

NOTE: This officer should not be assigned duties other than those of the safety officer.

NONCOMMISSIONED OFFICER IN CHARGE

D-6. The NCOIC assists the OIC and safety officer by performing duties as required; for example, he might supervise enlisted personnel who are supporting the LFX.

AMMUNITION DETAIL

D-7. This detail is composed of one or more ammunition handlers. The ammunition detail—

- Breaks down, issues, receives, accounts for, and safeguards live ammunition.
- Collects expended ammunition casings and other residue.

UNIT ARMORER

D-8. The unit armorer—

- Repairs rifles.
- Replaces parts.

ASSISTANT INSTRUCTOR

D-9. One assistant instructor is assigned for each one to ten firing points. Each assistant instructor—

- Ensures that all firers observe safety regulations and procedures.
- Assists firers having problems.

Medical Personnel

D-10. Medical personnel provide medical support, as required by regulations governing LFXs.

Control Tower Operators

D-11. Control tower operators—

- Raise and lower the targets.
- Time the exposures.
- Sound the audible signal.
- Give the fire commands.

NOTE: If possible, two men should be chosen to perform these functions.

Maintenance Detail

D-12. This detail should be composed of two segments: one to conduct small arms repair and one to perform minor maintenance on the target-holding mechanisms.

AMMUNITION POSITIONING AND ISSUANCE

D-13. To provide a safe and operational range, the following are recommended procedures for handling ammunition:

- Locate all ammunition for the weapons involved at firing sites outside of the backblast area (when applicable). Store ammunition at a position that minimizes the potential for ignition, explosion, or rapid burning.
- Issue ammunition to firing units immediately before scheduled training exercises. Distribute small arms ammunition to troops only when they are on the ready line or firing line.
- Cover all ammunition to protect it from the elements and direct rays of the sun. For proper ventilation, provide air circulation between the ammunition and cover.
- Limit the unpacking of ammunition at the firing line to the minimum number of rounds needed. Retain packaging material until firing is complete. Units do not burn wooden containers or indiscriminately fire ammunition to preclude return to a storage site.

COMPOSITE RISK MANAGEMENT

D-14. Composite risk management (CRM) is a decision-making process used to mitigate risks associated with all hazards that have the potential to injure or kill personnel, damage or destroy equipment, or otherwise impact mission effectiveness. The guiding principles of CRM are as follows:

- Integrate CRM into all phases of the mission and operations.
- Make risk decisions at the appropriate level.
- Accept no unnecessary risk.
- Apply the process cyclically and continuously.
- Do not be risk averse.

D-15. CRM is a five-step process:

(1) Identify hazards.
(2) Assess hazards to determine risk.
(3) Develop controls, and make risk decisions.
(4) Implement controls.
(5) Supervise and evaluate.

NOTE: Steps 1 and 2 are assessment steps; Steps 3 through 5 are management steps.

STEP 1–IDENTIFY HAZARDS

D-16. Hazards may arise from any number of areas. They can be associated with enemy activity, accident potential, weather or environmental conditions, health, sanitation, behavior, and/or materiel or equipment. Mission, enemy, terrain and weather, troops and support available, time available, civil considerations (METT-TC) factors serve as a standard format for identifying hazards, on- or off-duty.

D-17. CRM does not differentiate between the sources of the hazard. The loss of personnel, equipment, or materiel due to any hazard has the same disruptive impact on readiness or mission capabilities, no matter what the source.

STEP 2–ASSESS HAZARDS TO DETERMINE RISK

D-18. During mission analysis, course of action (COA) development, or analysis, rehearsal, and execution steps of the military decision-making process (MDMP), commanders assess hazards and assign risk in terms of probability and severity of adverse impact. During their assessment, they must consider both mission- and non-mission-related aspects that may have an impact. The end result of this assessment is an initial estimate of risk for each identified hazard as determined from the standardized application of the risk assessment matrix.

D-19. There are three substeps in this step:

(1) Assess the probability of an event or occurrence.

(2) Estimate the expected result or severity of an event or occurrence.

(3) Determine the level of risk using the standard risk assessment matrix.

Assess the Probability of an Event or Occurrence

D-20. Probability is the likelihood an event will occur based on prior experience. The probability levels estimated for each hazard are based on the mission, COA, or frequency of a similar event. There are five levels of probability (Table D-1):

- Frequent.
- Likely.
- Occasional.
- Seldom.
- Unlikely.

Table D-1. Five levels of probability.

LEVEL OF PROBABILITY	EXPLANATION
Frequent	Occurs very often, known to happen regularly.
Likely	Occurs several times, a common occurrence.
Occasional	Occurs sporadically, but is not common.
Seldom	Remotely possible, could occur at some time.
Unlikely	Can assume will not occur, but not impossible.

Estimate the Expected Result or Severity of an Event or Occurrence

D-21. Severity is the degree to which an incident will impact combat power, mission capability, or readiness. The degree of severity estimated for each hazard is based on the results of similar events. Severity is addressed in the four levels used on the risk assessment worksheet (Table D-2):

- Catastrophic.
- Critical.
- Marginal.
- Negligible.

Table D-2. Four levels of severity.

LEVEL OF SEVERITY	EXPLANATION
Catastrophic	• Complete mission failure or loss of ability to accomplish a mission. • Death or permanent total disability. • Loss of major or mission-critical systems or equipment. • Major property or facility damage. • Severe environmental damage. • Mission-critical security failure. • Unacceptable collateral damage.
Critical	• Severely degraded mission capability or unit readiness. • Permanent partial disability or temporary total disability exceeding three months time. • Extensive major damage to equipment or systems. • Significant damage to property or the environment. • Security failure. • Significant collateral damage.
Marginal	• Degraded mission capability or unit readiness. • Minor damage to equipment or systems, property, or the environment. • Lost days due to injury or illness, not exceeding three months. • Minor damage to property or the environment.
Negligible	• Little or no adverse impact on mission capability. • First aid or minor medical treatment. • Slight equipment or system damage, but fully functional or serviceable. • Little or no property or environmental damage.

Determine the Level of Risk Using the Standard Risk Assessment Matrix

D-22. Using the standard risk assessment matrix (Table D-3), commanders convert probability and severity for each identified hazard into a specified level of risk. All accepted residual risk must be approved at the appropriate level of command.

> **NOTE:** This assessment is an estimate, not an absolute. It may or may not be indicative of the relative danger of a given operation, activity, or event.

Table D-3. Risk assessment matrix.

RISK ASSESSMENT MATRIX						
Severity		Probability				
		Frequent A	Likely B	Occasional C	Seldom D	Unlikely E
Catastrophic	I	E	E	H	H	M
Critical	II	E	H	H	M	L
Marginal	III	H	M	M	L	L
Negligible	IV	M	L	L	L	L
E – Extremely High		H – High		M – Moderate	L – Low	

D-23. Risk is addressed in the four levels listed in the lower left corner of the matrix (Table D-4):

- Extremely high risk.
- High risk.
- Moderate risk.
- Low risk.

Table D-4. Four levels of risk.

LEVEL OF RISK	EXPLANATION
Extremely High Risk	Loss of ability to accomplish the mission if hazards occur. In the example noted in Table D-3, a frequent or likely probability of catastrophic loss (IA or IB) or a frequent probability of critical loss (IIA) exists. This implies that the risk associated with this mission, activity, or event may have severe consequences. The decision to continue must be weighed carefully against the potential gain to be achieved by continuing this COA.
High Risk	Significant degradation of mission capabilities (in terms of the required mission standard), inability to accomplish all parts of the mission, or inability to complete the mission to standard will result if hazards occur during the mission. Occasional to seldom probability of catastrophic loss (IC or ID) exists. A likely to occasional probability of a critical loss (IIB or IIC) exists. Frequent probability of marginal losses (IIIA) exists. This implies that if a hazardous event occurs, serious consequences will occur. The decision to continue must be weighed carefully against the potential gain to be achieved by continuing this COA.
Moderate Risk	Expected degraded mission capabilities (in terms of the required mission standard) will result if hazards occur during the mission. An unlikely probability of catastrophic loss (IE) exists. The probability of a critical loss is seldom (IID). Marginal losses occur with a likely or occasional probability (IIIB or IIIC). A frequent probability of negligible (IVA) losses exists.
Low Risk	Expected losses have little or no impact on accomplishing the mission. The probability of critical loss is unlikely (IIE), while that of marginal loss is seldom (IIID) or unlikely (IIIE). The probability of a negligible loss is likely or less (IVB through (IVE). Expected losses have little or no impact on accomplishing the mission. Injury, damage, or illness are not expected, or may be minor and have no long-term impact or effect.

STEP 3 – DEVELOP CONTROLS, AND MAKE RISK DECISIONS

D-24. In this step, commanders develop and apply controls, reassess the hazard to determine a residual risk, and make risk decisions. This process continues until an acceptable level of risk is achieved or until all risks are reduced to a level where benefits outweigh the potential cost. This step is accomplished during the COA development, COA analysis, COA comparison, and COA approval of the MDMP.

STEP 4 – IMPLEMENT CONTROLS

D-25. Leaders and staffs ensure that controls are integrated into SOPs, written and verbal orders, mission briefings, and staff estimates. The critical check for this step is to ensure that controls are converted into clear and simple execution orders. This step includes coordination and communication with—

- Appropriate superior, adjacent, and subordinate units, organizations, and individuals.
- Logistics Civil Augmentation Program (LOGCAP) organizations and civilian agencies that are part of the force or may be impacted by the activity, hazard, or its control.
- The media and nongovernmental organizations (NGO) when their presence impacts or is impacted by the force.

STEP 5 – SUPERVISE AND EVALUATE

D-26. This step involves implementing risk controls and enforcing them to standard, and validating the adequacy of selected control measures in supporting the objectives and desired outcomes. This continuous process provides the ability to identify weaknesses and to change or adjust controls based on performance, changing situations, conditions, or events.

RESPONSIBILITIES

D-27. CRM responsibilities are spread across three levels:

- Commander.
- Leaders.
- Individual.

COMMANDER

D-28. During implementation of the CRM process, the commander—

- Ensures that warfighting functions (WFF) are performed to standard to minimize human error, materiel failure, and environmental effects.
- Establishes a force protection policy and publishes a safety philosophy with realistic safety goals, objectives, and priorities.
- Ensures that his training assessment considers the WFF's ability to protect the force. Selects long-, short-, and near-term control actions and ensures implementation to improve force protection.
- Ensures that his staff integrates risk management into the planning and execution of training and operational missions.
- Makes risk decisions. Selects, monitors, and enforces implementation of controls for hazards most likely to result in loss of combat power. After implementing controls, if risk remains above the tolerance level established by higher command, he must elevate the risk decision to the appropriate command level.
- Ensures that the CRM process is evaluated during all AARs.
- Determines if unit performance meets force protection guidance. Determines effectiveness of hazard controls and necessary changes to guidance and controls. Ensures that these changes are fed back into the training management cycle and guidance for operational missions, including unit SOPs.

LEADERS

D-29. During implementation of the CRM process, leaders—

- Enforce METL task performance to standard. Adopt the crawl—walk—run approach in planning and executing training.
- Make use of automated on- and off-duty CRM tools and surveys available from the United States Army Combat Readiness Center (USACRC).
- Execute risk reduction controls selected by the commander by developing and implementing supporting leader level controls. Apply risk management procedures to each.

INDIVIDUAL

D-30. All Soldiers must understand how to use the CRM process to enhance mission success and to reduce or eliminate loss.

D-31. During implementation of the CRM process, Soldiers—

- Support commanders and leaders in the rapid identification and communication of hazards and associated risks that may impact on the mission.
- Provide immediate feedback to the leader as the mission progresses and hazards are encountered. Use short written messages, hand and arm signals, or radio transmissions to communicate first-hand information to leaders.
- In extreme situations, act alone or make risk decisions within the context of orders.

DA FORM 7566 (COMPOSITE RISK MANAGEMENT WORKSHEET)

D-32. DA Form 7566 (Composite Risk Management Worksheet, Figure D-1a and D-1b) provides a starting point to logically track the CRM process. It can be used to document risk management steps taken during planning, preparation, and execution of training and combat missions.

NOTE: For detailed CRM procedures, see FM 5-19.

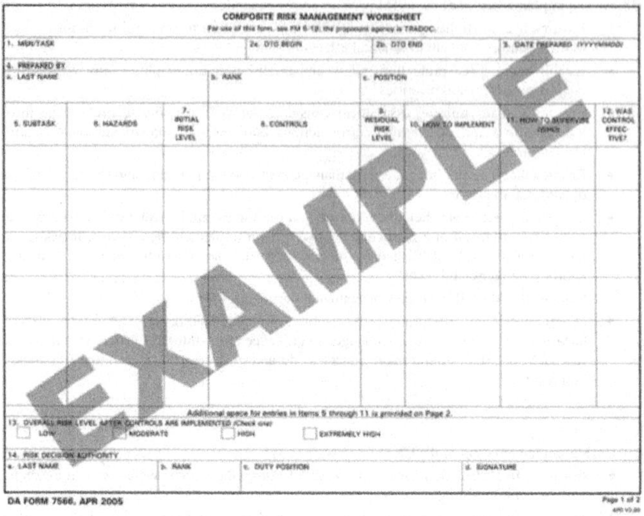

COMPOSITE RISK MANAGEMENT WORKSHEET

For use of this form, see FM 5-19; the proponent agency is TRADOC.

1. MSN/TASK	2a. DTG BEGIN	2b. DTG END	3. DATE PREPARED (YYYYMMDD)

4. PREPARED BY

a. LAST NAME	b. RANK	c. POSITION

5. SUBTASK	6. HAZARDS	7. INITIAL RISK LEVEL	8. CONTROLS	9. RESIDUAL RISK LEVEL	10. HOW TO IMPLEMENT	11. HOW TO SUPERVISE (WHO)	12. WAS CONTROL EFFEC-TIVE?

Additional space for entries in Items 5 through 11 is provided on Page 2.

13. OVERALL RISK LEVEL AFTER CONTROLS ARE IMPLEMENTED (Check one)

☐ LOW ☐ MODERATE ☐ HIGH ☐ EXTREMELY HIGH

14. RISK DECISION AUTHORITY

a. LAST NAME	b. RANK	c. DUTY POSITION	d. SIGNATURE

DA FORM 7566, APR 2005 Page 1 of 2

EXAMPLE

Figure D-1a. Sample DA Form 7566 (Composite Risk Management Worksheet).

ITEMS 5 THROUGH 12 CONTINUED:

5. SUBTASK	6. HAZARDS	7. INITIAL RISK LEVEL	8. CONTROLS	9. RESIDUAL RISK LEVEL	10. HOW TO IMPLEMENT	11. HOW TO SUPERVISE (WHO)	12. WAS CONTROL EFFECTIVE?

EXAMPLE

DA FORM 7566, APR 2005 — Page 2 of 2

Figure D-1b. Sample DA Form 7566 (Composite Risk Management Worksheet) (page 2).

D-33. See Table D-5 for instructions on completing DA Form 7566.

Table D-5. Worksheet instructions

ITEM	INSTRUCTION
1 through 4	Self-explanatory.
5	Subtask relating to the mission or task in Block 1.
6	Hazards–Identify hazards by reviewing METT-TC factors for the mission or task. Additional factors include historical lessons learned, experience, judgment, equipment characteristics and warnings, and environmental considerations.
7	Initial Risk Level–Factors include historical lessons learned, intuitive analyses, experience, judgment, equipment characteristics and warnings, and environmental considerations. Determine the initial risk for each hazard by applying the risk assessment matrix. Enter the risk level for each hazard.
8	Controls– For each hazard, develop one or more controls that will eliminate the hazard or reduce the risk of a hazardous incident. Specify who, what, where, why, when, and how for each control. Enter controls.
9	Residual Risk Level–Determine the residual risk for each hazard by applying the risk assessment matrix. Enter the residual risk level for each hazard.
10	How to Implement–Decide how each control will be put into effect or communicated to the personnel who will make it happen (written or verbal instruction; tactical, safety, garrison SOPs; rehearsals). Enter controls.
11	How to Supervise (Who)–Plan how each control will be monitored for implementation (continuous supervision, spot-checks) and reassess hazards as the situation changes. Determine if the controls worked and if they can be improved. Pass on lessons learned.
12	Was Control Effective–Indicate "Yes" or "No." Review during AAR.
13	Overall Risk Level–Select the highest residual risk level and circle it. This becomes the overall mission or task risk level. The commander decides whether the controls are sufficient to accept the level of residual risk. If the risk is too great to continue the mission or task, the commander directs development of additional controls or modifies, changes, or rejects the COA.
14	Risk Decision Authority–Signed by the appropriate level of command.

Appendix E

Range Procedures and Range Operations Checklist

This appendix contains the procedures used to conduct live-fire training exercises. These procedures support Army regulations (TCs 7-9 and 25-1), local range regulations, and established unit training SOPs. Conduct of the training exercise should clearly define and establish details and equipment needed to open and operate the range so that these actions do not adversely impact Soldiers' training time. The procedures outlined in this appendix should be followed in order to open the range and conduct effective training.

RANGE PROCEDURES

E-1. Before beginning an LFX, all personnel must receive an orientation on range operations. The orientation should outline the procedures for conducting the exercise, to include the duties of the nonfiring orders.

PERSONNEL AND DUTIES

D-2. To provide a safe and efficient range operation and effective instruction, certain duties may be required of personnel. The personnel may include—

- OIC.
- RSO.
- NCOIC.
- Ammunition detail.
- Unit armorer.
- Assistant instructor.
- Medical personnel.
- Control tower operators.
- Maintenance detail.

OFFICER IN CHARGE

D-3. The OIC is responsible for the overall operation of the range before, during, and after live-firing.

RANGE SAFETY OFFICER

D-4. The RSO—

- Is responsible for the safe operation of the range.
- Conducts a safety orientation before each scheduled LFX.
- Ensures that a brass and ammunition check is made before the unit leaves the range.
- Ensures that all personnel comply with the safety regulations and procedures prescribed for the conduct of an LFX.
- Ensures that all left-handed firers use left-handed firing devices.

NOTE: This officer should not be assigned duties other than those of the safety officer.

Noncommissioned Officer In Charge

D-5. The NCOIC assists the OIC and safety officer by performing duties as required; for example, he might supervise enlisted personnel who are supporting the LFX.

Ammunition Detail

D-6. This detail is composed of one or more ammunition handlers. The ammunition detail—

- Breaks down, issues, receives, accounts for, and safeguards live ammunition.
- Collects expended ammunition casings and other residue.

Unit Armorer

D-7. The unit armorer—

- Repairs rifles.
- Replaces parts.

Assistant Instructor

D-8. One assistant instructor is assigned for each one to ten firing points. Each assistant instructor—

- Ensures that all firers observe safety regulations and procedures.
- Assists firers having problems.

Medical Personnel

D-9. Medical personnel provide medical support, as required by regulations governing LFXs.

Control Tower Operators

D-10. Control tower operators—

- Raise and lower the targets.
- Time the exposures.
- Sound the audible signal.
- Give the fire commands.

NOTE: If possible, two men should be chosen to perform these functions.

Maintenance Detail

D-11. This detail should be composed of two segments: one to conduct small arms repair and one to perform minor maintenance on the target-holding mechanisms.

FIRING ORDER LINEUP

E-12. After the range cadre have given the safety and range briefings, they assemble the Soldiers in firing orders in correlation with the number of firing points on that range. After the firing order has been determined, firers have their weapons rodded, move to the firing line, and proceed to their assigned firing points, keeping their weapons pointed up and downrange at all times.

TOWER COMMANDS

E-13. Simple, standard fire commands are needed to avoid confusion and misunderstanding during LFXs.

GENERAL COMMANDS

E-14. The following general commands may be altered when necessary:

- "Firers, assume the ________ position."
- "Issue the firer ________ rounds of ammunition."
- "Coach, secure ________ rounds of ammunition."
- "Lock one round. Load."
- "Ready on the right?"
- "Ready on the left?"
- "Ready on the firing line?"
- "Commence firing when your targets appear."
- "Cease firing. Lock and clear all weapons."

Grouping Commands

E-15. Grouping commands include the following:

- "Firers, assume a good supported prone position."
- "Lock one of three single rounds. Load."
- "Ready on the right?"
- "Ready on the left?"
- "The firing line is ready."
- "Place your selector lever on SEMIAUTOMATIC."
- "Commence firing."
- "Cease fire. Lock and clear your weapons."
- "Clear on the right?"
- "Clear on the left?"
- "The firing line is clear."
- "Move down to your targets, and triangulate your shot group."
- "After all personnel have triangulated their targets, move back to the firing line."
- "At this time, make adjustments to your sights."
- "Repeat all firing commands until grouping standards are met."

ZERO COMMANDS

E-16. Tower commands are the same as grouping commands. Repeat all firing commands until zeroing standards are met.

Field Firing Exercises

E-17. Simple, standard fire commands are needed to avoid confusion during field firing exercises. Commands for exercises from stationary positions are as follows:

- "Firers, assume a good ________ position."
- "Lock one magazine of ________ rounds. Load."
- "Ready on the right?"
- "Ready on the left?"
- "The firing line is ready."
- "Place your selector lever on SEMIAUTOMATIC."
- "Scan your sector."
- "Cease fire. Lock and clear your weapon." (Place the selector lever in the SAFE position.)

E-18. The range officer relays his commands to the pit NCOIC by radio or telephone so he can keep abreast of the conduct of fire. Before each firing exercise, the range officer informs the pit NCOIC of the next exercise and any special instructions for target operation (for example, "The next firing will be for zero. Mark targets after each three-round shot group," or for slow fire, "The next firing will be ten rounds. Slow fire. Mark targets after each shot.").

E-19. Radio telephone operators (RTOs) relay commands to the pit and pass on special instructions to target operators as requested by assistant instructors.

NOTES: 1. Radio telephone operators never identify a firer on a particular firing point.

2. The command "MARK TARGET NUMBER ________" indicates that the target has been fired upon but has not withdrawn for marking.

PRACTICE RECORD FIRE AND RECORD FIRE

E-20. Simple, standard fire commands are needed to avoid confusion and misunderstanding during practice record fire and record fire.

Practice Record Fire

E-21. Practice record fire commands include the following:

- "Firers, assume a good supported (prone unsupported) position."
- "Scorers, point out the limits of your lane."
- "Firers, lock your first magazine. Load."
- "Scan your sector."
- "Cease fire."
- "Lock and clear all weapons."

Record Fire

E-22. Record fire commands include the following:

- "Firers, assume a good supported (prone unsupported) position."
- "Scorers, point out the limits of your sector."
- "Firers, lock your first 20-round magazine. Load."
- "Scan your sector."
- "Cease fire."
- "Lock and clear all weapons."

Rapid Fire Exercises

E-23. The following commands are used for rapid fire exercises:

- "Lock and clear all weapons."
- "Clear on the right?"
- "Clear on the left?"
- "The firing line is clear."
- "Firers, assume the ________ position."
- "Assistants, secure two magazines of five rounds each." ("Issue the firer one magazine of five rounds.")
- "Lock one magazine. Load."
- "Ready on the right?"
- "Ready on the left?"
- "Ready on the firing line?"
- "Watch your targets." ("Firers, assume the appropriate firing position and commence firing when the targets are presented.")

E-24. When all of the targets are withdrawn, the range officer checks for slow firers or malfunctions and then allows them to fire.

E-25. The pit NCOIC organizes, orients, and provides safety for the pit detail.

E-26. The success of KD firing depends on efficient operation of the targets and the close coordination between the pit NCOIC and range officer. All operators must know the proper procedure for operating and marking the target.

Marking Targets for Zeroing and Slow Fire

E-27. Targets are marked quickly after each shot or group of shots without command. During slow fire, the firer has a time limit of one minute for each shot. Twenty seconds is considered the maximum time limit for marking. A marker (spotter) is placed in the hit regardless of its location on the target. Each time the target is marked, the marker is removed from the previous hit, and the hole is pasted.

NOTE: 3-inch markers are used for 100, 200, and 300 meters; 5-inch markers are used for 500 meters.

Using Disk Markers

E-28. The target markers are painted black on one side and white on the opposite side. They are available in three dimensions:

- 1 1/2 inches (NSN 6920-00789-0864).
- 3 inches (NSN 6920-00-713-8255).
- 5 inches (NSN 6920-00-713-8254).

E-29. The disk spindle (NSN 6920-00-7138257) may also be procured through the supply channels.

RANGE OPERATIONS CHECKLIST

E-30. This checklist consists of nine sections, each covering a topic relating to range operations:

- Mission analysis.
- Double check.
- Become an expert.
- Determine requirements.
- Determine available resources.
- Foolproof.
- Occupy the range and conduct training.
- Close the range.
- KD range.

NOTE: The checklist should be modified to include local policy changes to regulations or SOPs.

E-31. The person responsible for the training must answer the questions in each section. To answer these questions, he should use the following procedures:

- Ask each question in order.
- Record a "Yes" answer by checking the GO column.
- Record a "No" or "Don't know" answer by checking the NO GO column.

- Refer to the checklist to find the GO and NO GO columns.
- After all of the questions in a section are answered, analyze the NO GOs.
  - Contact the people who reported them and ask if they have corrected each problem. If so, change the answer to GO.
  - If any NO GO remains, analyze it and implement a countermeasure for the shortfall. Afterward, check to ensure that the countermeasures work.
  - Before range operations start, be sure that a workable countermeasure is implemented for each safety hazard presented by a NO GO answer.

Mission Analysis

E-32. Mission analysis includes identifying the following:

- Who will be firing on the range? ________
- Number of personnel________ Units________
- What weapons and course will be used? Weapons________ Course________
- Where will the training be conducted? Range________
- When is the range scheduled for operations? Date________ Opens________ Closes________

Double Check

	GO	NO GO	REMARKS
1. Has sufficient ammunition been requested for the number of personnel?			
2. Are the range facilities adequate for the type of training to be conducted?			
3. Has enough time been scheduled to complete the training?			
4. Have conflicts that surfaced been resolved?			

Become an Expert

	GO	NO GO	REMARKS
1. Review TMs and FMs on the weapons to be fired.			
2. Talk with the armorer and other personnel experienced with the weapons to be fired.			
3. Review AR 385-63.			
4. Visit range control and read the installation's range instructions.			
5. Reconnoiter the range (preferably while it is in use).			
6. Check ARTEPs to see if training tasks can be integrated into the range training plan.			

Determine Requirements

E-33. Determining the requirements involves identifying the equipment and personnel necessary to conduct the training.

Personnel

E-34. The following personnel are required for range operations:

- OIC.
- Safety officer.
- Assistant safety officer.
- NCOIC.
- Ammunition NCO.

- Ammunition personnel (determined by type of range).
- Target detail and target operators.
- Tower operator.
- Concurrent training instructors.
- Assistant instructors.
- Radio telephone operators.
- Guards (range requirements).
- Medic(s).
- Air guard.
- Armorer.
- Truck driver (range personnel and equipment).
- Mechanic for vehicles.

E-35. Further, the following must be checked:

- Have you overstaffed your range?

Equipment

E-36. The following equipment is required for range operations:

- Range packet and clearance form.
- Safety fan and diagram (if applicable).
- Other safety equipment (for example, aiming circle, compass).
- Publications pertaining to the training that will be conducted.
- Lesson plans, status reports, and reporting folder.
- Range flag and light (night firing).
- Radios.
- Field telephone and wire.
- 292 antenna (if necessary).
- PA set with backup bullhorn(s).
- Concurrent training markers.
- Training aids for concurrent training stations.
- Sandbags.
- Tentage (briefing tent, warm-up tent).
- Space heaters (if needed).
- Colored helmets for control personnel.
- Safety paddles and vehicle flag sets or lights.
- Ambulance or designated vehicle.
- Earplugs.
- Water for drinking and cleaning.
- Scorecards.
- Master score sheet.
- Armorer's tools and cleaning equipment for weapons.
- Brooms, shovels, and other cleaning supplies and equipment.
- Tables and chairs (if needed).
- Target accessories.
- Fire extinguishers.
- Tarp, stakes, and rope to cover the ammunition.
- Toilet paper.
- Spare weapons and repair parts (as needed).

- Tow bar and slave cables for vehicles.
- Fuel and oil for vehicles and target mechanisms.

DETERMINE AVAILABLE RESOURCES

E-37. Determining available resources involves the following:

- Fill personnel spaces.
- Keep unit integrity.
- Utilize NCOs.
- Coordinate with supporting organizations:
  - Ammunition.
  - Transportation.
  - Training aids.
  - Medics.
  - Weapons.
  - Other equipment.

FOOLPROOF

E-38. To foolproof range operations—

- Write an overall lesson plan for the range.
- Organize a plan for firing:
  - Determine the range organization.
  - Outline courses of fire to be used.
  - Have fire commands typed for use on the range.
  - Set the rotation of stations.
- Rehearse concurrent training instructors and assistants.
- Brief the RTO on unique range control radio procedures.
- Brief and rehearse the reporting NCO on range operation and all of his duties.
- Collect and concentrate equipment for use on the range in one location.
- Obtain training aids.
- Pick up targets from the range warehouse (if required).
- Report to range control for a safety briefing (if required) and sign for any special items.
- Publish the LOI:
  - Uniform of range and firing personnel (helmets and earplugs).
  - Mode of transportation, departure times, and places.
  - Methods of messing to be used.
  - Any special requirements being placed on units.

OCCUPY THE RANGE AND CONDUCT TRAINING

E-39. Certain actions must be performed to properly occupy the range and conduct training.

Occupy the Range

E-40. When occupying the range, perform the following actions:

- Request permission to occupy the range.
- Establish good communications.
- Have designated areas prepared:
  - Parking.
  - Ammunition point.

    - Medical station.
    - Water point.
    - Concurrent training.
    - Mess.
    - Helipad.
    - Armorer.
- Inspect the range for operational condition.
- Raise the flag when occupying or firing according to the local SOP.
- Check the ammunition to ensure that it is the correct type and quantity.
- Ensure that range personnel are in the proper uniform and the equipment is in position.
- Receive firing units.
- Conduct safety checks on weapons.
- Check for clean, fully operational weapons.
- Conduct a safety briefing (to include administrative personnel on the range).
- Organize personnel into firing orders (keep unit integrity, if possible).
- Request permission to commence firing from range control.

Conduct of Firing

E-41. When conducting firing, the following must be checked:

- Are communications to range control satisfactory?
- Are commands from the tower clear and concise?
- Are range areas policed?
- Is ammunition accountability maintained?
- Is the master score sheet updated?
- Is personnel accountability maintained?
- Are vehicles parked in the appropriate areas?
- Is the air guard on duty and alert?
- Are personnel in proper uniform?
- Are earplugs in use?
- Are troops responding properly to commands?
- Are on-the-spot corrections being made when troops use poor techniques or fail to hit the target?
- Is conservation of ammunition being enforced?
- Are weapons cleared before they are taken from the firing line?
- Are personnel checked for brass or ammunition before they leave the range?
- Is anyone standing around not involved in training or support?

CLOSE THE RANGE

E-42. To properly close the range, the following must be performed:

- Close downrange according to the local SOP.
- Remove all equipment and ammunition from the range.
- Police the range.
- Repaste and resurface targets as required by range instructions
- Perform other maintenance tasks as required by the local SOP.
- Request a range inspector from range control when ready to be cleared.
- Submit an after-action report to headquarters.
- Report any noted safety hazards to the proper authorities.

KNOWN DISTANCE RANGE

E-43. Events held on the KD range require additional personnel and equipment.

Personnel

E-44. In addition to the personnel previously identified, the KD range requires—

- NCOIC of pit detail.
- Assistant safety officer for the pit area.

Equipment

E-45. In addition to the equipment previously identified, the KD range requires—

- A sound set for the pit area.
- Positive communication from the firing line to the pit area.
- Pasters.
- Glue and brushes for resurfacing targets.
- Lubricant for target frames.
- Proper targets mounted in target frames.
- Briefing on how to operate a KD range.
- Procedure for marking targets.
- Procedure for pit safety.

COMPUTER CONTROLLED RANGES

E-46. Modern computer-controlled ranges allow trainers to develop scenarios and control targets and battlefield simulation devices. This permits Soldiers and units to practice mission-essential tasks in a stressful environment. Computerized systems also provide performance feedback. Using data recorded during training, the computer generates AARs. This accurate feedback allows commander to assess the unit's performance and leaders to assess their unit's mission status and design training programs to overcome the identified shortcomings. The performance feedback highlights positive actions to reinforce correct procedures and to foster Soldiers' confidence, enabling Soldiers and leaders to recognize and correct their shortcomings.

RANGE SELECTION

E-47. This section identifies primary and alternate ranges used for training and qualification, with specific weapon systems and weapons based on applicable FMs. The following table lists primary and alternate ranges to satisfy weapon system training events.

Table E-1. Primary/alternate range selection.

WEAPON SYSTEM	TRAINING EVENT	RANGE	
		PRIMARY	ALTERNATE
M16-Series and M4	Zero	25-m range	AFF/ARF/MRF with 25m boots
	Sustainment/Record	AFF/ARF/MRF	25-m range
	Qualification	ARF/MRF/QTR	25-m range
	Night	AFF/ARF/MRF Night	25-m range
	CBRN Conditions	AFF/ARF/MRF	25-m range
AFF: Automated Field Fire Range ARF: Automated Record Fire Range MRF: Modified Record Fire Range QTR: Qualification Training Range			

Rifle/Machine Gun Zero Range (17801)

E-48. This range (Figure E-1) is used to train individual Soldiers on the skills necessary to align the sights and practice basic marksmanship techniques against stationary targets. The range is designed for conducting shot-grouping and zeroing exercises with the M16- and M4-series weapons, as well as crew-served machine guns. Its primary features include—

- 32 target frames at 25 meters.
- 16 target frames at 10 meters.
- 32 foxholes.

E-49. This range requires no automation. All targets are fixed at 25 meters from the firing line for M16- and M4-series weapons and 10 meters for machine gun.

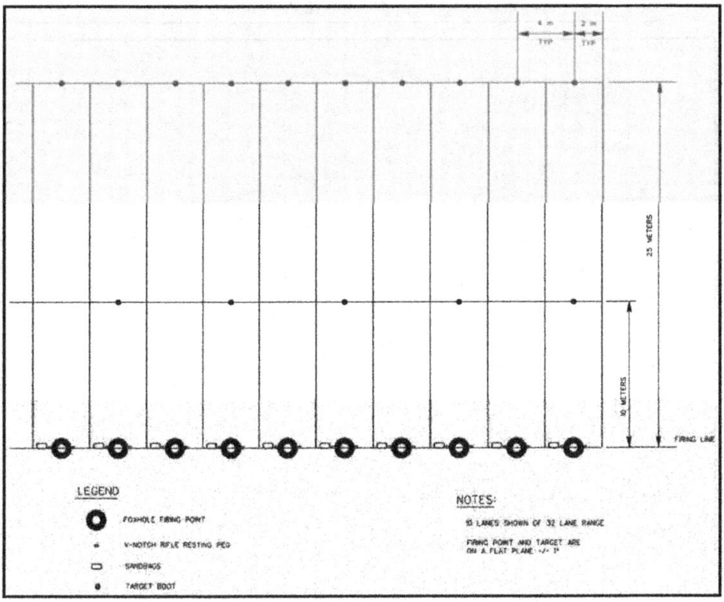

Figure E-1. Rifle/machine gun zero range (17801).

Automated Field Fire Range (17803)

E-50. The AFF range (Figure E-2) is used to train and familiarize Soldiers on the skills necessary to identify, engage, and hit stationary Infantry targets with M16- and M4-series weapons. Its primary features include—

- 96 stationary Infantry targets.
- 32 foxhole positions.

E-51. All targets are fully automated, and the event-specific target scenario is computer-driven and scored from the range operations center. The range operating system is fully capable of providing immediate performance feedback to the participants.

NOTE: This range can be used for automatic rifle practice.

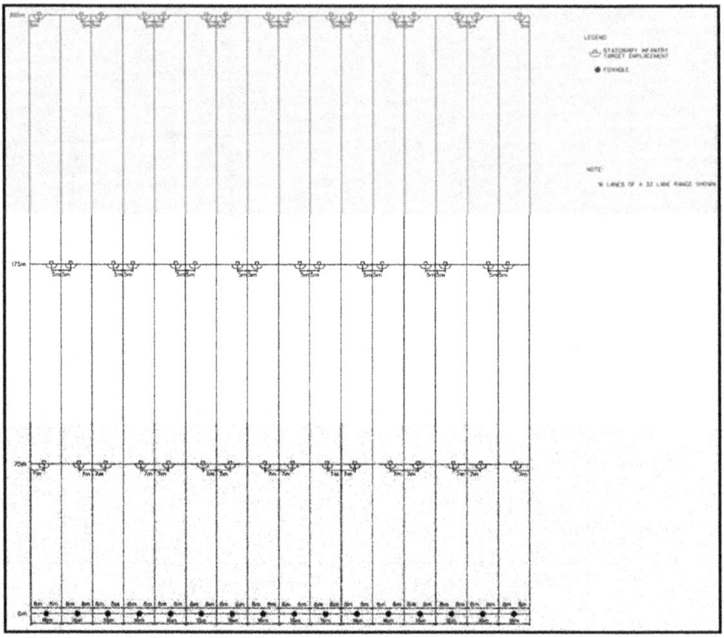

Figure E-2. Automated field fire range (17803).

Automated Record Fire Range (17805)

E-52. The ARF range (Figure E-3) is used to train and test individual Soldiers on the skills necessary to identify, engage, and defeat stationary Infantry targets for day/night qualification requirements with M16- and M4-series weapons. Its primary features include—

- 112 stationary Infantry targets.
- 16 foxholes.

E-53. All targets are fully automated, and the event-specific target scenario is computer-driven and scored from the range operations center. The range operating system is fully capable of providing immediate performance feedback to the participants.

NOTE: 1. To perform night fire, Soldiers fire at 50-meter targets from the night-fire line and baseline.

2. Range Set-up: Replace one of the 50m F-type silhouettes with an E- type silhouette. Low light illumination capability is required in both of the 50-meter target emplacements. Engage the F- type from the 25-meter night-firing line, and the E- type from the baseline.

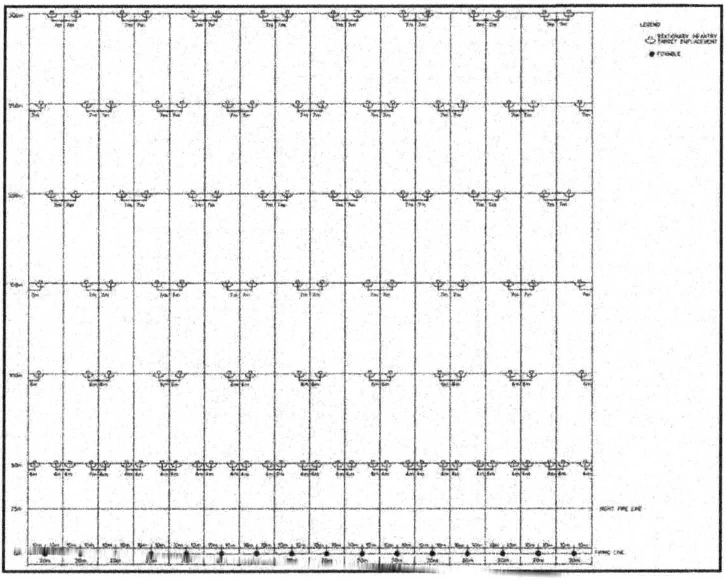

Figure E-3. Automated record fire range (17805).

Modified Record Fire Range (17806)

E-54. The modified record fire (MRF) range (Figure E-4) is used to train and test individual Soldiers on the skills necessary to identify, engage, and defeat stationary Infantry targets for day/night qualification requirements with M16- and M4-series weapons. This range combines the capabilities of the AFF (17803), ARF (17805), and automated night fire (17808) ranges to reduce land and maintenance requirements and increase efficiencies. Its primary features include—

- 144 stationary Infantry targets.
- 16 foxholes

E-55. All targets are fully automated, and the event-specific target scenario is computer-driven and scored from the range operations center. The range operating system is fully capable of providing immediate performance feedback to the participants.

NOTE: 1. To perform night fire, Soldiers fire at 50-meter targets from the night-fire line and baseline.

2. Range Set-up: Replace one of the 50m F-type silhouettes with an E- type silhouette. Low light illumination capability is required in both of the 50-meter target emplacements. Engage the F- type from the 25-meter night-firing line, and the E- type from the baseline.

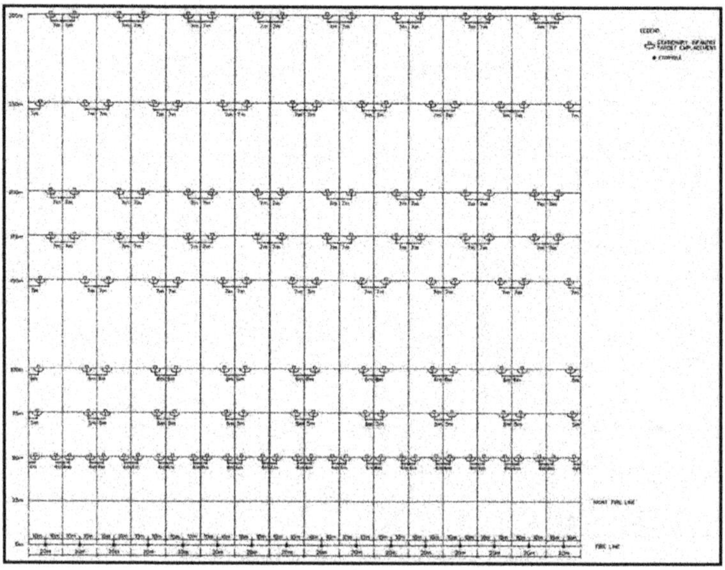

Figure E-4. Modified record fire range (17806).

Qualification Training Range (17809)

E-56. The qualification training (QTR) range (Figure E-5) is used to train and test Soldiers on the skills necessary to detect, identify, engage, and defeat stationary and moving Infantry targets in a tactical array with their prescribed weapons. This range combines the capabilities of the MRF range (17806), automated sniper field fire range (17812), combat pistol/military police firearms qualification course (17822), and multipurpose machine gun range (17833) to centralize training and reduce land, maintenance, and unit overhead requirements. Its primary features include—

- 429 stationary Infantry targets.
- 20 stationary armor targets.
- 20 moving Infantry targets.
- 10 stationary Infantry emplacements.
- 32 lanes rifle/machine gun zero.
- 15 lanes combat pistol qualification.
- 10 lanes sniper field fire.
- 16 lanes modified record fire.
- 10 lanes multipurpose machine gun.
- 2 target mechanisms each (400-meter and 700-meter).

E-57. All targets are fully automated, and the event-specific target scenario is computer-driven and scored from the range operations center. The range operating system is fully capable of providing immediate performance feedback to the participants.

NOTE: This range enhances throughput capability for units with multiple weapon densities by consolidating unit efforts to operate one training facility.

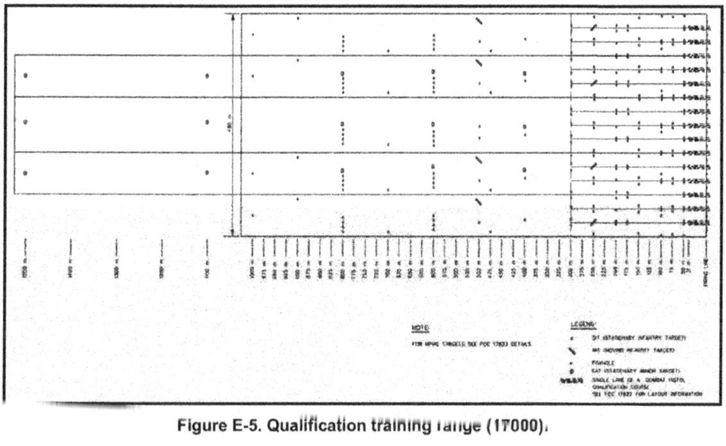

Figure E-5. Qualification training range (17000).

Appendix F

10-Meter Target Offsets and 25-Meter Zero Offsets

This section provides the 10-meter target offsets and the 25-meter zero offsets for M16- and M4-series weapons mounted with iron sights, optics, MILES, or aiming lasers. A blank, reproducible 10-meter target offset (Figure F-2) and an example of each weapon configuration (Figures F-3 through F-6) are provided. The M16A2 300-meter zeroing target is used for 25-meter zeroing with all weapon configurations.

MARKING 10-METER TARGET OFFSETS

F-1. To mark the proper 10-meter target offsets—

(1) Find the correct template for the weapon configuration.

(2) Starting from the center of the borelight circle on the offset, count the number of squares to the desired point of aim.

EXAMPLE

L2.0, U2.4

Starting from the center of the borelight circle (0.0, 0.0), move left 2 squares and up 2.4 squares.

NOTE: Each template also provides a number formula for the proper offset.

---

(3) Place the appropriate symbol or mark (Figure F-1).

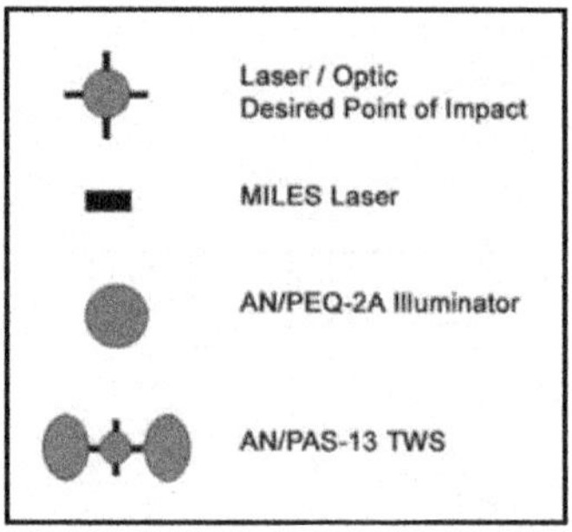

Figure F-1. 10-meter target offset symbols.

MARKING 25-METER ZERO OFFSETS

F-2. To mark the proper 25-meter zero offsets—

(1) Use only a M16A2 300-meter zeroing target.

(2) Find the correct target template for the weapon configuration.

(3) Count the number of squares, starting from the center of the 300-meter zeroing silhouette.

(4) Mark the designated strike point by drawing a small circle at the appropriate number of squares from the center of the 300-meter zeroing silhouette.

(5) Draw a 4-centimeter by 4-centimeter square, keeping the designated strike point at the center.

NOTES: 1. To reproduce the 10-meter target offset, copy the blank 10-meter target offset and place the example of the weapon being used on the back. This reproducible copy can be laminated and used repeatedly.

2. Table F-1 provides offset mounting information for various weapon configurations.

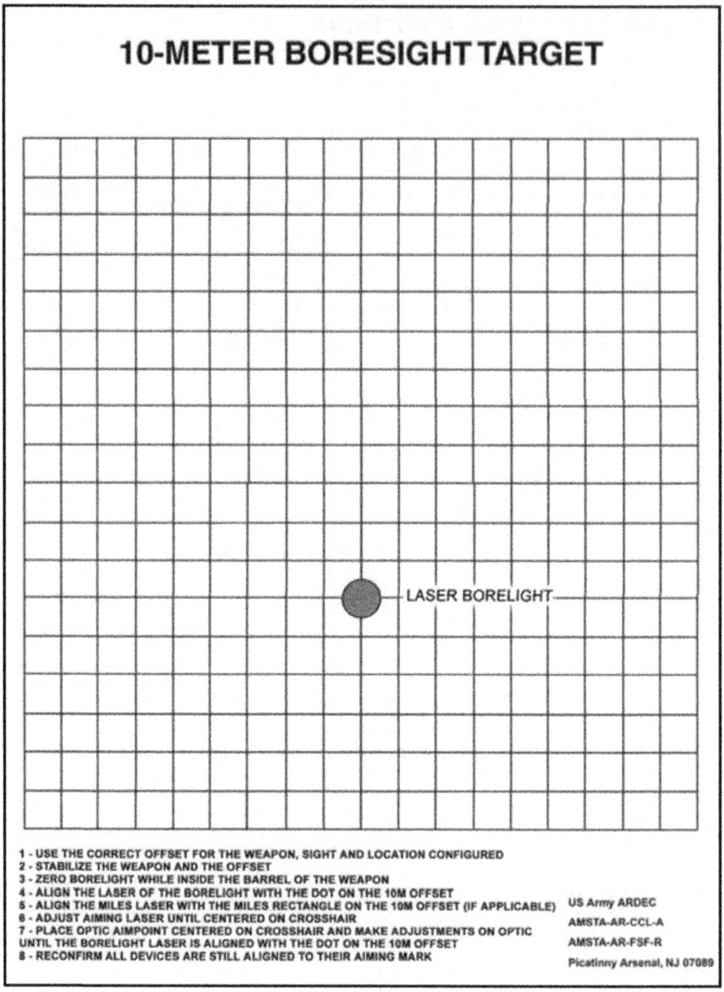

Figure F-2. Blank 10-meter target offset.

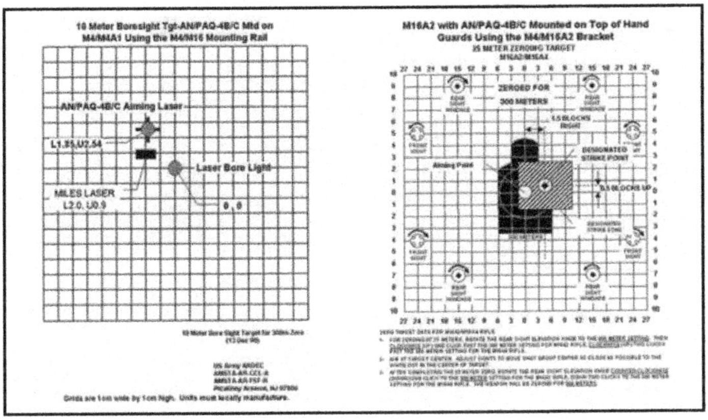

Figure F-3. M16A2 10-meter boresighting target/25-meter zeroing target offsets.

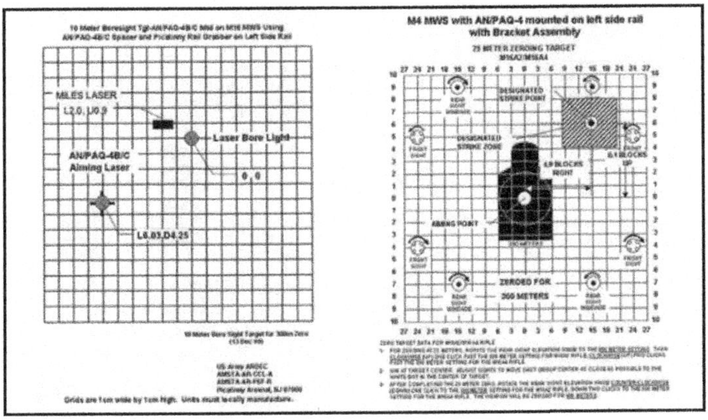

Figure F-4. M4 MWS 10-meter boresighting target/25-meter zeroing target offsets.

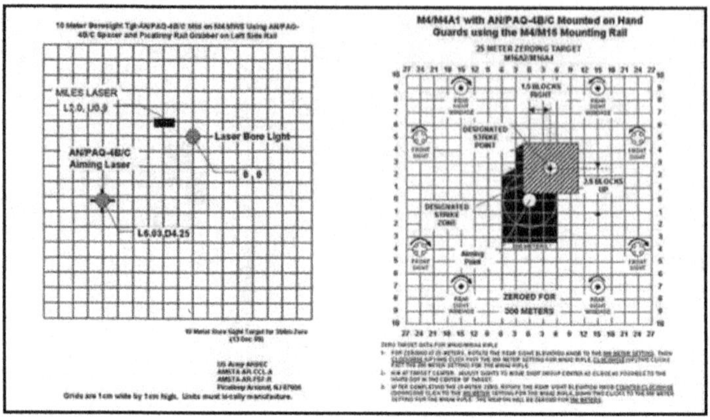

Figure F-5. M4/M4A1 10-meter boresighting target/25-meter zeroing target offsets.

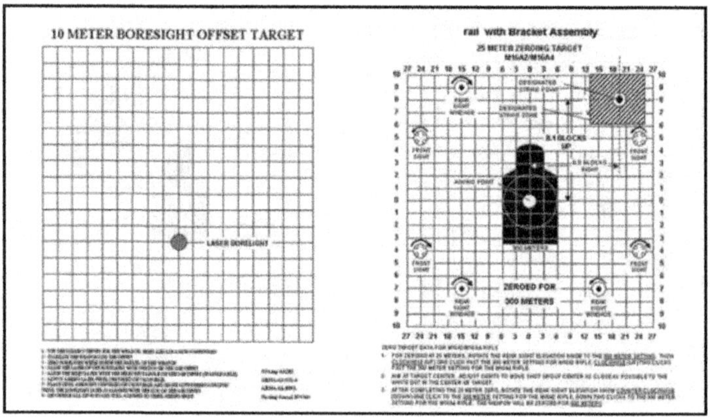

Figure F-6. M16A4 MWS 10-meter boresighting target/25-meter zeroing target offsets.

Table F-1. Offset mounting.

WEAPON	ACCESSORY	RAIL GRABBER	MOUNT	LOCATION	RANGE TO ZERO	ZERO OFFSET	BORESIGHT TARGET	MILES OFFSET
M16A2	Iron sight	N/A	N/A	N/A	300 m	0.0 0.0	0.0 4.2U	2.0L 0.9U
M16A2	M68	N/A	M68 gooseneck bracket	Carrying handle	300 m	0.0 1.4 cm DN	0.0 5.2U	2.0L 2.4U
M16A2	LTWS	TWS	TWS bracket assembly	Carrying handle	300 m	0.0 10D	0.0 13.4U	2.0L 2.4U
M16A2	TWS	N/A	TWS bracket assembly	Carrying handle	300 m	0.0 8.1D	0.0 11.5U	2.0L 2.4U
M16A2	AN/PVS-4	N/A	AN/PVS-4 mounting knob	Carrying handle	300 m	0.0 7.0D	0.0 9.4U	2.0L 0.9U
M16A2	AN/PAQ-4B/C	N/A	M4/M16 bracket	Hand guards	300 m	1.5R 0.5U	1.85L 2.54U	2.0L 0.9U
M16A2	AN/PEQ-2A/B	N/A	M4/M16 bracket	Hand guards	300 m	1.5L 0.5U	1.8R 2.4U	2.0L 0.9U
M16/M203	AN/PAQ-4B/C	N/A	Spacer and AN/PVS-4 mounting knob	Carrying handle	300 m	1.85R 2.6D	1.85L 8.6U	2.0L 3.9U
M16/M203	AN/PVS-4	N/A	AN/PVS-4 mounting bracket assembly	Carrying handle	300 m	4.2R 9.8D	TBD	2.0L 0.9U
M4/M4A1	BUIS	N/A	N/A	Upper receiver	300 m	0.0 0.0	0.0 4.01U	2.0L 0.9U
M4/M4A1	AN/PAQ-4B/C	N/A	M4/M16 bracket	Hand guards	300 m	1.5R 2.5U	1.85L 2.54U	2.0L 0.9U
M4/M4A1	LTWS	TWS	N/A	Upper receiver	300 m	0.0 4.5D	0.0 7.9U	TBD
M4/M4A1	TWS	Picatinny	TWS spacer and rail grabber	Upper receiver	300 m	0.0 5.7D	0.0 9.4U	2.0L 2.4U
M4/M4A1	AN/PEQ-2A/B	N/A	M4/M16 bracket	Hand guards	300 m	1.0L 0.3U	1.8R 2.4U	2.0L 0.9U
M4/M4A1	AN/PVS-4	Picatinny	Spacer and rail grabber	Upper receiver	300 m	0.0 3.4D	0.0 7.6U	2.0L 0.9U
M4/M4A1	M68	M68	Half-moon spacer	Upper receiver	300 m	0.0 1.4 cm DN	0.0 5.63U	2.0L 2.4U
M4/M203	BUIS	N/A	N/A	Upper receiver	300 m	0.0 0.0	0.0 6.01U	2.0L 0.9U
M4/M203	AN/PAQ-4B/C	N/A	Spacer and AN/PVS-4 mounting knob	Carrying handle	300 m	1.3R 1.9D	1.85L 8.6U	2.0L 0.9U
M4/M203	AN/PVS-4	Picatinny	Spacer and rail grabber	Upper receiver	300 m	0.0 3.4D	0.0 9.6U	2.0L 3.9U
NOTE: Target offsets not yet developed are indicated by TBD (to be developed).								

Table F-1. Offset mounting (continued).

WEAPON	ACCESSORY	RAIL GRABBER	MOUNT	LOCATION	RANGE TO ZERO	ZERO OFFSET	BORESIGHT TARGET	MILES OFFSET
M4 MWS	BUIS	N/A	N/A	Upper receiver	300 m	0.0 0.0	0.0 4.01U	2.0L 0.9U
M4 MWS	AN/PVS-4	Picatinny	Spacer and rail grabber	Upper receiver	300 m	0.0 3.4D	0.0 7.6U	2.0L 0.9U
M4 MWS	M68	M68	Rail grabber	Upper receiver	300 m	0.0 1.4 cm DN	0.0 5.63U	2.0L 2.4U
M4 MWS	LTWS	TWS	N/A	Upper receiver	300 m	0.0 4.5D	0.0 7.9U	2.0L 2.4U
M4 MWS	TWS	TWS	Spacer	Upper receiver	300 m	0.0 5.7D	0.0 9.4U	2.0L 2.4U
M4 MWS	ANPEQ-2A	Insight	N/A	Left	300 m	TBD	4.5L 1.0D	2.0L 0.9U
M4 MWS	AN/PEQ-2A/B	Insight	N/A	Right	300 m	N/A	5.5R 5.4D	2.0L 0.9U
M4 MWS	AN/PEQ-2A/B	Insight	N/A	Top	300 m	1.5L 0.5D	2.9R 2.3U	2.0L 0.9U
M4 MWS	AN/PEQ-2A/B	Picatinny	Spacer	Top	300 m	N/A	1.95R 4.1U	2.0L 0.9U
M4 MWS	AN/PEQ-2A/B	Picatinny	Spacer	Right	300 m	N/A	6.35R 4.4D	2.0L 0.9U
M4 MWS	AN/PEQ-2A/B	Picatinny	Spacer	Left	300 m	6.9R 2.0U	6.2L 0.60D	2.0L 0.9U
M4MWS	AN/PEQ-2A/B	Insight	Training adapter	Top	300 m	2.0L 1.5D	N/A	2.0L 0.9U
M4 MWS	AN/PAQ-4B/C	Picatinny	AN/PAQ-4B/C bracket adapter	Top	300 m	4.9R 6.1U	1.75L 3.9U	2.0L 0.9U
M4 MWS	AN/PAQ-4B/C	Picatinny	AN/PAQ-4B/C bracket adapter (spacer)	Right	300 m	N/A	6.9R 0.9D	2.0L 0.9U
M4 MWS	AN/PAQ-4B/C	Insight	N/A	Top	300 m	N/A	1.75L 2.15U	2.0L 0.9U
M4MWS	AN/PAQ-4B/C	Insight	N/A	Right	300 m	N/A	4.35R 0.65D	2.0L 0.9U
M4MWS	AN/PAQ-4B/C	Insight	N/A	Left	300 m	N/A	4.30L 4.25D	2.0L 0.9U
M4 MWS M203	BUIS	N/A	N/A	Upper receiver	300 m	0.0 0.0	0.0 6.01U	2.0L 0.9U
M4 MWS M203	AN/PAQ-4B/C	Picatinny	Bracket adapter (spacer	Left	300 m	4.9R 6.1U	6.0L 4.0D	2.0L 3.9U
M4 MWS M203	AN/PVS-4	Picatinny	Spacer	Upper receiver	300 m	0.0 3.4D	0.0 9.6U	2.0L 3.9U
M16A4 MWS	BUIS	N/A	N/A	Upper receiver	300 m	0.0 0.0	0.0 4.01U	2.0L 0.9U
M16A4 MWS	AN/PAQ-4B/C	Picatinny	AN/PAQ-4B/C bracket adapter (spacer)	Left	300 m	6.5R 8.1U	6.03L 4.25D	2.0L 0.9U
M16A4 MWS	TWS	TWs	Spacer	Upper receiver	300 m	0.0 6.0D	0.0 9.4U	2.0L 2.4U
M16A4 MWS	M68	M68	N/A	Upper receiver	300 m	0.0 0.0	0.0 5.63U	2.0L 2.4U
M16A4 MWS	AN/PEQ-2A/B	Insight	N/A	Left	300 m	3.0R 3.0U	4.5L 1.0D	2.0L 0.9U
M16A4 MWS	AN/PVS-4	Picatinny	Spacer	Upper receiver	300 m	0.0 4.6D	0.0 7.6U	2.0L 0.9U

NOTE: Target offsets not yet developed are indicated by TBD (to be developed).

Table F-1. Offset mounting (continued).

WEAPON	ACCESSORY	RAIL GRABBER	MOUNT	LOCATION	RANGE TO ZERO	ZERO OFFSET	BORESIGHT TARGET	MILES OFFSET
M16A4 MWS M203	BUIS	N/A	N/A	Upper receiver	300 m	0.0 0.0	0.0 6.01U	2.0L 0.9U
M16A4 MWS M203	AN/PAQ-4B/C	Picatinny	AN/PAQ-4B/C bracket adapter (spacer	Left	300 m	6.5R 8.1U	6.0L 4.0D	2.0L 3.9U
M16A4 MWS M203	AN/PVS-4	Picatinny	Spacer	Upper receiver	300 m	0.0 4.6D	0.0 9.6U	2.0L 3.9U
M16A4 MWS	AN/PEQ-2A/B	Picatinny	Spacer	Left	300 m	6.0R 2.0U	6.2L 0.60D	2.0L 0.9U
M16A4 MWS	AN/PEQ-2A/B	Picatinny	Spacer	Right	300 m	TBD	6.35R 4.4D	2.0L 0.9U
M16A4 MWS	AN/PEQ-2A/B	Picatinny	Spacer	Top	300 m	TBD	1.95R 4.1U	2.0L 0.9U
M16A4 MWS	AN/PEQ-2A/B	Insight	N/A	Right	300 m	TBD	5.5R 5.4 D	2.0L 0.9U
M16A4 MWS	AN/PEQ-2A/B	Insight	N/A	Top	300 m	1.5L 0.5D	2.0R 2.3U	2.0L 0.9U
M16A4 MWS	AN/PEQ-2A/B	Insight	Training adapter	Top	300 m	2.0L 1.5D	TBD	2.0L 0.9D
M16A4 MWS	AN/PAQ-4B/C	Picatinny	AN/PAQ-4B/C bracket adapter	Top	300 m	4.9R 6.1U	1.75L 3.9U	2.0L 0.9U
M16A4 MWS	AN/PAQ-4B/C	Picatinny	AN/PAC-4B/C bracket adapter	Right	300 m	N/A	6.0R 0.9D	2.0L 0.9U
M16A4 MWS	AN/PAQ-4B/C	Insight	N/A	Top	300 m	N/A	1.75L 2.15U	2.0L 0.9U
M16A4 MWS	AN/PAQ-4B/C	Insight	N/A	Right	300 m	N/A	4.35R 0.65D	2.0L 0.9U
M16A4 MWS	AN/PAQ-4B/C	Insight	N/A	Left	300 m	N/A	4.30L 4.25D	2.0L 0.9U
NOTE: Target offsets not yet developed are indicated by TBD (to be developed).								

Glossary

SECTION I – ACRONYMS AND ABBREVIATIONS

AAR	after-action review
ACH	Army combat helmet
ACOG	advanced combat optical gunsight
AFF	automated field fire
AMU	Army Marksmanship Unit
AP	armor-piercing
AR	Army regulation
ARF	automated record fire
ARM	advanced rifle marksmanship
ARS	adapter rail system
ATC	Army training center
ATPIAL	advanced target pointer/illuminator aiming light
BDC	bullet drop compensated
BFA	blank firing attachment
BOLC	Basic Officer Leaders Course
BRM	basic rifle marksmanship
BUIS	backup iron sight
CBRN	chemical, biological, radiological, and nuclear
CCO	close combat optic
COA	course of action
cm	centimeter or centimeters
CRM	composite risk management
DA Pam	Department of the Army pamphlet
EST	Engagement Skills Trainer
FM	field manual
FOV	field of view
fps	feet per second
ft	foot or feet
HWTS	heavy weapon thermal sight
I2	image-intensifier
IBA	interceptor body armor
IET	initial entry training
in	inch or inches
IR	infrared
KD	known distance
lb	pound or pounds
LBE	load-bearing equipment
LFX	live-fire exercise
LMTS	Laser Marksmanship Training System
LOGCAP	Logistics Civil Augmentation Program
LOMAH	location of misses and hits

LWTS	light weapon thermal sight
m	meter or meters
MDMP	military decision-making process
METL	mission-essential task list
METT-TC	mission, enemy, terrain and weather, troops and support available, time available, civil considerations
min	minute or minutes
mm	millimeter or millimeters
MOA	minute of angle
MOPP	mission-oriented protective posture
mph	miles per hour
MRF	modified record fire
MWS	modular weapon system
NCOES	Noncommissioned Officers Education System
NCOIC	noncommissioned officer in charge
NFOV	narrow field of view
NGO	nongovernmental organizations
NOD	night observation devices
NSN	national stock number
NVD	night vision device
OIC	officer in charge
OSUT	one-station unit training
OTV	outer tactical vest
PASGT	personnel armor system for ground troops
PH	probability of hit
PMCS	preventive maintenance checks and services
PMI	preliminary marksmanship instruction
POI	program of instruction
QTR	qualification training
RETS	Remote Electronic Target System
RFA	rimfire adapter
RSO	range safety officer
RTO	radio telephone operator
SAPI	small arms protective insert
SAW	squad automatic weapon
SDM	squad designated marksman
SDZ	surface danger zone
sec	second or seconds
SRC	short-range combat
SRM	short-range marksmanship
SRT	special reaction team
SRTA	short-range training ammunition
STRAC	Standards in Training Commission (DA Pam 350-38)
STX	situation training exercise
TADSS	training aids, devices, simulators, and simulations
TC	training circular
TSC	training support centers
TM	technical manual

TPIAL	target pointer illuminator/aiming light
TWS	thermal weapon sight
UO	urban operations
USACRC	United States Army Combat Readiness Center
WFF	warfighting functions
WFOV	wide field of view

SECTION II – TERMS

Active Army: All Regular Army (RA) forces in the Active Army.

adjusted aiming point: An aiming point that allows for gravity, wind, target movement, zero changes, and MOPP firing.

advanced marksmanship: Normally refers to marksmanship skills taught during ARM.

advanced rifle marksmanship: Normally refers to the formal marksmanship instruction received by infantrymen upon completion of BRM during OSUT.

aiming: A marksmanship fundamental; refers to the precise alignment of the rifle sights with the target.

aiming card: The M15A1 aiming card is a cardboard sleeve with a moveable insert. The rear sight aperture, front sight post, and target are pictured. This training device is used in conjunction with aiming instructions.

aiming point: A place on a target in which the rifle sights are aligned normally the target center of mass.

alibi target: A target or additional target a soldier is allowed to engage during qualification firing when unable to complete a record fire scenario due to circumstances beyond his control; for example, a target mechanism, weapon, or ammunition malfunction.

alternate course: Alternatives to standard qualification courses.

ammunition lot: A quantity of cartridges, each of which is made by one manufacturer under uniform conditions and is expected to work in a uniform manner.

ammunition lot number: Code number that identifies a particular quantity of ammunition from one manufacturer.

aperture: The hole in the rear sight.

armorer: One who services and makes repairs on small arms and performs similar duties to keep small arms ready for use.

Army Training and Evaluation Program: A guide for the training and evaluation of critical unit combat missions – crew/squad through battalion/task force echelon.

Army Training Center: Conducts OSUT and BRM. Locations are Fort Benning, GA; Fort Jackson, SC; Fort Knox, KY.

artificial illumination: Any light from a man-made source.

assault course: An area of ground used for training soldiers in attacking an enemy in close combat.

automatic fire: A firing mode that causes the weapon to continue firing as long as the trigger is held or until all ammunition has been expended.

ball: The projectile; the bullet.

ball ammunition: General-purpose standard service ammunition with a solid core bullet.

ball and dummy: An exercise that substitutes a dummy round for a live round without the firer knowing it. An excellent exercise for identifying and correcting trigger jerks.

ballistics: A science that deals with the motion and flight characteristics of projectiles.

barrel erosion: Wearing away of the surface of the bore due to the combined effects of gas washing, coring, and mechanical abrasion.

basic marksmanship: Fundamental marksmanship skills taught in BRM during IET and OSUT.

basic rifle marksmanship: The formal course of marksmanship instruction received by all soldiers.

battlesight zero: A sight setting that soldiers keep on their weapons. It provides the highest probability of hitting most high-priority combat targets with minimum adjustment to the aiming point, a 250 meter sight setting as on the M16A1 rifle, and a 300 meter sight setting as on the M16A2 rifle.

blank ammunition: A complete cartridge without the bullet used to simulate weapon firing.

blank firing adapter: A device that fits in the muzzle of the rifle; used only with blank ammunition.

brass: An alloy of copper and zinc used to make cartridge cases and bullet jackets. Also, a common name for expended cases.

breath control: The third marksmanship fundamental; refers to the control of breathing to help keep the rifle steady during firing.

bullet: The projectile or ball; the part that goes downrange. It may also be used to refer to the complete cartridge.

bull's-eye target: Any target with a round black circle and scoring rings. Normally used in competitive marksmanship training.

buttplate: Metal or rubber covering of the end of the stock on the rifle.

cadre coach: A trainer with expertise and knowledge exceeding that of the firer.

caliber: Diameter of the bore; for example, the M16-series rifle bore is 5.56mm (.223 inch).

cartridge: A complete round of ammunition.

center of mass: A point that is horizontally (left and right) and vertically (up and down) at the center of the target.

chambering: The step in the cycle of operation that refers too fully seating the round in the chamber of the rifle.

chamber plug: A range safety device that is a small plastic plug designed to fit into the chamber of the M16. A handle extends out the ejection port so safety personnel can see at a glance that the rifle is clear of ammunition.

clock method: Method of calling shots by referring to the figures on an ordinary clock dial assumed to have the target at its center. Also a method of determining the strength and direction of wind.

coach: Any individual who assists firers on the firing line.

coach-and-pupil method: Method of training in which pairs of pupils take turns practicing a procedure explained by the instructor/trainer.

cocking: The step in the cycle of operation that refers to the rearward movement of the bolt riding over the hammer, resetting the weapon for subsequent firing.

collective firing proficiency: Units delivering effective fire in a tactical setting. It requires individual skill plus command and control to engage all targets within an assigned sector.

concurrent training: Training that occurs at the same time that other unit members are using the primary training facilities.

cookoff: A round that fires as a result of a hot chamber without the trigger being pulled. It can occur any time until the weapon is cooled.

crack and thump: A method to determine the general direction and distance to an enemy firer who is shooting at you.

cradle: A vise-like mechanism that holds a weapon in a secure position for test firing.

cross dominance: A soldier with a dominant hand and a dominant eye that are not the same; for example, a right-hander firer with a dominant left eye.

cycle of operation: The eight steps involved in firing a round of ammunition: feeding, chambering, locking, firing, unlocking, extracting, ejecting, and cocking.

cyclic rate of fire: The maximum rate at which a weapon will fire in the automatic mode.

dime-washer exercise: A dry-fire exercise used to practice trigger squeeze.

downrange feedback: Used to describe any training technique that provides precise knowledge of bullet strike (whether hit or miss).

dry fire: A technique used to simulate the firing of a live round with an empty weapon. Any application of the fundamentals of marksmanship without live ammunition may be referred to as dry fire.

dry-fire moving target trainer: A small-motorized scaled target device used to teach the engagement of moving personnel targets.

dummy ammunition: A cartridge without a primer or powder. Primarily used for ball-and-dummy exercises on the live-fire line.

effective wind: The average of all the varying winds encountered.

ejection: The step in the cycle of operation that removes the expended cartridge from the weapon out of the ejection port.

elevation adjustment: Rotating the front sight post to cause the bullet to strike higher or lower on the target.

expert: The highest qualification rating.

external ballistics: What happens to the bullet between the time it leaves the rifle and the time it arrives at the target.

extraction: The step in the cycle of operation that pulls the round from the chamber.

eye relief: The distance from the firing eye to the rear sight. Eye relief is a function of stock weld.

feedback: Obtaining knowledge of performance.

feedback target: Targets designed for use at 75, 175, or 300 meters; includes an overprinted grid similar to a zero target.

feeding: The step in the cycle of operation that is the forward movement of the bolt, stripping the top round from the magazine and moving it toward the chamber.

field firing: Training on the standard field firing range with target banks at 75, 175, and 300 meters.

firing: The step in the cycle of operation that refers to pulling the trigger, releasing the hammer to strike the firing pin, which strikes the primer. The primer ignites and, in turn, ignites the powder charge within the cartridge case.

firing hand: The right hand of a right-handed firer. The left hand of a left-handed firer.

firing pin: Plunger in the bolt of a rifle that strikes the primer.

fleeting target: A moving target remains within observing or firing distance for such a short period that it affords little time for deliberate adjustment and fire against it.

functioning: (See cycle of operation.)

fundamentals of rifle marksmanship: The four essential elements needed to hit targets: steady position, aiming, breath control, and trigger squeeze.

gravity: The natural pull of all objects to the center of the earth.

grouping: A live-fire exercise with the objective of shooting tight shoot groups.

gun bore line: A reference line established by the linear extension of the bore axis of a gun.

headspace: The distance between the face of the bolt(fully closed) and the face of a fully chambered cartridge.

hold-off: (See adjusted aiming point.)

horizontal dispersion: The left-to-right displacement of bullets on a target.

immediate action: procedures applied to rapidly reduce any rifle stoppage without determining its cause.

individual firing proficiency: Individual firing skills; for example, an individual's performance on the record fire course.

Infantry Remoted Target System (IRETS): (See RETS.)

infrared aiming light: A unique night sighting system that uses infrared light to assist in the aiming process.

initial entry training: Indicates the first training received by a new soldier, includes the MOS-producing portion of his training such a one-station unit training (OSUT).

initial pressure: The applications of about half of the total trigger pressure it takes to fire a rifle.

instructor-trainer ratio: The number of soldiers for which each instructor/trainer is responsible.

internal ballistics: What happens to the bullet before it leaves the muzzle of the rifle.

known distance: Describes the older range complexes with large target frames behind a large berm and firing lines at 100 yards or 100-meter increments. (See FM 25-7.)

laser: Light amplification by simulated emission of radiation.

lead: Distance ahead of a moving target that a rifle must be aimed to hit the target.

lead rule: Provides the soldier guidance on how to adjust his aiming point to hit moving targets.

line of sight: A line between the rifle and the aiming point, extending from the firing eye through the center of the rear aperture, across the tip of the front sight post, and onto the target.

location of misses and hits : A projectile location system that provides immediate and precise information to the firer concerning bullet strike (hit or miss).

locking: The step in the cycle of operation that is a counterclockwise rotation of the bolt, securing it into the barrel locking lugs.

long-range sight: The aperture marked L on the M16A1 rifle equipped with standard sights; provides for a zero at 375 meters. The M16A1 rifle equipped with LLLSS has an aperture marked L, but it is a regular sight.

Low-Light Level Sight System (LLLSS): A sighting system for low visibility firing that replaces the standard front and rear sights on the M16A1 rifle.

marksman: The designation given to the lowest qualification rating.

maximum effective range: The greatest distance at which a soldier may be expected to deliver a target hit.

maximum effective rate of fire: The highest rates of fire that can be maintained and still achieve target hits.

maximum range: The longest distance a projectile will travel when fired from a weapon held at the optimum angle.

minute of angle: A angle that would cover 1 inch at a distance of 100 yards, 2 inches at 200 yards, and so on. Each click of sight adjustment on the M16A1 rifle with standard sights is equal to one minute of angle.

Multiple Integrated Laser Engagement System (MILES): A tactile shooting device that uses a low-powered laser to activate detectors placed on people and vehicles.

muzzle velocity: The speed of a projectile as it leaves the muzzle of the weapon.

natural point of aim: The direction of the body/rifle combination is oriented while in a stable, relaxed firing position.

natural respiratory pause: The temporary cessation of breathing between an exhale and inhale.

night firing: Firing performed under all conditions of limited visibility.

nonfiring hand: The opposite of the firing hand.

optical sight: Sight with lenses, prisms, or mirrors used in lieu of iron sights.

Paige sighting device: A device with a small-scaled target that fits into the muzzle of the weapon, allowing the soldier to practice aiming.

pasters: Small white or black gum-backed paper used for covering bullet holes.

peep sight: The rear sight; a sight with a small aperture (hole).

peer coach: A soldier with shooting experience and knowledge equal to that of the firer he is coaching.

pit: The target area behind the large berm of a KD range.

plastic practice ammunition: Ammunition with a plastic projectile, high-muzzle velocity (the light weight causes it to lose velocity rapidly with a maximum range of 250 meters or less) designed for use in close-in training areas; frangible bullet.

point of aim: The exact spot on a target the rifle sights are aligned with.

point of impact: The point that a bullet strikes; usually considered in relation to point of aim.

pop, no kick: A firing condition when the primer ignites and the powder charge does not. This normally results in lodging the bullet inside the barrel.

pop-up target: A silhouette target that is activated remotely so it can suddenly appear and fall when struck by a bullet.

practice record: Firing conducted on a qualification course for practice.

predetermined fire: A technique of aligning the rifle during good visibility so the rifle can be aligned and fired on designated areas when they cannot be seen due to darkness, smoke, or fog.

preparatory marksmanship training: All marksmanship training that takes place before live fire.

primer: A small explosive device in the center base of the cartridge case that is struck by the firing pin to fire the round.

probability of hit: Ranging from 0 to 1.0, it refers to the odds of a given round hitting the target at a given range.

qualification firing: Firing on any authorized course that results in meeting qualification requirements; may also be called record fire. (See record fire.)

quick fire: A technique of fire used to engage surprise targets at close range.

range card: Small chart on which ranges and directions to various targets and other important points in the area under fire are recorded.

rapid semiautomatic fire: A firing procedure that results in an accurate shot being fired every one or two seconds.

receiver: That portion of a firearm that holds the barrel and houses the bolt and firing mechanism.

recoil: The rearward motion or kick of a gun upon firing.

record fire: Any course of fire used to determine if qualification standards are met. The standard record fire course consists of 40 target exposures at ranges between 50 and 300 meters. The standard course requires 23 hits to qualify as marksman, 30 for sharpshooter, and 36 for expert.

reduced range ammunition: Ammunition that is designed to be a ballistic match with service ammunition to an appropriate range for training (may be less than maximum effective range) and a reduced maximum range.

regular rear sight: The M16A1 rifle rear sight that is zeroed for 250 meters (the unmarked aperture on rifles with standard sights and the aperture marked L on rifles equipped with LLLSS).

reinforcement training: Training conducted that is over and above scheduled training.

remedial action: A procedure applied after immediate action has failed to correct a malfunction, which determines the cause of the malfunction.

remedial training: Additional training presented to soldiers who have demonstrated special shooting problems.

Remote Electronic Target System: Range complexes. Some ranges include moving targets.

Reserve Components: Includes Army National Guard and Army Reserve forces.

ricochet fire: Fire in which the projectile glances from a surface after impact.

Riddle sighting device: A small magnetic device with a scaled target that attaches to the front sight assembly, allowing the soldier to practice aiming.

rifle cant: Any leaning of the rifle to the left or right from a vertical position during firing.

rim-fire adapter: The caliber .22-rim fire adapter (M261) consists of a bolt and a magazine insert, which allows standard .22 caliber ammunition to be fired in the M16 rifle.

round: May refer to a complete cartridge or to a bullet.

scaled-silhouette target: Any target that is reduced in size. When it is observed from 25 meters, it looks the same size as though at a greater range.

sector of fire: An area assigned to an individual, weapon, or unit to be covered by fire.

semiautomatic fire: A mode of fire that allows one round to be fired each time the trigger is pulled.

serviceability checks: A technical inspection of the rifle to determine if it is safe to fire and in working condition. (May not ensure accuracy.)

service ammunition: Standard ammunition used by the military. Ammunition designed for combat.

service rifle: The primary rifle of a military force.

service school: Branch schools such as the US Army Infantry School at Fort Benning, Ga. and the Armor School at Fort Knox, Ky.

sharpshooter: The middle rating of qualification.

shot group: A number of shots fired using the same aiming point, which accounts for rifle, ammunition, and firer variability. Three shots are enough, but any number of rounds may be fired in a group.

shot group analysis: A procedure for analyzing the size of shot groups on a target to determine firer error.

sight alignment: Placing the center tip of the front sight post in the exact center of the rear aperture.

sighter rounds: Rounds fired that allow the bullet strike to be observed in relation to the aiming point.

sight picture: Placing correct sight alignment on a selected aiming point on a target.

sight radius: The distance from the front sight post to the rear sight aperture of a rifle.

sighting device (M16): A small metal device with a tinted square of glass that is placed on the carrying handle, allowing a coach to see what the firer sees through the sights.

silhouette target: A target that represents the outline of a man.

spotters: A round cardboard disk placed in bullet holes with a small wooden peg so the bullet strike can be observed from the firing line.

squad automatic weapon: A lightweight, one-man, 5.56mm machine gun.

starlight scope: A weapon scope that amplifies ambient light so targets can be seen and effectively engaged during darkness. The AN/PVS-2 and AN/PVS-4 are used on the M16 rifle.

steady position: The first marksmanship fundamental, which refers to the establishment of a position that allows the weapon to be held still while it is being fired.

stock weld: The contact of the cheek with the stock of the weapon.

supported position: Any position that uses something other than the body to steady the weapon (artificial support).

suppressive fire: Any engagement that does not have a definite or visible target. Firing in the general direction of known or suspected enemy location.

sustained rate of fire: Rate of fire that a weapon can continue to deliver for an indefinite period without overheating.

terminal ballistics: What happens to the bullet when it comes in contact with the target.

tight shot group: A shot group with all bullet holes close together.

tracer ammunition: Ammunition with a substance at the rear of the bullet that ignites soon after firing. It burns brightly so the trajectory of the bullet can be seen.

tracking: Engaging moving targets where the lead is established and maintained; moving with the target as the trigger is squeezed.

train the trainer: Describes any training that is designed to train marksmanship instructors or coaches.

trainfire: A marksmanship program using pop-up targets in a realistic environment.

trajectory: The flight path the bullet takes from the rifle to the target.

trapping: A technique for engaging moving targets. The aiming point is established forward of the target. The rifle is held stationary and fired as the target approaches the aiming point.

trigger squeeze: The fourth fundamental; squeezing the trigger so that the movement of firing is a surprise, the lay of the weapon is not disturbed, and a large target hit can be expected.

unit marksmanship: All marksmanship training that is conducted by units.

unlocking: The step in the cycle of operation that refers to the clockwise rotation of the bolt after firing, freeing the bolt from the barrel locking lugs.

unsupported position: Any position that requires the firer to hold the weapon steady using only his body (bone support).

vertical dispersion: The up-and-down displacement of bullets on a target.

Weaponeer: A training device that simulates the firing of the M16 rifle to provide performance feedback.

windage adjustment: Moving the rear sight aperture to cause the bullet to strike left or right on the target.

wind value: The effect the wind will have on the trajectory of the bullet.

wobble area: The natural movement of the weapon/sight on and around an aiming point when the weapon is being held in a steady position.

zero criterion: The standard or requirement for zeroing; 4cm or smaller group at 25 meters.

zeroing: Adjusting the rifle sights so bullets hit the aiming point at a given range.

zero target: A scaled-silhouette target with a superimposed grid for use at 25 meters.

References

SOURCES USED

These sources were used to prepare this publication.

AR 385-10, The Army Safety Program. 23 August 2007.

AR 385-63, Range Safety. 19 May 2003.

DA Pam 350-38, Standards in Weapons Training. 16 August 2004.

DA Pam 385-63, Range Safety. 10 April 2003.

FM 5-19, Composite Risk Management. 21 August 2006.

TM 9-1005-249-10, , Operator's Manual for Rifle, 5.56-mm, M16 (NSN 1005-00-856-6885), Rifle 5.56-mm, M16A1 (1005-00-073-9421). 11 May 1990.

TM 9-1005-319-10, Operator's Manual for Rifle, 5.56 MM, M16A2 W/E (NSN 1005-01-128-9936) (EIC: 4GM); Rifle, 5.56 MM, M16A3 (1005-01-357-5112); Rifle, 5.56 MM, M16A4 (1005-01-383-2872) (EIC: 4F9); Carbine, 5.56 MM, M4 W/E (1005-01-231-0973) (EIC: 4FJ); Carbine, 5.56 MM, M4A1 (1005-01-382-0953) (EIC: 4GC), 1 October 1998. 01 October 1998.

DOCUMENTS NEEDED

These documents must be available to the intended users of this publication.

ARMY PUBLICATIONS

AR 350-1, Army Training and Leader Development. 03 August 2007.

FM 7-0, Training the Force. 22 October 2002.

FM 7-1, Battle Focused Training. 15 September 2003.

TC 7-9, Infantry Live-Fire Training. 30 September 1993.

TC 25-1, Training Land. 15 March 2004.

TC 25-8, Training Ranges. 5 April 2004.

TM 9-1005-319-23&P, Unit and Direct Support Maintenance Manual (Including Repair Parts and Special Tools List) FOR RIFLE, 5.56MM, M16A2 W/E (NSN 1005-01-128-9936) (EIC: 4GM) Rifle, 5.56MM, M16A3, (1005-01-357-5112) Rifle, 5.56MM, M16A4, W/E (1005-01-383-2872) (EIC: 4F9) Carbine, 5.56MM, M4 (1005-01-231-0973) (EIC: 4FJ) and Carbine, 5.56MM, M4A1 (1005-01-382-0953) (EIC: 4GC), 1 May 1991.

TM 9-1240-413-13&P, Operator and Field Maintenance Manual Including Repair Parts and Special Tools List for M68 Sight, Reflex, w/Quick Release and Sight Mount (COMP M2: NSN 1240-01-411-1265) (COMP M4: NSN 1240-01-540-3690). 19 March 2008.

TM 9-5855-1914-13&P, Operator and Field Maintenance Manual Including Repair Parts and Special Tools List for the Advanced Target Pointer Illuminator Aiming Light (ATPIAL) AN/PEQ-15 (NSN 5855-01-534-5931). 16 February 2007.

TM 9-6920-746-12&P, Operator's and Organizational Maintenance Manual (Including Repair Parts and Special Tools List) M2 Practice Bolt Plastic Ammunition for Rifle 5.56mm, M16 Series (NSN 1005-01-184-4041). 09 September 1986.

TM 11-5855-213-10, Operator's Manual for Night Vision Sight, Individual Served Weapon, AN/PVS-4 (NSN 5855-00-629-5334) (EIC: IPJ). 01 February 1993.

TM 11-5855-261-10, Operator's Manual for Infrared Aiming Light, AN/PAQ-4 (NSN 5855-01-107-5925). 28 May 1981.

TM 11-5855-301-12&P, Operator's and Unit Maintenance Manual (Including Repair Parts and Special Tools List) Light, Aiming, Infrared, AN/PAQ-4B (NSN 5855-01-361-1362), AN/PAQ-4C (5855-01-398-4315).15 May 2000.

TM 11-5855-306-10, Operator's Manual for Monocular Night Vision Device, AN/PVS-14 (NSN 5855-01-432-0524) (EIC: IPX). 15 May 2007.

TM 11-5855-308-12&P, Operator's and Unit Maintenance Manual (Including Repair Parts and Special Tools List), Target Pointer Illuminator/Aiming Light AN/PEQ-2A (NSN 5855-01-447-8992). 15 May 2000.

TM 11-5855-312-10, Operator's Manual Sight, Thermal AN/PAS-13B(V)2 (NSN 5855-01-464-3152); AN/PAS-13B(V)3 (5855-01-464-3151). 15 February 2005.

TM 11-5855-316-10, Operator's Manual AN/PAS-13C(V)1 Sight, Thermal (NSN 5855-01-523-7707); AN/PAS-13C(V)2 Sight, Thermal (NSN 5855-01-523-7713) (MWTS); AN/PAS-13C(V)3 Sight, Thermal (NSN 5855-01-523-7715). 15 June 2006.

TM 11-5855-317-10, Operator's Manual For Sight, Thermal AN/PAS-13D(V)2 (NSN 5855-01-524-4313); AN/PAS-13D(V)3 NSN 5855-01-524-4314) (HWTS). 15 February 2007.

ARMY FORMS

DA Form 2028, Recommended Changes to Publications and Blank Forms.

DA Form 3595-R, Record Fire Scorecard.

DA Form 3601-R, Single Target—Field Fire I Scorecard.

DA Form 5239-R, 100-, 200-, and 300-Meter Downrange Feedback Scorecard.

DA Form 5241-R, Single and Multiple Targets—Field Fire II Scorecard.

DA Form 5789-R, Record Fire Scorecard—Known-Distance Course (LRA).

DA Form 5790-R, Record Fire Scorecard—Scaled Target Alternate Course.

DA Form 7489-R, Record Night Fire Scorecard (LRA).

DA Form 7566, Composite Risk Management Worksheet.

DA Form 7649-R, Squad Designated Marksman—Record Fire I and II.

DA Form 7650-R, Squad Designated Marksman—Position Evaluation.

READINGS RECOMMENDED

These sources contain relevant supplementary information.

AR 190-11, Physical Security of Arms, Ammunition, and Explosives. 15 November 2006.

AR 350-38, Training Device Polices and Management. 15 October 1993.

DA Pam 350-9, Index and Description of Army Training Devices. 03 September 2002.

DA Pam 385-64, Ammunition and Explosive Safety Standards. 15 December 1999.

FM 1-02, Operational Terms and Graphics. 21 September 2004.

FM 3-06, Urban Operations. 26 October 2006.

FM 3-06.11, Combined Arms Operations in Urban Terrain. 28 February 2002.

FM 3-11.4, Multiservice Tactics, Techniques, and Procedures for Nuclear, Biological, and Chemical (NBC) Protection. 02 June 2003.

FM 4-25.11, First Aid. 23 December 2002.

FM 20-3, Camouflage, Concealment, and Decoys. 30 August 1999.

FM 21-10, Field Hygiene and Sanitation. 21 June 2000

FM 3-21.75, The Warrior Ethos and Soldier Combat Skills. 28 January 2008.

FM 22-6, Guard Duty. 17 September 1971.

FM 25-4, How to Conduct Training Exercises. 10 September 1984.

STP 21-1-SMCT, Soldier's Manual of Common Tasks, Skill Level 1. 14 December 2007.

STP 21-24-SMCT, Soldier's Manual of Common Tasks (SMCT), Skill Levels 2-4. 2 October 2006.

TC 25-20, A Leader's Guide to After-Action Reviews. 30 September 1993.

TC 90-1, Training for Urban Operations. 19 May 2008.

TM 9-1005-249-23&P, Unit and Direct Support Maintenance Manual (Including Repair Parts and Special Tools List) for Rifle, 5.56-mm, M16 (NSN 1005-00-856-6885) (EIC: 4F7); Rifle, 5.56-mm, M16A1 (1005-00-073-9421) (EIC: 4FC). 19 June 1991.

TM 9-6920-363-12&P, Operator's and Organizational Maintenance Manual (Including Repair Parts and Special Tools List) for Conversion Kit (Cal .22 Rimfire Adapter) M261 (NSN 1005-01-010-1561) for Rifle: 5.56-mm, M16 and M16A1. Reprinted with Changes 1 and 2. 21 August 1984.

INTERNET WEBSITES

U.S. Army Publishing Directorate, http://www.apd.army.mil/

Reimer Doctrine and Training Digital Library, http://www.train.army.mil

Index

B

M

SINGLE TARGET
FIELD FIRE I SCORECARD

For use of this form see FM 3-22.9; the proponent agency is TRADOC

ID CODE	UNIT	DATE (YYYYMMDD)	EVALUATOR'S ID CODE

TABLE 1 — SUPPOR¯ED FIRING POSITION

RD	RANGE (m)	TIME (sec)	HIT	MISS
1	75	6	☐	☐
2	175	8	☐	☐
3	300	10	☐	☐
4	175	8	☐	☐
5	75	6	☐	☐
6	300	10	☐	☐
7	300	10	☐	☐
8	75	75	☐	☐
9	175	175	☐	☐
10	175	8	☐	☐
11	300	10	☐	☐
12	175	8	☐	☐
13	75	6	☐	☐
14	300	10	☐	☐
15	175	8	☐	☐
16	75	6	☐	☐
17	300	10	☐	☐
18	75	6	☐	☐
TOTAL				

TABLE 2 — SUPPORTED FIRING POSITION

RD	RANGE (m)	TIME (sec)	HIT	MISS
1	75	6	☐	☐
2	175	8	☐	☐
3	300	10	☐	☐
4	175	8	☐	☐
5	75	6	☐	☐
6	300	10	☐	☐
7	300	10	☐	☐
8	75	6	☐	☐
9	175	8	☐	☐
10	175	8	☐	☐
11	300	10	☐	☐
12	175	8	☐	☐
13	75	6	☐	☐
14	300	10	☐	☐
15	175	8	☐	☐
16	75	6	☐	☐
17	300	10	☐	☐
18	75	8	☐	☐
TOTAL				

TABLE 3 — PRONE FIRING POSITION

RD	RANGE (m)	TIME (sec)	HIT	MISS
1	75	6	☐	☐
2	175	8	☐	☐
3	300	10	☐	☐
4	175	8	☐	☐
5	75	6	☐	☐
6	300	10	☐	☐
7	300	10	☐	☐
8	75	6	☐	☐
9	175	8	☐	☐
TOTAL				

TABLE 4 — KNEELING FIRING POSITION

RD	RANGE (m)	TIME (sec)	HIT	MISS
1	75	6	☐	☐
2	175	8	☐	☐
3	75	6	☐	☐
4	175	8	☐	☐
5	75	6	☐	☐
6	175	8	☐	☐
7	75	6	☐	☐
8	75	6	☐	☐
9	175	8	☐	☐
TOTAL				

SCORE

TABLE	HIT	MISS	NO FIRE
1			
2			
3			
4			

REMARKS

DATE INITIALED (YYYYMMDD)	SCORER'S INITIALS

100-, 200- AND 300-METER DOWNRANGE FEEDBACK SCORECARD

For use of this form see FM 3-22.9; the proponent agency is TRADOC.

ID CODE	UNIT	DATE *(YYYYMMDD)*	EVALUATOR'S ID CODE

ZERO CONFIRMATION		
RANGE (m)	HIT	MISS
300		
ZERO CONFIRMATION REFIRE		
RANGE (m)	HIT	MISS
300		

TABLE 1 -- DOWNRANGE SILHOUETTES, SUPPORTED FIRING POSITION			
RDS	RANGE (m)	HIT	MISS
5	100		
10	200		
5	300		
TOTAL			

TABLE 2 -- DOWNRANGE SILHOUETTES, PRONE FIRING POSITION			
RDS	RANGE (m)	HIT	MISS
5	100		
10	200		
5	300		
TOTAL			

ZERO-SHOT LOCATIONS

300M

WIND SPEED (mph):

WIND DIRECTION (use arrow):

SHOT LOCATIONS

SUPPORTED FIRING POSITION

TABLE 1

PRONE UNSUPPORTED FIRING POSITION

TABLE 2

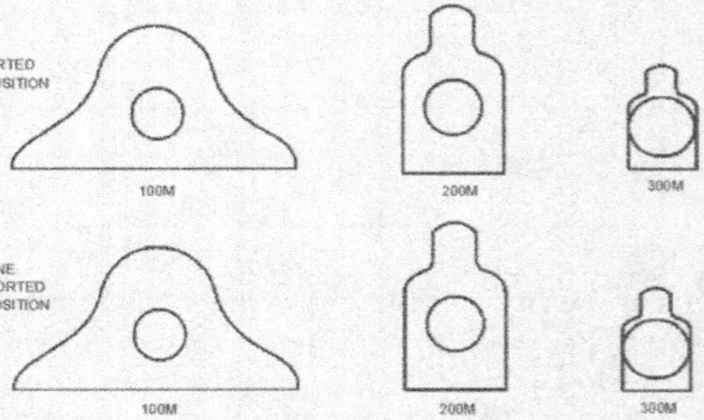

REMARKS

DATE INITIALED *(YYYYMMDD)*	SCORER'S INITIALS

SINGLE AND MULTIPLE TARGETS FIELD FIRE II SCORECARD

For use of this form see FM 3-22.9; the proponent agency is TRADOC.

ID CODE	UNIT	DATE *(YYYYMMDD)*	EVALUATOR'S ID CODE

TABLE 1 SUPPORTED FIRING POSITION

RD	RANGE (m)	TIME (sec)	HIT	MISS
1	75	5	☐	☐
2	175	7	☐	☐
3	75	11	☐	☐
4	300		☐	☐
5	75	9	☐	☐
6	175		☐	☐
7	75	10	☐	☐
8	300		☐	☐
9	175	11	☐	☐
10	300		☐	☐
		TOTAL		

TABLE 2 SUPPORTED FIRING POSITION

RD	RANGE (m)	TIME (sec)	HIT	MISS
1	175	7	☐	☐
2	75	10	☐	☐
3	300		☐	☐
4	75	9	☐	☐
5	175		☐	☐
6	300	9	☐	☐
7	75	9	☐	☐
8	175		☐	☐
9	175	11	☐	☐
10	300		☐	☐
11	75	9	☐	☐
12	175		☐	☐
13	175	11	☐	☐
14	300		☐	☐
15	75	5	☐	☐
16	175	11	☐	☐
17	300		☐	☐
18	75	9	☐	☐
19	175		☐	☐
20	75	10	☐	☐
21	300		☐	☐
22	175	7	☐	☐
		TOTAL		

TABLE 3 PRONE FIRING POSITION

RD	RANGE (m)	TIME (sec)	HIT	MISS
1	75	6	☐	☐
2	175	8	☐	☐
3	75	13	☐	☐
4	300		☐	☐
5	75	11	☐	☐
6	175		☐	☐
7	75	12	☐	☐
8	300		☐	☐
9	175	13	☐	☐
10	300		☐	☐
11	75	11	☐	☐
12	175		☐	☐
13	175	8	☐	☐
14	75	6	☐	☐
15	75	11	☐	☐
16	175		☐	☐
17	75	12	☐	☐
18	300		☐	☐
19	75	11	☐	☐
20	175		☐	☐
21	175	13	☐	☐
22	300		☐	☐
		TOTAL		

SCORE

TABLE	HIT	MISS
2		
3		
TOTAL		

REMARKS

DATE INITIALED *(YYYYMMDD)*	SCORER'S INITIALS

DA FORM 5241-R, SEP 2008 — DA FORM 5241-R JUN 1989, IS OBSOLETE. — APD PE v1.00

RECORD FIRING SCORECARD -- KNOWN DISTANCE COURSE

For use of this form see FM 3-22.9; the proponent agency is TRADOC

ID CODE	UNIT	DATE *(YYYYMMDD)*	EVALUATOR'S ID CODE

TABLE 1
PRONE SUPPORTED OR FOXHOLE SUPPORTED FIRING POSITION
(TIME: 120 SECONDS)

RD	RANGE (m)	HIT	MISS	RD	RANGE (m)	HIT	MISS
1		☐	☐	11		☐	☐
2		☐	☐	12		☐	☐
3		☐	☐	13		☐	☐
4	300	☐	☐	14	300	☐	☐
5		☐	☐	15		☐	☐
6	E-SIL	☐	☐	16	E-SIL	☐	☐
7		☐	☐	17		☐	☐
8		☐	☐	18		☐	☐
9		☐	☐	19		☐	☐
10		☐	☐	20		☐	☐
TOTAL				TOTAL			

TABLE 2
PRONE UNSUPPORTED FIRING POSITION
(TIME: 60 SECONDS)

RD	RANGE (m)	HIT	MISS
1		☐	☐
2		☐	☐
3		☐	☐
4	300	☐	☐
5		☐	☐
6	E-SIL	☐	☐
7		☐	☐
8		☐	☐
9		☐	☐
10		☐	☐
TOTAL			

TABLE 3
KNEELING FIRING POSITION
(TIME: 60 SECONDS)

RD	RANGE (m)	HIT	MISS
1		☐	☐
2		☐	☐
3		☐	☐
4	300	☐	☐
5		☐	☐
6	F-SIL	☐	☐
7		☐	☐
8		☐	☐
9		☐	☐
10		☐	☐
TOTAL			

SCORE

TABLE	HIT	MISS	NO FIRE
1			
2			
3			

FIRER'S QUALIFICATION SCORE

QUALIFICATION SCORE RATINGS *(Check One)*

☐ 38-40 -- EXPERT ☐ 26-32 -- MARKSMAN

☐ 33-37 -- SHARPSHOOTER ☐ 25 AND BELOW -- UNQUALIFIED

LIGHT

WIND

NIGHT FIRE EXERCISE

DATE *(YYYYMMDD)*	HIT	MISS	GO	NO GO
			☐	☐

CBRN FIRE EXERCISE

DATE *(YYYYMMDD)*	HIT	MISS	GO	NO GO
			☐	☐

THE FIRER WILL BE ISSUED 40 ROUNDS. THE ROUNDS WILL BE PRELOADED IN FOUR 10-ROUND MAGAZINES - TWO FOR TABLE 1, ONE FOR EACH REMAINING TABLE.

ZERO			
ELEV			
WIND			

DATE INITIALED *(YYYYMMDD)*	SCORER'S INITIALS
DATE INITIALED *(YYYYMMDD)*	OFFICER'S INITIALS

This scorecard is used to score known distance course record fire qualification when the known distance range is used. This course is used only when the standard record fire course is not available.

CONDUCT OF FIRE

The uniform for qualification ia a helmet, LBE/LBV, and interceptor body armor with front and back SAPI plates (if available). No other armor is required.

(1) **Table 1 -- Supported Firing Position.** The firer is given two 10-round magazines to engage an E-silhouette at 300 yards within 120 seconds.

(2) **Table 2 -- Unsupported Firing Position.** The firer is given a 10-round magazine to engage an E-silhouette at 200 yards within 60 seconds.

(3) **Table 3 -- Unsupported Firing Position.** The firer is given a 10-round magazine to engage an F-silhouette at 100 yards within 60 seconds.

SCORING

Scoring is conducted in the pits, with the results provided after each table. A hit is scored for any bullet hole that is within or touches some part of the silhouette facing.

RECORD FIRING SCORECARD -- SCALED TARGET ALTERNATE COURSE

For use of this form see FM 3-22.9; the proponent agency is TRADOC.

ID CODE	UNIT	DATE (YYYYMMDD)	EVALUATOR'S ID CODE

TABLE 1 PRONE SUPPORTED OR FOXHOLE SUPPORTED FIRING POSITION (TIME: 120 SECONDS)								TABLE 2 PRONE UNSUPPORTED FIRING POSITION (TIME: 60 SECONDS)				TABLE 3 KNEELING FIRING POSITION (TIME: 60 SECONDS)			
RD	RANGE (m)	HIT	MISS	RD	RANGE (m)	HIT	MISS	RD	RANGE (m)	HIT	MISS	RD	RANGE (m)	HIT	MISS
1	300	☐	☐	11	150	☐	☐	1	300	☐	☐	1	150	☐	☐
2	300	☐	☐	12	150	☐	☐	2	250	☐	☐	2	150	☐	☐
3	250	☐	☐	13	100	☐	☐	3	200	☐	☐	3	100	☐	☐
4	250	☐	☐	14	100	☐	☐	4	200	☐	☐	4	100	☐	☐
5	200	☐	☐	15	100	☐	☐	5	150	☐	☐	5	100	☐	☐
6	200	☐	☐	16	100	☐	☐	6	150	☐	☐	6	100	☐	☐
7	200	☐	☐	17	100	☐	☐	7	100	☐	☐	7	100	☐	☐
8	200	☐	☐	18	100	☐	☐	8	100	☐	☐	8	100	☐	☐
9	150	☐	☐	19	50	☐	☐	9	100	☐	☐	9	50	☐	☐
10	150	☐	☐	20	50	☐	☐	10	50	☐	☐	10	50	☐	☐
TOTAL				TOTAL				TOTAL				TOTAL			

SCORE			
TABLE	HIT	MISS	NO FIRE
1			
2			
3			
FIRER'S QUALIFICATION SCORE			

QUALIFICATION SCORE RATINGS *(Check One)*

☐ 36-40 -- EXPERT ☐ 23-29 -- MARKSMAN

☐ 30-35 -- SHARPSHOOTER ☐ 22 AND BELOW -- UNQUALIFIED

FIRER ISSUED 40 ROUNDS TO ENGAGE 10 TARGETS. NO MORE THAN 4 ROUNDS PER TARGET. THE ROUNDS WILL BE PRELOADED IN ONE 20-ROUND MAGAZINE FOR TABLE 1, ONE 10-ROUND MAGAZINE FOR TABLE 2, AND ONE 10-ROUND MAGAZINE FOR TABLE 3.

NIGHT FIRE EXERCISE

DATE *(YYYYMMDD)*	HIT	MISS	GO	NO GO
			☐	☐

CBRN FIRE EXERCISE

DATE *(YYYYMMDD)*	HIT	MISS	GO	NO GO
			☐	☐

REMARKS

DATE INITIALED *(YYYYMMDD)*	SCORER'S INITIALS
DATE INITIALED *(YYYYMMDD)*	OFFICER'S INITIALS

This scorecard is used to score alternate course record fire qualification when the 25-meter scaled silhouette target (NSN 6920-01-167-1396) is used. The alternate course is used only when standard record fire and known distance ranges are unavailable.

NOTE: If zeroing/grouping exercises are *not performed* on the *day of record fire*, six rounds of training/sustainment ammunition will be fired for 25-meter zero confirmation prior to conducting the qualification course.

CONDUCT OF FIRE

Alternate course qualification firers will have one 20-round magazine and two 10-round magazines. To ensure that firers do not forget which targets they engaged and shoot a given target more than the prescribed number of times, firers should adhere to the following guideline: Engage targets on the sheet from left to right and nearest to farthest. Engagement should follow this order: 50m, 100m left, 100m center, 100m right, 150m left, 150m right, 200m left, 200m right, 250m, and 300m.

Though the time between each firing position is not specified, enough time should be allotted to allow the firer to clear his weapon, quickly change firing positions, and reload before beginning the next firing table. The range RSO ensures that enough time is given between each change in firing positions to facilitate the timely flow of the record fire qualification table.

(1) Table 1 -- Prone Supported Firing Position or (at the unit commander's discretion) Foxhole Supported Firing Position. The firer is given one 20-round magazine to engage 10 silhouettes on the target sheet. Table 1 includes 2 rounds for each silhouette. Firing must be completed in 120 seconds. No more than 2 hits are scored for each silhouette.

(2) Table 2 -- Prone Unsupported Firing Position. The firer is given one 10-round magazine to engage 10 silhouettes on the same target sheet. Table 2 includes 1 round for each silhouette. Firing must be completed in 60 seconds. No more than 1 hit is scored for each silhouette.

(3) Table 3 -- Kneeling Firing Position. The firer is given one 10-round magazine to engage 10 silhouettes on the same target sheet. Table 3 includes 2 rounds for each silhouette positioned 50 to 150 meters away. Firing must be completed in 60 seconds. No more than 2 hits are scored for each silhouette.

SCORING

The same target sheet is used for every 40-round qualification table that a firer completes. One hit is awarded for each round that strikes within or touches some part of the silhouette. A maximum of four hits for each silhouette on the same target sheet are scored.

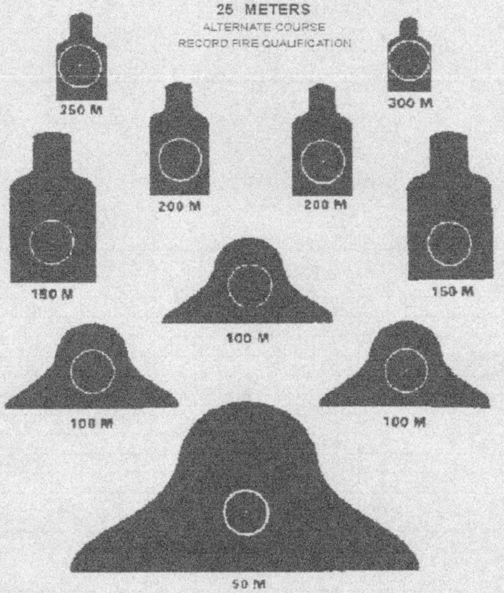

THE WHITE DOT ON EACH TARGET SHOWS THE CENTER OF MASS AIMING POINT. BULLETS SHOULD HIT WITHIN THE CIRCLE, BUT ARE SCORED AS HITS IF THEY HIT ANY PART OF THE SILHOUETTE.

RECORD NIGHT FIRE SCORECARD

For use of this form see FM 3-22.9; the proponent agency is TRADOC.

ID CODE	UNIT	DATE (YYYYMMDD)	EVALUATOR'S ID CODE

TABLE 1
FOXHOLE SUPPORTED FIRING POSITION

RD	RANGE (m)	TIME (sec)	HIT	MISS	NO FIRE	RD	RANGE (m)	TIME (sec)	HIT	MISS	NO FIRE
1	50L	3				11	100	8			
2	200	6				12	200				
3	100	4				13	150	8			
4	150	5				14	50R				
5	100	8				15	100	8			
6	150	5				16	150				
7	50R	3				17	200	6			
8	200	6				18	150	5			
9	150	5				19	50L	9			
10	250	7				20	250				
TOTAL						TOTAL					

TABLE 2
PRONE UNSUPPORTED FIRING POSITION

RD	RANGE (m)	TIME (sec)	HIT	MISS	NO FIRE	RD	RANGE (m)	TIME (sec)	HIT	MISS	NO FIRE
1	100	5				11	150	8			
2	200	8				12	50R	5			
3	150	6				13	100	12			
4	50L	12				14	200				
5	200					15	150	12			
6	150	12				16	50L				
7	100					17	100	12			
8	50R	12				18	150				
9	150					19	200	12			
10	100	8				20	100				
TOTAL						TOTAL					

SCORE

TABLE	HIT	MISS	NO FIRE
1			
2			
TOTAL			

FIRER'S QUALIFICATION SCORE

QUALIFICATION SCORE RATINGS *(Check One)*

- ☐ 36-40 – EXPERT
- ☐ 17-23 – MARKSMAN
- ☐ 24-34 – SHARPSHOOTER
- ☐ 16 AND BELOW – UNQUALIFIED

AIMING DEVICE USED *(Check One)*

- ☐ AN/PEQ-2A/B
- ☐ AN/PAQ-4B/C
- ☐ AN/PVS-4

REMARKS

DATE INITIALED (YYYYMMDD)	SCORER'S INITIALS
DATE INITIALED (YYYYMMDD)	OFFICER'S INITIALS

SQUAD DESIGNATED MARKSMAN
RECORD FIRE I AND II

For use of this form, see FM 3-22.9; the proponent agency is TRADOC.

ID CODE	UNIT	DATE *(YYYYMMDD)*	EVALUATOR'S ID CODE

RECORD FIRE I - MECHANICAL SIGHT ADJUSTMENT			RECORD FIRE II - ADJUSTED AIMING POINTS/OPTIC		
TARGET RANGE/POSITION	HIT	MISS	TARGET RANGE/POSITION	HIT	MISS
100m unsupported	☐	☐	100m unsupported	☐	☐
100m unsupported	☐	☐	100m unsupported	☐	☐
100m unsupported	☐	☐	100m unsupported	☐	☐
100m unsupported	☐	☐	100m unsupported	☐	☐
200m supported	☐	☐	200m supported	☐	☐
200m supported	☐	☐	200m supported	☐	☐
200m unsupported	☐	☐	200m unsupported	☐	☐
200m unsupported	☐	☐	200m unsupported	☐	☐
300m supported	☐	☐	300m supported	☐	☐
300m supported	☐	☐	300m supported	☐	☐
300m unsupported	☐	☐	300m unsupported	☐	☐
300m unsupported	☐	☐	300m unsupported	☐	☐
400m supported	☐	☐	400m supported	☐	☐
400m supported	☐	☐	400m supported	☐	☐
400m supported	☐	☐	400m supported	☐	☐
400m supported	☐	☐	400m supported	☐	☐
500m supported	☐	☐	500m supported	☐	☐
500m supported	☐	☐	500m supported	☐	☐
500m supported	☐	☐	500m supported	☐	☐
500m supported	☐	☐	500m supported	☐	☐
TOTAL			**TOTAL**		
☐ **PASS (14-20)** ☐ **FAIL (13 and below)**			☐ **PASS (14-20)** ☐ **FAIL (13 and below)**		
DATE INITIALED *(YYYYMMDD)*			SCORER'S INITIALS		
DATE INITIALED *(YYYYMMDD)*			OFFICER'S INITIALS		

DA FORM [illegible]649-R, SEP 2008 APD PE v1.00

SQUAD DESIGNATED MARKSMAN
POSITION EVALUATION

For use of this form, see FM 3-22.9; the proponent agency is TRADOC.

☐ FOXHOLE SUPPORTED ☐ PRONE UNSUPPORTED		ID CODE	UNIT	DATE (YYYYMMDD)	EVALUATOR'S ID CODE
PHASE	EYE RELIEF	TRIGGER FINGER	ELBOWS	NON-FIRING HAND	LEGS
EXAMPLE	How far is nose from charging handle?	What part of 1st joint (tip, middle, joint)?	Is platform stable?	Where is it located (Soldier's choice, stable)?	Where are they located (stability for upper body)?
Phase I – Steady Position					
Phase II – Borelight Dry Fire					
Phase V – Practice Qualification					
100m					
200m					
300m					
400m					
500m					
Phase VI – Record Qualification					
100m					
200m					
300m					
400m					
500m					
DATE INITIALED (YYYYMMDD)			SCORER'S INITIALS		

DA FORM 7650-R, SEP 2008 APD PE v1.00

Made in the USA
Monee, IL
05 November 2026

67808158R00295